3 Volumes

CASUAL CALCULUS

A Friendly Student Companion

CASUAL CALCULUS

3 Volumes

A Friendly Student Companion

Kenneth H Luther

Valparaiso University, USA

World Scientific

NEW JERSEY • LONDON • SINGAPORE • BEIJING • SHANGHAI • HONG KONG • TAIPEI • CHENNAI • TOKYO

Published by

World Scientific Publishing Co. Pte. Ltd.
5 Toh Tuck Link, Singapore 596224
USA office: 27 Warren Street, Suite 401-402, Hackensack, NJ 07601
UK office: 57 Shelton Street, Covent Garden, London WC2H 9HE

British Library Cataloguing-in-Publication Data
A catalogue record for this book is available from the British Library.

CASUAL CALCULUS: A FRIENDLY STUDENT COMPANION
(In 3 Volumes)

ISBN 978-981-124-263-2 (set_hardcover)
ISBN 978-981-124-264-9 (set_paperback)
ISBN 978-981-124-265-6 (set_ebook for institutions)
ISBN 978-981-124-266-3 (set_ebook for individuals)

ISBN 978-981-122-392-1 (vol. 1_hardcover)
ISBN 978-981-122-488-1 (vol. 1_paperback)
ISBN 978-981-122-393-8 (vol. 1_ebook for institutions)
ISBN 978-981-122-394-5 (vol. 1_ebook for individuals)

ISBN 978-981-124-197-0 (vol. 2_hardcover)
ISBN 978-981-124-198-7 (vol. 2_paperback)
ISBN 978-981-124-199-4 (vol. 2_ebook for institutions)
ISBN 978-981-124-211-3 (vol. 2_ebook for individuals)

ISBN 978-981-122-395-2 (vol. 3_hardcover)
ISBN 978-981-122-489-8 (vol. 3_paperback)
ISBN 978-981-122-396-9 (vol. 3_ebook for institutions)
ISBN 978-981-122-397-6 (vol. 3_ebook for individuals)

For any available supplementary material, please visit
https://www.worldscientific.com/worldscibooks/10.1142/11927#t=suppl

Printed in Singapore

Dedicated to the bands that kept me sane during the years this material accumulated:

Anathema, Angra, Avantasia, Black Sabbath, Blue Oyster Cult, Circus Maximus, Dream Theater, Dynazty, Edenbridge, Epica, Evergrey, Fates Warning, Iron Maiden, Judas Priest, Kamelot, Kingcrow, Masterplan, Mournful Congregation, Nightingale, Nightwish, Nocturnal Rites, Novembers Doom, Opeth, Pain of Salvation, Power Quest, Pyramaze, Rhapsody of Fire, Rush, Saga, Serenity, Seventh Wonder, Shape of Despair, Signum Regis, Sonata Arctica, Stratovarius, Styx, Sunn O))), Symphony X, Tangerine Dream, Theocracy, Threshold, Toto, Voyager, Yes

Preface

Who the Heck Needs This Book?

Oh great, another Calculus textbook! Sort of. This is not intended to be a completely stand-alone Calculus textbook; rather, I see it more as a supplement. Some people will want more exercises than are presented here. Some will want more theoretical development of the concepts. But some, I hope *you*, will find a good balance between discussion of ideas and demonstration of those ideas in action via lots of solved problems. This book is intended to fill that middle ground between a regular text and a solution guide.

My goal is to serve students who are looking for a "second voice" to speak the content at them. This book contains narrative sections similar to a bazillion other texts, but I hope you find it to be conversational, fun, and totally unlike a stereotypical stuffy math text. When I was an undergraduate student at Mount Union College (now the University of Mount Union), one of my Physics & Astronomy Professors, Dr. James Rodman, used to advocate for "the half-drunken approach." He was not suggesting students should bring booze to class, but rather, he was emphasizing a state of mind: try to stay relaxed and at ease, and don't be intimidated by the content coming at you. This advice didn't work so well in his Celestial Mechanics class, which met at 8am, and in which I was the only student, but otherwise I've always appreciated the sentiment. My aim is to maintain a casual and informal tone of conversation as we work our way through the material. I may "speak" to you directly in second person, as I recognize and see you as an individual trying to tackle some sophisticated math. I may refer to myself in first person, if I have some experience or advice that may help. I may refer to you and I together as "we," because I recognize that we are all life-long learners, and we are experiencing the content together. The author promises that he will never refer to himself in the third person, or refer haughtily to "the author" anywhere in this text. Except for right there.

This work contains worked solutions to *all* of the exercises. My own students' feedback makes it clear that the availability of worked solutions is very helpful. But to get the most out of what's here, you have to meet me half-way: this text is intended to be *fully consumed.* It is not intended to be a text where each section has 120 exercises from which an instructor assigns a dozen. If you dive into this book, I hope you read almost every section and work each and every exercise.

If you're a math major or STEM student needing a rigorous treatment of Calculus, then once this gets you past the initial problem solving phases, its job will be done — and I hope it gets you interested in and motivated for a deeper treatment and further study. If all you need is a boost getting started, this is for you. For those who are taking Calculus from a more utilitarian point of view, or who just need an alternate take on some of the subjects, this is for you. If you're one of the growing number of students taking Calculus in a hybrid or exclusively on-line format, either from a traditional institution or from one of the aggregate course sites, this may help you. If you are returning to a study of Calculus after a long gap, this may serve as a good review. It is not a "normal" Calculus text. It is not going to appeal to the instructor who adjusts his bowtie and monacle before discussing math-em-a-tics (always pronounced fully with four crisp syllables, you know), but I would be as happy as my dogs lying in their favorite sunny spot on the deck if students recommended it to each other.[1]

What Is the Structure of This Book?

First of all, this "book" is split into three volumes, each corresponding to one semester of a typical three semester Calculus sequence. Volume 1 (Chapters 1–6) corresponds to a standard Calculus I, Volume 2 (Chapters 7–12) goes with Calculus II, and then Volume 3 (Chapters 13–18) matches what is often Calculus 3 (Multivariable Calculus). In most cases we will

[1]You get to see a picture of my dogs in Chapter 8.

follow the usual pattern of traditional Calculus texts (you know the ones, where if you line up ten of them in a row, nine of them will have "The Fundamental Theorem of Calculus" as Chapter 5 and "Sequences and Series" as Chapter 11). However, in some places, the order is rearranged a bit. In particular, antiderivatives are introduced at the same time as derivatives, and as a derivative technique (such as the power rule) is introduced, the corresponding antiderivative technique is seen as well. Also, in Volume 2, the topic of approximation and Taylor polynomials comes before discussion of infinite series and a return to Taylor Series; I find it easier to introduce the big ideas this way. So far, the Calculus Police have not come to get me.

The section-by-section set up of each Volume is as follows: Each topic of content begins with a narrative section that leads you through the main ideas and presents examples along the way. After each Example is a "You Try It" problem that's very similar to the example. Stop after each Example and try your hand at the associated YTI problem. The solutions to all YTI problems are at the end of the very section they're shown in — so, if you think you've succeeded at the YTI problem, go check the solution to be sure. Or, if you get stuck on the YTI problem, then go look at its solution to get a hint. Once you've completed a section, you'll see the YTI problems collected, along with a set of Practice Problems and Challenge Problems. The Practice Problems should be similar to the YTI Problems, but you're getting them all at once, and so you don't necessarily know which specific technique to use or which Example to follow — you have to think about it! The solutions to the Practice Problems are at the back end of the book. So while they're all available, they are more physically separated from the section they come from; the idea is that you might be inclined to rely on them a bit less, although they are still there when you need them. Then finally, the Challenge Problems are a bit tougher than the others, and you can use those to see if you're successfully synthesizing the ideas you've seen in their section. Solutions to Challenge Problems come after those to the Practice Problems. If you like video games, think of the Practice Problems as gaining experience points, and the Challenge Problems as leveling up.

If this book does to appeal to some instructors as a resource, the structure of the book lends itself to those who like to have more active learning or flipped classrooms. For example, a section and its You Try It problems could be assigned as class preparation activities, with the Practice Problems following as in-class group exercises, and the Challenge Problems as individual homework. Or, the Practice Problems could be done before class, and the Challenge Problems left as group work in class.

Two minor contextual items to look for are these:

(1) Each section in Volume 1 ends with a summary of key ideas labeled, "Have You Learned?" Beyond that, I assume you've become mature enough in your reading that you are able to determine key ideas on your own.
(2) At many locations in the text, I will pose some "Food For Thought" (FFT) based on an open question left unanswered. These little puzzlers are bracketed by the symbol 🍽 (it's supposed to be a fork, a plate, and a knife).

To keep you focused on problem solving, some derivations or other more theoretical discussions are held off until the end of a section. I have always been a fan of heavy metal, and I see jumping into a derivation or proof as the mathematical version of jumping into a mosh pit: you're mostly there to sing along, but every once in a while, you have to wade in and get a bit bruised. So, each of these more mathematically violent discussions are set off with a subheading of, "Into the Pit!!"

But before we get to any of the content, I want to give you a motivational speech based on my two decades (sigh) of teaching and observing learners. This comes next in the Prelude. Don't worry, there are cartoons!

Contents

Chapter 0

Prelude: Your Mathematics Pep Talk

When taking a difficult math class, students often struggle with motivation as related to the future utility of the material, the necessity of retaining apparent minutia for the purposes of exam taking, and the philosophical question of why we have to be here when reruns of The Office are on. As a pep talk for students about to embark on a journey through Calculus, here are the "Ten Most Important Things A Math Student Should Remember".

(10) Sometimes, the answer is NEVER.

Fig. 0.1 Never.

The question is, of course, "*When am I ever going to use this stuff in real life?*". You may be astounded to learn that you just might go through the remainder of your existence without ever having used implicit differentiation or integration by partial fractions again. Shocking, I know. But if you restricted your curriculum to only those classes in which all of the material is directly used in real life, your transcripts would contain entries such as "DRV 101: Red Lights" and "BR 110: Flushing Techniques"; academic majors would be described by such titles as "Lawn Care" and "Personal Hygiene". The fact is that in many courses — not just math — you are required to read, learn, and act out topics which will never darken your doorstep again. Yet, it is predominantly math courses in which this creates the biggest objections. Has anyone ever raised their hand in a Spanish class and asked, "When will I ever use this in real life?"

Remember that in math, what you are being taught is, hopefully, not the entirety of what you are learning. I know this is a tired cliche, but math teachers can be considered personal trainers for the brain. Do you lift weights? Jog or run on a treadmill? Are you doing these things because you plan on getting a job that requires you to repeatedly run in circles all day? Or to lie on you back and lift heavy, circular slabs of iron up and down several times? If not, then why do it? It's because you know that by exercising in this fashion (or any other), you are making your body healthy, and training your body to physically respond quickly and efficiently in other situations. Well, math classes do the same thing by training your analytical processes to respond quickly and efficiently in other situations. There is a reason that problems you solve are often called "exercises" ... they're just like burpees! They raise your threshold of what you consider difficult; after Calculus, you'll look back at a lot of algebraic processes you faced in the past and say, "Gosh, I can't believe I used to think this was the hardest thing ever." Also, some processes in math — even when you never encounter them again — will improve your intuition for quantitative outcomes and behaviors. We need more people in this world who are quantitatively literate.

(9) If You Don't Ask, I Won't Tell.

Fig. 0.2 Can I help?

Ask questions. Get help. It's as simple as that. If you don't let your instructor know that you need help with something, then he or she can't help you. Often students will approach me with, "I'm sorry to bother you, but...". Or, "I know you're busy, but...". I am sorry if you have other professors who make you feel as if you're intruding, but I can't control them. My personal philosophy, and one I hope other instructors share, is this: taking time to help you with class material is my job. It's why I am here. If you are interrupting me while I'm doing something else, chances are it is some dull exercise in tedium related to a campus committee or departmental assessment — from which I am glad to be delivered even for a few short minutes! And sure, maybe sometimes you'll catch me when I'm on the way out for something else, or when I'm trying to prepare for a class that starts in ten minutes, but that should not be the norm. If your university is primarily residential, then the campus is your home. I come to your home for many hours per day for the primary reason of helping you learn. Why in the world would you not take advantage of that? In Image 0.2, the goof whose bathroom is flooding due to plumbing problems is ignoring the available help. The plumber is standing right there and is waiting to help, but the goof is trying to pretend the problem doesn't exist, or that eventually he'll be able to solve it all himself. Don't be that goof.

(8) Flatten the Curve.

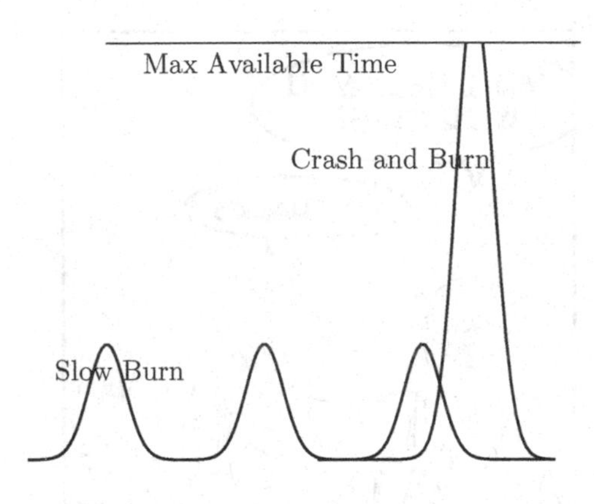

Fig. 0.3 Time on task.

Your assignment is due at 9am tomorrow. It's now 8pm the night before. Have you started it yet? No? Well, Survivor Season 182 starts tonight at 9pm, so you'll start your work at 10pm when the show over, and you'll be fine, right? I think we all know the answer to that.

Your work on problems should begin soon after they're assigned. There are several reasons for this. First, if you wait to begin until the night before the problems are due, it could be several days since you heard about the material in class. How much are you going to remember? Second, if you start early, you can pace yourself. Maybe just doing one or two exercises a day will be enough! Third, when a problem stumps you, you can set it aside and think about it in the shower the next morning. Or you can go to see your instructor about it, or ask a question in class. You won't be scrambling to make them all work the night before they're due. As an instructor, there's nothing more face-palm inducing than to see a student's assignment contain an answer to problem 17 that reads: "17) ??". Because that "??" gets interpreted as, "I tried this problem but didn't know what to do, and since I left myself no time to ask you about it, I can't get it done." Waiting to the last minute and doing a rush job doesn't do anyone any good.

Suppose someone gave you a fine glass of Cabernet Sauvignon to try. (Bear with me if you're not old enough to drink yet!) Sure, you could just gulp it down quickly. But what did you get out of it? Did you learn how it tasted? Do you understand its flavor? If someone served you another glass of a red wine, would you be able to tell if it was the same or not? Can you now tell the difference between Cabernet Sauvignon and Merlot? That's the difference between simply going through the motions of quickly getting the wine from your mouth to your stomach, and actually drinking the wine. Treat your homework solutions like a glass of fine wine. Savor them. Understand them. Learn what they teach you about new problems.

This isn't just about math, you know. Time management is an essential self-defense mechanism in college. Image 0.3 shows a hypothetical graph of the time a student spends on a given class during a portion of the semester. There is some minimum amount of time spent on the class simply by the act of going to class; that's the "In Class" time. The rest of the time devoted to the class comes from what the student does out of class. In the "Slow Burn" curve, the student is devoting a small amount of time each day to the class — by reading the text and doing homework on a daily basis. When the work is due, not much extra time is needed to assemble the solutions to be handed in because many of them are already done. Further, because the student is involved in the subject daily, that student will have a better change to learn the material well. Now consider the other curve; on this one, the student waits until the night before each assignment is due and expends a lot of time getting it done. For any one class, this might not be a problem all the time. But somewhere up the vertical axis is the total amount of time available for classwork during a day, i.e. 24 hours minus sleeping and eating. And suppose now that assignments from several classes are due the same day. To get all the assignments done, this student will have to add together the peaks of activity from all the classes at once, and the total time needed may exceed the time available! That's why this curve is called the "Crash and Burn Curve". Even if only one class is involved, in this method of time management, the student is not engaged in the subject on a daily basis and is less likely to learn the material well.

This book is being written in 2019, 2020, and 2021. By now, we are all familiar with the phrase "flatten the curve". So: flatten your time management curve by spreading out your work; don't allow things to collect until you need spikes of activity to get things done, when you may not have enough time to deal with that spike.

(7) Be Bob Ross, not Jackson Pollock.

Fig. 0.4 Wow, that's a mess.

Remember in grade school when you got all excited because you started using college-ruled paper with the skinnier lines on it? This meant you could squeeze more work onto one page! How awesome was that?!

Well ... STOP IT. Let your work breathe. Spread it out. Don't cram two columns on one page. Don't write really small for most of the page, and then start writing sideways up the margin at the bottom of the page to fit in the rest of that solution. Don't allocate space for your problems before you even start working on them. Write a lot things down! Use that paper! Kill those trees! Don't ask, "Do I have to show these steps?", but rather ask, "Why shouldn't I show these steps? It's my work, after all." Leave yourself a trail of bread crumbs (math crumbs?) in your work, so that *when* (not if) you have to go back to refer to your work to see how you did something previously, you can make sense of your own work. You put a lot of time into your work, why not make your own product useful to you when it's time to study for a test?

Having said all that, there's a difference between splatting your work all over the page in a completely disorganized fashion, and laying it out in a clear and deliberate design.

Each solution you create tells a story; be sure you carefully show the beginning, middle, and end. Your solutions should be self-contained, so that a grader does not have to have the original questions in order to tell whether your answers are right. They should stand on their own. Look at it this way: your work should be so complete that a semi-informed reader, who has a general familiarity with the content, would be able to understand from your work alone what the question was, what work you did to address the question, and what your final answer is. Because when it comes time to study for a test, that semi informed reader is going to be you. It's likely you will have forgotten the fine details of what went into the work you're using as a study guide. Did you leave yourself enough information to make your work useful, or did you leave yourself nothing useful?

When you look at your completed work, can you see and understand the details of the landscape that has arisen in your work, or does you page look like you ingested little bits of math and then sneezed onto the page?

When it comes to presenting your mathematics, be Bob Ross, not Jackson Pollack.

(6) Math be like a languaj: it gots propar gramer and spelin.

Fig. 0.5 Efficient communication requires proper use of language.

The notation that we today use in mathematics has developed over centuries. Granted, a lot of it might not make sense, but hey, it's what we have. So learn to love it, and learn to use it.

If started I paragraph a two rite, and used whales improperly by in rong putting ones and order wrong in and leafing out some ... you'd have a really hard time reading and understanding it. Math is the same. When writing math, each symbol has a purpose. The absence of a symbol can be just as important as its presence. Are you using symbols properly? Do you know what the symbols actually mean? Are you forming proper mathematical grammar? One of the most egregious violations against mathematical grammar is the abuse of the poor equals sign. That little guy gets no respect. Let's say the question is, "What are the roots of $f(x) = x^2 - 4x + 3$?" What do you think of this answer?

$$f(x) = x^2 - 4x + 3 = (x - 1)(x - 3) = 1, 3$$

This is very representative of how results will come in on many submissions. But it's awful. Felony homicide has been committed on the third equals sign. It is not being used to connect two things that are equal; rather, it's being used to mean "here was one result, and it sort of leads to this next thing." The term $(x-1)(x-3)$ is not *equal* to the list of two numbers 1 and 3. Now, surely $x=1$ and $x=3$ are the correct answers to the question, but the presentation is poor. Rather, we should use words to make transitions between mathematical phrases. A better response would be,

$$\text{Since } f(x) = x^2 - 4x + 3 = (x-1)(x-3),$$

$$\text{then } f(x) = 0 \text{ when } x = 1 \text{ or } x = 3.$$

Does that take a couple more seconds to write? Yep. Does it use a bit more paper? Yep. Is it more clear, and does it use proper notation? Yep. This goes back to Item 7 above: use the paper. Space things out. Let them breathe. Each solution tells a story, be sure you show the beginning, middle, and end. In fact, the better version of this answer is so self-contained that you could show it to someone else who had not seen the original question, and they would know what the question was, what work you did to respond, and what your answer is. I'm repeating these phrases from before because they are so important.

You want your instructor to be clear and precise in her mathematical communications with you — when creating notes, or when providing feedback on work. Return the favor.

(5) "Almost" only counts in Horseshoes and hand-grenades.

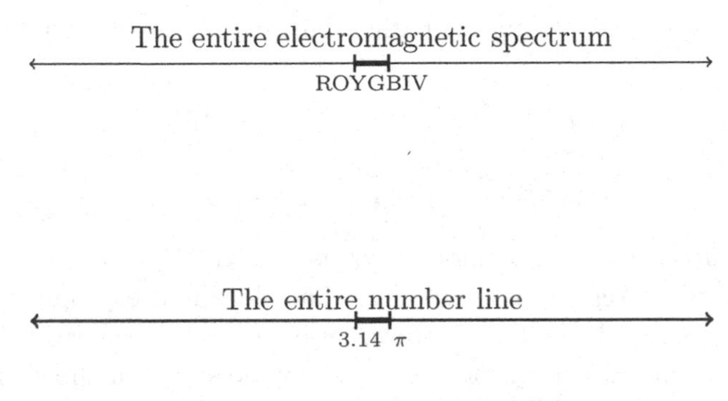

Fig. 0.6 π is equal to 3.14, right?

You know, the entire electromagnetic spectrum is really, really big. The visible portion of the spectrum is, relatively, very, very small. Image 0.6 shows a representation of this. Since the visible portion of the spectrum is such a tiny part of the whole thing, why are we making such a fuss about minor differences within the visible spectrum? Instead of assigning different colors and inventing our complicated friend Roy G. Biv, (Red, Orange, Yellow, Green, Blue, Indigo, Violet), why don't we just assign all the colors as BBBBBBB?

This is ridiculous, right? We assign different colors because we know there are different colors. We see them! So let's shift to the numerical spectrum, which is also really, really big. And on that scale, surely representing the number π as 3.14, or the number $\sqrt{2}$ as 1.57 is no big deal, right? ... Wrong, it is a big deal. We know that π and 3.14 are not the same number. Sure, they are really close to each other, but just like B is different than I on the visible spectrum, π and 1.57 are different on the numerical spectrum.

When possible, keep your answers exact. If your result should be π^2, write it as π^2, not 9.87. If the correct value is $\sqrt{3} + 1/2$, then write it that way, not as 2.2. Every time you round numbers off, you introduce error. Be as precise as you can. Now clearly, this will require judgment calls. If an exact answer turns out to be something hideous, and writing the exact version will obscure any intuitive understanding of what the number really is, then sure, approximate away. For example, one of the roots of the polynomial $f(x) = x^4 - 2x^3 + 2x^2 - 1$ is

$$x = -\frac{\sqrt[3]{17 + 3\sqrt{33}}}{3} + \frac{2}{3\sqrt[3]{17 + 3\sqrt{33}}} + \frac{1}{3}$$

It would be ridiculous to report this exact value, and we'd instead report it as approximately -0.544. On the other hand, one of the roots of the polynomial $f(x) = x^4 - 2x^2 - 1$ is $x = \sqrt{1 + \sqrt{2}}$. Is it *not* ridiculous to report the value this way. You can still have intuition about the number. Is it negative? No. Is it larger than 10? Clearly, no. Is it larger than 5? Clearly, no. It's certainly larger than 1, but likely not larger than 2. Why spoil this perfectly good number by writing a less accurate version as 1.55?

(4) You Can't Have It Both Ways.

Fig. 0.7 But I hate word problems!.

You can't simultaneously complain about not being shown practical uses of things you're learning, while at the same time balking at being posed questions that put the content into practical terms.

Okay, we know that many word problems are not that practical. The results of that race, between a car and a train moving perpendicularly away from each other at two different speeds, have never been that useful. But if you can't respond to the simple context-based questions which are — like hotel toilet seats — "sanitized for your protection", then how are you ever going to be able to respond to truly practical questions that reflect the complexities and uncertainties of the real world? Start small. Learn how to apply the content you're learning to pseudo-practical scenarios, then move on to the big leagues later.

(3) If You Have to Cheat, You'd Better Cheat on the Cheating, Too.

Fig. 0.8 If you need to cheat, you probably won't be able to cheat effectively.

Some students in math classes will cheat. There's no way around that fact. Usually cheating takes the form of borrowing answers from another student's work — either with or without that person's permission, and either by being in the same room or by looking something up on-line. But here's the thing: if you are so uncomfortable with the content you're being asked about that you feel the need to borrow someone else's work, how in the world do you know your content well enough that you can tell if what you're copying is actually correct?

In my history as a math professor, the vast majority of cheating I've had to report was discovered because students shared answers that were ridiculously incorrect. That could mean they shared "answers" among themselves. It could mean they both found the same, but incorrect, on-line "resource." But certainly, the borrower was completely unable to diagnose whether what he or she was copying was reasonable, or total malarkey. Correct answers are expected to be similar. Incorrect answers that are identical because of the ridiculous similarities they share stand out like a marshmallow in a bowl of raisins.

(2) Three Little Letters.

Fig. 0.9 The difference three little letters make.

Here's the harsh truth: your teacher wants you to fail. Can you believe that? It's true. But I'm not talking about your grade. I'm talking about your first try on a hard problem. Look, if you can do all your problems correctly on your first try, you probably signed up for the wrong class. You should be challenged by the problems you do. Very often, they won't go well the first time. But that's good! Your teacher wants you to fail to get the problem right on that first try because that means you are challenging yourself and you're about to learn something. A set-back does not come because you didn't understand a topic or get a problem right the first time, the set-back comes if you don't try again. People who run a four-minute mile didn't do that the first time they tried. Professional pitchers in baseball did not throw a 90 mile-per-hour fastballs the first time out. They tried, and tried again. So should you. Be determined.

This idea is not original to me, I've heard it come out at professional conferences, and I have no idea where it originated, but it's a solid truth: there are three letters that can change your outlook when you're getting frustrated. Have a look at the two panels in Image 0.9; the extra three letters in the second panel of makes that image tell an entirely different story than the first panel. Imagine how different the mindset and motivation is for the person on the right as compared to the person on the left. Which one are you?

(1) Knowledge is Not a Toxin.

Fig. 0.10 Knowlege is not a toxin.

Let's say that every negative stereotype of a math class is true. It's too hard. You'll never get it. It makes you miserable. Worst of all, you'll never use any of it in real life. Well ... suck it up, clench your teeth, and turn the page. Just remember — if you accidentally learn something, it won't hurt you.

Chapter 1

Years of Work in One Chapter

1.1 The Basics of Functions

Introduction

Learning calculus is a lot like learning a new language. When you learn a new language, you start by learning some vocabulary. But you can't stop there, you have to learn how those vocabulary words get put together to make sentences. Then you learn how sentences get put together to make paragraphs. You have to learn to spell the words right, and make sure the sentences and paragraphs are grammatically correct. Calculus, or math in general, is no different. We take lots of individual things, put them together into larger constructions, and in order for those constructions to mean anything, we have to create and spell them correctly. In math, our "words" are functions, and our "sentences" can be equations or other expressions. And we put these sentences together into solutions to problems, or demonstrations (proofs) of new mathematical concepts.

As they are the building blocks of mathematics, in both the "grammatical" and computational senses, we must become adept at their use. Computationally, each individual function is like a little engine that takes input, churns up that input, and then gives output. Different engines, or functions, do different things to the input. We can take many individual functions and combine them into more complicated functions. Fortunately, there's a relatively small number of basic function types we have to get used to in order to make lots of headway into applicable mathematics. In order to discuss and use functions, though, we have to learn more related language.

So let's start off with a review of the language of functions. Without a firm grasp of what functions are and how we talk about them, life in a calculus class is difficult. Because you've made it this far in your personal mathematical trajectory, you surely have used functions before, and so here we just need a recap.

Functions in Algebraic Form

Functions are associations between sets of numbers. The sets of numbers we're concerned with will be part, or all, of the real numbers. You probably think of functions in their algebraic form, where a function is seen as a recipe that takes input, does something to it, and returns the resulting output. For example, I'm sure you recognize this as a function:

$$f(x) = x^2 + 5 \tag{1.1}$$

Pause for a minute to appreciate that there's quite a story in that little expression. On the left side alone, we get the name of the function (f for function, we're clever like that) and a name (x) for the input on which f acts. The right side gives a recipe for how the function acts on that input. We're told that our input will be squared, and then 5 will be added to that result. The input might be be a number ($f(1) = 6$), or it could be a replacement ($f(a) = a^2 + 5$), or it could be something more complicated: $f(x + 4) = (x + 4)^2 + 5$.

There are plenty of times we get tired of using f and x, and there is no reason we can't use other letters, even from different alphabets. The expression $h(t)$ identifies a function named h, with input named t. The expression $\sigma(\alpha)$ identifies a function named σ with input named α. Let's do better than describing the input as "input", though. Using the traditional notation $f(x)$, the input x can be appropriate numbers or expressions; since x can vary, we call it a *variable*. By "appropriate numbers", we mean those which are allowed to be used according to the nature of the function; as you know, the numbers we can use as input for $f(x) = x^2 + 5$ and those we can use for $f(x) = \sqrt{x - 2}$ are different. We might also mean that certain numbers are ruled out explicitly for some reason; for example, if you had a bad experience with numbers larger than 10, you so might invent a function $f(x) = (x - 6)^3$ for $x \leq 10$. There may be implicit or explicit rules about what values x can take on, but we still call it an *independent* variable.

One of the reasons you need to know how to read and write functional notation is that it helps you recognize and describe constructions for comparison. In the example $f(x) = x^2 + 5$, do you know that $f(2x)$ and $2f(x)$ are *not* the same thing? But in other cases, $f(2x)$ and $2f(x)$ might indeed be equal. To have conversations like this, the ability to understand and handle functional notation is crucial. You have to know that "$f(2x)$" means that we want to replace any instance of x in the recipe with $2x$; here, that means $f(2x) = (2x)^2 + 5$. But "$2f(x)$" means to simply take the original function and multiply the whole thing by 2; here, that means $2f(x) = 2(x^2+5)$. And $4x^2+5$ is certainly not equal to $2x^2+10$, so indeed, $f(2x)$ is not equal to $2f(x)$.

A very common mistake, that you have surely made before, is to assume that $f(a+b)$ is equal to $f(a) + f(b)$. For some functions, that might be true. But for most functions, it's *not* true. And again, this conversation can only happen when you understand functional notation. The first expression means take your function f, go to wherever the variable sits, and replace the variable with the expression $a + b$. The second means make TWO items, one with a plugged into the function and one with b plugged into the function, and then add the two items together. You may not get the same thing as in the first expression! If you ever are in a position to be grading math problems, charge students a dollar each time they claim that $\sqrt{a+b} = \sqrt{a} + \sqrt{b}$. This is definitely not true (try it with $a = 4$ and $b = 5$), but using that step could very well be the most common algebraic mistake on Earth.

Here's a bit of practice with functional notation.

You Try It

(1) Given the function $f(x) = \sqrt{x+5}$, identify the name of the function and the name of the independent variable, and then find (simplify where possible) $f(3)$, $f(x-1)$, $f(4+11)$, and $f(4) + f(11)$.

You Try It

(2) Given the function $b(w) = 1/(w^2 - 4)$, identify the name of the function and the name of the independent variable, and then find (simplify where possible) $b(3)$, $b(-w)$, $b(0+1)$, and $b(0) + b(1)$.

You Try It

(3) If $f(x) = 2x + 4$, and a, b are constants, which of the following statements are true?

(a) $f(x - 4) = 2x$
(b) $f(x) - 4 = 2x$
(c) $f(a + b) = f(a) + f(b)$
(d) $f(a \cdot b) = f(a) \cdot f(b)$

Let's go back to our sample function in Eq. (1.1), $f(x) = x^2 + 5$, and see what we else can learn from the right side of the expression:

- The set of numbers we are allowed to use for the independent variable is called the *domain* of the function. In this case, we can plug any number we want into $x^2 + 5$, and so the domain of this function is all real numbers — or, in symbols, $\mathbb{R}$.
- The set of numbers we can get in return from a function is called the *range* of the function. In this case, no matter what we put in for x, we will always get back a positive real number that's 5 or larger — so, the range of this function is that set of numbers. It's cumbersome to write "all real numbers greater than or equal to five" in words, so we can use various notations to shorten this up. In "interval notation" for sets of numbers, we could write $[5, \infty)$ - this collects all numbers between 5 and infinity, including 5 itself (remember, a square bracket includes that endpoint, while a round bracket would not). In set notation, we can write $\{y : y \geq 5\}$ where we have now named the output of the function by the name y. Or, we could write $f(x) \geq 5$. All of these describe the same set / range.

Since I'm sure you've played the domain / range game before, this is a good time to review it:

You Try It

(4) Find the domain and range of $f(x) = \sqrt{x + 5}$.
(5) Find the domain and range of $f(x) = 1/(x^2 - 4)$.

Functions in Graphical Form

Ever since you saw $f(x) = x^2 + 5$ in the paragraphs above, you've probably had a burning desire to graph it. I'm sure you recognize it as a parabola, with vertex at $(0, 5)$, that opens upwards. The graph of a function is a visual representation of the association between the domain and range of the function. If we have the algebraic form of a function, we get the graph quickly by either plotting points (where we make a list of a bunch of x-values, compute the corresponding functional values, and plot the resulting points) or by using past experience and adapting already known graphs to a new function. You probably don't need to plot points to know the graph of $f(x) = x^2 + 5$; rather, you already know the graph of $f(x) = x^2$, and you understand what adding 5 to that function does to its graph.

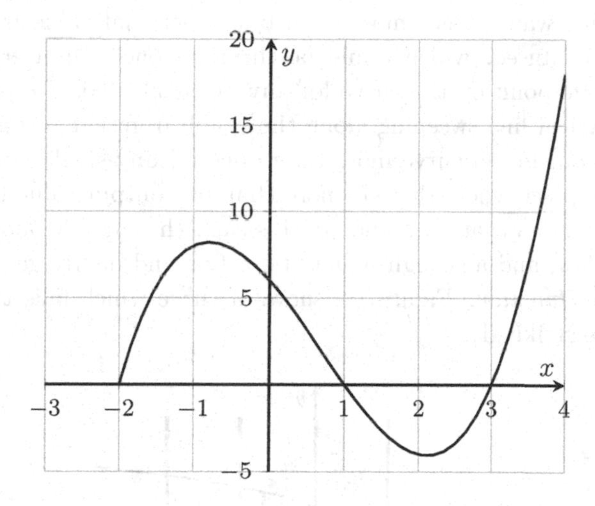

Fig. 1.1 There's no "formula", but it's still a function!

Now it's nice to have the algebraic form of a function, but it isn't completely necessary. A graph alone is a perfectly good representation of a function. Consider the graph shown in Fig. 1.1. We do not know the algebraic form of this function, but who cares? We still know lots of things. We know that the name of the independent variable is t (from the horizontal axis). The name of the function itself is g (from the vertical axis). Therefore, this graph represents a function $g(t)$. We know that if $t = 1$ then g is 0, i.e. $g(1) = 0$. We know that $g(3) = 0$. We can estimate any other value we need. From what we see of the graph, it looks like only real numbers larger than -2 are used as values of the independent variable, and so the domain of this function appears to be $t \geq -2$. If we guess that that low point around $t = 2$ has a functional value of about -4, and if we assume the graph just keeps going to the upper right, then we have a range of approximately $[-4, \infty)$. This graph is a perfectly good function even though we don't know its recipe!

On the other hand, not any ol' squiggly line on a graph is considered an honest function. A curve has to obey certain restrictions to be allowed in the exclusive "function" club. For one thing, there can be no confusion about associations. If we have a member of the domain selected, i.e. an input value, there can be no confusion about which output value gets associated with it. In the algebraic sense, that's a no-brainer. If we plug in a value for x, we don't want to get more than one functional value back. If we did, we'd be confused: which would be the right one? On a graph, there can be only one point on the curve for any given x value. To test this, we imagine a vertical line sweeping along the graph from left to right; if that line ever hits our graph curve more than once for any single x-value, then we've found a place where there is more than one output value for a single input (x) value, and that's a no-no. In this case, the curve is said to fail the *vertical line test*, and although it may be a fine and dandy generic curve, it's not a true *function*. Figure 1.2 shows a curve which fails the vertical line test quite strikingly.

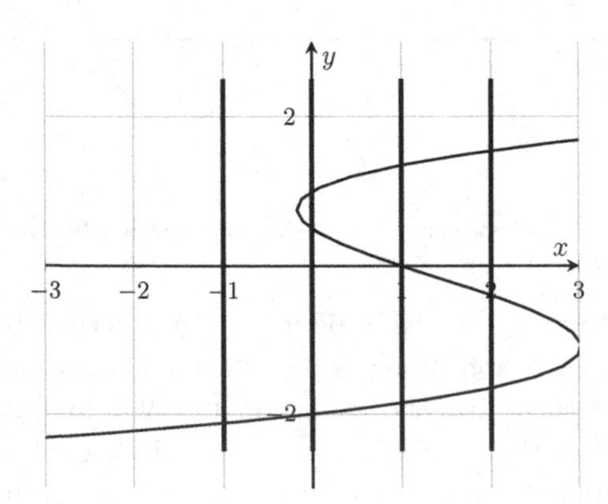

Fig. 1.2 Vertical lines at many x-values cross the graph more than once. This graph fails the vertical line test.

The obvious follow-up question is: Given that there's a vertical line test, is there a *horizontal line test*? The answer is yes, and a function that passes the horizontal line test is super-special. Let's call our vertical axis the y-axis, since we're used to that. Consider a horizontal line sweeping across a curve / function from bottom to top. If this curve never crosses the function more than once for any single y value, what does that mean? It means that for any y value in the range of the function, there is only one x-value that produces it. In this case the function is called *one-to-one*: for any x-value, there's only one possible y-value, and for any possible y-value there's only one x-value that gives it. This is actually a pretty cool thing. Our example from above, $f(x) = x^2 + 5$, is NOT a one-to-one function, since its graph fails the horizontal line test, as shown in Fig. 1.3.

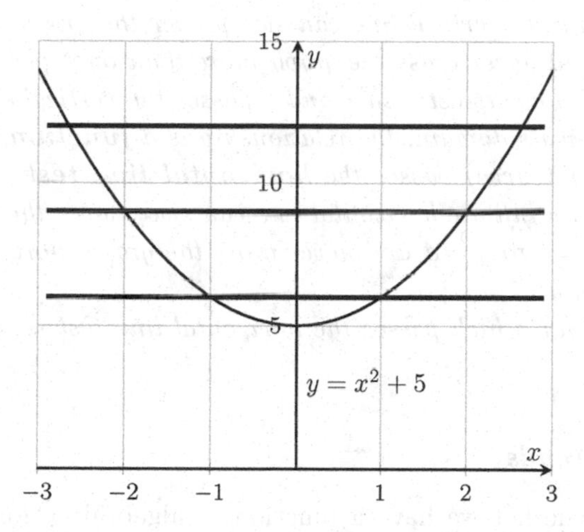

Fig. 1.3 Horizontal lines at many y-values cross the graph twice. This graph fails the horizontal line test.

What are the consequences of failing the horizontal line test? Well, for example, suppose we wanted to know where the value $y = 6$ comes from. We now have *two* possible origins: we get $y = 6$ from both $x = -1$ and $x = 1$. There's nothing wrong with that, it's just super-special when there's only one x value for every possible y-value.

You Try It

(6) Four curves are shown in Fig. 1.5 (at the end of this section). Only two of them are actually functions. Identify them (using labels A, B, C, D). Of those two, only one of them is one-to-one. Identify it.

Here is a summary of definitions related to relationships and functions, so far:

Definition 1.1.

(1) The ***domain*** *of a relationship $y = f(x)$ is the set of all values of x for which y is defined.*
(2) The ***range*** *of $y = f(x)$ is the collected set of all values of y generated by all values of x in the domain.*
(3) A graph (curve) passes the ***vertical line test*** *on an interval $[a, b]$ if an infinite vertical line can sweep over the graph from $x = a$ to $x = b$ and never cross the graph more than once per value of x.*
(4) If a relationship between x and y passes the vertical line test everywhere on its domain, the relationship is a ***function****.*
(5) A graph (curve) passes the ***horizontal line test*** *on an interval $[c, d]$ if an infinite horizontal line can sweep over the graph of $f(x)$ from $y = c$ to $y = d$ and never cross the graph more than once per value of y.*
(6) A function which passes the horizontal line test is called ***one-to-one****.*

Identifying Trends

Regardless of whether we have a function in algebraic ("formula") form, graphical form, or both, sometimes there are deeper patterns to be observed, and being able to express these patterns using functional notation is important. Here is an example of categorization that is very visually evident on a graph, and somewhat easily assessed algebraically.

Functions can be categorized as *even* functions, *odd* functions, or neither. The algebraic definition of these is as follows:

Definition 1.2.

- *$f(x)$ is an even function if $f(-x) = f(x)$.*
- *$f(x)$ is an odd function if $f(-x) = -f(x)$.*

Clearly this is a test of your ability to read and interpret functional notation. What the heck does "$f(-x) = f(x)$" mean? Don't panic, it's pretty simple, and we can look at it on a value by value basis as well as a global basis.

On a value by value basis, $f(-x) = f(x)$ means that if we evaluate the function any one value and also its opposite, we'll get the same result back both times. That is, $f(-1) = f(1)$, or $f(-3/2) = f(3/2)$, or $f(-\sqrt{5}) = f(\sqrt{5})$, and so on. Now, we can't sit here and test out every possible value in the domain of a function, so then we move to the more global inspection of this definition. The notation is telling us that we can test every value in the domain simultaneously by replacing every instance of "x" in a function's recipe with $-x$ (that's what $f(-x)$ means) and seeing if we return to the original function $f(x)$ (so that $f(-x) = f(x)$). These situations are not rare, and we've been working with such a function all along: $f(x) = x^2 + 5$ is an even function. When we construct $f(-x)$, we get $f(-x) = (-x)^2 + 5$. Then when we simplify that, we have $(-x)^2 + 5 = x^2 + 5$. And now we're back to the original function. Put into a "stream" of equalities,

$$f(-x) = (-x)^2 + 5 = x^2 + 5 = f(x)$$

So for $f(x) = x^2 + 5$, we have $f(-x) = f(x)$, meaning $f(x) = x^2 + 5$ is an even function.

There is also a nice interplay between even functions and their graphs. If $f(x)$ is even, then we get the same value back from both $x = -1$ and $x = 1$. The values of $f(-2)$ and $f(2)$ must the same, and similarly for $f(-0.775)$ and $f(0.775)$, etc. The visual result of these equalities among pairs of points is that the left and right halves of the graph (separated by the y-axis) will be mirror images of each other. This is called being *symmetric about the y-axis.*

You Try It

(7) Test algebraically whether $g(t) = \sqrt{t^2 + t}$ is an even function or not.

(8) One of the two graphs in Fig. 1.6 (at the end of this section) shows an even function. Identify it.

The definition of an odd function is similar. Writing that $f(-x) = -f(x)$ is, in functional language, the most efficient way of writing that if we plug in $-x$ for x, we'll get the *negative* of the original function back. When can this happen? How about $f(x) = x^3$?

$$f(-x) = (-x)^3 = -x^3 = -f(x)$$

So $f(x) = x^3$ is an odd function. In contrast, $f(x) = x^3 + 1$ is NOT an odd function. It's not an even function either. Make sure you see why! Graphically, oddness is seen as a sort of upside down flip across the origin. For an odd function, $f(-1)$ is the same as $-f(1)$, or $f(-3)$ is the same as $-f(3)$; there is an upside down symmetry across the origin. A function is odd if you can take the right half of its graph, reflect it across the y-axis, and then flip it upside down across the x-axis to end up with the left half of its graph. If you know the graph of $f(x) = x^3$, you can see this upside down symmetry there; this is called being *symmetric about the origin*. Figure 1.4 shows samples of the symmetry shown by even and odd functions; can you identify the functions used in those illustrations?

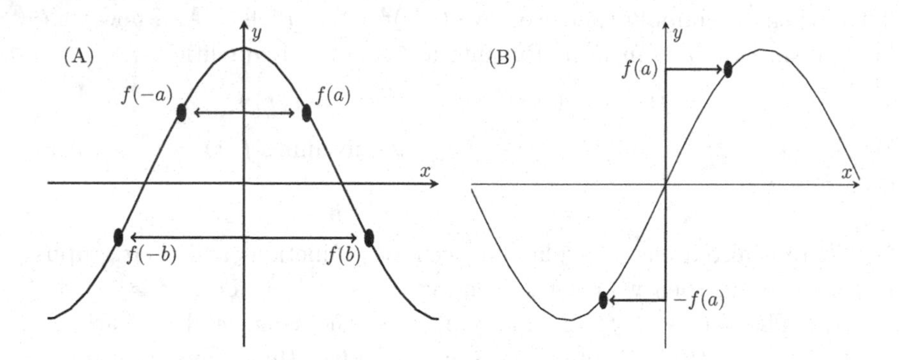

Fig. 1.4 Graphical symmetry for (A) even and (B) odd functions.

You Try It

(9) One of the two graphs in Fig. 1.7 (at the end of this section) shows an odd function. Identify it.

(10) Test algebraically whether $g(t) = 1/(t^3 - t)$ is an odd function or not.

Functions In Tabulated Form

Math classes are, like hotel toilet seats, sanitized for your protection. Mathematically, it is easist and most efficient to have functions in algebraic and graphical form, and so that's how we deal with most of them. But the real world doesn't always work like that. In most "real world" settings, relationships are not found or given so neatly. Most relationships come from observation. Most observation is not done continuously. For example, if you are monitoring the temperature throughout the day, you likely have a table of data:

t	0	0.5	1.0	1.5	2.0	...
T	75	77	78	78	76	...

There's no recipe. The graph is just a bunch of points, not a nice curve. But your data still represents a perfectly good function, and you'd probably name it $T(t)$, temperature T as a function of time t.

We will often work with tabulated functions. Don't freak out.

Have You Learned...

- How to read and use functional notation?
- How to find the domain and range of a function?
- How to spot even and odd functions, both algebraically and graphically?
- That functions can be represented graphically and in tabular form as well as by a formula?

The Basics of Functions — Problem List

The Basics of Functions — You Try It

Try these problems, appearing above, for which solutions are available at the end of this section.

(1) Given the function $f(x) = \sqrt{x+5}$, identify the name of the function and the name of the independent variable, and then find and simplify $f(3)$, $f(x-1)$, $f(4+11)$, and $f(4) + f(11)$.
(2) Given the function $b(w) = 1/(w^2-4)$, identify the name of the function and the name of the independent variable, and then find and simplify $b(3)$, $b(-w)$, $b(0+1)$, and $b(0) + b(1)$.
(3) If $f(x) = 2x + 4$, and a, b are constants, which of the following statements are true?

(a) $f(x-4) = 2x$
(b) $f(x-4) = 2x$
(c) $f(x) - 4 = 2x$
(d) $f(a+b) = f(a) + f(b)$
(e) $f(a \cdot b) = f(a) \cdot f(b)$

(4) Find the domain and range of $f(x) = \sqrt{x+5}$.
(5) Find the domain and range of $f(x) = \dfrac{1}{x^2-4}$.

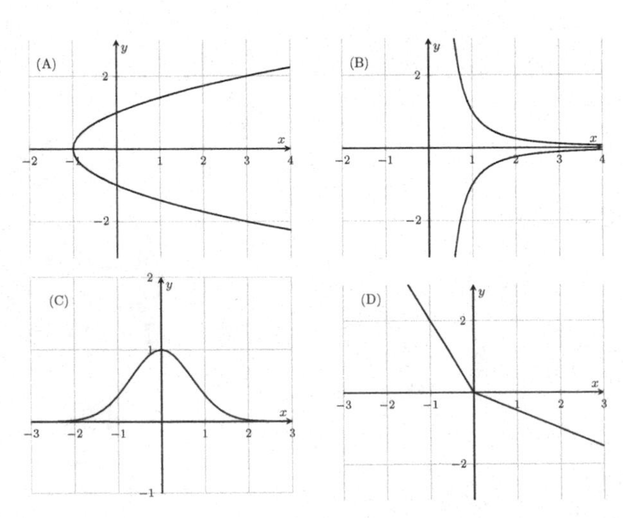

Fig. 1.5 Find the functions, as well as the one-to-one function (w/ YTI 6).

(6) Four curves are shown in Fig. 1.5. Only two of them are actually functions. Identify them (using labels A, B, C, D). Of those two, only one of them is one-to-one. Identify it.

(7) Test algebraically whether $g(t) = \sqrt{t^2 + t}$ is an even function or not.

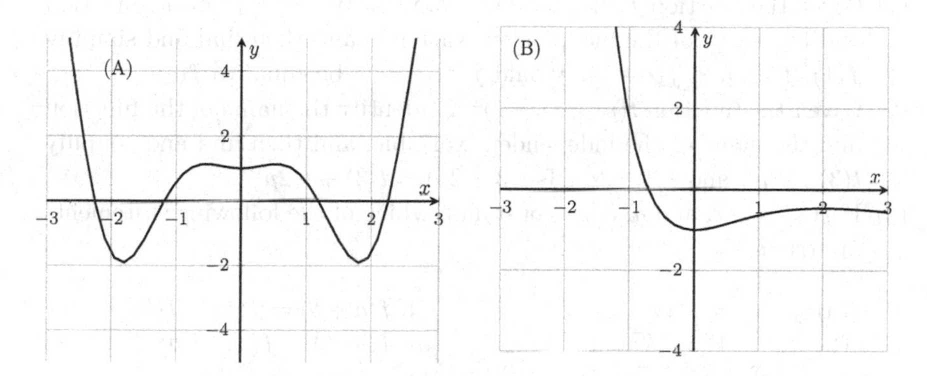

Fig. 1.6 Find the even function (w/ YTI 8).

(8) One of the two graphs in Fig. 1.6 shows an even function. Identify it.

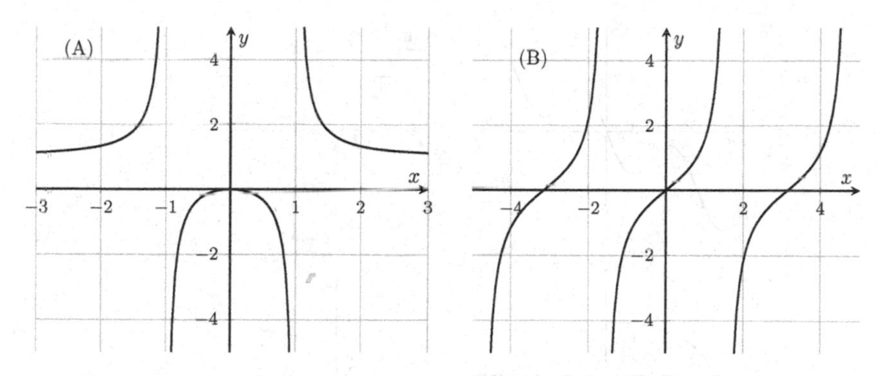

Fig. 1.7 Find the odd function (w/ YTI 9).

(9) One of the two graphs in Fig. 1.7 shows an odd function. Identify it.

(10) Test algebraically whether this is an odd function or not: $g(t) = \dfrac{1}{t^3 - t}$

The Basics of Functions — Practice Problems

Try these as you get the hang of the You Try It problems. Solutions to these problems are available in Sec. A.1.1.

(1) Given the function $f(x) = x/\sqrt{x^2 + c}$, identify the name of the function and the name of the independent variable, and then find and simplify $f(1)$, $f(c)$, and $f(c-1)$. Would $f(c) - f(1)$ be equal to $f(c-1)$?

(2) Given the function $t(p) = (p-2)^2/2$, identify the name of the function and the name of the independent variable, and then find and simplify $t(3)$, $t(2p)$, and $t(3+2p)$. Is $t(3+2p) = t(3) + t(2p)$?

(3) If $f(x) = \sqrt{x}$, and a, b are constants, which of the following statements are true?

(a) $f(2x) = 2f(x)$
(b) $f(x-1) = \sqrt{x} - 1$
(c) $f(x) - 1 = \sqrt{x} - 1$
(d) $f(a+b) = f(a) + f(b)$
(e) $f(a \cdot b) = f(a) \cdot f(b)$

(4) Find the domain and range of $h(p) = \dfrac{1}{p} - \sqrt{2p}$.

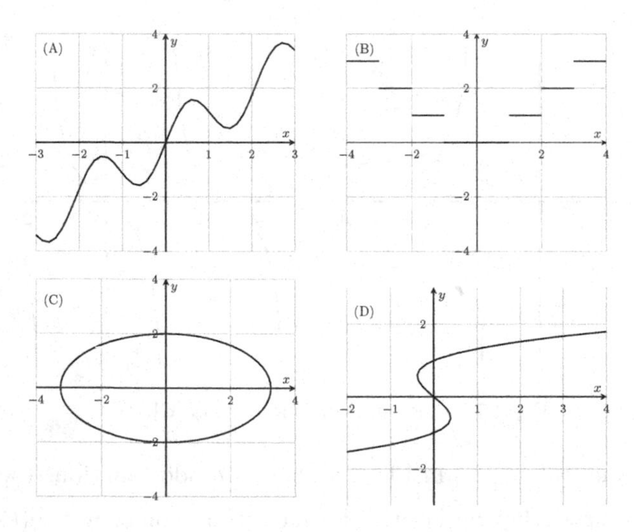

Fig. 1.8 Find the functions, and in particular, identify the even and odd functions (w/ PP 6).

(5) Four curves are shown in Fig. 1.8. Only two of them are actually functions. Identify them (using labels A,B,C,D). Of those two, one is odd, and one is even. Identify them.

(6) Determine algebraically whether $f(x) = (x - 1)(x^2 - 2)^{3/2}$ is an even function, an odd function, or neither.

The Basics of Functions — Challenge Problems

Try these problems to test your skills with the ideas in this section. Solutions to these problems are available in Sec. B.1.1.

(1) It's your turn to design some functions.
   (a) Give a function $f(x)$ that has not yet appeared in this section for which $f(a+b) = f(a) + f(b)$, and demonstrate that this expression is true for your function. Give another function $f(x)$ that has not appeared for which $f(a + b) \neq f(a) + f(b)$, and demonstrate that this expression is true for your function.
   (b) Give a function $f(x)$ that has not yet appeared in this section for which $f(2a) = 2 * f(a)$, and demonstrate that this expression is true for your function. Give another function $f(x)$ that has not appeared for which $f(b/2) = f(b)/f(2)$, and demonstrate that this expression is true for your function.

(2) For $g(t) = t^2 + 1$, create and simplify the expressions $g(t + 1)$, $g(2t)$, and $\dfrac{g(t) - g(1)}{t - 1}$.

(3) Find the domain and range of this function, and show whether it is even, odd, or neither: $f(t) = \dfrac{t^4 - 4}{t^2 + 2}$

The Basics of Functions — You Try It — Solved

(1) Given the function $f(x) = \sqrt{x+5}$, identify the name of the function and the name of the independent variable, and then find and simplify $f(3)$, $f(x-1)$, $f(4+11)$, and $f(4)+f(11)$.

☐ The name of the function is f. The independent variable is x. The expressions are as follows:

$$f(3) = \sqrt{3+5} = \sqrt{8} = 2\sqrt{2}$$
$$f(x-1) = \sqrt{(x-1)+5} = \sqrt{x+4}$$
$$f(4+11) = \sqrt{(4+11)+5} = \sqrt{20} = 2\sqrt{5}$$
$$f(4)+f(11) = \sqrt{4+5}+\sqrt{11+5} = \sqrt{9}+\sqrt{16} = 3+4 = 7$$

Note that $f(4+11)$ is not the same as $f(4)+f(11)$. ■

(2) Given the function $b(w) = 1/(w^2-4)$, identify the name of the function and the name of the independent variable, and then find and simplify $b(3)$, $b(-w)$, $b(0+1)$, and $b(0)+b(1)$.

☐ The name of the function is b. The independent variable is w. The expressions are as follows:

$$b(3) = \frac{1}{3^2-4} = \frac{1}{5}$$
$$b(-w) = \frac{1}{(-w)^2-4} = \frac{1}{w^2-4}$$
$$b(0+1) = \frac{1}{(0+1)^2-4} = \frac{1}{1^2-4} = -\frac{1}{3}$$
$$b(0)+f(1) = \frac{1}{0^2-4} + \frac{1}{1^2-4} = -\frac{1}{4} - \frac{1}{3} = -\frac{7}{12}$$

Note that $b(-w) = b(w)$ but $b(0+1)$ is not the same as $b(4)+b(1)$. ■

(3) If $f(x) = 2x+4$, and a, b are constants, which of the following statements are true?

(a) $f(x-4) = 2x$
(b) $f(x)-4 = 2x$
(c) $f(a+b) = f(a)+f(b)$
(d) $f(a \cdot b) = f(a) \cdot f(b)$

□ Make sure you understand the difference between expressions like "$f(x-4)$" and "$f(x)-4$". The former means: go into the recipe for the function, and wherever you see x, replace it by $x-4$. The latter means take the entirety of the function recipe for $f(x)$ and subtract 4. Therefore, with $f(x) = 2x+4$:

(a) $f(x-4) = 2(x-4)+4 = 2x-8+4 = 2x-4$, so "$f(x-4) = 2x$" is *false*.
(b) $f(x)-4 = (2x+4)-4 = 2x$, so "$f(x)-4 = 2x$" is *true*.
(c) $f(a+b) = 2(a+b)+4 = 2a+2b+4$, so "$f(a+b) = f(a)+f(b)$" is *false*.
(d) $f(a \cdot b) = 2(a \cdot b)+4 = 2ab+4$, so "$f(a \cdot b) = f(a) \cdot f(b)$" is *false*.

■

---

(4) Find the domain and range of $f(x) = \sqrt{x+5}$.

□ Since $x+5$ can't be negative, the domain of f is $x \geq -5$. Since we will never get a negative result, the range of f is all non-negative numbers, or $[0, \infty)$. ■

---

(5) Find the domain and range of $f(x) = \dfrac{1}{x^2-4}$.

□ Since we can't have $x^2-4 = 0$, the domain of f is all real numbers except -2 and 2. We can write this as $x \neq \pm 2$, or $x \in \mathbb{R} \setminus \{-2, 2\}$, etc. We can get pretty much any number except 0 as output of this function, so the range is all real numbers except 0. ■

---

(6) Four curves are shown in Fig. 1.9. Only two of them are actually functions. Identify them (using labels A,B,C,D). Of those two, only one of them is one-to-one. Identify it.

□ (A) and (B) fail the vertical line test, so (C) and (D) are the two functions. Since (C) fails the horizontal line test, then it is not one-to-one. (D) passes the horizontal line test, and is the one-to-one function.

■

---

(7) Test algebraically whether $g(t) = \sqrt{t^2+t}$ is an even function or not.

□ If g is even, we need $g(-t) = g(t)$. But

$$g(-t) = \sqrt{(-t)^2+(-t)} = \sqrt{t^2-t}$$

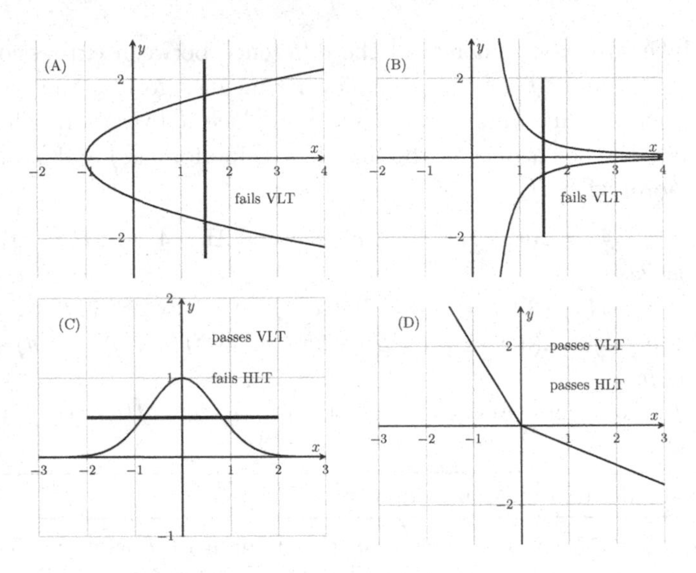

Fig. 1.9 Find the functions, as well as the one-to-one function (w/ YTI 6).

and so $g(-t)$ is not the same as $g(t)$. The function is not even. Note that this doesn't automatically make it an odd function, a function can be neither even nor odd. ■

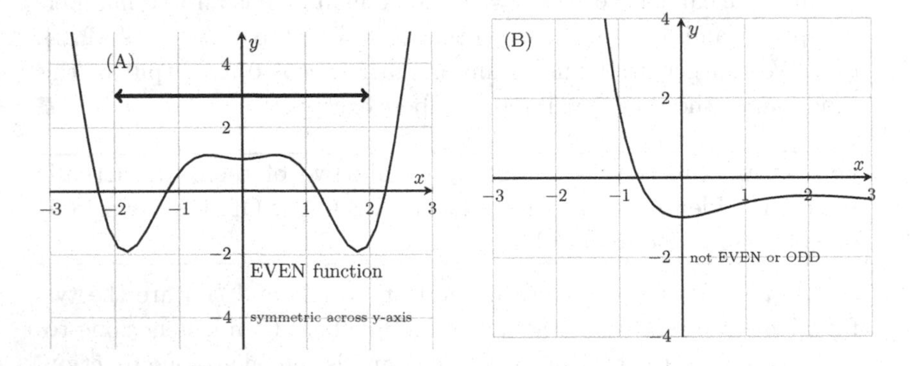

Fig. 1.10 Find the even function (w/ YTI 8).

---

(8) One of the two graphs in Fig. 1.10 shows an even function. Identify it.

□ Function (A) is symmetric about the y-axis (i.e. it has left / right symmetry) and therefore it is even. ■

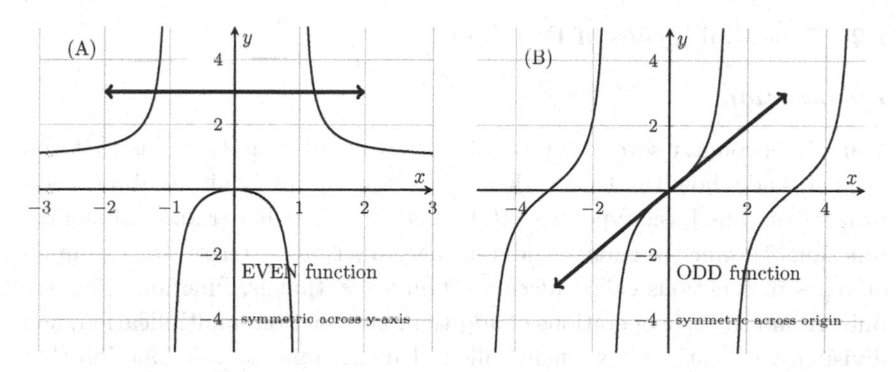

Fig. 1.11 Find the odd function (w/ YTI 9).

---

(9) One of the two graphs in Fig. 1.11 shows an odd function. Identify it.

☐ Function (A) is symmetric about the y-axis and is therefore even. Function (B) is symmetric across the origin, i.e. $f(-x)$ is the negative image of $f(x)$ for each x. So, (B) is odd. ■

---

(10) Test algebraically whether this is an odd function or not: $g(t) = \dfrac{1}{t^3 - t}$.

☐ If g is odd, we need $g(-t) = -g(t)$. Testing this,

$$g(-t) = \frac{1}{(-t)^3 - (-t)} = \frac{1}{-t^3 + t} = \frac{1}{-(t^3 - t)} = -\frac{1}{t^3 - t} = -g(t)$$

and therefore the function is odd. ■

1.2 Essential Types of Functions

Introduction

You will encounter several types of functions so often in Calculus that you need to know how to identify them, and develop an intuition about their properties. The basic types we will look at here are: linear and polynomial functions, power functions, and rational functions. These are examples of types of functions called *algebraic functions*, that is, functions that are built from the basic operations of addition, subtraction, multiplication, and division, with rational exponents allowed in the mix, too. A function that is not an algebraic function is called a *trascendental function*; this category includes exponenential functions, logarithmic functions, and trigonometric functions. We will look individually at the transcendental functions after reviewing the algebraic functions.

Linear Functions

A linear function is a specific example of a polynomial, which is defined more generally next. A linear function is a relation that can be written as $y = a_1x + a_0$, where x is the independent variable, y is the dependent variable, and a_0 and a_1 are constants.

The graph of a linear function is a straight line, which is no surprise given its name. In fact, the form of a linear function is more familiar as $y = mx + b$, where m is the slope of the line and b is its y-intercept. (I'll assume these terms are familiar, if they're not, you have a lot of review to do!)

By the time your experience in a first semester of Calculus is done, you'll be so tired of finding the equations of lines that it won't even be funny. In Calculus I, we will spend a good amount of effort writing equations of lines; however, the information we can provide about the lines will get more and more ... um ... *interesting.*

Just to brush the dust off, let's review a traditional problem involving linear functions — finding the equation of a line. To do this, we need a point (x_0, y_0) on the line and the line's slope m, so that we can fill in the *point-slope formula*

$$y - y_0 = m(x - x_0) \tag{1.2}$$

If we are given two points (x_0, y_0) and (x_1, y_1) on the line, that's just as good, since we can immediately compute the slope from those two points:

$$m = \frac{y_1 - y_0}{x_1 - x_0} \tag{1.3}$$

I can't stress strongly enough that when using the point-slope formula, we plug numbers into the spots held by x_0, y_0 and m. The y and x themselves are *left alone.* They are the variables that must appear in the final form of the equation of the line.

EX 1 Find the equation of the line connecting the points $(1, 2)$ and $(3, -4)$.

We need to compute the slope,

$$m = \frac{y_1 - y_0}{x_1 - x_0} = \frac{3 - 1}{-4 - 2} = \frac{2}{-6} = -\frac{1}{3}$$

And then we can pick one of the two points, say $(1, 2)$, to use in the point-slope formula (1.2):

$$y - y_0 = m(x - x_0)$$
$$y - 2 = -\frac{1}{3}(x - 1) \quad \blacksquare$$

Your past experience may be urging you to multiply this out and rearrange it to the *slope-intercept form* $y = mx + b$. You don't always need to do this, since the point-slope form of the line's equation is perfectly good. In fact, I'd argue it's better than the slope-intercept form. When you look at the slope-intercept form, you can see the line's slope, but you can no longer tell what point was used to create the equation. If you leave your equation in point-slope formula, we can still see both m AND the point (x_0, y_0). Of course, if you need quick reference to the y-intercept, then go ahead and rearrange as needed.

For your entertainment, here are a couple of problems that are more interesting. They are examples of what will become quite common: needing to find the equation of a line for which seemingly strange information is given.

You Try It

(1) Find the equation of the line connecting the point $(-1, -1)$ to the vertex of the parabola $y = 2 - x^2$.
(2) The point $(1/\sqrt{2}, 1/\sqrt{2})$ is on the unit circle. Find the equation of the line that just skims the circle at this point.

Finally, note that a line whose equation looks like $y = b$ will be a horizontal line, with slope 0, while a line whose equation looks like $x = a$ is a vertical line, with undefined (infinite) slope.

Polynomial Functions

Polynomial functions are functions of the form

$$y = a_n x^n + a_{n-1} x^{n-1} + a_{n-2} x^{n-2} + \ldots + a_2 x^2 + a_1 x + a_0$$

where all of the a_i's are constants called *coefficients*. The highest power of x appearing in the equation is the *degree* of the polynomial. Linear functions have degree 1; quadratic functions have degree 2; cubic functions have degree 3. The domain of a polynomial is all real numbers. The range can vary depending on the coefficients and may or may not be easy to determine. The graph of a polynomial will have "bends" in it, the number of which is no more than one less than the degree - meaning, a quadratic polynomial's graph has no more than one bend in it, a cubic polynomial's will have no more than two, and so on. The number of *roots* (places where we get $y = 0$) is no larger than the degree. Usually, when presented with a polynomial, we are going to get asked to do any or all of the following:

(1) Graph it
(2) Factor it
(3) Find its roots

Usually, factoring a polynomial leads to finding its roots, and knowing its roots makes it relatively easy to graph. Other than for quadratic polynomials, though, we have no quick and easy way to factor them, so trial and error is usually required. When you have a quadratic polynomial, you can factor and determine roots by examining the coefficients, or by using the *quadratic formula*. Just in case you have forgotten this gem, it says that the two roots of a polynomial $y = ax^2 + bx + c$ are given by

$$x_1, x_2 = \frac{-b \pm \sqrt{b^2 - 4ac}}{2a} \tag{1.4}$$

EX 2 Compare the polynomials $y = x^4 + 1$ and $y = x^4 - 1$ by factoring them, determining their roots, and finding how many bends appear in their graphs.

The polynomial $y = x^4 + 1$ does not factor. It has no roots. (Well, actually it has roots that are complex numbers, but we don't care about those right now.) There is only one bend in its graph, at $(0, 1)$. The graph opens upward on both sides; it's an even function, and so symmetric across the y-axis. ■

The polynomial $y = x^4 - 1$ factors as $y = (x^2 - 1)(x^2 + 1)$ which in turn factors to

$$y = (x - 1)(x + 1)(x^2 + 1)$$

There are two roots, at $x = -1$ and $x = 1$. (The term $x^2 + 1$ contributes no roots). Its graph has to cross the x-axis twice, at the roots. It will also have only one bend, since it has exactly the same shape as $y = x^4 - 1$, just shifted down. These curves are shown in Fig. 1.12.

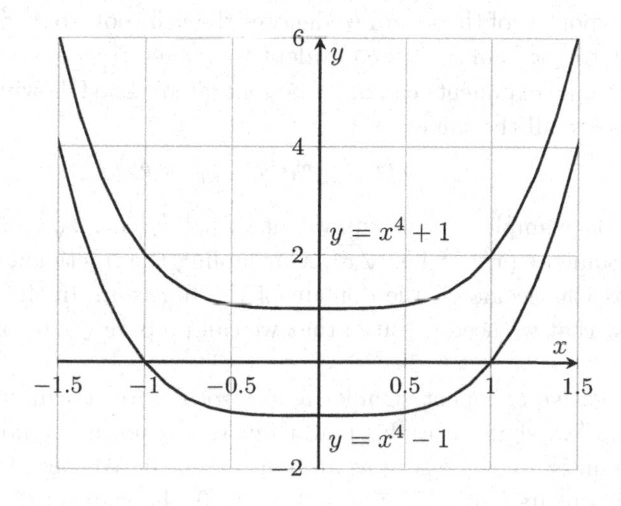

Fig. 1.12 Behavior of graphs.

You Try It

(3) Categorize the polynomial $y = (x-1)^5$ by providing its degree, domain, range, and roots. How many bends are in the graph?
(4) Categorize the polynomial $y = x^3 + x^2 - 2x$ by providing its degree, domain, range, and roots. How many bends are in the graph?

Power Functions

A power function is any functions that contains a term of the form $x^{p/q}$, where p and q are integers (I'll let you guess the one integer value q is not allowed to have). These are all power functions:

$$x^{3/2} \quad ; \quad 3x^{-2} \quad ; \quad x^{1/5} + x$$

and so on. Note that functions of the form x^n, where n is a positive integer, are power functions, but they're also polynomials. They're dull as power functions and even more boring as polynomials. The second example above, $3x^{-2}$, is a power function but not a polynomial (it is, though, both a power function and a rational function, as we'll see below).

The domains, ranges and graphs of power functions can vary based on the exponents. To decrypt a power function and start assessing its properties, you have to remember a few basic principles of exponents:

- An exponent of the form $1/n$ denotes the nth root, so $x^{1/2}$ is equivalent to $\sqrt{x}$, and $x^{1/5}$ is equivalent to $\sqrt[5]{x}$, etc.
- Fractional exponents can often be separated. The following expressions are all the same:

  $$x^{p/q} = (x^p)^{1/q} = (x^{1/q})^p$$

  So, for example, $x^{3/2}$ is equivalent to $(x^{1/2})^3$ i.e. $(\sqrt{x})^3$. It's also the same as $(x^3)^{1/2}$ i.e. $\sqrt{x^3}$. Uncoupling the fractional exponent helps you decide on the domain of the function. In this case, we know that we need $x \geq 0$ so that we can prepare $\sqrt{x}$ before cubing it.
- A negative exponent signifies a reciprocal. So we can write $3x^{-2}$ as $3/x^2$ ... that's why this power function is not a polynomial. It's domain can now be seen as all reals except 0. We can also think a bit about its graph; as x gets closer to 0, the value of the function gets larger and larger (for both positive and negative x), and as x

gets larger, the value of the function gets smaller and smaller and closer to 0.

Remember to be very careful when dealing with reciprocals. For example, only one of the following two expressions is equivalent to $(x + x^2)^{-1}$. Which one?

$$\frac{1}{x + x^2} \quad \text{or} \quad \frac{1}{x} + \frac{1}{x^2}$$

You Try It

(5) Find the domain and range of $f(x) = x^{2/3} + x$.
(6) Simplify the expression $1 - (x - 4)^{-2}$.

Rational Functions

Rational functions are functions written as the quotient of two polynomials. If $P(x)$ and $Q(x)$ are two polynomials, then $f(x) = P(x)/Q(x)$ is a rational function. These functions are lots of fun, since their domains, ranges, and graphs can be quite varied. The sample power function $3x^{-2}$ above is a rational function since it can be rewritten as $3/x^2$, which is a quotient of the two polynomials 3 and x^2. Other examples of rational functions are:

$$\frac{1}{x^2 - 2} \; ; \; \frac{x^2 - 4}{x^4 + 1} \; ; \; \frac{2x^7 + 5}{2x(x^2 + 1)}$$

Determination of the properties and behavior of a rational function is usually all about:

- Simplifying and canceling as much as possible
- Determining when the denominator would become zero, both before and after simplification
- Determining what happens as x gets big, both in the positive and negative directions

Graphs of rational functions often have vertical and horizontal asymptotes. Please note the spelling and pronounciation of that word: asympTOTE, not asympTOPE. And it's not spelled with two *s*'s, either. If you spell it with two *s*'s, your reader will think you are calling him or her a name.

EX 3 Find the domain of $g(x) = (x^2 - x)/(x^2 + x)$, and describe how the graph behaves as x gets larger and larger in the positive direction.

This is an example of keeping an eye on the denominator both before and after simplification. At first, we can see this obvious factoring:

$$\frac{x^2 - x}{x^2 + x} = \frac{x(x-1)}{x(x+1)}$$

At this point, we have locked in the fact that we cannot use either $x = 0$ or $x = -1$, no matter how the expression simplifies further. The domain of this function is all reals except 0 and -1. Now, the next obvious thing to do is cancel the x's:

$$\frac{x^2 - x}{x^2 + x} = \frac{x(x-1)}{x(x+1)} = \frac{x-1}{x+1}$$

An important thing to remember here is that even though it looks like $x = 0$ would be OK in the simplified form of the function, we have already removed $x = 0$ from the domain; we can't magically start using it again. The domain remains all reals except 0 and -1. While the graph of this function will go crazy near $x = -1$ (it has a vertical asymptote there), values around $x = 0$ cause no problems — therefore, the graph will behave well in the vicinity of $x = 0$, but will have a hole in the graph at $x = 0$ since we can't plot a point there. In all, this graph would have a hole at $x = 0$, a vertical asymptote at $x = -1$; as x gets larger and larger in the positive direction, the value of the function gets closer and closer to 1. The graph is shown in Fig. 1.13. ■

You Try It

(7) Find the domain of $f(x) = (5x+4)/(x^2+3x+2)$. What odd things (asyptotes, holes) occur in its graph?
(8) Is $f(x) = (x^2 - 4)/(x^4 + 1)$ even, odd, or neither?

Piecewise Functions

Piecewise functions are functions in which the recipe you use depends on the value of the independent variable. The various recipes, and corresponding ranges of use, are usually given in brackets. For example, consider the piecewise function

$$f(x) = \begin{cases} x^2 & \text{for} \quad x < 0 \\ -2x + 2 & \text{for} \quad 0 \leq x < 1 \\ x - 1 & \text{for} \quad x \geq 1 \end{cases}$$

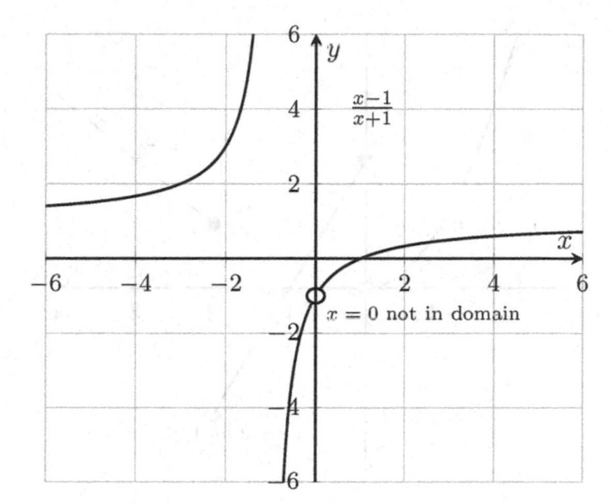

Fig. 1.13 Algebraic "memory" of domains.

If we want $f(-2)$, then we use the first recipe since $x = -2$ falls within the bucket "$x < 0$". So, $f(-2) = (-2)^2 = 4$. If we want $f(1/2)$, then we use the middle recipe:

$$f\left(\frac{1}{2}\right) = -2\left(\frac{1}{2}\right) + 2 = 1$$

And similarly, $f(5) = 5 - 1 = 4$. The graph of this function (shown in Fig. 1.14) is made of three pieces, which may or may not connect. From left to right along the x-axis, the graph starts as a parabola, then becomes a line, then a different line; these transitions happen at $x = 0$ at $x = 1$, and the pieces may or may not connect at the junctions.

The primary example of a common piecewise function is the absolute value function, $f(x) = |x|$. We all know what the bars mean, but the formal definition of this function is given in piecewise form:

$$f(x) = \begin{cases} x & \text{for } x \geq 0 \\ -x & \text{for } x < 0 \end{cases}$$

You probably remember what the graph of this looks like, it's an upright V-shape with the apex at the origin, where the lines $y = -x$ and $y = x$ join. There are two "pieces" to any absolute value function. So when you really want to do something with such a function, you need to be able to extract its piecewise representation — trust me on this, it will come back in the not so distant future.

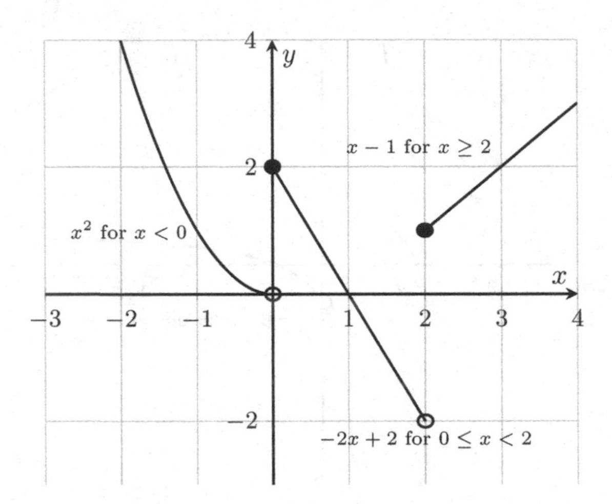

Fig. 1.14 Pieces of a piecewise function.

EX 4 Write $g(x) = |x - 5|$ in piecewise form and describe the two parts of its graph.

When $x - 5 \geq 0$, i.e. when $x \geq 5$, the absolute value bars do nothing. When $x - 5 < 0$, the absolute value bars change the sign of the result, and we can write $-(x - 5)$ as $-x + 5$. So, the piecewise representation of this function is

$$g(x) = \begin{cases} x - 5 & \text{for } x \geq 5 \\ -x + 5 & \text{for } x < 5 \end{cases}$$

The graph of this function (see Fig. 1.15) will show two lines $x - 5$ and $-x + 5$, joining at the point $(5, 0)$. ■

You Try It

(9) Write $f(x) = |4 - x^2|$ in piecewise form and describe its graph.

Have You Learned...

- The names of many different types of functions?
- The characteristics of polynomial, power, and rational functions?
- That piecewise functions should not be scary?
- That the absolute value function is a piecewise function in disguise?

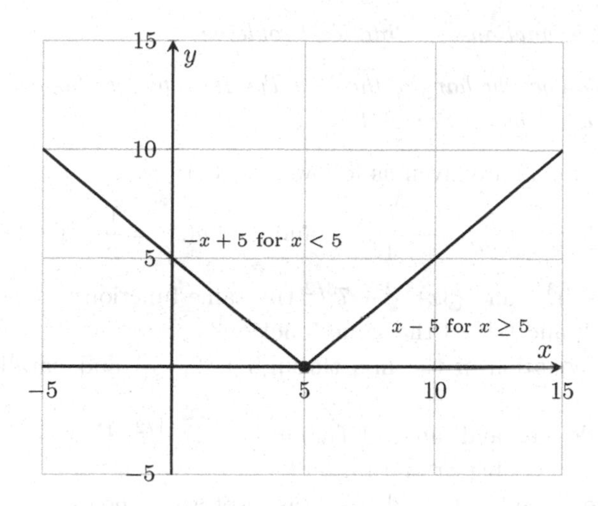

Fig. 1.15 The function $g(x) = |x - 5|$.

Basic Types of Functions — Problem List

Basic Types of Functions — You Try It

These appeared above; solutions begin on the next page.

(1) Find the equation of the line connecting the point $(-1, -1)$ to the vertex of the parabola $y = 2 - x^2$.
(2) The point $(1/\sqrt{2}, 1/\sqrt{2})$ is on the unit circle. Find the equation of the line that just skims the circle at this point.
(3) Categorize the polynomial $y = (x-1)^5$ by providing its degree, domain, range, and roots. How many bends are in the graph?
(4) Categorize the polynomial $y = x^3 + x^2 - 2x$ by providing its degree, domain, range, and roots. How many bends are in the graph?
(5) Find the domain and range of $f(x) = x^{2/3} + x$.
(6) Simplify the expression $1 - (x - 4)^{-2}$.
(7) Find the domain of $f(x) = (5x + 4)/(x^2 + 3x + 2)$. What odd things (asyptotes, holes) occur in its graph?
(8) Is $f(x) = (x^2 - 4)/(x^4 + 1)$ even, odd, or neither?
(9) Write $f(x) = |4 - x^2|$ in piecewise form and describe its graph.

Basic Types of Functions — Practice Problems

Try these as you get the hang of the You Try It problems. Solutions to these problems are available in Sec. A.1.2.

(1) If $f(x)$ and $g(x)$ are given as follows, is $f(x) = g(x)$?

$$f(x) = \frac{1}{x+1} \quad \text{and} \quad g(x) = \frac{1}{x} + 1$$

(2) Are $(3x+7)^{3/2}$ and $(3x)^{3/2} + 7^{3/2}$ the same function?
(3) Are $|x+4|$ and $|x|+4$ the same function?
(4) Find the equation of the line through $(-2,-3)$ and parallel to $y = 2x+7$.
(5) Find the domain and range of $f(x) = (x+2)^{-1/2}$. How does the graph behave as x gets bigger and bigger?
(6) Find the graphical oddities (asymptotes, holes) of $h(t) = \dfrac{t^3+t^2}{t^3+t^2-2t-2}$.
(7) Put the function $f(x) = 3 - |x+2|$ into piecewise form and sketch its graph.

Basic Types of Functions — Challenge Problems

Try these problems to test your skills with the ideas in this section. Solutions to these problems are available in Sec. B.1.2.

(1) Find the equation of the line that connects the cusp of the graph of $f(x) = 4 - |x-1|$ and the hole in the graph of

$$g(x) = \frac{x^2+x-2}{x^2-x}$$

(2) Find the domain and graphical oddities (asymptotes, holes) of $g(x) = \dfrac{x^2-5x+4}{\sqrt{x}+1}$. (Hint: You should try the technique of "rationalizing the denominator" ... how's that for a blast from the past?)
(3) Put the function $f(x) = |x^3 - 2x^2 - 3x|$ into piecewise form and make a reasonable sketch of its graph.

Essential Types of Functions — You Try It — Solved

(1) Find the equation of the line connecting the point $(-1,-1)$ to the vertex of the parabola $y = 2 - x^2$.

☐ The vertex of the parabola $y = 2 - x^2$ is at $(0, 2)$. Therefore, the slope of the line connecting $(-1,-1)$ to this vertext is

$$m = \frac{2-(-1)}{0-(-1)} = 3$$

and the equation of the line comes from Eq. (1.2) as:

$$\begin{aligned} y - y_0 &= m(x - x_0) \\ y - 2 &= 3(x - 0) \\ y &= 3x + 2 \quad \blacksquare \end{aligned}$$

(2) The point $(1/\sqrt{2}, 1/\sqrt{2})$ is on the unit circle. Find the equation of the line that just skims the circle at this point.

☐ This point is exactly in the middle of the arc of the circle in the first quadrant. The line that skims the circle there is on a perfect downward diagonal, and so has a slope of -1. With that slope and the given point, we have

$$\begin{aligned} y - y_0 &= m(x - x_0) \\ y - \frac{1}{\sqrt{2}} &= (-1)\left(x - \frac{1}{\sqrt{2}}\right) \quad \blacksquare \end{aligned}$$

(3) Categorize the polynomial $y = (x-1)^5$ by providing its degree, domain, range, and roots. How many bends are in the graph?

☐ This polynomial has degree 5. It's domain is all real numbers. We can also get all numbers back from the function (an odd exopnent returns negative numbers, too), so the range is also all reals. There is only one root, at $x = 1$. There are no real bends in the graph, only a slight kink, much like that on $y = x^3$. ■

(4) Categorize the polynomial $y = x^3 + x^2 - 2x$ by providing its degree, domain, range, and roots. How many bends are in the graph?

☐ This polynomial has degree 3. It's domain and range are all real numbers. To see the roots, it's better to factor it as $y = x(x^2+x-2) = x(x-1)(x+2)$; thus, we see there are three roots, at $x = -2, 0, 1$. Since the curve must pass through the x-axis three times, there must be two bends in the graph. ■

---

(5) Find the domain and range of $f(x) = x^{2/3} + x$.

□ The domain and range are really controlled by the $x^{2/3}$. If we consider this term as $(x^{1/3})^2$, or $(\sqrt[3]{x})^2$, we can see that the domain is all real numbers (since we can take the cube root of anything), and the range is all non-negative reals (since we square everything). ■

---

(6) Simplify the expression $1 - (x-4)^{-2}$.

□ The point of this exercise is to make sure that you do NOT try to simplify this as:

$$1 - (x-4)^{-2} = 1 - \frac{1}{(x-4)^2} = 1 - \frac{1}{x^2} + \frac{1}{4^2}$$

Don't do that!! Rather, do this:

$$\begin{aligned} 1 - (x-4)^{-2} &= 1 - \frac{1}{(x-4)^2} = \frac{(x-4)^2}{(x-4)^2} - \frac{1}{(x-4)^2} \\ &= \frac{(x-4)^2 - 1}{(x-4)^2} = \frac{x^2 - 8x + 15}{(x-4)^2} \\ &= \frac{(x-3)(x-5)}{(x-4)^2} \end{aligned}$$ ■

---

(7) Find the domain of $f(x) = (5x+4)/(x^2+3x+2)$. What odd things (asyptotes, holes) occur in its graph?

□ By factoring, we get

$$f(x) = \frac{5x+4}{x^2+3x+2} = \frac{5x+4}{(x+1)(x+2)}$$

so we can see that the domain is all reals except $x = -1$ and $x = -2$. Vertical asymptotes occur at these locations, and there are no holes in the graph since no terms in the denominator cancel out. ■

---

(8) Is $f(x) = (x^2-4)/(x^4+1)$ even, odd, or neither?

□ Since

$$f(-x) = \frac{(-x)^2 - 4}{(-x)^4 + 1} = \frac{x^2 - 4}{x^4 + 1} = f(x)$$

then this is an even function. ■

(9) Write $f(x) = |4 - x^2|$ in piecewise form and describe its graph.

□ Consider the quantity $4 - x^2$ without the absolute values. This quantity will be negative when $x < -2$ and $x > 2$, so these are the regions in which the absolute value bars do their job. On $-2 \le x \le 2$, the bars do nothing. So, in piecewise form, we have

$$f(x) = \begin{cases} -(4 - x^2) & \text{for} \quad x < -2 \\ 4 - x^2 & \text{for} \ 0 \le x \le 2 \\ -(4 - x^2) & \text{for} \quad x > 2 \end{cases}$$

The graph will have 3 parts. The middle part, on $-2 \le x \le 2$, is the regular upside down parabola $4 - x^2$. On the wings, for $x < -2$ and $x > 2$, we have the negative portion of $4 - x^2$ flipped upside down to become positive. The whole graph looks like a big curvy W. ■

1.3 Exponential and Logarithmic Functions

Introduction

Having reviewed many types of algebraic functions, it's now time to review a few of the more common transcendental functions. The first ones we'll look at are exponential and logarithmic functions. Note that the information contained here is intended to be a *review* of things you already know about these functions from a pre-calculus course. The problems in this problem set are NOT going to be enough for you to feel comfortable with the topic if you need more than just a review. If that's the case, dive into a pre-calc book.

Exponential Functions

Consider the sequence of numbers $2^0, 2^1, 2^2, 2^3, 2^4 \ldots$. When written out, this list becomes $1, 2, 4, 8, 16, \ldots$. These numbers are growing bigger, and the farther we go out in the list, the faster the numbers grow. The number 2 is the *base* of these expressions, and the changing values are the *exponent.* In fact, since the exponent varies, we can just replace it with a variable, and recognize this collection as a set of values of the function 2^x. What happens when x is negative? Well, remember that a negative exponent signifies a reciprocal, so $2^{-1}, 2^{-2}, 2^{-3}, \ldots$ are the same as $1/2, 1/4, 1/8, \ldots$. Can x be a fraction? Sure! We already know that $2^{1/2}$ is the same as $\sqrt{2}$, and $2^{-1/3}$ is the same as $1/\sqrt[3]{2}$, etc. We can even use the number 2^π if we want to. So, since we can use any real number for x, the domain of the function $y = 2^x$ is all real numbers. The range, though, is not all real numbers. Did you notice that all of the values of 2^x listed in this paragraph are positive? And did you notice that there is no way to get a 0 out of 2^x? The range of 2^x is all positive real numbers.

The graph of $y = 2^x$ is shown in Fig. 1.16 along with a related graph.

There is nothing special about 2 as the base. We can select any base b we would like, as long as $b > 0$. And the graph of any $y = b^x$ would be similar to the graph of 2^x with the following stipulations:

- All graphs of $y = b^x$ (where $b > 0$) go through the point $(0, 1)$ since $b^0 = 1$ for all b.
- If $b > 2$, then the graph of b^x is steeper than the graph of 2^x

- If $1 < b < 2$, then the graph of b^x is less steep than the graph of 2^x
- If $0 < b < 1$, the graph still goes through $(0, 1)$ but is sort of flipped upside down relative to the graph of 2^x. Why? Well, consider $b = 1/2$, and get a few sample points: $(1/2)^2 = 1/4$, $(1/2)^3 = 1/8$ and so on. So whereas larger x gives larger output for $y = 2^x$, larger x will given *smaller* output for $y = (1/2)^x$. The graph of $y = (1/2)^x$ is shown in Fig. 1.16. FFT: What happens when x is negative in $y = (1/2)^x$?

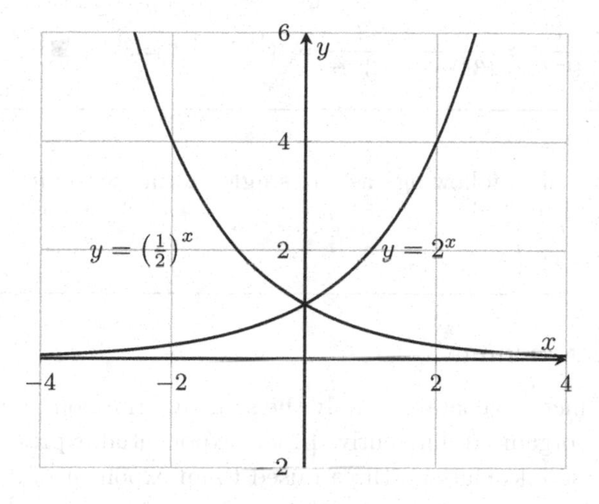

Fig. 1.16 Sample exponential curves.

You Try It

(1) On the same set of axes, sketch the graphs of $3^x, 4^x, (1/3)^x$ and $(1/4)^x$.

In the event that we want to manipulate exponential functions, we can use well-known properties of exponents, such as:

- $b^a \cdot b^c = b^{a+c}$
- $b^a / b^c = b^{a-c}$
- $(b^a)^c = b^{ac}$

EX 1 Simplify the expression $2^{2x}4^{-3x}$ into a single term.

Since $2^{2x} = (2^2)^x = 4^x$, we can write this as $4^x 4^{-3x} = 4^{x-3x} = 4^{-2x}$. This last value is also known as 16^{-x} or even $(1/16)^x$. ■

EX 2 Write $81^x/9^{-x}$ as a single term with a base of 3.

Since $81 = 3^4$ and $9 = 3^2$, then we can rewrite the function as

$$\frac{81^x}{9^{-x}} = \frac{(3^4)^x}{(3^2)^{-x}} = \frac{3^{4x}}{3^{-2x}} = 3^{4x-(-2x)} = 3^{6x} \quad ■$$

You Try It

(2) Write the following as a single term with a base of 2: $\left(\frac{1}{4}\right)^x \left(\frac{1}{8}\right)^{-2x}$

Logarithmic Functions

Logarithmic functions contain exactly the same information as exponential functions, just organized differently. In the exponential expression $3^2 = 9$, we are given the information "three raised to an exponent of 2 gives a result of 9". There are three pieces of data: the base, the exponent, and the result. In this expression, the result (9) is emphasized by being put on a side by itself.

There are some cases when we are more interested in the exponent in one of these relations. If you are familiar with the decibel, Richter, or pH scales, you are familiar with scales that reflect the exponent of a value rather than the value itself. For example, if $[H^+]$ represents the concentration of hydrogen ions in a solution, then the solution's pH value depends on the exponent of that concentration: a solution where $[H^+] = 10^{-7}$ has a pH value of 7.

Logarithms are functions that highlight the exponent in a relation between base, exponent, and result. Given the same information as above, that it takes an exponent of 2 to change a base of 3 into a result of 9, we write $\log_3(9) = 2$. In this expression, it is the exponent that is emphasized by being put on a side by itself. It's the same information, organized differently.

For example, the expression $\log_2(8)$ is asking, "What exponent of 2 is required to get a result of 8?" The answer is 3, and so we'd write $\log_2(8) = 3$. Similarly, $\log_2(32) = 5$ and $\log_2(1/2) = -1$.

You Try It

(3) Evaluate these expressions and write the known information in its exponential form:

$$\log_4 2 \quad ; \quad \log_{10} 1000 \quad ; \quad \log_3 \tfrac{1}{9} \quad ; \quad \log_5 \sqrt{5}$$

When put in more general form, we write logarithmic functions like this: $y = \log_b(x)$. The base is b, and x is a result that happens when y is the exponent of b. These two expressions mean exactly the same thing:

$$y = \log_b(x) \qquad \leftrightarrow \qquad b^y = x$$

Since we count in base 10, a common logarithm is the logarithm with a base of 10. The graph of $y = \log_{10}(x)$ is shown in Fig. 1.17. How do we know the curve is really $y = \log_{10}(x)$ and not that of some other base, like $\log_5(x)$? Easy: it goes through the point $(10, 1)$. We know that $10^1 = 10$ and so also $1 = \log_{10}(10)$. Thus, the graph of $y = \log_{10}(x)$ needs to use the point $(10, 1)$. This leads to a general rule as shown below.

Given that the curve shown is $y = \log_{10}(x)$, any other logarithmic curve will have the same shape with the following stipulations:

- All graphs of $y = \log_b(x)$ (where $b > 0$) go through the point $(1, 0)$ since $0 = \log_b(1)$ for all b. All graphs of $y = \log_b(x)$ go through the point $(b, 1)$ since $1 = \log_b(b)$. Thus,
- If $b > 10$, then the graph of $y = \log_b(x)$ is "flatter" than the graph of $y = \log_{10}(x)$
- If $1 < b < 10$, then the graph of $y = \log_{10}(x)$ is "flatter" than the graph of $y = \log_b(x)$
- If $0 < b < 1$, then the graph of $y = \log_b(x)$ is flipped over the x-axis, but still goes through $(1, 0)$ and $(b, 1)$.

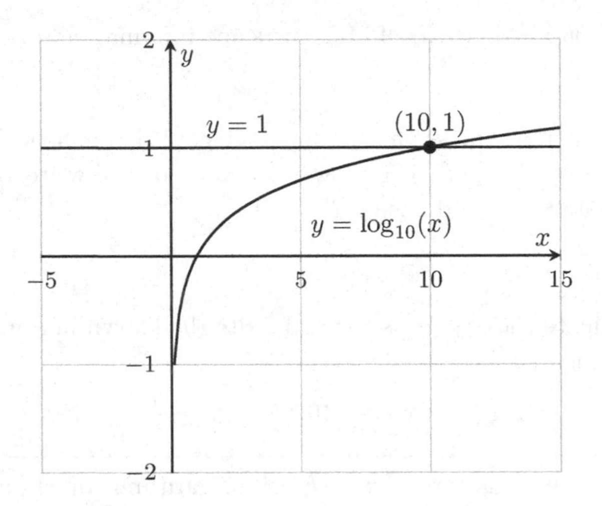

Fig. 1.17 Sample logarithmic curve and identifying point.

> **You Try It**
>
> (4) Figure 1.18 (at the end of the section) shows two curves of the form $y = \log_b(x)$. Identify the base b in each.

Some of the algebraic properties of logarithmic functions are similar to those of exponential functions:

- $\log_b(ac) = \log_b(a) + \log_b(c)$
- $\log_b(a/c) = \log_b(a) - \log_b(c)$
- $\log_b(a^p) = p\log_b(a)$

EX 3 Write $\log_{10}(25) + \log_{10}(125)$ as a single logarithm of the form $a\log_{10}(5)$.

Using the properties of logarithms, we can write this as

$$\log_{10}(25) + \log_{10}(125) = \log_{10}(5^2) + \log_{10}(5^3)$$
$$= 2\log_{10}(5) + 3\log_{10}(5) = 5\log_{10}(5) \quad \blacksquare$$

> **You Try It**
>
> (5) Write $\log_3 16 + \log_3(1/4)$ as a single logarithm of the form $a\log_3 2$.

Solving Equations Containing Exponential and Logarithmic Functions

What happens when we start combining exponential and logarithmic functions? For example, what is the value of $\log_b(b^2)$? Without knowing what b is, we know that this value is 2. Remember, that logarithmic function is asking, "What exponent do we put on b to get a result of b^2?" That's clearly 2, and it doesn't matter what b is. Similarly, $\log_b(b^3) = 3$, $\log_b(b^7) = 7$, $\log_b(b^{-4}) = -4$, and so on. In general,

$$\log_b(b^x) = x$$

This means that if we apply a logarithm of base b to an exponent with base b, we get the exponent back. So in an equation that contains a variable up in an exponent, we can get it out of the exponent by applying a logarithmic function; this is similar to how we square something to free a variable trapped inside a square root. Many functions have other functions that "undo" them; these are called *inverse functions* and you'll see those more generally in another setting.

EX 4 Solve the equation $3^{4x-2} = 7$ for x.

The variable we need to get at is in an exponent, so let's apply a logarithm (remember to do it to both sides). Follow the steps in order *(i)* through *(v)*:

i) $3^{4x-2} = 7$

ii) $\log_3 3^{4x-2} = \log_3 7$ $\longrightarrow$ *iv)* $4x = \log_3 7 + 2$

iii) $4x - 2 = \log_3 7$ *v)* $x = \frac{1}{4}(\log_3 7 + 2)$

To see a decimal representation of that value, we'd need a calculator. But we prefer exact answers when we can get 'em! ■

You Try It

(6) Solve the equation $10^{x/3+5} = 2$ for x.

We can combine the exponential and logarithmic functions in the other order, too:

$$b^{\log_b(x)} = x$$

This means that if we apply an exponent of base b to a logarithm of base b, the two functions eliminate each other again.

EX 5 Solve the equation $3\log_2(5x) = 8$ for x.

We want to apply the exponential function 2^x here to remove the logarithm, but the coefficient of 3 is in the way. We need to move it first:

$i)\quad 3\log_2(5x) = 8$

$ii)\quad \log_2(5x) = \dfrac{8}{3} \qquad \longrightarrow \qquad iv)\quad 5x = 2^{8/3}$

$iii)\quad 2^{\log_2(5x)} = 2^{8/3} \qquad\qquad v)\quad x = \dfrac{2^{8/3}}{5}$ ■

You Try It

(7) Solve the equation $\log_{10}(x+5) = 2$ for x.

The Special Functions e^x and $\ln(x)$

The exponential and logarithmic functions that are needed most often in applications are $y = e^x$ and $y = \log_e(x)$. Since that logarithm is so important, it gets its own name: it is the "natural logarithm" and is abbreviated $y = \ln(x)$. The number e is a transcendental number, like π, which is not a rational number and has no repeating portions in its decimal representation. Its value is approximately $2.718\ldots$. If you want to see the value of e computed to one million decimal places, just do an internet search for the phrase "e to one million decimal places"; at the time of the writing of this text, this search brings up a link to a text file hosted by NASA which contains the very cool thing you're looking for.

At the basic level we're at now, these functions are treated no differently (and behave no differently) than all the other exponential and logarithmic functions. Therefore, these functions will appear in the practice and challenge problems without further discussion. You'll get your fill of them later in the course, we don't need to dwell on them too much right now, except to know how to use them.

Exponential Functions in Action

Exponential functions are most commonly used in problems related to growth and decay of some quantity. These can be population problems, interest / principal problems, radioactive decay problems, etc. Here are a couple of typical examples:

EX 6 The balance B in an idle bank account doubles every 10 years. If the initial balance is \$1000, write a general formula for the balance as a function of time $B(t)$ and predict the balance after 55 years and 3 months.

The balance increases per year like this:

$$B(0) = 1000 \quad ; \quad B(10) = 2000 \quad ; \quad B(20) = 4000 \quad ; \quad B(30) = 8000$$

and so on. Let's break these numbers down to see any patterns. First, we write the list again as

$$B(0) = 1000\,;\ B(10) = 2 * 1000\,;\ B(20) = 4 * 1000\,;\ B(30) = 8 * 1000$$

and then as

$$\begin{aligned} B(0) = 2^0 * 1000 \quad &; \quad B(10) = 2^1 * 1000 \\ B(20) = 2^2 * 1000 \quad &; \quad B(30) = 2^3 * 1000 \end{aligned}$$

The only value changing is the the exponent of 2. With each term giving a value of $B(t)$, we can link the exponent to t as follows:

$$\begin{aligned} B(0) = 2^{0/10} * 1000 \quad &; \quad B(10) = 2^{10/10} * 1000 \quad ; \\ B(20) = 2^{20/10} * 1000 \quad &; \quad B(30) = 2^{30/10} * 1000 \end{aligned}$$

and now we recognize the overall form of the function as

$$B(t) = 1000 \cdot 2^{t/10}$$

Now, since 55 years and 3 months is 55.25 years, we can predict a value of

$$B(55.25) = 1000 \cdot 2^{55.25/10} = 1000 \cdot 2^{5.525} \approx \$46,046$$

(where I used a CAS to estimate the value). ■

You Try It

(8) A radioactive isotope has a half-life of 15 years (this means half of the current amount is gone after 15 more years). Suppose there are initially 20 grams present. Find a general formula for the amount remaining as a function of time, $m(t)$.

Have You Learned...

- The common features shared by all exponential functions?
- The common features shared by all logarithmic functions?
- That exponential and logarithmic functions give exactly the same information, just in different formats?
- To algebraically manipulate exponential and logarithmic functions?
- To solve equations involving exponential and logarithmic functions?

Exponential and Logarithmic Functions — Problem List

Exponential and Logarithmic Functions — You Try It

These appeared above; solutions begin on the next page.

(1) On the same set of axes, sketch the graphs of $3^x, 4^x, (1/3)^x$, and $(1/4)^x$.

(2) Write the following as a single term with a base of 2: $\left(\frac{1}{4}\right)^x \left(\frac{1}{8}\right)^{-2x}$

(3) Evaluate these expressions and write the known information in its exponential form:

$$\log_4 2 \quad ; \quad \log_{10} 1000 \quad ; \quad \log_3 \tfrac{1}{9} \quad ; \quad \log_5 \sqrt{5}$$

(4) Figure 1.18 shows two curves of the form $y = \log_b(x)$. Identify the base b in each.

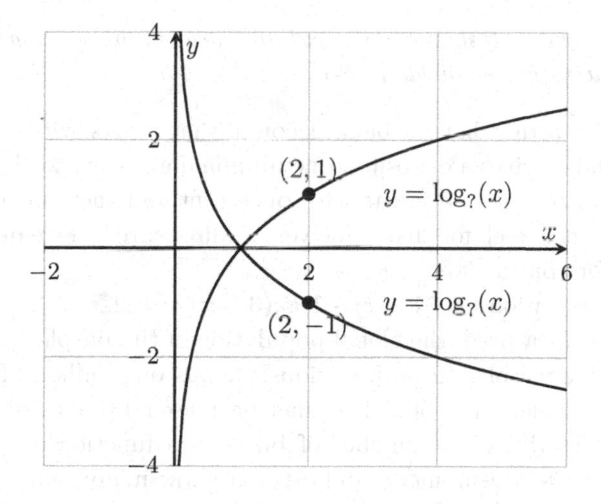

Fig. 1.18 Guess the logarithmic bases!

(5) Write $\log_3 16 + \log_3(1/4)$ as a single logarithm of the form $a \log_3(2)$.

(6) Solve the equation $\log_{10}(x + 5) = 2$ for x.

(7) Solve the equation $10^{x/3+5} = 2$ for x.

(8) A radioactive isotope has a half-life of 15 years (this means half of the current amount is gone after 15 more years). Suppose there are initially 20 grams present. Find a general formula for the amount remaining as a function of time, $m(t)$.

Exponential and Logarithmic Functions — Practice Problems

Try these as you get the hang of the You Try It problems. Solutions to these problems are available in Sec. A.1.3.

(1) Sketch the graph of $y = 4^x - 3$; identify at least two points.
(2) What are the domain and range of the function $f(x) = 3e^{-2x}$?
(3) Write $2\ln 4 - \ln 2$ as a single logarithm.
(4) Find the value of $2^{\log_2 3 + \log_2 5}$.
(5) Solve the equation $e^{2x+3} - 7 = 0$.
(6) Solve the equation $\ln(4 - 2x) - \ln(2) = -3$.
(7) If as population $P(t)$ starts at 100 and doubles every 3 hours, write a general formula for the population as a function of time, $P(t)$. What will the population be after 15 hours?

Exponential and Logarithmic Functions — Challenge Problems

Try these problems to test your skills with the ideas in this section. Solutions to these problems are available in Sec. B.1.3.

(1) Would you rather have a bank account that starts with one million dollars and receives a deposit of one million dollars per week for a year, or a bank account that starts with once cent and then has its balance doubled every week for that same year? Make sure to give quantitative reasons for your answer.
(2) Solve the equation $\log_9(1 - x) + \log_9(3 - x) = 1/2$.
(3) When first measured, the global population of the purple nosed pointy headed pine warbler (a bird, obviously), was one million. Every year since that time, the population has decreased by a third. Write a function describing the number of birds as a function of time, $B(t)$, since the first measurement, and estimate how many years are needed for the population to drop to 2500.

Exp. and Log. Functions — You Try It — Solved

(1) On the same set of axes, sketch the graphs of $3^x, 4^x, (1/3)^x$ and $(1/4)^x$.

□ Figure 1.19 shows these curves. ■

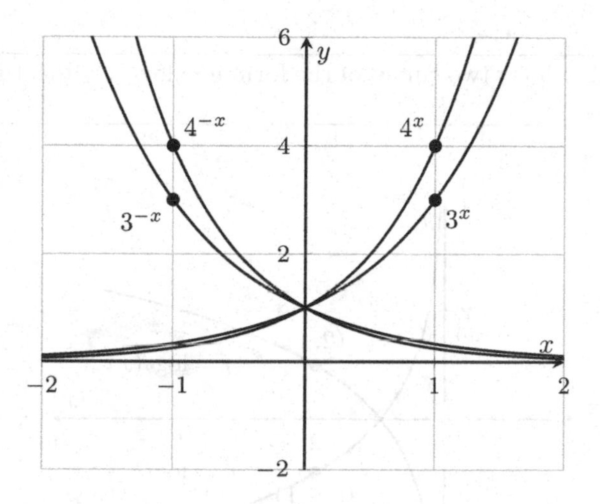

Fig. 1.19 Plotting 3^x, 4^x, $(1/3)^x$, $(1/4)^x$.

(2) Write the following as a single term with a base of 2: $\left(\frac{1}{4}\right)^x \left(\frac{1}{8}\right)^{-2x}$

$$\square \left(\frac{1}{4}\right)^x \left(\frac{1}{8}\right)^{-2x} = \left(\frac{1}{2^2}\right)^x \left(\frac{1}{2^3}\right)^{-2x} = \left(2^{-2}\right)^x \left(2^3\right)^{2x}$$
$$= 2^{-2x}2^{6x} = 2^{-2x+6x} = 2^{4x} \quad ■$$

(3) Evaluate these expressions and write the known information in its exponential form:

$$\log_4 2 \quad ; \quad \log_{10} 1000 \quad ; \quad \log_3 \tfrac{1}{9} \quad ; \quad \log_5 \sqrt{5}$$

$$\square \quad \log_4 2 = \log_4(\sqrt{4}) = \log_4(4^{1/2}) = \frac{1}{2}$$

$$\log_3 \frac{1}{9} = \log_3 3^{-2} = -2$$

$$\log_{10} 1000 = \log_{10} 10^3 = 3$$

$$\log_5 \sqrt{5} = \log_5 5^{1/2} = \frac{1}{2} \quad \blacksquare$$

---

(4) Figure 1.20 shows two curves of the form $y = \log_b(x)$. Identify the base b in each.

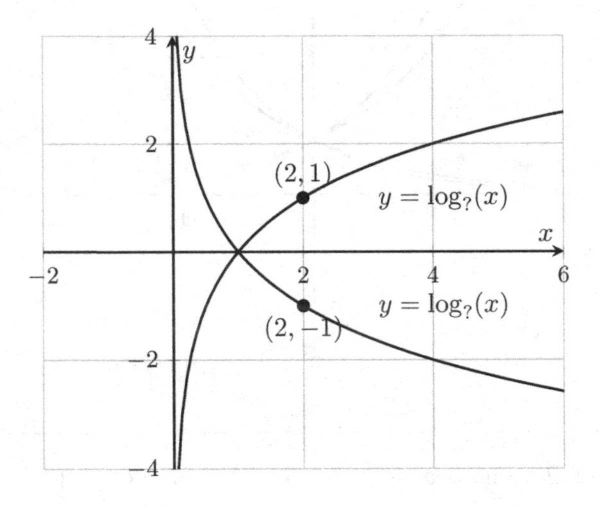

Fig. 1.20 Guess the logarithmic bases!

□ Curve (A) goes through the point $(2, 1)$, so we know that this curve shows $1 = \log_b 2$, or in exponential form, $b^1 = 2$. So b must be 2, and this is the curve $y = \log_2(x)$. Curve (B) goes through the point $(1/2, 1)$ and so by similar reasons is the curve $y = \log_{1/2}(x)$. ■

---

(5) Write $\log_3 16 + \log_3 \frac{1}{4}$ as a single logarithm of the form $a \log_3(2)$.

$$\square \ \log_3 16 + \log_3 \frac{1}{4} = \log_3 2^4 + \log_3 2^{-2} = 4\log_3 2 - 2\log_3 2 = 2\log_3 2 \quad \blacksquare$$

(6) Solve the equation $\log_{10}(x+5) = 2$ for x.

☐ Remember that logarithms simply relate a base, an exponent, and the result. The information given as $\log_{10}(x+5) = 2$ tells us that if we have a base of 10, the result of $x+5$ is generated by an exponent of 2. That same relation can be written as $x+5 = 10^2$. And so, $x = 95$. ■

(7) Solve the equation $10^{x/3+5} = 2$ for x.

☐ Since we are given an exponential relationship, our strategy is to employ a logarithm to get into the exponent and dig out the x: Starting with $10^{x/3+5} = 2$, we can apply the logarithm to both sides, to set of this sequence:

$$\begin{aligned} \log_{10} 10^{x/3+5} &= \log_{10} 2 \\ \frac{x}{3} + 5 &= \log_{10} 2 \\ \frac{x}{3} &= \log_{10} 2 - 5 \\ x &= 3(\log_{10} 2 - 5) \quad \blacksquare \end{aligned}$$

(8) A radioactive isotope has a half-life of 15 years (this means half of the current amount is gone after 15 more years). Suppose there are initially 20 grams present. Find a general formula for the amount remaining as a function of time, $m(t)$.

☐ If the isotope has a half life of 15 years and $m(t)$ starts at 20, we have

$$\begin{aligned} m(0) &= 20 \\ m(15) &= \frac{1}{2}(20) \\ m(30) &= \frac{1}{2} \cdot \frac{1}{2}(20) = \frac{1}{2^2}(20) \\ m(45) &= \frac{1}{2} \cdot \frac{1}{2^2}(20) = \frac{1}{2^3}(20) \quad \cdots \end{aligned}$$

After any t years, the pattern above shows that

$$m(t) = \frac{1}{2^{t/15}}(20) = 20 \cdot 2^{-t/15} \quad \blacksquare$$

1.4 Compositions and Inverses

Introduction

In the last section, a subtlety snuck in that deserves mention. Did you notice that when expressions like

$$10^{x/3+5} \quad ; \quad \log_{10}(x+5) \quad ; \quad e^{2x+3} \quad ; \quad \ln(4-2x)$$

arose, we had situations in which a simple, basic function was made more complicated by the "insertion" of another function? For example, the first one in the sample list shows a basic function 10^x in which $x/3+5$ has been placed into the exponent. This type of combination of functions is called *composition* of functions.

We also saw these two results:

$$\log_b(b^x) = x \quad \text{and} \quad b^{\log_b(x)} = x$$

These are instances of *composition* in which the combination of the two functions (here $\log_b x$ and b^x) creates an expression that simplifies all the way down to just x.

In general, we can select almost any two functions for the purpose of "plugging" one function into the other (composition), but when we spot a pair of functions which essentially annihilate each other upon composition, we've identified a very special pair. (I'm sure you can think of other instances of this — how about $\sqrt{x}$ and x^2?)

Let's explore these concepts, starting simple, and building up to the important results.

Combining Functions To Build New Functions

We have discussed several types of functions — linear, polynomial, power, rational, exponential, and logarithmic. While these types of functions are fun when taken by themselves, they are even more entertaining when we combine them!

Functions can be combined with the basic algebraic operations of addition, subtraction, multiplication, and division. These methods of combination are, hopefully, self-explanatory. We can also combine functions through *composition*. Loosely (and poorly) put, this is when you "plug one function into another function". All of these combinations can have implications for the domain, range, and graph of the resulting expression.

To be more specific, two functions $f(x)$ and $g(x)$ can be combined algebraically in these ways:

$$f(x) \pm g(x) \quad ; \quad f(x)g(x) \quad ; \quad \frac{f(x)}{g(x)}$$

The domains of $f(x) \pm g(x)$ and $f(x)g(x)$ will be restricted to the common regions of the domains of both original functions. The domain of $f(x)/g(x)$ will start with such a restricted set accounting for both $f(x)$ and $g(x)$, from which any roots of $g(x)$ are then removed. Do you see why?

EX 1 If $f(x) = \ln(x)$ and $g(x) = \sqrt{x-2}$, find the domain of $f(x)/g(x)$.

The domain of $f(x)$ is $x > 0$. The domain of $g(x)$ is $x \geq 2$. Note that the interval $(0, 2)$ is in the domain of $f(x)$, but not $g(x)$ — so those numbers cannot be in the domain of the quotient. At this point, the domain of the quotient is evolving toward $[2, \infty)$. But further, $x = 2$ will make $g(x) = 0$, and so $x = 2$ cannot be in the new domain. Therefore, the domain of $f(x)/g(x)$ is $x > 2$, or $(2, \infty)$. ■

You Try It

(1) If $f(x) = \ln(x)$ and $g(x) = \sqrt{2-x}$, create and find the domains of $f + g$, $f - g$, fg, and f/g.

Composition and Domains

Composition of functions is often (poorly) described as "plugging one function into another"; that phrase captures the mechanics of symbol manipulation on paper, but misses the heart of what happens behind the scenes. For example, the composition of the functions $f(x) = x^2$ and $g(x) = x - 1$ can proceed in two ways: we can get either "plug" f into g, yielding $x^2 - 1$, or we can "plug" g into f, yielding $(x-1)^2$. But the action of composition which is going on "behind the scenes" is more of a filtering process that links input to output. Understanding this filtering process goes hand in hand with understanding the notation (yes, again).

Given two functions $f(x)$ and $g(x)$, the two different compositions can be indicated notationally as $f(g(x))$ and $g(f(x))$. Note how these are completely different than $f(x) \cdot g(x)$ and $g(x) \cdot f(x)$. We can also form $f(f(x))$ or $g(g(x))$. This notation helps us see the two levels at which composition operates: the level of an individual value traveling from input to output, and the level of algebraic manipulation (this latter is the "plug one function into another" version).

At the level of individual values, composition means that a value is handed to one function, and the result of that operation is handed to the second function. Notationally, the order of custody of the input value is followed from the inside, out. If $f(x) = 3x - 1$ and $g(x) = 1 + x^2$, then to evaluate $f(g(2))$, we first find $g(2) = 5$, then pass that 5 to the function f; in all, $f(g(2)) = f(5) = 14$. Similarly, to get the value of $g(f(-1))$, we first find that $f(-1) = -4$, so that $g(f(-1)) = g(-4) = 17$.

The domain of a composition is build from the domains of the individual functions. We build the domain of $f(g(x))$ through the following filtering process:

(1) We can start only with values allowed by $g(x)$.
(2) We must ensure any item produced by $g(x)$ is then allowed in $f(x)$.
(3) So, if a given starting value can pass successfully through $g(x)$, but the result of that operation cannot then pass through $f(x)$, that individual starting value is not allowed in the domain of $f(g(x))$.

Similarly, to find the domain of $g(f(x))$, we must

(1) start with values allowed by $f(x)$.
(2) ensure any item produced by $f(x)$ is then allowed in $g(x)$.

Domains found this way will hold regardless of what happens during algebraic simplification of the composition.

At the algebraic level, then okay, we can form a composition by "plugging one function into the other" and shaking out the result. But, while a simplified composition may *appear* to have a domain of all real numbers, that might be misleading — we must consider the above rules no matter what the simplified composition looks like! This is why saying composition is just "plugging one function into another" misses the whole point — we have to keep the filtering process of the domain in mind.

EX 2 If $f(x) = \sqrt{x-1}$ and $g(x) = 3x + 2$, create and find the domains of $f(g(x))$, $g(f(x))$, $f(f(x))$, $g(g(x))$.

The domain of g is all reals; the domain of f is $x \geq 1$. Then,

- $f(g(x)) = \sqrt{(3x+2)-1} = \sqrt{3x+1}$. We can use all reals for g, but then must ensure f can still work, so we need $x \geq -1/3$.
- $g(f(x)) = 3\sqrt{x-1}+1$. The domain for this is $x \geq 1$.
- $f(f(x)) = \sqrt{\sqrt{x-1}-1}$. We first start with $x \geq 1$. Then we must ensure that the interior of the outer square root remains positive or zero. So we need $\sqrt{x-1} \geq 1$, or $x - 1 \geq 1$ or $x \geq 2$. This requirement supercedes the other, so the domain of $f(f(x))$ is $x \geq 2$.
- $g(g(x)) = 3(3x+2) + 2 = 9x + 8$. Its domain is all reals. ■

You Try It

(2) If $f(x) = 1/x$ and $g(x) = 1 - \sqrt{x}$, then create and find the domains of $f(g(x))$, $g(f(x))$, $f(f(x))$, $g(g(x))$.

Later in this content, we will need to investigate the reverse question: not "what do you get when you form the composition of these two functions?" but rather "what two functions were used to form this expression by composition?"

EX 3 What two functions $f(x)$ and $g(x)$ form $h(x) = \ln(x^2 + 2)$ as the composition $f(g(x))$?

If we use $f(x) = \ln(x)$ and $g(x) = x^2 + 2$, then $h(x) = (g(x)) = \ln(x^2 + 2)$. ■

You Try It

(3) What two functions $f(x)$ and $g(x)$ form $G(x) = x^2/(x^2 + 4)$ as the composition $f(g(x))$?

(4) What three functions $f(x)$, $g(x)$, and $h(x)$ form $H(x) = 1 - 3^{x^2}$ when composed as $f(g(h)))$?

Variations of Functions and Graphs

Some variations of functions have a systematic impact on graphs:

- The graph of $f(x) + c$ is the graph of $f(x)$ shifted up by c units.
- The graph of $f(x) - c$ is the graph of $f(x)$ shifted down by c units.
- The graph of $f(x + c)$ is the graph of $f(x)$ shifted left by c units.
- The graph of $f(x - c)$ is the graph of $f(x)$ shifted right by c units.
- The graph of $cf(x)$ is the graph of $f(x)$ stretched vertically by a factor of c. If $c < 0$, the graph flips upside down, too.
- The graph of $f(cx)$ is the graph of $f(x)$ contracted horizontally by a factor of c.

These moves are known as translations (the first four) or contractions (the last two). FFT: Three of the six items above involve compositions. Can you identify them?

EX 4 I'll bet you remember what the graph of $y = \sin(x)$ looks like. (If not, look it up in preparation for the next section!) How can you adapt that graph to create the graphs $y = \sin(x) + 2$, $y = \sin(x + \pi/2)$, $y = 2\sin(x)$ and $y = \sin(2x)$.

- $y = \sin(x) + 2$ is the graph of $\sin(x)$ shifted up by 2.
- $y = \sin(x + \pi/2)$ is the graph of $\sin(x)$ shifted left by $\pi/2$; it actually then matches the graph of $\cos(x)$, which explains the identity $\sin(x + \pi/2) = \cos(x)$!
- $y = 2\sin(x)$ is twice as tall as $\sin(x)$: the same period, but twice the amplitude.
- $y = \sin(2x)$ is the graph of $\sin(x)$ contracted by a factor of two; therefore, an entire period of $\sin(2x)$ occurs on $[0, \pi]$ rather than the original $[0, 2\pi]$. ■

You Try It

(5) How can the graph of $y = x^2$ be adapted to create the graph of $y = (x + 1)^2$?

(6) How can the graph of $y = \sqrt{x}$ be adapted to create the graph of $y = \sqrt{x} + 3$?

You Try It

(7) How can the graph of $y = \log_{10}(x)$ be adapted to form the graph of $y = \log_{10}(x-5)$?
(8) How are the graphs of (a) $y = e^x + 2$, (b) $y = e^{x-2}$, (c) $y = -e^x$, and (d) $y = e^{-x}$ related to the graph of $y = e^x$?

Inverse Functions and Restricted Domains

It's time to think about domains and ranges again. Recall that the domain of a function is the set of all numbers allowed as input to the function, and the range is the set of all possible outputs. The function itself leads an input value from the domain towards the corresponding output value in the range. Often a more interesting problem is the reverse; the *inverse* of a function takes an element of the function's range and "returns" it to the element of the domain that produced it. (That is, using the inverse of a function is like playing a Reverse card from the game Uno on the function itself!) Notationally, if the function is f, its inverse is denoted f^{-1}. Please be aware that this does not mean reciprocal! This is an unfortunate criss-cross of notation, but we're stuck with it. When the exponent of -1 appears, context is everything.

The simplest level to examine the inverse of a function is on a point by point basis: if $f(3) = 8$, we'd also say $f^{-1}(8) = 3$.

EX 5 (a) If $f(-4) = 17$, what is $f^{-1}(17)$? (b) If $g(x) = 2x - 1$, what is $g^{-1}(3)$?

(a) By applying the very little we know about inverses already, if $f(-4) = 17$, then $f^{-1}(17) = -4$. To answer (b), we need to know which value in the domain of g leads to a value of 3 in the range; if $g(x) = 2x - 1$, then $g(x) = 3$ when $2x - 1 = 3$, or when $x = 2$. Then $g(2) = 3$ and $g^{-1}(3) = 2$. ■

So far, this business of inverse functions does not seem so bad, but let's complicate it by trying the strategy of part (b) in EX 5 again, to search for the value of $h^{-1}(1)$ in the case that $h(x) = x^2$. To do this, we need to know which value in the domain of h leads to a value of 1 in the range, so that we can reverse it. However, there are two such values: both $h(-1) = 1$

and $h(1) = 1$. So do we report $h^{-1}(1) = -1$ or $h^{-1}(1) = 1$? That's quite a little pickle, but there's a fix.

Given any function $f(x)$ and its entire associated domain, we cannot automatically conclude that there will be an inverse function partner to f. There might be, but there might not be. In the case that there is not, we still might get somewhere by considering the same function, but over a smaller portion of the original domain. In order to discuss an inverse, we need to ensure we are operating in a portion of the domain where a given output value in the range is produced by one, and only one, input value. When a particular output value of the function is produced by more than one input, this makes it hard for the inverse function to do its job, as we just saw. We want to avoid this situation, and there is a visual (graphical) way to do so; do you remember the *horizontal line test* from Def. 1.1? When we extend a horizontal line across the graph of a function, that line has an associated y value, and places where the horizontal line crosses the graph are locations where a given x value from the domain produces this y value in the range. If a horizontal line intersects the graph of $f(x)$ in more than one location, then the output value indicated by that line is produced by more than one input; the function fails the horizontal line test, and a supposed inverse function would get confused, as above. In this case, we say the inverse function will not exist.

If a function passes the horizontal line test everywhere on its graph, its inverse function will exist; each output is associated with one unique input, and vice versa. If a function does not pass the horizontal line test at first, we can try to spot a smaller portion of the graph (i.e. restrict the domain) such that the portion we're now considering does pass the horizontal line test.

Writing "passes the horizontal line test" can get tiring, but recall from Def. 1.1 that we can simply say the function is *one-to-one*.

Useful Fact 1.1. *If a function is one-to-one on an interval $[a, b]$, then it has an inverse there. It is also true that if a function is known to have an inverse on $[a, b]$, then it must be one-to-one. Together, we say that a function has an inverse on $[a, b]$ if and only if it is one-to-one there.*

The phrase "if and only if" is like mathematical superglue that binds together two conditions. The structure "(Condition A) if and only if

(Condition B)" means that Condition A implies Condition B, and also Condition B implies Condition A. If either condition is true, then both are true. FFT: Which of the following three statements are true?

- $ab = 0$ if and only if $a = 0$ or $b = 0$.
- $x^2 - 1 = 0$ if and only if $x = 1$.
- $a - b = b - a$ if and only if $a = b = 0$.

(Hint: two of the three statements are true.)

EX 6 The graphs of $x^3 + 5x$ and $\sqrt{x^2+1}$ are shown in Fig. 1.21 as curves (A) and (B), respectively. (The entire graph of $\sqrt{x^2+1}$ is made of both the dashed and solid parts of curve (B)). Which curve is one-to-one?

The graph of $x^3 + 5x$ passes the horizontal line test (at least in the region shown), and so is one-to-one. The entire graph of $\sqrt{x^2+1}$ (both dashed and solid together) does not pass the horizontal line test, and so is not one-to-one. However, if we restricted the domain of $\sqrt{x^2+1}$ to only $[0, \infty)$ (so that now we have only the thick solid portion of the graph in (B)), the resulting curve is one-to-one. ■

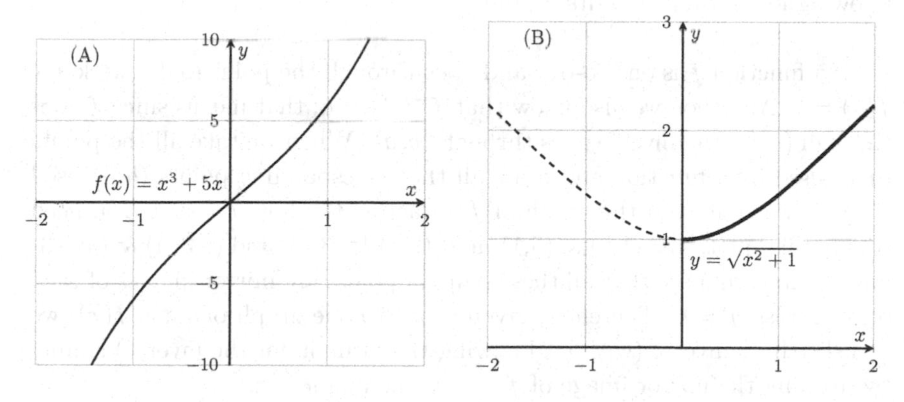

Fig. 1.21 Diagnosing where a function is one-to-one.

Where a function is one-to-one, and therefore has an inverse, there is an interplay between domains and ranges. A function takes an element of its domain and returns an element of its range. A function's inverse takes an element of the function's range and returns an element of the function's domain when that's possible (i.e. when the function is one-to-one). So, the inverse f^{-1} of a function f is a function itself; the domain of f^{-1} is the range of f, and the range of f^{-1} is the (restricted?) domain of f. (And here's another reminder that the notation f^{-1} does NOT indicate a reciprocal.)

EX 7 The functions (a) $f(x) = x^5$, (b) $g(x) = -\sqrt{x}$, and (c) $h(x) = \sqrt{x^3+1}$ are all one-to-one in their natural domains. What are the domains of their inverses?

The domain of each inverse is the range of the corresponding function. (a) The range of $f(x) = x^5$ is all real numbers, and so the domain of $f^{-1}(x)$ is also all real numbers. (a) The domain of $g(x) = -\sqrt{x}$ is $x \geq 0$ and its range is all negative real numbers (and zero), and so the domain of $g^{-1}(x)$ is $x \leq 0$. (c) The domain of $h(x) = \sqrt{x^3+1}$ is $x \geq -1$, and over that domain, the range is all non-negative real numbers, and so the domain of $h^{-1}(x)$ is $x \geq 0$. ■

Have you noticed that we're able to determine a lot about inverse functions (sample values, domain, and range) without even having recipes for them? There's one other thing we can generate for an inverse without even knowing its "formula" — its graph!

If a function f is one-to-one and goes through the point (a, b), we know $f(a) = b$. And then we also know that $f^{-1}(b) = a$; that means since f goes through (a, b), the inverse goes through (b, a). When we take all the points (a, b) used by a function and draw all the corresponding points (b, a) used by f^{-1}, we've flipped the graph of $f(x)$ across the line $y = x$. If you need to, try this: plot, say, (1,5) & (5,1), and (2,3) & (3,2), and $(-3, 4)$ & $(4, -3)$ and so on. You'll see that all these pairs of points are mirror images of each other across $y = x$. Therefore, given a one-to-one graph of some $f(x)$, we can sketch its inverse (without knowing the formula for the inverse) simply by drawing the mirror image of f across the line $y = x$.

Useful Fact 1.2. *If a function $f(x)$ has an inverse $f^{-1}(x)$, then the graphs of $f(x)$ and $f^{-1}(x)$ are symmetric across the line $y = x$.*

EX 8 A function $g(x)$, represented by the dashed line in Fig. 1.22, is one-to-one for $x \geq 1$. Which of the other two graphs (1) and (2)

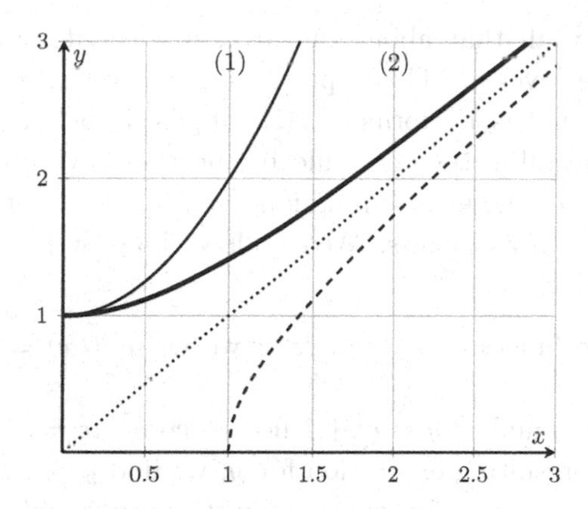

Fig. 1.22 Identifying an inverse by symmetry across $y = x$.

represent its inverse function? (Note $y = x$ is shown as the straight line.)

Because the thicker curve, (2), is the curve that is symmetric with $g(x)$ across the line $y = x$, it represents the inverse function g^{-1}. It turns out that $g(x) = \sqrt{x^2 - 1}$ and its inverse function is $g^{-1}(x) = \sqrt{x^2 + 1}$. ■

You Try It

(9) The function $h(t) = 2/t^2$ is shown together as both the dashed and thick solid lines in Fig. 1.23 (at the end of this section). The thick solid half for $t > 0$ is one-to-one. Which of the other two graphs (1, marked with o) and (2, marked with x) represent the inverse function for the restricted half of $h(t) = 2/t^2$? (Note $y = t$ is shown for reference.)

So far, we've seen that if we're given a function which is one-to-one, or can be made one-to-one upon a restriction of its domain, then:

- Knowing individual points (a, b) on the graph of the function allow us to know individual points (b, a) on the graph of the inverse.
- The domain of $f^{-1}(x)$ is the range of $f(x)$ and the range of $f^{-1}(x)$ is the domain of $f(x)$.
- If we know the graph of the function, we can construct the graph of the inverse by drawing the function's image across the line $y = x$.

And we can know all that about the inverse without even knowing the "formula" for the inverse. That's pretty impressive! But if the original one-to-one function has a "formula" (as opposed to being tabulated or known only graphically), then it would be more satisfying if we could find the representation of the inverse function, too. That is merely a matter of algebra, and it works as follows. We'll follow along by finding the inverse of $f(x) = x^3 + 1$:

(1) Write the function as "$y = \ldots$"; we write $f(x) = x^3 + 1$ as $y = x^3 + 1$.
(2) Exchange y and x; $y = x^3 + 1$ now becomes $x = y^3 + 1$.
(3) Solve the resulting expression for y; we find $y = \sqrt[3]{x-1}$.
(4) This is the inverse function, name it appropriately; we now know that for $f(x) = x^3 + 1$, we have $f^{-1}(x) = \sqrt[3]{x-1}$.

EX 9 Find a domain over which $f(x) = (x-3)^4$ is one-to-one, and find its inverse function there.

The shape of this graph is very parabola-like (although it's steeper since the exponent is 4), with a vertex at $(3, 0)$. (Sketch it!) Every output for an x which is larger than 3 is duplicated by another x which is less than 3. So to eliminate duplication and leave a one-to-one function, we can restrict the domain to $x \geq 3$. With that done, we know the inverse exists and is found this way:

$$i) \quad y = (x-3)^4$$

$$ii) \quad x = (y-3)^4 \qquad \longrightarrow \qquad iv) \quad y = x^{1/4} + 3$$

$$iii) \quad x^{1/4} = y - 3 \qquad\qquad v) \quad f^{-1}(x) = \sqrt[4]{x} + 3 \quad \blacksquare$$

You Try It

(10) Find a domain over which $f(x) = e^{x^2}$ is one-to-one, and find its inverse function there.

Finally, here's a note about why we want to worry about inverse functions to begin with. Knowing the inverse of a given function is the key to unlocking that function in the process of solving an equation. You have already known that to solve for a variable tied up in, say, an expression that's squared, we would need to apply a square root. Well, that's the magic of inverse functions. Formally written, given a function $f(x)$ and its inverse $f^{-1}(x)$, the following two statements are true:

$$f(f^{-1}(x)) = x \quad \text{and} \quad f^{-1}(f(x)) = x$$

For example, in the last section, we noted that exponential and logarithmic functions can be applied to one another to force an elimination:

$$\log_b(b^x) = x \quad \text{and} \quad b^{\log_b(x)} = x$$

Although we didn't use the term in those notes, it turns out that exponential functions and logarithmic functions are inverses! The function e^x is the inverse of $\ln(x)$, and vice versa. $\log_{10}(x)$ is the inverse of 10^x. And so on. So more generally, if *anything* is trapped inside an exponential or logarithmic function, we apply the appropriate inverse to free the inside:

$$\log_b(b^{g(x)}) = g(x) \quad \text{and} \quad b^{\log_b g(x)} = g(x)$$

From now on, any time you're introduced to a new type of function — or when you're asked to review a certain type of function, such as trigonometric functions in the next section! — you should also expect to learn about the inverses as well. Believe me, I take no joy in the fact I'm about to remind you of trigonometric functions and their inverses.[1]

Have You Learned...

- How to assess domains of constructions like $f \pm g$, fg, f/g.
- How to read and use and functional notation involving composition of functions?
- How to determine the domain of a composition of functions?
- The graphical and algebraic relationships between a function and its inverse?
- When a function even has an inverse?
- How functions and their inverses are used when solving equations?
- That some already-familiar functions are actually inverses of each other?
- That inverse notation f^{-1} does *not* imply a reciprocal?

[1]Well, maybe a little.

Compositions and Inverses — Problem List

Compositions and Inverses — You Try It

These appeared above; solutions begin on the next page.

(1) If $f(x) = \ln(x)$ and $g(x) = \sqrt{2-x}$, create and find the domains of $f+g$, $f-g$, fg, and f/g.
(2) If $f(x) = 1/x$ and $g(x) = 1 - \sqrt{x}$, then create and find the domains of $f(g(x))$, $g(f(x))$, $f(f(x))$, $g(g(x))$.
(3) What two functions $f(x)$ and $g(x)$ form $G(x) = x^2/(x^2+4)$ as the composition $f(g(x))$?
(4) What three functions $f(x)$, $g(x)$, and $h(x)$ form $H(x) = 1 - 3^{x^2}$ when composed as $f(g(h)))$?
(5) How can the graph of $y = x^2$ be adapted to create the graph of $y = (x+1)^2$?
(6) How can the graph of $y = \sqrt{x}$ be adapted to create the graph of $y = \sqrt{x+3}$?
(7) How can the graph of $y = \log_{10}(x)$ be adapted to form the graph of $y = \log_{10}(x-5)$?
(8) How are the graphs of (a) $y = e^x + 2$, (b) $y = e^{x-2}$, (c) $y = -e^x$, and (d) $y = e^{-x}$ related to the graph of $y = e^x$?
(9) The function $h(t) = 2/t^2$ is shown together as both the dashed and thick solid lines in Fig. 1.23. The thick solid half for $t > 0$ is one-to-one. Which of the other two graphs (1, marked with o) and (2, marked with x) represent the inverse function for the restricted half of $h(t) = 2/t^2$? (Note $y = t$ is shown for reference.)
(10) Find a domain over which $f(x) = e^{x^2}$ is one-to-one, and find its inverse function there.

Compositions and Inverses — Practice Problems

Try these as you get the hang of the You Try It problems. Solutions to these problems are available in Sec. A.1.4.

(1) If $f(x) = \sqrt{1+x}$ and $g(x) = \sqrt{1-x}$, then create and find the domains of $f+g$, $f-g$, fg, f/g.
(2) If $f(x) = 1 - 3x$ and $g(x) = 5x^2 + 3x + 2$, then create and find the domains of $f(g(x))$, $g(f(x))$, $f(f(x))$, $g(g(x))$.
(3) If $f(x)2/(x+1)$, $g(x) = x^3$, and $h(x) = \sqrt{x+3}$, what is $f(g(h(x)))$?

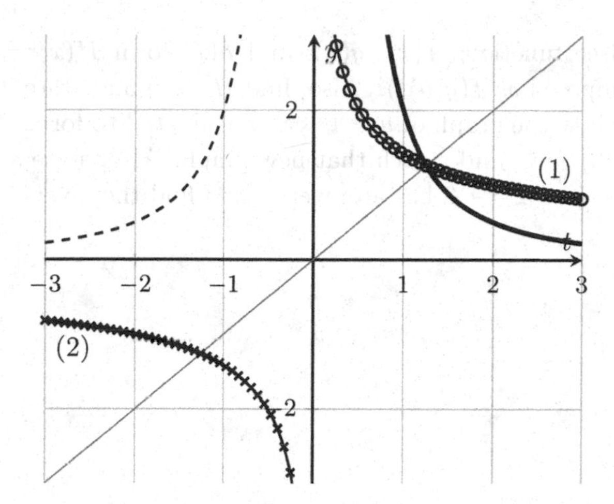

Fig. 1.23 $2/t^2$ is made of, at first, both the dashed and thick solid curves. The other curves (marked with o or x) are the candidate inverses.

(4) What two functions $f(t)$ and $g(t)$ form $u(t) = e^t/(1+e^t)$ as the composition $f(g(t))$?
(5) What three functions $f(x)$, $g(x)$, and $h(x)$ form $H(x) = \log_2(\sqrt{x^4 - 1})$ when composed as $f(g(h)))$?
(6) How can the graph of $y = x^2$ be adapted to create the graph of $y = 1 - x^2$?
(7) How can the graph of $y = \sqrt[3]{x}$ be adapted to create the graph of $y = 1 + \sqrt[3]{x - 1}$?
(8) How can the graph of $y = e^x$ be adapted to create the graph of $y = 1 + 2e^x$?
(9) How can the graph of $y = \ln(x)$ be adapted to create the graphs (a) $y = -\ln(x)$, (b) $y = \ln(-x)$, (c) $y = \ln|x|$?
(10) Find a domain over which $f(x) = 2(x-1)^4$ is one-to-one, and find its inverse function there.

Compositions and Inverses — Challenge Problems

Try these problems to test your skills with the ideas in this section. Solutions to these problems are available in Sec. B.1.4.

(1) If $f(x) = \sqrt{2x+3}$ and $g(x) = x^2 + 1$, then create and find the domains of $f(g(x))$, $g(f(x))$, $f(f(x))$, $g(g(x))$. Then, pick the function from $f(x)$ and $g(x)$ which is already one-to-one, and find its inverse.

(2) What three functions $f(x)$, $g(x)$, and $h(x)$ form $H(x) = \sqrt[3]{\sqrt{x} - 1}$ when composed as $f(g(h)))$? Also, find $H^{-1}(x)$ and state its domain.
(3) Describe how the graph of $y = x^4$ can be adapted to form the graph of $y = (x+2)^4 + 3$, and sketch that new graph. Also, state a domain on which $y = (x+2)^4 + 3$ has an inverse, and find the inverse.

Compositions and Inverses — You Try It — Solved

(1) If $f(x) = \ln(x)$ and $g(x) = \sqrt{2-x}$, create and find the domains of $f+g$, $f-g$, fg, and f/g.

□ The domain of $f(x)$ is $x > 0$. The domain of $g(x)$ is $x \leq 2$. The only values allowed in both sets are $0 < x \leq 2$. Thus,

- $f(x) + g(x) = \ln(x) + \sqrt{2-x}$; the domain of this is $0 < x \leq 2$.
- $f(x) - g(x) = \ln(x) - \sqrt{2-x}$; the domain of this is $0 < x \leq 2$.
- $f(x)g(x) = \ln(x)\sqrt{2-x}$; the domain of this is $0 < x \leq 2$.
- $f(x)+g(x) = \ln(x)/\sqrt{2-x}$; the domain of this starts as $0 < x \leq 2$, but since $g(2) = 0$, we must remove $x = 2$ from the new domain. Therefore the domain of $f(x)/g(x)$ is $0 < x < 2$.

■

(2) If $f(x) = 1/x$ and $g(x) = 1 - \sqrt{x}$, then create and find the domains of $f(g(x))$, $g(f(x))$, $f(f(x))$, $g(g(x))$.

□ Note that the domains of $f(x)$ and $g(x)$ alone are $x \neq 0$ and $x \geq 0$, respectively. Then,

$$f(g(x)) = \frac{1}{1-\sqrt{x}}$$

The domain of $f(g(x))$ must ensure that $g(x) = 1 - \sqrt{x}$ is never zero. We can't use $x < 0$, and we must also note that $g(1) = 0$. So the domain of $f(g(x))$ is $x \geq 0$ AND $x \neq 1$. In set notation, this is $[0,1) \cup (1,\infty)$. In bracket notation, we could write this domain as $\{x : x \geq 0, x \neq 1\}$.

$$g(f(x)) = 1 - \sqrt{\frac{1}{x}} = 1 = \frac{1}{\sqrt{x}}$$

The domain of $g(f(x))$ is $x > 0$, i.e. $(0,\infty)$.

$$f(f(x)) = \frac{1}{(1/x)} = x$$

Since $f(f(x)) = x$, it is easy to jump to the conclusion that the domain of $f(f(x))$ is all reals. But, we could not use $x = 0$ originally in $f(x)$, so we can't suddenly start using it now. The domain of $f(f(x))$ is all reals except 0, i.e. $\{x : x \neq 0\}$, i.e. $(-\infty,0) \cup (0,\infty)$.

$$g(g(x)) = 1 - \sqrt{1-\sqrt{x}}$$

The domain of $g(g(x))$ begins with $x \geq 0$ because of the interior $\sqrt{x}$, but we also now have to ensure that $1 - \sqrt{x} \geq 0$; this corresponds to $1 \geq \sqrt{x}$ or $x \leq 1$. So all together, when we put $x \geq 0$ together with $x \leq 1$, the domain of $g(g(x))$ is $0 \leq x \leq 1$. ■

---

(3) What two functions $f(x)$ and $g(x)$ form $G(x) = x^2/(x^2+4)$ as the composition $f(g(x))$?

□ $G(x)$ can be decomposed into $f \circ g$ as follows:

$$G(x) = \frac{x^2}{x^2+4} = f(g(x)) \quad \text{for } f(x) = \frac{x}{x+4} \text{ and } g(x) = x^2$$ ■

---

(4) What three functions $f(x)$, $g(x)$, and $h(x)$ form $H(x) = 1 - 3^{x^2}$ when composed as $f(g(h(x)))$?

□ With $f(x) = 1-x$, $g(x) = 3^x$, and $h(x) = x^2$, we get $g(h(x)) = 3(x^2)$, and then $H(x)$ is $f(g(h(x)))$. ■

---

(5) How can the graph of $y = x^2$ be adapted to create the graph of $y = (x+1)^2$?

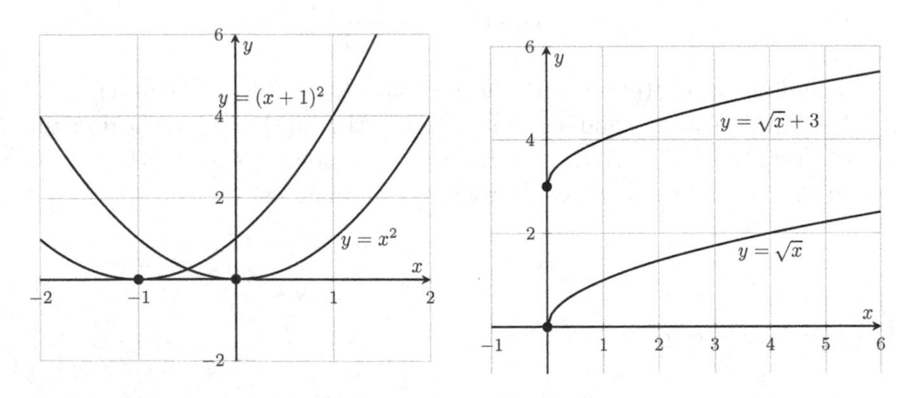

Fig. 1.24 Horizontal and vertical translation in action.

□ The graph of $y = (x+1)^2$ can be found by taking the graph of $y = x^2$ and shifting it to the left by 1 unit, so that the vertex moves from $(0,0)$ to $(-1,0)$ — see Fig. 1.24(A). ■

(6) How can the graph of $y = \sqrt{x}$ be adapted to create the graph of $y = \sqrt{x} + 3$?

□ The graph of $y = \sqrt{x}+3$ can be found by taking the graph of $y = \sqrt{x}$ and shifting it up by 3 units, so that the left endpoint moves from $(0, 0)$ to $(0, 3)$ — see Fig. 1.24(B). ■

(7) How can the graph of $y = \log_{10}(x)$ be adapted to form the graph of $y = \log_{10}(x - 5)$?

□ The graph of $y = \log_{10} x$ is $y = \log_{10}(x)$ shifted to the right by 5 units, so that the x-intercept moves from $(1, 0)$ to $(6, 0)$ — see Fig. 1.25(A). ■

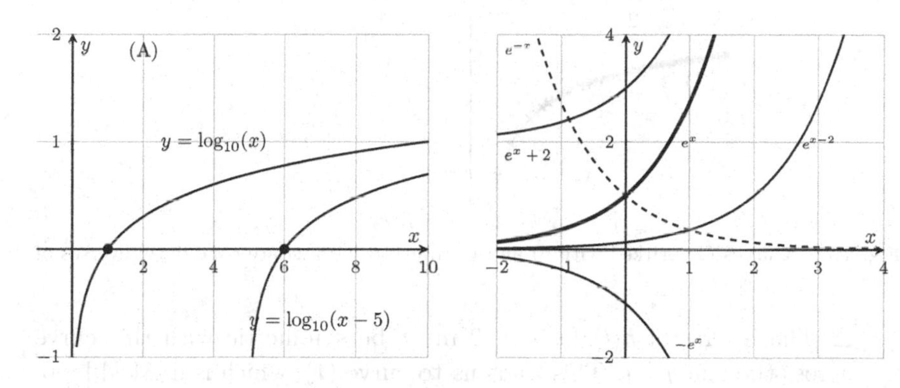

Fig. 1.25 Variations on $\log_{10} x$ and e^x.

(8) How are the graphs of (a) $y = e^x + 2$, (b) $y = e^{x-2}$, (c) $y = -e^x$, and (d) $y = e^{-x}$ related to the graph of $y = e^x$?

□ Relative to the graph of $y = e^x$, we get (a) $y = e^x + 2$ is shifted 2 upwards; (b) $y = e^{x-2}$ is shifted 2 right; (c) $y = -e^x$ is reflected across the x-axis; (d) $y = e^{-x}$ is reflected across the y-axis; (e) $y = -e^{-x}$ is reflected around both axes. These are shown in Fig. 1.25(B). ■

(9) The function $h(t) = 2/t^2$ is shown together as both the dashed and thick solid lines in Fig. 1.26. The thick solid half for $t > 0$ is one-to-one. Which of the other two graphs (1, marked with o) and (2, marked with x) represent the inverse function for the restricted half of $h(t) = 2/t^2$? (Note $y = t$ is shown for reference.)

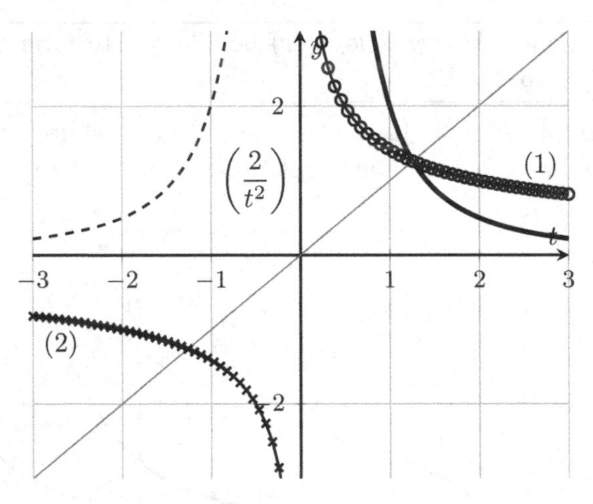

Fig. 1.26 Curves (1, marked with o) and (2, marked with x) are candidate inverses of $2/t^2$.

☐ The inverse of $h(t)$ for $t > 0$ must be symmetric with this curve across the line $y = t$. This leads us to curve (1), which is marked by o. ■

(10) Find a domain over which $f(x) = e^{x^2}$ is one-to-one, and find its inverse function there.

☐ This is an even function, for which $f(x) = f(-x)$. Figure 1.27 shows that if we select a domain $[0, \infty)$, the resulting curve (thick solid) passes the horizontal line test, and so is one-to-one. On that domain, the inverse is found as follows:

$$i)\quad y = e^{x^2}$$

$$ii)\quad x = e^{y^2} \qquad \longrightarrow \qquad iv)\quad \ln x = y^2$$

$$iii)\quad \ln x = \ln e^{y^2} \qquad\qquad v)\quad y = \sqrt{\ln(x)}$$

The inverse function $f^{-1}(x) = \sqrt{\ln(x)}$ is also shown in Fig. 1.27.

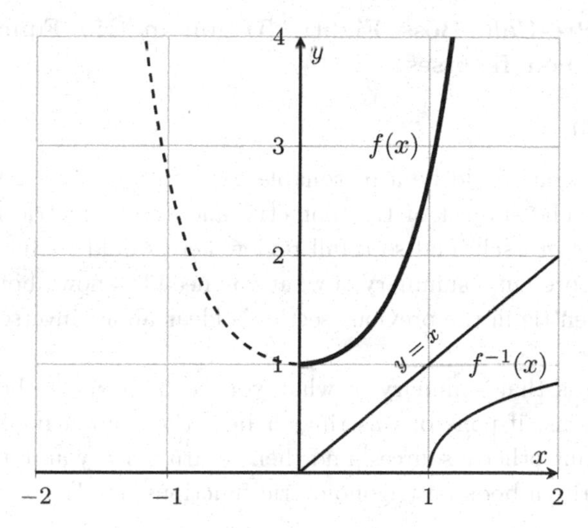

Fig. 1.27 Restricting the domain of $y = e^{x^2}$ to make it one-to-one.

1.5 The Pre-Calc Boss Fight: Trigonometric Functions (and Their Inverses)

Introduction

Contrary to what would be a reasonable expectation, this section will be quite short. The subject of trigonometry and trigonometric functions is often a course in itself, and so a full review here would be futile. Rather, my plan is to present a summary of what you need to know about trig functions, and then tie in the previous section's ideas about inverse functions.

And here is that summary of what you need to know about trigonometric functions; if none of this rings a bell, then put a bookmark here, go explore some other resources, and then return when you're ready. (Hey, maybe I'll write a book on trigonometric functions next!)

- There are six trigonometric functions: the "main three" are $\sin(x)$, $\cos(x)$, $tan(x)$, and the three reciprocal functions are $\sec(x)$, $\csc(x)$, $\cot(x)$.
- The independent variable in each represents an angle (in radians, please ... degrees should be banished).
- Each has a definition based on ratios of sides from a right triangle. You should know what these are.
- Each is periodic.
- Each has a graph which you should be able to sketch.
- Each has some commonly occurring values which you should know.

There you have everything important from course on trigonometric functions! Now let's move on to talking about inverse trigonometric functions.

Inverse Trigonometric Functions

Given that you should know what the graphs of the six trig functions look like, you should know that not a single one of them is one-to-one! For example, $f(x) = \sin(x)$ fails the horizontal line test big-time, as seen in Fig. 1.28.

So, how in the world are trig functions going to have inverses? Life would be easier if they didn't. Unfortunately, life isn't always fair, and trig functions do have inverses. The secret is in restriction of domains. Note that for $f(x) = \sin(x)$, if we restrict our domain to the interval $[-\pi/2, \pi/2]$, we

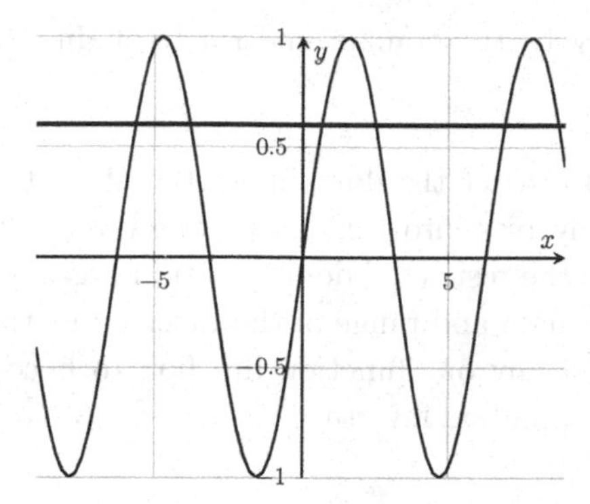

Fig. 1.28 $\sin(x)$ fails the horizontal line test, so how can it have an inverse?

end up covering the entire spread of possible output of the function (everything from -1 to 1) in a one-to-one fashion. If we take any x values less than $\pi/2$ or greater than $\pi/2$, we duplicate values of $f(x)$ that are already known. This restricted domain is shown with the thick portion of the curve in Fig. 1.29.

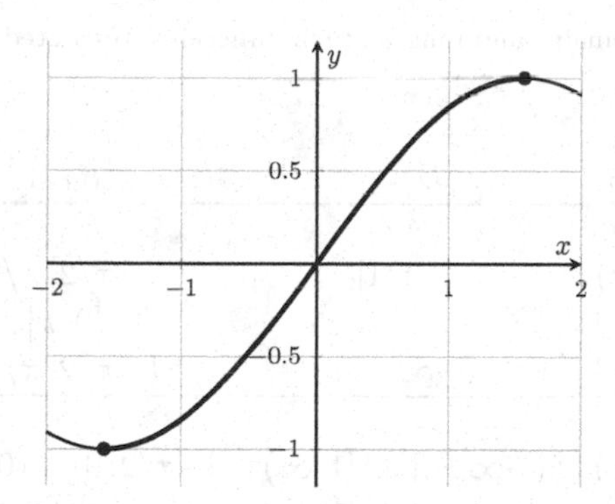

Fig. 1.29 Restricting the domain of $\sin(x)$ to a one-to-one portion.

On this restricted domain, we can expect an inverse of $\sin(x)$ to exist. We cleverly name this function $\sin^{-1}(x)$; the graph of this inverse sine function will be the (restricted) graph of $y = \sin(x)$ flipped around the line $y = x$. (See if you can take a pencil and sketch in the graph of $\sin^{-1}(x)$ into Fig. 1.29.) Since the restricted domain and range of $\sin(x)$ are $[-\pi/2, \pi/2]$

and $[-1, 1]$ respectively, the domain and range of $\sin^{-1}(x)$ are $[-1, 1]$ and $[-\pi/2, \pi/2]$.

We also need to restrict the domains of the other trig functions in order to bring their inverses into existence. The tables shown in Figs. 1.30 and 1.31 summarize the restricted domains and ranges of all trig functions, and the resulting domain and range of the inverse functions. Note how the domain and range of any one function flip flop to become the range and domain of the corresponding inverse.

$f(x)$	D_f	R_f
$\sin(x)$	$[-\pi/2, \pi/2]$	$[-1, 1]$
$\cos(x)$	$[0, \pi)$	$[-1, 1]$
$\tan(x)$	$(-\pi/2, \pi/2)$	$(-\infty, \infty)$
$\csc(x)$	$[-\pi/2, 0) \cup (0, \pi/2]$	$(-\infty, -1] \cup [1, \infty)$
$\sec(x)$	$[0, \pi/2) \cup (\pi/2, \pi]$	$(-\infty, -1] \cup [1, \infty)$
$\cot(x)$	$(0, \pi)$	$(-\infty, \infty)$

Fig. 1.30 Domains and ranges of trig functions, restricted for inverses.

$f^{-1}(x)$	$D_{f^{-1}}$	$R_{f^{-1}}$
$\sin^{-1}(x)$	$[-1, 1]$	$[-\pi/2, \pi/2]$
$\cos^{-1}(x)$	$[-1, 1]$	$[0, \pi]$
$\tan^{-1}(x)$	$(-\infty, \infty)$	$(-\pi/2, \pi/2)$
$\csc^{-1}(x)$	$(-\infty, -1] \cup [1, \infty)$	$[-\pi/2, 0) \cup (0, \pi/2]$
$\sec^{-1}(x)$	$(-\infty, -1] \cup [1, \infty)$	$[0, \pi/2) \cup (\pi/2, \pi]$
$\cot^{-1}(x)$	$(-\infty, \infty)$	$(0, \pi)$

Fig. 1.31 Domains and ranges of inverse trig functions.

Figure 1.32 shows the graphs of the first three primary trig functions on their restricted domains, with their corresponding inverse functions. Can you tell which function is which? I'll tell you this much: curves A1 and A2 are the sines, B1 and B2 are the cosines, and C1 and C2 are the tangents.

But which is the restricted trig function, and which is the inverse? This leads to YTI 1:

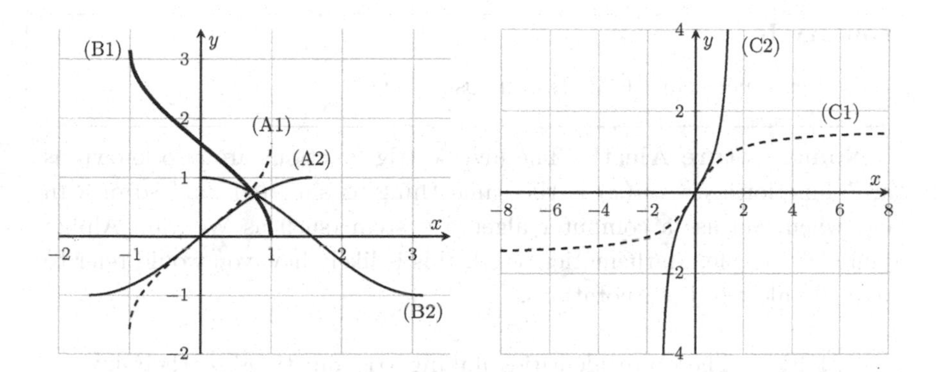

Fig. 1.32 **Identifying (restricted) trig functions and their inverses.**

> **You Try It**
>
> (1) Refer to Fig. 1.32. Of curves A1 and A2, which is $\sin(x)$ and which is $\sin^{-1}(x)$? Of curves B1 and B2, which is $\cos(x)$ and which is $\cos^{-1}(x)$? Of curves C1 and C2, which is $\tan(x)$ and which is $\tan^{-1}(x)$?

With these restricted domains, we can begin evaluating the inverse trig functions. But to do that, you have to know what they really mean. It's well and good to name them and give their domains and ranges, but what are they, really? Well, consider the usual definition of inverse function applied to this case. If $\sin(\pi/4) = 1/\sqrt{2}$, then $\sin^{-1}(1/\sqrt{2}) = \pi/4$. Or if $\csc(\pi/6) = 2$, then $\csc^{-1}(2) = \pi/6$. In general, an expression like $\sin^{-1}(x)$ should be read as "what angle, from within the appropriate range, has a sine of x?"

EX 1 Evaluate $\sin^{-1}(1/2)$ and $\sin^{-1}(-1/2)$.

The phrase $\sin^{-1}(1/2)$ defined itself, but you can think of it like a query: it's asking for the angle from within the interval $[-\pi/2, \pi/2]$ that has a sine of $1/2$. Whether by calculator or table, you can find that this angle is $\pi/6$. Next, $\sin^{-1}(-1/2)$ is asking for the angle from within the interval $[-\pi/2, \pi/2]$ that has a sine of $-1/2$. By the known symmetries of the sine function, we know from the first result that this will be $-\pi/6$. Together,

$$\sin^{-1}\left(\frac{1}{2}\right) = \frac{\pi}{6} \quad \text{and} \quad \sin^{-1}\left(-\frac{1}{2}\right) = -\frac{\pi}{6} \quad \blacksquare$$

You Try It

(2) Evaluate $\sin^{-1}(\sqrt{3}/2)$ and $\cos^{-1}(-1)$.

Nomenclature Alert! The inverse trig functions are also known as "arc" functions; *arcsin*(x) is the same thing as $\sin^{-1}(x)$, and so on. In fact, when you use a computer algebra system (such as Wolfram Alpha, Symbolab, Maple, Mathematica, etc.), this is likely how you would refer to them. Be alert to both notations.

FFT: There are identities linking trig functions to their inverse functions, such as this one (with a mix of notation for practice):

$$\sin^{-1}(a) = \csc^{-1}\left(\frac{1}{a}\right) \quad \text{and} \quad \arcsin(b) = \left(sin^{-1}\frac{1}{b}\right)$$

(for $a, b \neq 0$). For example, $sin^{-1}(1/\sqrt{2}) = \csc^{-1}(\sqrt{2})$. There are five other similar expressions for each other trig function & inverse pair, which say that a trig function of a given value is equal to the inverse trig function of the reciprocal value. Why do these work?

Combining Trig Functions and Their Inverses

The examples above were about individual values of the inverse trig functions. We can also systematically analyze compositions of these inverse trig functions with their regular trig counterparts, in the same way we can combine e^x and $\ln(x)$ or x^2 and $\sqrt{x}$. For example, consider the expression $\sin(\sin^{-1}(x))$. Decrypting this expression, we see that it is asking for "the sine of the angle whose sine is x". Well, that's x. Similarly, the expression $\sin^{-1}(\sin(x))$ is asking for "the angle whose sine is $\sin(x)$?" Well, that's x again. Together, assuming we are working in the proper domains,

$$\sin(\sin^{-1}(x)) = \sin^{-1}(\sin(x)) = x$$

Any similar combination of a trig function with its own inverse will result in elimination of the functions. But also, it turns out that when you combine any one trig function with a different trig function's inverse, the two functions will eliminate each other, leaving behind a more complicated algebraic expression than just x, as this next example shows.

EX 2 Simplify $\sin(\cos^{-1}(x))$ into its algebraic form.

This expression is seeking "the sine of the angle whose cosine is x". Well, let's call this angle A and set it inside a right triangle; if the cosine of the angle A is x, which can also be written $x/1$, then from the basic definition of cosine (adjacent over hypotenuse) we can start labelling a right triangle as shown in the left half of Fig. 1.33. Then by the Pythagorean Theorem, we can fill in the length of the remaining (opposite) side, as shown in the right half of Fig. 1.33.

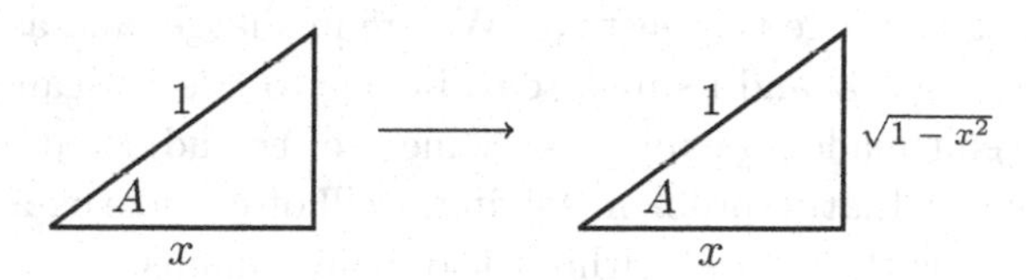

Fig. 1.33 Information helpful for simplifying $\sin(\cos^{-1}(x))$.

And now, since we know that $\sin(A)$ comes from opposite over hypotenuse,

$$\sin(A) = \sin(\cos^{-1}(x)) = \frac{\sqrt{1-x^2}}{1} = \sqrt{1-x^2}$$

Even better, now that the triangle is complete, we know all of the following compositions:

$$\sin(\cos^{-1}(x)) = \sqrt{1-x^2} \qquad \csc(\cos^{-1}(x)) = \frac{1}{\sqrt{1-x^2}}$$

$$\cos(\cos^{-1}(x)) = x \qquad \sec(\cos^{-1}(x)) = \frac{1}{x}$$

$$\tan(\cos^{-1}(x)) = \frac{\sqrt{1-x^2}}{x} \qquad \cot(\cos^{-1}(x)) = \frac{x}{\sqrt{1-x^2}} \quad \blacksquare$$

Since there are 36 different ways to combine a trig function with an inverse trig function, there are still plenty of others to discover!

You Try It

(3) Simplify $\tan(\sin^{-1}(x))$ into its algebraic form.

FFT: Any relations which mix trig functions and inverses, such as $\sin^{-1}(a) = \csc^{-1}(1/a)$, $\sin(\sin^{-1}(x)) = \sin^{-1}(\sin(x)) = x$, and

$\sin(\cos^{-1}(x)) = \sqrt{1-x^2}$, are actually a bit trickier than they appear; technically, they hold only when we're working in the proper domains. For example, the statement $\sin^{-1}\sin(11\pi/2) = 11\pi/2$ is technically *false*, based on Table 1.31. Why is it false, and how can it be corrected?

Whenever we blend trig functions and their inverses in a single expression, we are (technically) restricting the values of x we can work with from that point forward. In practice, though, we may or may not acknowledge this issue, because we can just say, "oh, we're redefining the inverse trig function so that its range is adjusted." We are in charge, and as long as we set up the trig function and its inverse to be one-to-one, we can proceed — even if we haven't made an explicit statement of the adjusted domain and range. This means that in problem solving, we'll often take expressions like $\sin^{-1}(\sin x) = x$ as they stand without too many qualms.

Solving Equations With Trig Functions and Their Inverses

We can use inverse trig functions to solve equations in the same way we use any other inverse function; that is, we use combinations of functions and their inverses to eliminate each other and leave variables behind. You have to know (or be able to find) special values of the trig functions when needed, and/or deal with reference angles.

EX 3 Solve $\sin(2x-1) = -\dfrac{1}{2}$.

If you have spectacular memory of special angles, you may remember that $-\pi/6$ is the angle within the restricted domain of $\sin(x)$ at which the sine is $-\pi/2$, and so we know that $2x - 1 = -\pi/6$. But if not, we can apply the inverse sine function to both sides (keeping in mind that we must be acting in the restricted domain / range of sine and its inverse), to get

$$\sin^{-1}(\sin(2x-1)) = \sin^{-1}\left(-\frac{1}{2}\right)$$

when then (once you come up with the value on the right hand side by table or computer algebra system) becomes

$$2x - 1 = -\frac{\pi}{6}$$

$$x = -\frac{\pi}{12} + \frac{1}{2} \quad \blacksquare$$

EX 4 Solve $\tan^{-1}(x+2) = -\frac{\pi}{3}$.

Again, if you have spectacular memory of special angles, you may recall that, within its restricted domain, the tangent of $-\pi/3$ is $-\sqrt{3}$, so that $x+2$ must be equal to $-\sqrt{3}$. But if not, we can apply the tangent function to both sides (keeping in mind that we must be acting in the restricted domain / range of tangent and its inverse), to get

$$\tan(\tan^{-1}(x+2) = \tan\left(-\frac{\pi}{3}\right)$$

which then simplifies as

$$x + 2 = -\sqrt{3}$$
$$x = -2 - \sqrt{3} \quad \blacksquare$$

You Try It

(4) Solve $2\sec(x/3) = \frac{4}{\sqrt{3}}$.

Have You Learned...

- How trigonometric functions can have inverses even though they're not one-to-one?
- How to evaluate (find values of) inverse trigonometric functions?
- The graphs of some inverse trigonometric functions?
- How to algebraically combine trigonometric functions and inverse trigonometric functions, even ones that don't match?
- How to solve equations using inverse trigonometric functions?
- How to be alert to issues related to domain and range when mixing trig functions and inverse trig functions?

Trig Functions & Their Inverses — Problem List

Trig Functions & Their Inverses — You Try It

These appeared above; solutions begin on the next page.

(1) Refer to Fig. 1.32 earlier in this section. Of curves A1 and A2, which is $\sin(x)$ and which is $\sin^{-1}(x)$? Of curves B1 and B2, which is $\cos(x)$ and which is $\cos^{-1}(x)$? Of curves C1 and C2, which is $\tan(x)$ and which is $\tan^{-1}(x)$?
(2) Evaluate $\sin^{-1}(\sqrt{3}/2)$ and $\cos^{-1}(-1)$.
(3) Simplify $\tan(\sin^{-1}(x))$ into its algebraic form.
(4) Solve $2\sec(x/3) = \dfrac{4}{\sqrt{3}}$.

Trig Functions & Their Inverses — Practice Problems

Try these as you get the hang of the You Try It problems. Solutions to these problems are available in Sec. A.1.5.

(1) Curves D1 and D2 in Fig. 1.34 are a trigonometric function (on its restricted domain) and its inverse. Identify them.

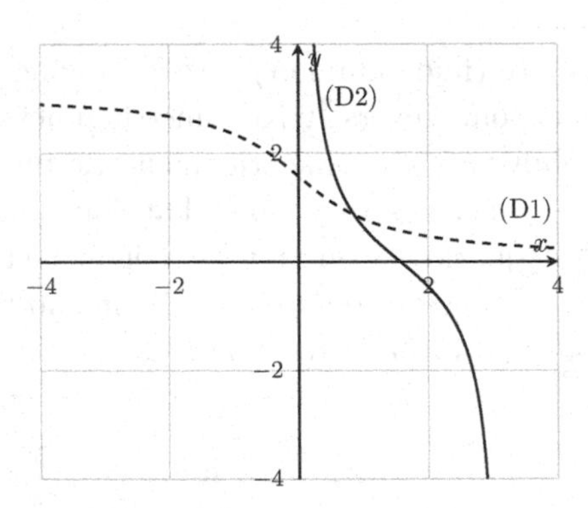

Fig. 1.34 Name the trig function and its inverse.

(2) Evaluate $\arctan(-1)$ and $\csc^{-1}(2)$.
(3) Simplify $\sin(\tan^{-1}(x))$ into its algebraic form.
(4) Solve $4\cos^{-1}(x^2) = \dfrac{2\pi}{3}$.

Trig Functions & Their Inverses — Challenge Problems

Try these problems to test your skills with the ideas in this section. Solutions to these problems are available in Sec. B.1.5.

(1) Sketch the graph of $f(x) = \sec(x)$ on its restricted domain, and $\sec^{-1}(x)$, on the same set of axes.
(2) Evaluate $\sec^{-1}(\sqrt{2})$ and $\sin^{-1}(1)$.
(3) Simplify $\cos(\cot^{-1}(1/x))$ into its algebraic form.

Trig Functions (and Their Inverses) — You Try It — Solved

(1) Refer to Fig. 1.35. Of curves A1 and A2, which is $\sin(x)$ and which is $\sin^{-1}(x)$? Of curves B1 and B2, which is $\cos(x)$ and which is $\cos^{-1}(x)$? Of curves C1 and C2, which is $\tan(x)$ and which is $\tan^{-1}(x)$?

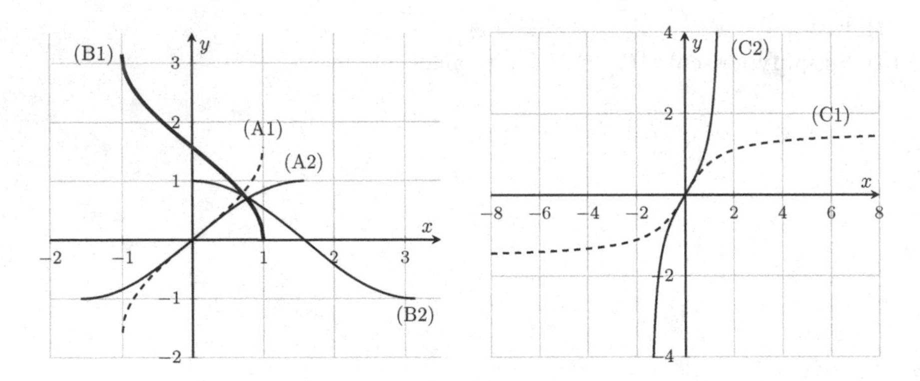

Fig. 1.35 Identifying (restricted) trig functions and their inverses.

☐ A1 is $\sin^{-1}(x)$ and A2 is $\sin(x)$. B1 is $\cos^{-1}(x)$ and B2 is $\cos(x)$. C1 is $\tan^{-1}(x)$ and C2 is $\tan(x)$. ■

(2) Evaluate $\sin^{-1}(\sqrt{3}/2)$ and $\cos^{-1}(-1)$.

☐ The expression $\sin^{-1}(\sqrt{3}/2)$ represents the angle between $-\pi/2$ and $\pi/2$ whose sine is $\sqrt{3}/2$. That angle is $\pi/3$. The expression $\cos^{-1}(-1)$ represents the angle between 0 and π whose cosine is -1. That angle is π. So:

$$\sin^{-1}\left(\frac{\sqrt{3}}{2}\right) = \frac{\pi}{3} \quad \text{and} \quad \cos^{-1}(-1) = \pi \quad ■$$

(3) Simplify $\tan(\sin^{-1}(x))$ into its algebraic form.

☐ $\tan(\sin^{-1}(x))$ wants the tangent of the angle whose sine is x. If the sine of this angle is x, then we have an opposite side of length x, a hypotenuse of length 1. This angle (A) and associated lengths are indicated in Fig. 1.36. We can then complete the triangle by labeling the adjacent side with length $\sqrt{1-x^2}$. So the tangent of this mystery angle is opp / adj $= \dfrac{x}{\sqrt{1-x^2}}$ and

$$\tan(\sin^{-1}(x)) = \frac{x}{\sqrt{1-x^2}} \quad \blacksquare$$

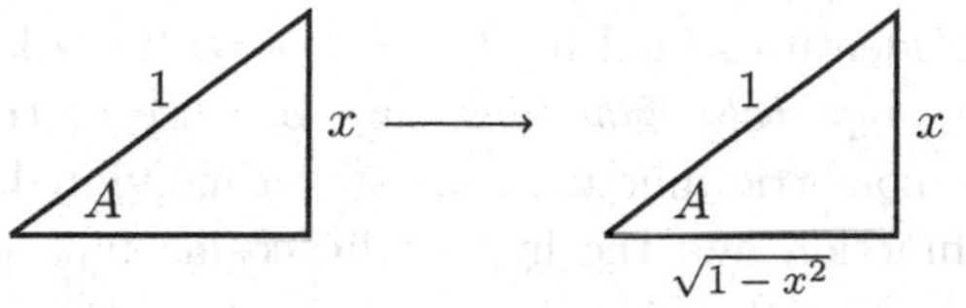

Fig. 1.36 Information helpful for simplifying $\tan(\sin^{-1}(x))$.

---

(4) Solve $2\sec\left(\dfrac{x}{3}\right) = \dfrac{4}{\sqrt{3}}$.

☐ Follow along in order *(i)* to *(vi)*:

i) $2\sec(x/3) = \dfrac{4}{\sqrt{3}}$

ii) $\sec(x/3) = \dfrac{2}{\sqrt{3}}$ $\longrightarrow$ *iv)* $\dfrac{x}{3} = \dfrac{\pi}{6}$

iii) $\sec^{-1}(\sec(x/3)) = \sec^{-1}\dfrac{2}{\sqrt{3}}$ *v)* $x = \dfrac{\pi}{2}$ $\blacksquare$

1.6 The Greatest Functions Ever!

Introduction

There is a collection of functions which are defined using the function e^x, but which behave similarly to trigonometric functions in how they relate to each other via identities (and in other ways we'll see later). They are collectively called *hyperbolic functions*, and because of their operational similarity to trigonometric functions, are given individual names like the hyperbolic sine function and the hyperbolic cosine function (I'll let you guess how many others there are besides these first two, and what their names might be.)

These functions do not show up nearly as much as exponential, logarithmic, or (regular) trigonometric functions, and so we'll give them a briefer treatment. However, you should maintain at least a passing familiarity with them, because they do become important from time to time. For example, here's a fun fact: if you freely hung a chain from two points at equal height, you might think the shape that the sagging chain traces out is a parabola, but it is not — it hangs in the shape of a hyperbolic cosine function. Similarly, the shape of the St. Louis Arch is properly described as a hyperbolic cosine function.

We will take a first look at these functions in this section. It will be a glancing look, kind of like when you go see a large display of Christmas lights just by driving through in your car. You certainly don't see it all, but it looks pretty anyway. Then these functions will be peppered into examples and exercises as appropriate in later sections.

The most important thing to remember about these functions is that they are named after trigonometric functions, but are actually defined using exponential functions. Because they relate to both important categories of functions, these are the greatest functions ever! OK, maybe describing hyperbolic functions that way is a bit of ... hyperbole?[2]

The Hyperbolic Sine and Cosine Functions

Let's get right to it.

---

[2]Yes, I worked that hard to squeeze in this bad joke.

Definition 1.3. *The hyperbolic sine and cosine functions are defined as follows:*

$$\sinh(x) = \frac{e^x - e^{-x}}{2} \qquad \textit{and} \qquad \cosh(x) = \frac{e^x + e^{-x}}{2}$$

From these simple definitions, we can determine that ...

- ... the domain of each is all real numbers
- ... the range of $\sinh(x)$ is all real numbers
- ... the range of $\cosh(x)$ is all real numbers greater than or equal to 1 (be sure you see why)

EX 1 Find the values of $\sinh(0)$, $\sinh(-1)$, and $\sinh(1)$. Is there a relation between $\sinh(-1)$ and $\sinh(1)$?

Based on Def. 1.3,

$$\sinh(0) = \frac{e^0 - e^{-0}}{2} = \frac{1-1}{2} = 0$$

$$\sinh(-1) = \frac{e^{-1} - e^{-(-1)}}{2} = \frac{1}{2}\left(\frac{1}{e} - e\right) \approx -1.175$$

$$\sinh(1) = \frac{e^1 - e^{-1}}{2} = \frac{1}{2}\left(e - \frac{1}{e}\right) \approx 1.175$$

And, we see that $\sinh(-1) = -\sinh(1)$. ■

You Try It

(1) Find the values of $\cosh(0), \cosh(-1)$, and $\cosh(1)$. Is there a relation between $\cosh(-1)$ and $\cosh(1)$?

At values of x which can be written as $a \ln b$, values of $\sinh(x)$ and $\cosh(x)$ will tidy up quite nicely. For example,

$$\sinh(\ln 2) = \frac{e^{\ln 2} - e^{-\ln 2}}{2} = \frac{1}{2}\left(2 - \frac{1}{2}\right) = \frac{3}{4}$$

$$\cosh(\ln 8) = \frac{e^{\ln 8} + e^{-\ln 8}}{2} = \frac{1}{2}\left(8 + \frac{1}{8}\right) = \frac{65}{16}$$

In general,

$$\sinh(\ln a) = \frac{e^{\ln a} - e^{-\ln a}}{2} = \frac{1}{2}\left(a - \frac{1}{a}\right) = \frac{a^2 - 1}{2a}$$

$$\cosh(\ln b) = \frac{e^{\ln b} + e^{-\ln b}}{2} = \frac{1}{2}\left(b - \frac{1}{b}\right) = \frac{b^2 + 1}{2b}$$

It helps to remember that $e^{\ln a} = a$ and $-\ln b = \ln b^{-1}$ so that $e^{-\ln b} = e^{\ln b^{-1}} = b^{-1}$.

You Try It

(2) Find the values of $\sinh(\ln 3), \cosh(2 \ln 4)$. Write the values as rational numbers.

The graphs of $\sinh(x)$ and $\cosh(x)$ are shown in Fig. 1.37. The trends we see in those graphs continue out of the frame, and there are no surprises. Both graphs rise very fast to the right, because e^x becomes the dominant contribution to the value of each function as x gets larger. The graph of $\sinh(x)$ falls fast to the left, because $-e^{-x}$ becomes the dominant contribution for $x < 0$. FFT: Can you explain the behavior of $\cosh(x)$ for $x < 0$?

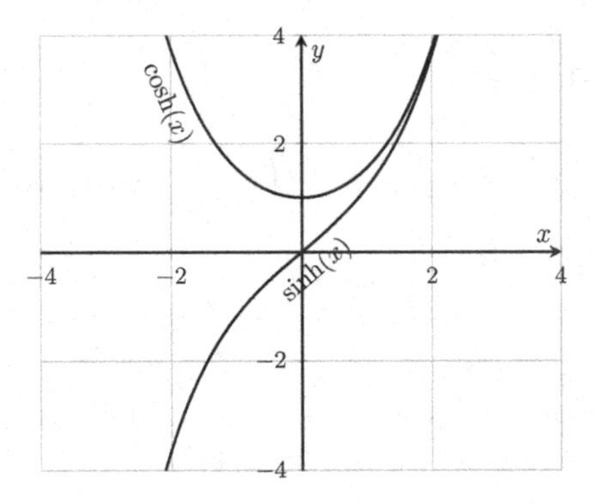

Fig. 1.37 Graphs of $\sinh(x)$ and $\cosh(x)$.

It's natural to look at these graphs and wonder why the names of these things reference trig functions and hyperbolas, since neither of those are evident in the graphs so far? Unfortunately, I have to keep you in suspense for those answers, because a better examination requires that we've seen (a) derivatives, and (b) parametric equations. However, here is a sneak preview.

EX 2 Use Def. 1.3 to simplify the expression $\sinh^2(x) - \cosh^2(x)$.

We have

$$\sinh^2(x) + \cosh^2(x) = \left(\frac{e^x - e^{-x}}{2}\right)^2 - \left(\frac{e^x + e^{-x}}{2}\right)^2$$

$$= \frac{1}{4}\left((e^{2x} + e^{-2x} - 2e^x e^{-x}) - (e^{2x} + e^{-2x} + 2e^x e^{-x})\right) = \frac{1}{4}\left(-4e^x e^{-x}\right) = -1$$

So altogether, we have $\sinh^2(x) - \cosh^2(x) = -1$. Alternately, we can write this as, $\cosh^2(x) - \sinh^2(x) = 1$. This is not *exactly* the same as everyone's trigonometric identity $\sin^2 x + \cos^2 x = 1$, but it's ... similar. ■

You Try It

(3) Use Def. 1.3 to compare the expressions $\sinh(2x)$ and $2\sinh(x)\cosh(x)$.

The Others

At this point, it may not surprise you to learn that there is a hyperbolic tangent function. Like its partners $\sinh(x)$ and $\cosh(x)$, it's fundamentally an exponential function:

Definition 1.4. *The hyperbolic tangent functions is defined as follows:*

$$\tanh(x) = \frac{e^x - e^{-x}}{e^x + e^{-x}}$$

A closer look at this definition will show, though, that we also have this relation:

$$\boxed{\tanh(x) = \frac{\sinh(x)}{\cosh(x)}}$$

We might as well dive all the way in and design three other functions, too!

$$\operatorname{sech}(x) = \frac{1}{\cosh(x)} \quad ; \quad \operatorname{csch}(x) = \frac{1}{\sinh(x)} \quad ; \quad \coth(x) = \frac{1}{\tanh(x)}$$

These hyperbolic secant, cosecant, and cotangent functions are included for completeness, we will not be using them later.

You Try It

(4) Use Def. 1.4 to find the values of $\tanh(0)$ and $\tanh(1)$.

Each of these six functions has an inverse function, given no other name than — for example — the "inverse hyperbolic tangent function". We know about these in the same way we know about any other inverse function. For example,

- Since $\sinh(x)$ is one-to-one everywhere, its inverse function exists anywhere.
- Since $\cosh(x)$ fails the horizontal line test, we have to select a restricted domain, such as $[0, \infty)$, in order to define an inverse hyperbolic cosine function.
- As an example for calculation, $\sinh^{-1}(4)$ would be the value of x for which $\sinh(x) = 4$. In order to determine the actual value of $\sinh^{-1}(4)$, we would have to solve the equation $(e^x - e^{-x})/2 = 4$. Solving equations like this for x can be done, and it leads to combinations of power functions and the natural logarithm; but, this is some heavy lifting that isn't really necessary for now. And so, the value of $\sinh^{-1}(4)$ can just be referred to in its pure exact form as $\sinh^{-1}(4)$, and that's it ... unless we want to use an approximate decimal value instead.

EX 3 Solve the equation $\sinh^2(x) - 4\sinh(x) + 4 = 0$. (Leave any inverse function value in its pure form.)

By factoring, we see that $\sinh^2(x) - 4\sinh(x) + 4 = (\sinh(x) - 2)^2$, so we are solving $(\sinh(x) - 2)^2 = 0$ or $\sinh(x) = 2$. The solution value for this is $x = \sinh^{-1}(2)$, and we have no other information about this value without getting an estimated solution to $(e^x - e^{-x})/2 = 2$. So, we just leave the value written in its pure form as $x = \sinh^{-1}(2)$. ■

You Try It

(5) Solve the equation $\cosh^2 x - 5\cosh(x) + 6 = 0$.

This is enough on these functions for now. They're sort of the mathematical equivalent of the state of Delaware. You've heard of it, you can find it on a map, but it's never your actual destination. It's good to know it exists, though, in case you ever meet anyone who says they are from that state.[3] And just like you might pass through a small piece of Delaware on your way to somewhere else, we'll drop in on these functions from time to

[3]Relax, Delaware citizens, I went to school there for 2.5 years!

time to help practice techniques we will be learning. In the meantime, just appreciate how cool it is that we can define a set of functions using the exponential function e^x, and have those functions relate to each other with identities that look an awful lot like trigonometric identities.

Have You Learned...

- ... the fundamental definitions of $\sinh(x)$ and $\cosh(x)$?
- ... how to define four other similar functions using these two primary functions?
- ... how to evaluate special values of these functions?
- ... a few identities relating some of these six functions?
- ... how to use inverse values of these functions to solve simple equations, when appropriate?

The Greatest Functions Ever! — Problem List

The Greatest Functions Ever! — You Try It

These appeared above; solutions begin on the next page.

(1) Find the values of $\cosh(0), \cosh(-1)$, and $\cosh(1)$. Is there a relation between $\cosh(-1)$ and $\cosh(1)$?
(2) Find the values of $\sinh(\ln 3), \cosh(2\ln 4)$. Write the values as rational numbers.
(3) Use Def. 1.3 to compare the expressions $\sinh(2x)$ and $2\sinh(x)\cosh(x)$.
(4) Use Def. 1.4 to find the values of $tanh(0)$, $tanh(-1)$, and $\tanh(1)$. Is there a relation between $\cosh(-1)$ and $\cosh(1)$?
(5) Solve the equation $\cosh^2 x - 5\cosh(x) + 6 = 0$.

The Greatest Functions Ever! — Practice Problems

Try these as you get the hang of the You Try It problems. Solutions to these problems are available in Sec. A.1.6.

(1) State the domain and range of $\tanh(x)$ and give an approximate sketch its graph. (You should be able to identify some trends by its definition in terms of $\sinh(x)$ and $\cosh(x)$, and can plot a few points, right?)
(2) Give the following values as rational numbers: $\tanh(-3)$, $\text{sech}(\ln 10)$, $\coth(3)$.
(3) Give a definition of $\text{sech}(x)$ using exponential functions, and see if there is a relationship between $\text{sech}^2(x)$ and $\tanh^2(x)$.
(4) Let $f(x) = \sinh(x)/(\cosh^2(x) + 1)$. Does the graph of $f(x)$ have any vertical asymptotes? How do you know?
(5) There is a trigonometric identity $\cos(A+B) = \cos A \cos B - \sin A \sin B$. Can you develop a similar identity for $\cosh(A + B)$?
(6) Does the equation $\tanh^2 x + 4\tanh(x) + 4 = 0$ have any solutions? Why or why not? (Hint: Use the result of Practice Problem 1.)

The Greatest Functions Ever! — Challenge Problems

Try these problems to test your skills with the ideas in this section. Solutions to these problems are available in Sec. B.1.6.

(1) Categorize each of $\sinh(x), \cosh(x)$, and $\tanh(x)$ as even or odd, and explain how you know.

(2) From EX 2, we know that there is no direct counterpart to $\sin^2 x + \cos^2 x = 1$, in that $\sinh^2(x)+\cosh^2(x) \neq 1$. But can $\sinh^2(x)+\cosh^2(x)$ be related to *anything* interesting? (Hint: In You Try It 2, we've already seen an identity involving $\sinh(2x)$, so maybe ....)

(3) If there are any solutions to $\sinh^2(x) - 8\cosh(x) + 16 = 0$, find them; write the values in their pure exact form.

The Greatest Functions Ever! — You Try It — Solved

(1) Find the values of $\cosh(0), \cosh(-1)$, and $\cosh(1)$. Is there a relation between $\cosh(-1)$ and $\cosh(1)$?

☐ These three values are:

$$\cosh(0) = \frac{e^0 + e^{-0}}{2} = \frac{1+1}{2} = 1$$

$$\cosh(-1) = \frac{e^{-1} + e^{-(-1)}}{2} = \frac{e^{-1} + e^1}{2} = \frac{1}{2}\left(\frac{1}{e} + e\right)$$

$$\cosh(1) = \frac{e^1 + e^{-1}}{2} = \frac{1}{2}\left(e + \frac{1}{e}\right)$$

And it looks like $\cosh(-1) = \cosh(1)$. ■

(2) Find the values of $\sinh(\ln 3), \cosh(2\ln 4)$. Write the values as rational numbers.

☐ By properties of logarithms, $2\ln 4 = \ln 4^2 = \ln 16$. So,

$$\sinh(\ln 3) = \frac{e^{\ln 3} - e^{-\ln 3}}{2} = \frac{1}{2}\left(3 - \frac{1}{3}\right) = \frac{4}{3}$$

$$\cosh(2\ln 4) = \cosh(\ln 4^2)\frac{e^{\ln 16} + e^{-\ln 16}}{2} = \frac{1}{2}\left(16 + \frac{1}{16}\right) = \frac{257}{32}$$ ■

(3) Use Def. 1.3 to compare the expressions $\sinh(2x)$ and $2\sinh(x)\cosh(x)$.

☐ First,

$$\sinh(2x) = \frac{1}{2}(e^{2x} - e^{-2x})$$

Also,

$$\begin{aligned} 2\sinh(x)\cosh(x) &= 2 \cdot \left(\frac{1}{2}(e^x - e^{-x})\right) \cdot \left(\frac{1}{2}(e^x + e^{-x})\right) \\ &= \frac{1}{2}\left((e^x - e^{-x})(e^x + e^{-x})\right) \\ &= \frac{1}{2}(e^{2x} - e^{-2x}) \end{aligned}$$

So that $\sinh(2x) = 2\sinh(x)\cosh(x)$. ■

(4) Use Def. 1.4 to find the values of $tanh(0)$, $tanh(-1)$, and $\tanh(1)$. Is there a relation between $\cosh(-1)$ and $\cosh(1)$?

□ By the fundamental definition of $\tanh(x)$, we have

$$\tanh(0) = \frac{e^0 - e^{-0}}{e^0 + e^{-0}} = \frac{1-1}{1+1} = 0$$

$$\tanh(-1) = \frac{e^{-1} - e^{-(-1)}}{e^{-1} + e^{-(-1)}} = \frac{e^{-1} - e^1}{e^{-1} + e^1} = \frac{1-e^2}{1+e^2}$$

$$\tanh(1) = \frac{e^1 - e^{-1}}{e^1 + e^{-1}} == \frac{e^2 - 1}{e^2 + 1}$$

And so $\tanh(-1) = -\tanh(1)$. ■

(5) Solve the equation $\cosh^2 x - 5\cosh(x) + 6 = 0$.

□ By factoring, we can rewrite $\cosh^2 x - 5\cosh(x) + 6 = 0$ to $(\cosh^2 x - 2)(\cosh(x) - 3) = 0$, which means we have solutions when $\cosh(x) = 2$ or $\cosh(x) = 3$. Therefore, the solutions are $x = \cosh^{-1}(2)$ and $x = \cosh^{-1}(3)$. ■

Chapter 2

Take It to the Limit

2.1 Introduction to Limits

Introduction

One of the fun things we do in Calculus is break out our "mathematical microscope" and examine very small scale behavior of functions, usually near a place where something interesting happens. For the moment, we forget the question, "What happens to $f(x)$ at a particular x value?" and instead concentrate on the question, "What happens to $f(x)$ as we get closer and closer to a particular x value?" For example, the question, "What happens to $f(x)$ as x gets closer and closer to 0?" has entirely different answers for $f(x) = x$ and $f(x) = 1/x$. And we even get different answers for $f(x) = 1/x$ depending on whether we're sneaking up on $x = 0$ from the left side or the right side. We might also ask the question, "What happens to $f(x) = 1/x$ as x gets bigger and bigger?" Would that depend on whether x is positive or negative?

These are examples of *limits*. Being able to understand and evaluating limits is critical for developing almost every mathematical procedure you'll do in the next two or three semesters.

Notation

The notation

$$\lim_{x \to a} f(x) = L \tag{2.1}$$

is translated as "the limit of $f(x)$ as x approaches a is L". In other words, as x gets closer and closer to a, the function $f(x)$ gets closer and closer to L. When we're investigating limits, we're usually told $f(x)$ and a, and need to discover L. Sometimes it's obvious from the function itself, but sometimes it's not so obvious. There's another level of detail, too: it's possible for x to approach its target a from either the left or the right. For example, x can approach 2 by passing through $1.8, 1.9, 1.99, 1.999, \ldots$ OR by passing through $2.2, 2.1, 2.01, 2.001, \ldots$. And it's possible for the function to approach different values from each side. So we define the "left hand limit" to be the limit obtained when x approaches a from the left, and it is denoted by

$$\lim_{x \to a^-} f(x)$$

If you haven't noticed the difference between this limit notation and what's shown in the original example of Eq. (2.1), look closely for the superscript of "$-$" on the a, which is the indicator that x is approaching a from the left side. (Yes, we are now using superscripts on subscripts.) It's important not to let that — be confused with a — which indicates a negative value; here, a can be positive or negative.

Do you care to take a guess as to the notation for a limit in which x approaches a from the right? We define the "right hand limit" to be the limit obtained when x approaches a from the right, and it is signified by

$$\lim_{x \to a^+} f(x)$$

Useful Fact 2.1. *The overall limit L shown in (2.1) exists only when the left and right hand limits agree on what the value should be. Otherwise, we say the limit does not exist.*

Visual Examples of Limits

Figure 2.1 shows a function with vertical asymptotes at $x = -2$ and $x = 2$. There is a lot of exciting action on this graph! To name just a few items,

- As x approaches 0 from the left, the function (as read off the y-axis) is approaching 0 too. As x approaches 0 from the right, the function is again approaching 0. The overall limit of $f(x)$ as x approached 0 here seems to be 0. In other words, we have a conjecture from the graph that

$$\lim_{x \to 0^-} f(x) = 0 \quad ; \quad \lim_{x \to 0^+} f(x) = 0 \quad \text{and so} \quad \lim_{x \to 0} f(x) = 0$$

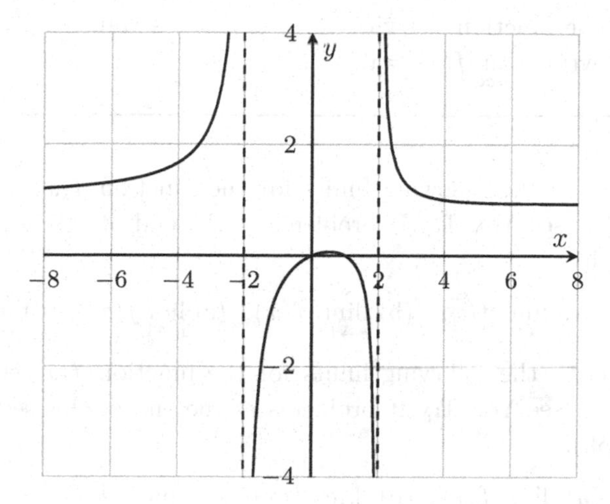

Fig. 2.1 Vertical asymptotes as indicators of limiting trends.

- As x approaches 1 from the left, the function is approaching 0. As x approaches 1 from the right, the function is again approaching 0. So we conjecture that

$$\lim_{x\to 1^-} f(x) = 0 \quad ; \quad \lim_{x\to 1^+} f(x) = 0 \quad \text{and so} \quad \lim_{x\to 1} f(x) = 0$$

- As $x \to 2$ from the left, we see the function heading down to negative infinity. But as $x \to 2$ from the right, we see the function heading up to positive infinity. Since these values don't agree, we write

$$\lim_{x\to 2^-} f(x) = -\infty \quad \text{and} \quad \lim_{x\to 2^+} f(x) = +\infty$$

  and so $\lim\limits_{x\to 2} f(x)$.
- The behavior near $x = -2$ is similar to that near $x = 2$.
- There are "normal" and uninteresting limits at other places, too. For example, see if you agree that $\lim\limits_{x\to -4} f(x) \approx 2$. The actual value of that limit is 5/3, but 2 is a pretty good estimate based on the graph itself. In fact, any behavior read from a graph is just a conjecture, even though it can be a pretty good one! To become firm on the value of a limit, we need to analyze the limit quantitatively, which we will see how to do soon.
- As x gets bigger (that is, as we move farther to the right along the x-axis) we see that $f(x)$ seems to be leveling off. We might guess

that the function is trying to level off at a value of $y = 1$, and so we'd write $\lim_{x\to\infty} f(x) = 1$.

You Try It

(1) Describe the following limits for the function $f(x)$ shown in the graph (see You Try It problems at the end of this section for the graph).

(a) $\lim_{x\to 1^-} f(x)$ (b) $\lim_{x\to 1^+} f(x)$ (c) $\lim_{x\to 1} f(x)$ (d) $\lim_{x\to 5} f(x)$

(2) Describe the following limits for the function $f(x)$ shown in the graph (see You Try It problems at the end of this section for the graph).

(a) $\lim_{x\to -7} f(x)$ (b) $\lim_{x\to -3} f(x)$ (c) $\lim_{x\to 0} f(x)$ (d) $\lim_{x\to \pi} f(x)$

Terminology issues: When reporting on limits that either don't match from the left and right, or which involve $\pm infty$, let's preserve as much information about the limit as we can. The phrase "the limit doesn't exist" will refer specifically to the case when the left and right hand limits do not agree, regardless of whether they are finite or infinite. Then, we can use a phrase like "the limit is $+\infty$" or "the limit is $-\infty$ to refer to the case when the left and right hand limits agree and are both either $+\infty$ or $-\infty$. For example, in Fig. 2.2, two different behaviors exhibited by the graph as x approaches -1 and as x approaches 3. To preserve that information, we can report that

$$\lim_{x\to -1} f(x) = +\infty \quad \text{and} \quad \lim_{x\to 3} f(x) \text{ does not exist}$$

Limits By Numerical Experimentation

Using a graph to investigate limits can lead to intuitive understanding of a function's behavior, but we are still just "eyeballing" the limits. A more substantive exploration of limits requires stronger quantitative techniques. A first entry into that is rather coarse, but simple. We can investigate a limit simply by calculating values of the function at points close to, and approaching, the target value $x = a$. For emphasis, remember that we don't care what the value of $f(x)$ AT $x = a$ is, we only care about the trend as x gets closer to a.

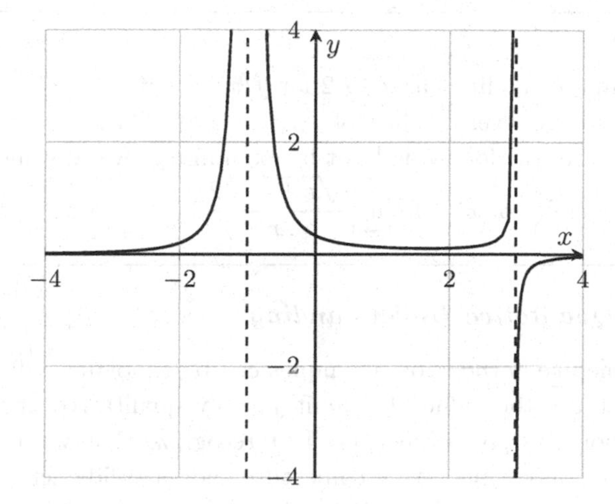

Fig. 2.2 Different limiting behaviors at vertical asymptotes.

EX 1 Estimate the limit $\lim_{x \to 1} \frac{x-1}{\sqrt{x}-1}$ by examining several values of $f(x)$ in the vicinity of $x = 1$.

First, note that this function is not defined at $x = 1$. A common error is to then say that the limit there does not exist. But in fact, it does. Here is a table of values of $f(x)$ calculated / approximated around $x = 1$ (note that values on both sides are included, to reveal left and right hand limits).

x	0.8	0.9	0.99	0.999	1.001	1.01	1.1
$f(x)$	1.8944	1.9486	1.9950	1.99950	2.0005	2.005	2.0488

so it seems that

$$\lim_{x \to 1} \frac{x-1}{\sqrt{x}-1} = 2$$

because the function approaches 2 from both the left and right of $x = 1$. We can't put a value into the table for $x = 1$ itself, but we don't need to, since the limit doesn't care what happens AT $x = 2$. (This table is evidence that there is a hole in the graph of $f(x)$ at $x = 1$. If you plot this curve in your favorite math platform — go ahead and try it, I'll wait! — you won't see the hole, but we know it's there.) ■

You Try It

(3) Estimate the limit as $x \to 2$ for $f(x) = (x^2 - 2x)/(x^2 - x - 2)$ by examining several values of $f(x)$ in the vicinity of $x = 2$.

(4) Estimate the following limit by examining several values of $f(x)$ in the vicinity of $x = 0$: $\lim_{x \to 0} \dfrac{\sqrt{x+4} - 2}{x}$

Limits By Qualitative Understanding

Making a table like in the above examples can be annoying. Often, though, we can conjecture the value of a limit just by qualitative understanding of the behavior of a given function. But recognize that we're still in the zone where our assessment is a conjecture, not a solid fact. Digging in with more quantitative tools is what we'll do in the next section. So, when you see "asses the limit", that means give a qualitative description of the limit process, based on trends. When you see something more precise, like "determine the limit", that means bring your quantitative tools to bear to develop an definitive result for the limit.

EX 2 Assess the limit $\lim_{x \to 1^-} \dfrac{3}{1-x}$.

The function is certainly undefined at $x = 1$. But, again, we don't really care. We want the limit as $x \to 1$ from the left, meaning we're coming in past $0.8, 0.9, 0.99, 0.999, \ldots$. As we pass these values, the denominator $f(x)$ stays positive, but gets smaller and smaller. Thus, the function itself stays positive and gets bigger and bigger. The closer x is to 1 (on the left), the larger $f(x)$ gets. So we write:

$$\lim_{x \to 1^-} \frac{3}{1-x} = \infty$$

(Note that the limit as $x \to 1$ from the right would be $-\infty$.) ■

EX 3 Assess the limit $\lim_{x \to \infty} \dfrac{x}{x+3}$.

As x gets bigger and bigger, the $+3$ in the denominator matters less and less, and the function "looks" more and more like just x/x. We are passing through values like $10/13, 100/103, \ldots, 10000/10003$, and so on. As x gets bigger, the function trends closer to 1. So we can write

$$\lim_{x \to \infty} \frac{x}{x+3} = 1$$

Again, note this is just a guess, although it's a pretty darn good one. In the next section, we'll see formal rules for evaluation of limits, then apply them to firmly determine limits such as this one. ■

You Try It

(5) Assess the limit $\lim_{x \to 5^+} \dfrac{6}{x-5}$.

(6) Assess the limit $\lim_{x \to 1} \dfrac{2-x}{(x-1)^2}$.

Have You Learned...

- How to read and use limit notation?
- The definition of a left and right sided (i.e. left and right hand) limits?
- The conditions under which a limit exists or does not exist?
- How to estimate a limit graphically?
- How to estimate a limit numerically?
- How to recognize limits of some simple functions just from your knowledge of how the function behaves?

Introduction to Limits — Problem List

Introduction to Limits — You Try It

These appeared above; solutions begin on the next page.

(1) Describe the following limits for the function $f(x)$ shown in Fig. 2.3.

(a) $\lim_{x\to 1^-} f(x)$ (b) $\lim_{x\to 1^+} f(x)$ (c) $\lim_{x\to 1} f(x)$ (d) $\lim_{x\to 5} f(x)$

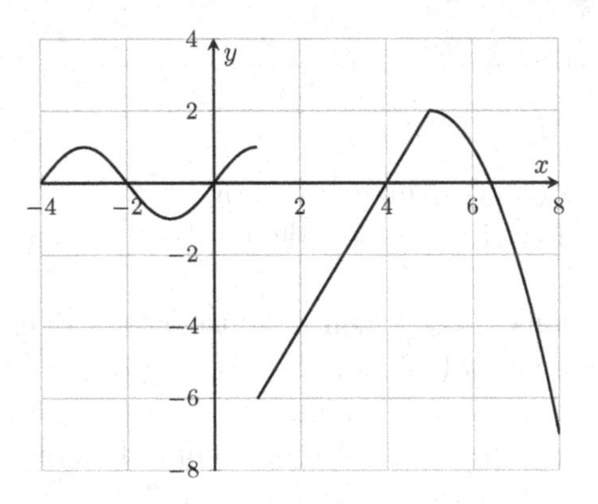

Fig. 2.3 A piecewise function for YTI 1.

(2) Describe the following limits for the function $f(x)$ shown in Fig. 2.4.

(a) $\lim_{x\to -7} f(x)$ (b) $\lim_{x\to -3} f(x)$ (c) $\lim_{x\to 0} f(x)$ (d) $\lim_{x\to \pi} f(x)$

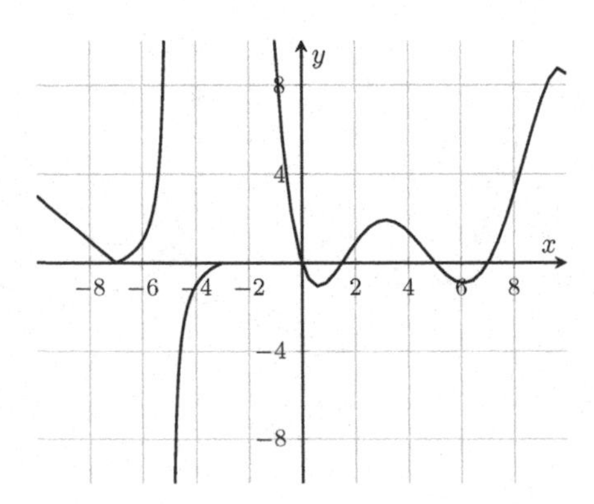

Fig. 2.4 A piecewise function for YTI 2.

(3) Estimate the limit as $x \to 2$ for $f(x) = (x^2 - 2x)/(x^2 - x - 2)$ by examining several values of $f(x)$ in the vicinity of $x = 2$.
(4) Estimate the following limit by examining several values of $f(x)$ in the vicinity of $x = 0$: $\lim_{x\to 0} \frac{\sqrt{x+4}-2}{x}$
(5) Assess the limit $\lim_{x\to 5^+} \frac{6}{x-5}$.
(6) Assess the limit $\lim_{x\to 1} \frac{2-x}{(x-1)^2}$.

Introduction to Limits — Practice Problems

Try these as you get the hang of the You Try It problems. Solutions to these problems are available in Sec. A.2.1.

(1) Describe the following limits for the function $g(t)$ shown in Fig. 2.5.

$$\text{(a)} \lim_{t\to 0^-} g(t) \quad \text{(b)} \lim_{t\to 0^+} g(t) \quad \text{(c)} \lim_{t\to 2^-} g(t) \quad \text{(d)} \lim_{t\to 2^+} g(t)$$

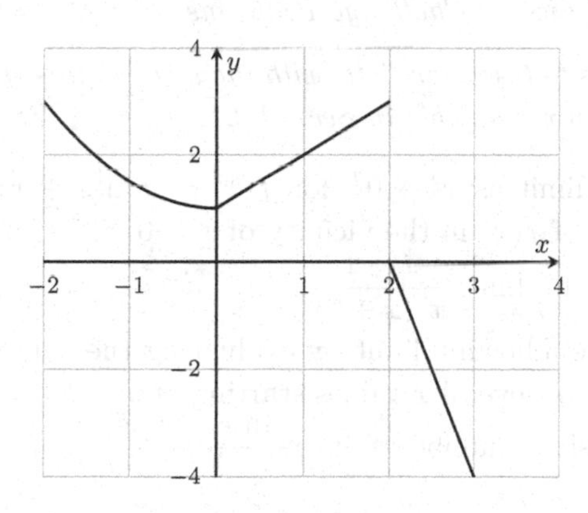

Fig. 2.5 A piecewise function for PP 1.

(2) Describe the following limits for the function $R(x)$ shown in Fig. 2.6.

$$\text{(a)} \lim_{x\to 2} R(x) = \quad \text{(b)} \lim_{x\to 5} R(x) \quad \text{(c)} \lim_{x\to -3^-} R(x) \quad \text{(d)} \lim_{x\to -3^+} R(x)$$

(3) Estimate the limit as $x \to -1$ for $f(x) = (x^2 - 2x)/(x^2 - x - 2)$ by examining several values of $f(x)$ in the vicinity of $x = -1$.
(4) Estimate the limit as $x \to 0$ for $f(x) = \tan 3x/\tan 5x$ by examining several values of $f(x)$ in the vicinity of $x = 0$.

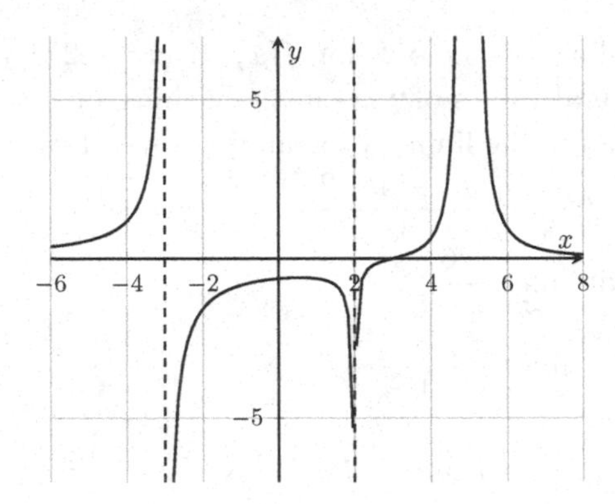

Fig. 2.6 A piecewise function for PP 2.

(5) Assess the limit $\lim_{x\to 5^-} \dfrac{6}{x-5}$.

(6) Assess the limit $\lim_{x\to 0} \dfrac{x-1}{x^2(x+2)}$.

Introduction to Limits — Challenge Problems

Try these problems to test your skills with the ideas in this section. Solutions to these problems are available in Sec. B.2.1.

(1) Estimate the limit as $x \to 0^+$ for $f(x) = x\ln(x + x^2)$ by examining several values of $f(x)$ in the vicinity of $x = 0$.

(2) Assess the limit $\lim_{x\to -2^+} \dfrac{x-1}{x^2(x+2)}$.

(3) Investigate the following limit by evaluating the value of the function for a sequence of several x values starting at $x = 1$ and decreasing to a very small positive number: $\lim_{x\to 0^+} \dfrac{\tan x - x}{x^3}$

Introduction to Limits — You Try It — Solved

(1) Describe the following limits for the function $f(x)$ shown in Fig. 2.7.

$$\text{(a)} \lim_{x\to 1^-} f(x) \quad \text{(b)} \lim_{x\to 1^+} f(x) \quad \text{(c)} \lim_{x\to 1} f(x) \quad \text{(d)} \lim_{x\to 5} f(x)$$

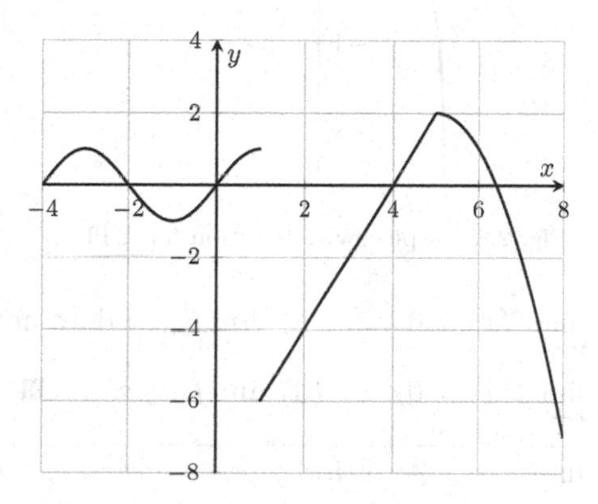

Fig. 2.7 A piecewise function for YTI 1.

☐ The limits are:

$$\text{(a)} \lim_{x\to 1^-} f(x) = 1 \quad \text{(b)} \lim_{x\to 1^+} f(x) = -6 \quad \text{(c)} \lim_{x\to 1} f(x) \text{ does not exist}$$

The overall limit as $x \to 1$ does not exist since the left and right hand limits are different there. On the other hand, the left and right hand limits as $x \to 5$ are both the same and so the overall limit there is equal to the common value:

$$\text{(d)} \lim_{x\to 5} f(x) = 2 \quad \blacksquare$$

(2) Describe the following limits for the function $f(x)$ shown in Fig. 2.4.

$$\text{(a)} \lim_{x\to -7} f(x) \quad \text{(b)} \lim_{x\to -3} f(x) \quad \text{(c)} \lim_{x\to 0} f(x) \quad \text{(d)} \lim_{x\to \pi} f(x)$$

☐ For three of these limits, we see the same limit from the left and right, and so the limits there are the common values. For one limit, we see different values from the left and right, and so the limit there does not exist. The limit in (d) needs to be estimated, since we can't tell the precise value from the graph — although we can make a good guess! The limits are:

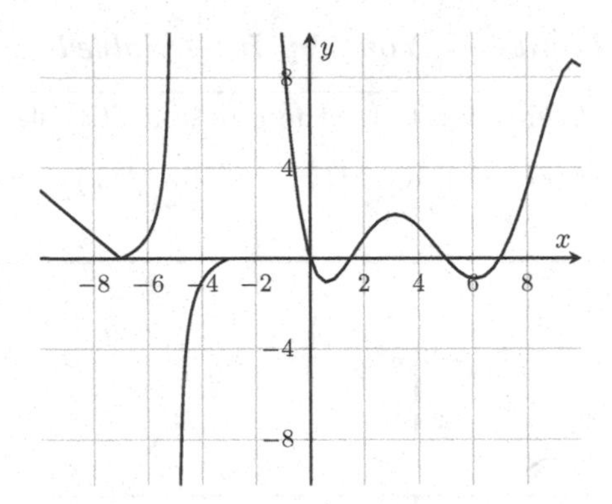

Fig. 2.8 A piecewise function for YTI 2.

$$\text{(a)} \lim_{x\to-7} f(x) = 0 \quad , \quad \text{(b)} \lim_{x\to-3} f(x) \text{ does not exist}$$

$$\text{(c)} \lim_{x\to 0} f(x) = 0 \quad , \quad \text{(d)} \lim_{x\to\pi} f(x) \approx \pi \quad \blacksquare$$

---

(3) Estimate the limit as $x \to 2$ for $f(x) = (x^2 - 2x)/(x^2 - x - 2)$ by examining several values of $f(x)$ in the vicinity of $x = 2$.

□ Here is a (split, for page formatting) table of values around $x = 2$; the split occurs where $x = 2$ would sit:

x	2.5	2.1	2.05	2.01	2.005	2.001
$f(x)$	.7143	.6774	.6721	.6678	.6672	0.6668
x	1.999	1.995	1.99	1.95	1.9	
$f(x)$	0.6666	0.6661	0.6656	0.6610	0.6552	

So it looks like

$$\lim_{x\to 2} f(x) \approx 0.6667 = \frac{2}{3} \quad \blacksquare$$

---

(4) Estimate the following limit by examining several values of $f(x)$ in the vicinity of $x = 0$: $\lim_{x\to 0} \frac{\sqrt{x+4}-2}{x}$

□ Here is a table of values around $x = 0$:

x	−0.2	−0.1	−0.01	0.01	0.1	0.2
$f(x)$	0.2532	0.2516	0.2502	0.2498	0.2484	0.2469

and so it looks like

$$\lim_{x\to 0} \frac{\sqrt{x+4}-2}{x} \approx 0.25 = \frac{1}{4} \quad \blacksquare$$

(5) Assess the limit $\lim_{x \to 5^+} \dfrac{6}{x-5}$.

☐ As x approaches 5 from the right, the quantity $x - 5$ approaches zero, but stays positive, so

$$\lim_{x \to 5^+} \frac{6}{x-5} = +\infty \quad \blacksquare$$

(6) Assess the limit $\lim_{x \to 1} \dfrac{2-x}{(x-1)^2}$.

☐ As x approaches 1, the quantity $(x-1)^2$ approaches zero, but stays positive. (Note that it doesn't matter whether x approaches 1 from the left or right.) The numerator, though, approaches 1. So

$$\lim_{x \to 1} \frac{2-x}{(x-1)^2} = +\infty \quad \blacksquare$$

2.2 Computation of Limits

Introduction

We've looked at the concept of limits and have evaluated some simple ones using graphs, numerical experiments, and some plain ol' know-how. But, to move along we need a systematic way of computing limits. We'll start with rules of limits and then see how to use these rules to evaluate limits of complicated functions.

Limits Laws

Finding limits of complicated functions requires knowing how limits of smaller, individual functions work together. Here are some of the basic rules. We start by assuming that the individual limits

$$\lim_{x \to a} f(x) \qquad \text{and} \qquad \lim_{x \to a} g(x)$$

both exist and are known. Then we can start putting them together in the following ways:

$$\boxed{\text{LL1)} \ \lim_{x \to a}(f(x) \pm g(x)) = \lim_{x \to a} f(x) \pm \lim_{x \to a} g(x)} \tag{2.2}$$

This law allows us to compute the limit of a sum as the sum of the limits, or the limit of a difference as the difference of the limits. For example, from simple know-how (or graphs), we know these limits:

$$\lim_{x \to 0} \cos(x) = 1 \qquad \text{and} \qquad \lim_{x \to 0} e^x = 1$$

and so we now also know that

$$\lim_{x \to 0}(\cos(x) + e^x) = \lim_{x \to 0} \cos(x) + \lim_{x \to 0} e^x = 1 + 1 = 2$$

by simply adding the results of the individual limits. Another limit law is:

$$\boxed{\text{LL2)} \ \lim_{x \to a}(f(x) \cdot g(x)) = \left(\lim_{x \to a} f(x)\right) \cdot \left(\lim_{x \to a} g(x)\right)} \tag{2.3}$$

In other words, the limit of a product is the product of the limits. Using the same two examples as above, we now know that

$$\lim_{x \to 0}(\cos(x) \cdot e^x) = \lim_{x \to 0} \cos(x) \cdot \lim_{x \to 0} e^x = 1 \cdot 1 = 1$$

If we have limit laws for addition, subtraction and multiplication of limits, then we probably have one for division, too:

$$\boxed{\text{LL3)} \ \lim_{x \to a} \frac{f(x)}{g(x)} = \frac{\lim_{x \to a} f(x)}{\lim_{x \to a} g(x)} \ \text{as long as} \ \lim_{x \to a} g(x) \neq 0} \tag{2.4}$$

This rule works like you think it should, with the warning that if the denominator is approaching zero, all bets are off. With our ongoing example functions and limits,

$$\lim_{x\to 0} \frac{\cos(x)}{e^x} = \frac{\lim_{x\to 0}\cos(x)}{\lim_{x\to 0} e^x} = \frac{1}{1} = 1$$

Law 2 can be extended in the case that one of the functions is a constant. First, realize that for the constant function $f(x) = c$, we have

$$\lim_{x\to a} c = c$$

This looks strange, but c is called a constant because it's, well, *constant*! Its value does not change. So c doesn't care that x is approaching a, c always stays c. And so as x approaches a, $f(x) = c$ approaches (in fact, is always equal to) c. Combining this with limit law 2 gives us:

$$\boxed{\text{LL4)}\ \lim_{x\to a}(c \cdot f(x)) = c \cdot \lim_{x\to a} f(x)} \tag{2.5}$$

Here's one that's so simple it's almost hard to get, see if you can figure it out:

$$\boxed{\text{LL5)}\ \lim_{x\to a} x = a} \tag{2.6}$$

Repeated combinations of laws (5) and (2) give us:

$$\boxed{\text{LL6)}\ \lim_{x\to a} x^n = a^n} \tag{2.7}$$

Combination of all these above rules let us evaluate many different kinds of limits in a systematic way. First, let's practice applying them generally.

EX 1 Given the values of the limits

$$\lim_{x\to 2} f(x) = 5 \quad ; \quad \lim_{x\to 2} g(x) = -1; \quad \lim_{x\to 2} h(x) = 0$$

determine the values of the following limits:

$$\text{a) } \lim_{x\to 2} f(x)g(x) \quad ; \quad \text{b) } \lim_{x\to 2} g(x) - h(x); \quad \text{c) } \lim_{x\to 2} \frac{f(x)}{h(x)}$$

Using appropriate limit laws from above, we have:

$$\text{a) } \lim_{x\to 2} f(x)g(x) = \lim_{x\to 2} f(x) \cdot \lim_{x\to 2} g(x) = (5)(-1) = 5$$

$$\text{b) } \lim_{x\to 2} g(x) - h(x) = \lim_{x\to 2} g(x) - \lim_{x\to 2} h(x) = -1 - 0 = -1$$

$$\text{c) } \lim_{x\to 2} \frac{f(x)}{h(x)} = \frac{\lim_{x\to 2} f(x)}{\lim_{x\to 2} h(x)} \text{ which does not exist}$$

Limit (c) does not exist since the denominator approaches zero while the numerator approaches 5. ■

We can also use these limit laws to evaluate the limit of a polynomial:

EX 2 Determine the limit $\lim_{x \to -1} (3x^4 - 2x^2 + 7)$.

We now have laws which govern how to take the limit of a power of x, the limit of sums / differences, the limit of a term with a constant coefficient, and the limit of a constant itself. Using all of these,

$$\begin{aligned}\lim_{x\to-1}(3x^4 - 2x^2 + 7) &= \lim_{x\to-1}(3x^4) - \lim_{x\to-1}(2x^2) + \lim_{x\to-1} 7 \\ &= 3 \cdot \lim_{x\to-1}(x^4) - 2 \cdot \lim_{x\to-1}(x^2) + \lim_{x\to-1} 7 \\ &= 3\left(\lim_{x\to-1} x\right)^4 - 2\left(\lim_{x\to-1} x\right)^2 + \lim_{x\to-1} 7 \\ &= 3(-1)^4 - 2(-1)^2 + (7) \\ &= 3 - 2 + 7 = 8 \quad \blacksquare\end{aligned}$$

A Note on Notation: In an early draft of this section, thankfully now edited, I had written the limit in EX 2 as

$$\lim_{x\to-1} 3x^4 - 2x^2 + 7$$

rather than

$$\lim_{x\to-1}(3x^4 - 2x^2 + 7)$$

Do you understand why that was awkward, and how to avoid the same mistake in your own work?

Note that the end result of that last example is what we would have found if we just plugged $x = -1$ into the polynomial in the first place. This leads to a new general rule, involving polynomials and rational functions (since rational functions are quotients of polynomials):

Useful Fact 2.2. *If $f(x)$ is a polynomial or rational function and $x = a$ is in the domain of $f(x)$, then*

$$\lim_{x\to a} f(x) = f(a)$$

In other words, if you can plug the value $x = a$ into a polynomial or rational function, then you can find the limit by doing just that.

We can continue with a few more general limit laws:

$$\boxed{\text{LL7)}\ \lim_{x\to a}(f(x))^n = \left(\lim_{x\to a} f(x)\right)^n} \tag{2.8}$$

So that, for example,

$$\lim_{x \to \pi/2} \sin^4(x) = \left(\lim_{x \to \pi/2} \sin(x) \right)^4 = (1)^4 = 1$$

Finally,

$$\boxed{\text{LL8)} \ \lim_{x \to a} \sqrt[n]{(f(x)} = \sqrt[n]{\lim_{x \to a} f(x)}} \tag{2.9}$$

So that, for example,

$$\lim_{x \to e} \sqrt[4]{\ln(x)} = \sqrt[4]{\lim_{x \to e} \ln(x)} = \sqrt[4]{1} = 1$$

You Try It

(1) If $\lim_{x \to a} f(x) = -3$, $\lim_{x \to a} g(x) = 0$, and $\lim_{x \to a} h(x) = 8$, then use the steps allowed by limit laws to determine the values of

$$\text{(a)} \ \lim_{x \to a} [f(x) + h(x)] \quad \text{(b)} \ \lim_{x \to a} [f(x)]^2$$

$$\text{(c)} \ \lim_{x \to a} \sqrt[3]{h(x)} \quad \text{(d)} \ \lim_{x \to a} \frac{1}{f(x)}$$

(2) Use the steps allowed by limit laws to determine

$$\lim_{x \to 3} (x^2 - 4)(x^3 + 5x - 1)$$

Confirm that this is what you'd get by evaluating $f(3)$.

(3) Determine $\lim_{x \to -1} (x^{15} - 12x^3 + x^2 + \pi)$.

General Computations

We saw in the above section that limits of simple functions are pretty easy to find. And we found out that if you have a polynomial or rational function for which $x = a$ is in the domain, then the limit of that function as x approaches a will be the same as the function's value at $x = a$.

But as is often the case, the easy cases are the boring ones. Consider the limit:

$$\lim_{x \to 2} \frac{x^2 - 4}{x - 2}$$

This does NOT follow the rule above that allows us to simply plug in $x = 2$ to find the limit, since $x = 2$ is not in the domain of the function. On the other hand, we can't leap right to the conclusion that the limit doesn't exist just because the denominator is going to zero. Sure, the denominator is going to zero, but so is the numerator. And when there's a race between the numerator and denominator to zero, you can't predict the outcome. We do know that if the denominator goes to zero and the numerator goes to a finite number other than zero, the limit will become ∞ or $-\infty$. But when we have a limit approaching $0/0$, we have no idea what's going to happen. This is called an *indeterminate form*, since we can't determine the outcome without further exploration.

Luckily, most of the "further exploration" in a limit like the one above involves simplification and/or qualitative reasoning. If we continue the limit above, we see that:

$$\lim_{x\to 2}\frac{x^2-4}{x-2}=\lim_{x\to 2}\frac{(x-2)(x+2)}{x-2}=\lim_{x\to 2}(x+2)=4$$

While the function is not defined at $x=2$, the limit as x *approaches* 2 is still just fine, and is equal to 4. (There is a hole in this graph at the point $(2,4)$.)

Here are a few more examples:

EX 3 Determine $\displaystyle\lim_{x\to -1}\frac{x^2+3x+2}{x^2+4x+3}$.

Even if we wanted to do this by plugging in $x=-1$, we can't, since the limit becomes $0/0$. So instead, we try to simplify:

$$\lim_{x\to -1}\frac{x^2+3x+2}{x^2+4x+3}=\lim_{x\to -1}\frac{(x+2)(x+1)}{(x+3)(x+1)}=\lim_{x\to -1}\frac{x+2}{x+3}=\frac{1}{2}\quad\blacksquare$$

EX 4 Determine $\displaystyle\lim_{h\to 0}\frac{(3+h)^2-3^2}{h}$.

This is in the form $0/0$. Our only hope is to multiply out the numerator and hoping something good happens:

$$\lim_{h\to 0}\frac{(3+h)^2-3^2}{h}=\lim_{h\to 0}\frac{9+6h+h^2-9}{h}=\lim_{h\to 0}\frac{h^2+6h}{h}=\lim_{h\to 0}\frac{h(h+6)}{h}$$

Something good has indeed happened, because now we get to cancel the h:

$$\ldots=\lim_{h\to 0}(h+6)=6$$

In this example, note that I could replace the 3 with a 4 or a -2 or any other number. We could actually do the limit with a variable in place of the 3:

$$\lim_{h\to 0} \frac{(x+h)^2 - x^2}{h} = \lim_{h\to 0} \frac{h^2 + 2xh}{h} = \lim_{h\to 0} h + 2x = 2x$$

Our answer is in the form of a function, too! I'm pointing this out now because your ability to do a limit like this one will have a big payoff when we discuss derivatives. ■

You Try It

(4) Determine $\displaystyle\lim_{x\to 2} \frac{x^2 - x + 6}{x - 2}$

(5) Determine $\displaystyle\lim_{h\to 0} \frac{(4+h)^2 - 16}{h}$

(6) Determine $\displaystyle\lim_{x\to 9} \frac{x^2 - 81}{\sqrt{x} - 3}$

For a final example of a case that combines simplification and/or general know-how, let's bring back in everyone's favorite function:

EX 5 Determine $\displaystyle\lim_{x\to 3} \frac{|x-3|}{2x-6}$

We need to check if the left and right hand limits are the same. Let's consider the right hand limit first. As x approaches 3 from the right, the term $x - 3$ is positive, and so the absolute value bars do nothing. So for $x \geq 3$, we can write this as:

$$\lim_{x\to 3^+} \frac{|x-3|}{2x-6} = \lim_{x\to 3^+} \frac{x-3}{2(x-3)} = \lim_{x\to 3^+} \frac{1}{2} = \frac{1}{2}$$

As x approaches 3 from the left, the term $x - 3$ is negative, and so the absolute value bars do their thing. So for $x < 3$, we can write this as:

$$\lim_{x\to 3^-} \frac{|x-3|}{2x-6} = \lim_{x\to 3^-} \frac{-(x-3)}{2(x-3)} = \lim_{x\to 3^+} \frac{-1}{2} = -\frac{1}{2}$$

Since the left and right hand limits do not exist, the overall limit does not exist. ■

You Try It

(7) Determine $\displaystyle\lim_{x\to -4} |x+4|$

Have You Learned...

- How to compute limits of polynomials?
- How to compute limits of rational functions, often through simplification?
- How to recognize when the limit of a rational function does not exist?
- How to compare left and right hand limits for functions involving absolute value?

Computation of Limits — Problem List

Computation of Limits — You Try It

These appeared above; solutions begin on the next page.

(1) If $\lim_{x\to a} f(x) = -3$, $\lim_{x\to a} g(x) = 0$, and $\lim_{x\to a} h(x) = 8$, then use the steps allowed by limit laws to determine the values of

(a) $\lim_{x\to a}[f(x)+h(x)]$ (b) $\lim_{x\to a}[f(x)]^2$ (c) $\lim_{x\to a}\sqrt[3]{h(x)}$ (d) $\lim_{x\to a}\dfrac{1}{f(x)}$

(2) Use the steps allowed by limit laws to determine

$$\lim_{x\to 3}(x^2-4)(x^3+5x-1)$$

Confirm that this is what you'd get by evaluating $f(3)$.

(3) Determine $\lim_{x\to -1}(x^{15}-12x^3+x^2+\pi)$.

(4) Determine $\lim_{x\to 2}\dfrac{x^2-x+6}{x-2}$.

(5) Determine $\lim_{h\to 0}\dfrac{(4+h)^2-16}{h}$.

(6) Determine $\lim_{x\to 9}\dfrac{x^2-81}{\sqrt{x}-3}$.

(7) Determine $\lim_{x\to -4}|x+4|$.

Computation of Limits — Practice Problems

Try these as you get the hang of the You Try It problems. Solutions to these problems are available in Sec. A.2.2.

(1) If $\lim_{x\to a} f(x) = -3$, $\lim_{x\to a} g(x) = 0$, and $\lim_{x\to a} h(x) = 8$, then use the steps allowed by limit laws to determine the values of

(a) $\lim_{x\to a}\dfrac{f(x)}{h(x)}$ (b) $\lim_{x\to a}\dfrac{g(x)}{f(x)}$ (c) $\lim_{x\to a}\dfrac{f(x)}{g(x)}$ (d) $\lim_{x\to a}\dfrac{2f(x)}{h(x)-f(x)}$

(2) Use the steps allowed by limit laws to determine $\lim_{t\to -1}(t^2+1)^3(t+3)^5$.

(3) Determine $\lim_{x\to 3}\dfrac{x^2+2}{x-4}$.

(4) Determine $\lim_{x\to 4}\dfrac{x^2-4x}{x^2-3x-4}$.

(5) Determine $\lim_{x\to 1}\dfrac{x^3-1}{x^2-1}$.

(6) Determine $\lim\limits_{h \to 0} \dfrac{\frac{1}{3+h} - \frac{1}{3}}{h}$.

(7) Determine $\lim\limits_{x \to -4^-} \dfrac{|x+4|}{x+4}$.

(8) If $F(x) = \dfrac{x^2-1}{|x-1|}$, determine

(a) $\lim\limits_{x \to 1^+} F(x)$ (b) $\lim\limits_{x \to 1^-} F(x)$ (c) $\lim\limits_{x \to 1} F(x)$

(9) Use the defining equations in Def. 1.3 to determine $\lim\limits_{x \to 0} \sinh(x)$ and $\lim\limits_{x \to 0} \cosh(x)$.

Computation of Limits — Challenge Problems

Try these problems to test your skills with the ideas in this section. Solutions to these problems are available in Sec. B.2.2.

(1) If $f(x) = \sqrt{x}$, build and evaluate the limit $\lim\limits_{h \to 0} \dfrac{f(1+h) - f(1)}{h}$. (Hint: Try rationalizing the numerator.)

(2) Determine $\lim\limits_{x \to 1.5} \dfrac{2x^2 - 3x}{|2x-3|}$.

(3) Invent an example of two functions $f(x)$ and $g(x)$ such that $\lim\limits_{x \to a} [f(x) + g(x)]$ exists even though neither $\lim\limits_{x \to a} f(x)$ nor $\lim_{x \to a} g(x)$ exist.

Computation of Limits — You Try It — Solved

(1) If $\lim_{x\to a} f(x) = -3$, $\lim_{x\to a} g(x) = 0$, and $\lim_{x\to a} h(x) = 8$, then use the steps allowed by limit laws to determine the values of

$$\text{(a)}\ \lim_{x\to a}[f(x)+h(x)] \quad \text{(b)}\ \lim_{x\to a}[f(x)]^2$$

$$\text{(c)}\ \lim_{x\to a}\sqrt[3]{h(x)} \quad \text{(d)}\ \lim_{x\to a}\frac{1}{f(x)}$$

☐ The limits are

$$\text{(a)}\ \lim_{x\to a}[f(x)+h(x)] = \lim_{x\to a} f(x) + \lim_{x\to a} h(x) = -3+8=5$$

$$\text{(b)}\ \lim_{x\to a}[f(x)]^2 = \lim_{x\to a} f(x)\cdot\lim_{x\to a} f(x) = (-3)(-3) = 9$$

$$\text{(c)}\ \lim_{x\to a}\sqrt[3]{h(x)} = \sqrt[3]{\lim_{x\to a} h(x)} = \sqrt[3]{8} = 2$$

$$\text{(d)}\ \lim_{x\to a}\frac{1}{f(x)} = \frac{1}{\lim_{x\to a} f(x)} = \frac{1}{-3} == -\frac{1}{3} \quad ■$$

(2) Use the steps allowed by limit laws to determine

$$\lim_{x\to 3}(x^2-4)(x^3+5x-1)$$

Confirm that this is what you'd get by evaluating $f(3)$.

☐
$$\begin{aligned}\lim_{x\to 3}(x^2-4)(x^3+5x-1) &= \lim_{x\to 3}(x^2-4)\cdot\lim_{x\to 3}(x^3+5x-1)\\ &= (\lim_{x\to 3} x^2 - \lim_{x\to 3} 4)\cdot(\lim_{x\to 3} x^3 + \lim_{x\to 3} 5x - \lim_{x\to 3} 1)\\ &= (9-4)\cdot(27+15-1)\\ &= (5)(41) = 205\end{aligned}$$

This is indeed what we'd get just by plugging in $x = 3$:

$$(3^2-4)(3^3+5(3)-1) = (5)(41) = 205 \quad ■$$

(3) Determine $\lim_{x\to -1}(x^{15} - 12x^3 + x^2 + \pi)$.

☐ This is a polynomial, and $x = -1$ is in its domain. So we're allowed to find the limit by direct substitution:

$$\begin{aligned}\lim_{x\to -1}(x^{15} - 12x^3 + x^2 + \pi) &= (-1)^{15} - 12(-1)^3 + (-1)^2 + \pi\\ &= -1 + 12 + 1 + \pi = 12 + \pi \quad ■\end{aligned}$$

(4) Determine $\lim_{x \to 2} \dfrac{x^2 - x + 6}{x - 2}$.

□

$$\lim_{x \to 2} \frac{x^2 - x + 6}{x - 2} = \lim_{x \to 2} \frac{(x+2)(x-3)}{x-2} \text{ which does not exist} \quad \blacksquare$$

(5) Determine $\lim_{h \to 0} \dfrac{(4+h)^2 - 16}{h}$.

□ $\lim_{h \to 0} \dfrac{(4+h)^2 - 16}{h} = \lim_{h \to 0} \dfrac{16 + 8h + h^2 - 16}{h} = \lim_{h \to 0} (8 + h) = 8$ ■

(6) Determine $\lim_{x \to 9} \dfrac{x^2 - 81}{\sqrt{x} - 3}$.

□ We can rationalize the denominator, or we can apply some clever factoring:

$$\begin{aligned}
\lim_{x \to 9} \frac{x^2 - 81}{\sqrt{x} - 3} &= \lim_{x \to 9} \frac{(x-9)(x+9)}{\sqrt{x} - 3} \\
&= \lim_{x \to 9} \frac{(\sqrt{x} - 3)(\sqrt{x} + 3)(x+9)}{\sqrt{x} - 3} \\
&= \lim_{x \to 9} (\sqrt{x} + 3)(x + 9) = (3+3)(9+9) = 108 \quad \blacksquare
\end{aligned}$$

(7) Determine $\lim_{x \to -4} |x + 4|$.

□ $\lim_{x \to -4} |x + 4| = 0$ since $\lim_{x \to -4^-} |x + 4| = 0$ and $\lim_{x \to -4+} |x + 4| = 0$ ■

2.3 Limits Involving Infinity

Introduction

We've seen some limits that ended up involving $+\infty$ or $-\infty$, so lets' put those in a larger context. There are two ways that ∞ or $-\infty$ can drop in and say hi in a limit process:

(1) When $f(x)$ approaches $\pm\infty$ as x approaches some finite value a. We saw a few of these in the last section, as we tried to gain some insight into how vertical asymptotes provide a visual guide to the behavior of a function for some $x \to a$. For this case, we'll have a recap.

(2) When x itself approaches ∞. In this case, $f(x)$ might also approach $\pm\infty$, but it might not. For this case, let's try to get some intuition about expected behaviors of $f(x)$, and then lock in some quantitative reasoning.

The Secret of Vertical Asymptotes

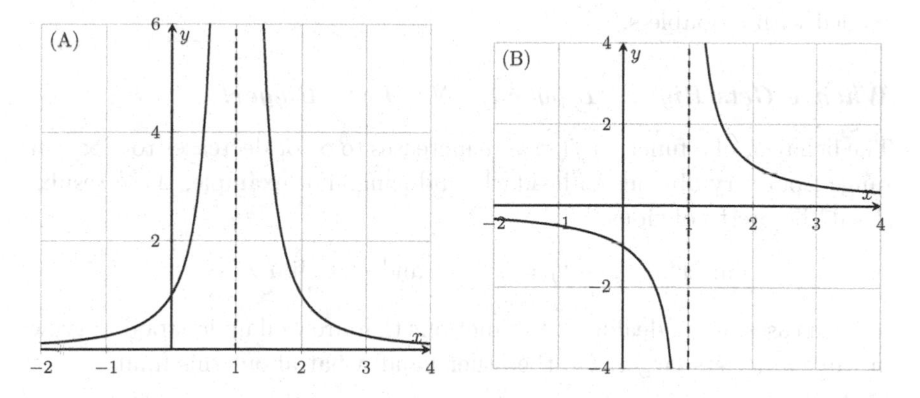

Fig. 2.9 Varieties of vertical asymptotes.

The graph of $f(x)$ has a vertical asysmptote when $f(x)$ itself heads off to either ∞ or $-\infty$ as x approaches some fixed $x = a$. The function can head to the same infinity on both sides of $x = a$ (as seen in Fig. 2.9A), or can head in opposite directions (as seen in Fig. 2.9B). Further, a function can have more than one vertical asymptote. Vertical asymptotes appear in all sorts of functions (remember the graph of $\tan x$?), but generally, if you're seeing them, there's a good chance you're looking at a rational function. At least, the patterns of predictability go best with those, and here's how they work:

Useful Fact 2.3. *The number of vertical asymptotes in the graph of a function $f(x)$ is equal to the number of zeros of the denominator of $f(x)$ (presuming $f(x)$ has been fully simplified), and the asymptotes are located at x-values corresponding to those zeros.*

(The need to have $f(x)$ fully simplified is reflected in EX 1 of the last section. If the denominator of $f(x)$ has a zero whose contributing term will eventually be canceled or eliminated by further algebra, then there will be a hole in the graph, not an asymptote.)

A function's graph can have multiple vertical asymptotes, and can have up to two horizontal asymptotes as well (one to the left, and one to the right).

Oh, by the way: please pay attention to the proper spelling and pronunciation of "asymptote". It is not "asymtope", and is it definitely not spelled with a double-s.

When x Gets Big .... Bigger ... No, Even Bigger!

The behavior of a function $f(x)$ as x increases to ∞ or decreases to $-\infty$ can range from very obvious to headache-inducing. For example, these results should be pretty obvious:

$$\lim_{x\to\infty}(x^2+3x+1)=\infty \qquad \text{and} \qquad \lim_{x\to\infty} e^x=\infty$$

In both cases, we're dealing with functions that are well understood to grow in value as x gets larger. On the other hand, what about this limit?

$$\lim_{x\to\infty}\sin(x)$$

This is not so obvious. It's tempting to call the limit 1 since $\sin x$ hits that value an infinite number of times as x makes its way to ∞. But $\sin(x)$ also hits every other number between -1 and 1 an infinite number of times. It's also tempting to call the limit 0 since that's the central value of the oscillations of $\sin x$. But since there is no one unique identifiable value that $\sin x$ tends to as $x\to\infty$, we say that this limit does not exist.

As with other limit problems, the interesting events occur in functions that are written as quotients. When a function is written as a quotient, then the ultimate fate of the function as $x \to \pm\infty$ is determined by a "battle" between the numerator and denominator. If the numerator gets bigger more quickly than the denominator, then the numerator "wins" and pulls the function's values larger and larger. If the denominator gets bigger more quickly than the numerator, then the denominator "wins" and pulls the function's values towards zero. Here are a couple of examples of these instances:

EX 1 Assess $\lim_{x\to\infty} \dfrac{x^4 - 2x}{x+1}$ and $\lim_{x\to-\infty} \dfrac{x^4 - 2x}{x+1}$

In either limit, the degree of the numerator is larger than the degree of the denominator; therefore, the numerator "wins" and wants to draw the function up to larger and larger values. But the signs of the results have to be inspected carefully. In the first limit, as $x \to \infty$, both the numerator and denominator end up being positive, and so

$$\lim_{x\to\infty} \frac{x^4 - 2x}{x+1} = +\infty$$

In the second limit, though, as $x \to -\infty$, the numerator becomes positive but the denominator stays negative; therefore,

$$\lim_{x\to-\infty} \frac{x^4 - 2x}{x+1} = -\infty \quad \blacksquare$$

EX 2 Assess $\lim_{x\to\infty} \dfrac{x+2}{x^2+3}$

In this limit, the the degree of the numerator is smaller than the degree of the denominator; therefore, the denominator "wins" and wants to draw the function to zero:

$$\lim_{x\to\infty} \frac{x+2}{x^2+3} = 0 \quad \blacksquare$$

You Try It

(1) Assess $\lim_{x\to\infty} \dfrac{1}{2x+3}$

Now, there is also a middle ground, when neither the numerator or denominator "wins" decisively. This happens when the degree of the numerator is the same as the degree of the denominator. It is these final cases where all the fun happens. For example,

EX 3 Assess $\lim_{x\to\infty} \dfrac{x^2+x}{2x^2-1}$

Intuitively, the argument might go like this: As x gets bigger and bigger, the $+x$ in the numerator and the -1 in the denominator get less and less important. So as $x \to \infty$, this function ends up looking more and more like $f(x) = x^2/2x^2$, which simplifies to $1/2$. Therefore, as $x \to \infty$, the function itself should be tending towards $1/2$. ■

These limit values are evident in the graph of $f(x)$ as *horizontal asymptotes.* If $f(x)$ is trending towards some constant value L as x goes to ∞ or $-\infty$, then the graph of $f(x)$ has a horizontal asymptote at $y = L$. Revisiting examples from above,

- in EX 1, $f(x)$ does not have a horizontal asymptote.
- in EX 2, $f(x)$ has a horizontal asymptote at $y = 0$, the x-axis.
- in EX 3, $f(x)$ has a horizontal asymptote at $y = 1/2$.

The story that's evolving can be summarized like this:

Useful Fact 2.4. *For a rational function $f(x) = P(x)/Q(x)$ (a ration of two polynomials),*

- *The x-axis ($y = 0$) will be a horizontal asymptote of $f(x)$ when $deg(P) < deg(Q)$.*
- *A horizontal asymptote will form (but will not be the x-axis) on the graph of $f(x)$ when $deg(P) = deg(Q)$, and that asymptote will be the line $\lim_{x\to\infty} f(x)$ and / or $\lim_{x\to-\infty} f(x)$.*

You Try It

(2) What is the horizontal asymptote of $f(x) = \dfrac{-2x^2+x+1}{x^2+3}$

Now let's see the quantitative technique that can be used to confirm all the intuitive results we just developed.

A Matter of Factoring

In limits where $x \to \pm\infty$, particularly those whose outcome is not certain from mere inspection of the function, a standard procedure is to locate the highest power of x from each of the numerator and denominator and factor those terms out; then, we hope for cancellation. The purpose of this is to eliminate, where possible, the "big bully" term that's trying to control everything and reveal the less important terms that can balance things out. Let's redo EXamples 1–3 with this procedure (note that now we will be *determining* the limits rather than just assessing the limit).

EX 4 Determine the following limits

$$\text{(A)} \lim_{x\to\infty} \frac{x^4 - 2x}{x+1} \qquad \text{(B)} \lim_{x\to\infty} \frac{x+2}{x^2+3} \qquad \text{(C)} \lim_{x\to\infty} \frac{x^2+x}{2x^2-1}$$

We identify and factor out the term with the highest power from each of the numerator and denominator. In (A), then, we factor out x^4 from the numerator and x from the denominator, and then apply known limit laws:

$$\frac{x^4 - 2x}{x+1} = \frac{x^4(1 - 2x/x^4)}{x(1+1/x)} = x^3 \cdot \frac{1 - 2/x^3}{1 + 1/x}$$

We see that the x^3 dominates everything else a $x \to \infty$; applying known limit laws, we get

$$\lim_{x\to\infty} \frac{x^4 - 2x}{x+1} = \lim_{x\to\infty} x^3 \cdot \frac{1 - 2/x^3}{1 + 1/x} = \infty$$

In (B), we factor out x from the numerator and x^2 from the denominator:

$$\frac{x+2}{x^2+3} = \frac{x(1+2/x)}{x^2(1+3/x^2)} = \frac{1}{x} \cdot \frac{1+2/x}{1+3/x^2}$$

Here, the leading $\frac{1}{x}$ controls the result, and

$$\lim_{x\to\infty} \frac{x+2}{x^2+3} = \lim_{x\to\infty} \frac{1}{x} \cdot \frac{1+2/x}{1+3/x^2} = 0$$

In (C), we'll factor out an x^2 from both the numerator and the denominator, then cancel it:

$$\frac{x^2+x}{2x^2-1} = \frac{x^2(1+1/x)}{x^2(2-1/x^2)} = \frac{1+1/x}{2-1/x^2}$$

The x^2 was controlling the conversation, but now that it's gone, we can see the subtlety of the function, and it is much more clear what will happen as $x \to \infty$. Applying limit laws,

$$\lim_{x\to\infty} \frac{x^2+x}{2x^2-1} = \lim_{x\to\infty} \frac{1+1/x}{2-1/x^2} = \frac{1+0}{2+0} = \frac{1}{2}$$

Hooray! Our intuition from Examples 1, 2, and 3 was correct! ■

You Try It

(3) Determine $\lim_{x\to\infty} \dfrac{x^3+5x}{2x^3-x^2+4}$

(4) Determine $\lim_{x\to\infty} \dfrac{x^4-2x+1}{x^3+3}$

Here's a summary of what we know so far. When $f(x)$ is a rational function, i.e. the ratio of two polynomials $f(x) = P(x)/Q(x)$, then $\lim_{x\to\infty} f(x)$ behaves according to these rules (in which "deg" means *degree of*):

- If $deg(P) > deg(Q)$ then the numerator "wins", and the limit is either ∞ or $-\infty$; the choice of sign must be found by careful inspection of the function. There is no horizontal asymptote for $f(x)$.
- If $deg(P) < deg(Q)$ then the denominator "wins", and the limit is zero — so that the x-axis is a horizontal asymptote for $f(x)$.
- If $deg(P) = deg(Q)$ then neither the denominator nor numerator really "wins"; the limit must be explored further, perhaps by factoring the highest powers of x from P and Q. The value of the limit represents the horizontal asymptote on the graph of $f(x)$.

But what if $f(x)$ is not technically a rational function, but is still a *ratio*? Well, the factoring business still works quite well, but you may have to do it more carefully.

EX 5 Evaluate $\lim_{x\to\infty} \dfrac{\sqrt{4x^4+x}}{3x^2-1}$

This one's a little harder to figure out at first sight. The polynomials involved are of different degree, but you don't want to leap to the conclusion that this limit will be ∞ because the degree of the polynomial in the numerator is bigger. Remember, there's a square root on the numerator, too — and if you ignore the $+x$ in the numerator, then $\sqrt{4x^4}$ will look an awful lot like $2x^2$! So it's not clear what's going on yet. What we do know, though, is that the highest power of x in the numerator is 4, so we'll factor out an x^4 from the numerator. The highest power of x in the denominator is 2, so we'll factor out an x^2 from there:

$$\frac{\sqrt{4x^4+x}}{3x^2-1} = \frac{\sqrt{x^4(4+1/x^3)}}{x^2(3-1/x^2)} = \frac{x^2\sqrt{4+1/x^3}}{x^2(3-1/x^2)} = \frac{\sqrt{4+1/x^3}}{3-1/x^2}$$

and now it's a lot more clear that

$$\lim_{x\to\infty}\frac{\sqrt{4x^4+x}}{3x^2-1}=\lim_{x\to\infty}\frac{\sqrt{4+1/x^3}}{3-1/x^2}=\frac{\sqrt{4+0}}{3-0}=\frac{2}{3}\quad\blacksquare$$

You Try It

(5) Determine $\lim_{x\to\infty}\frac{\sqrt{9x^6-x}}{x^3+1}$

Other Asymptotes

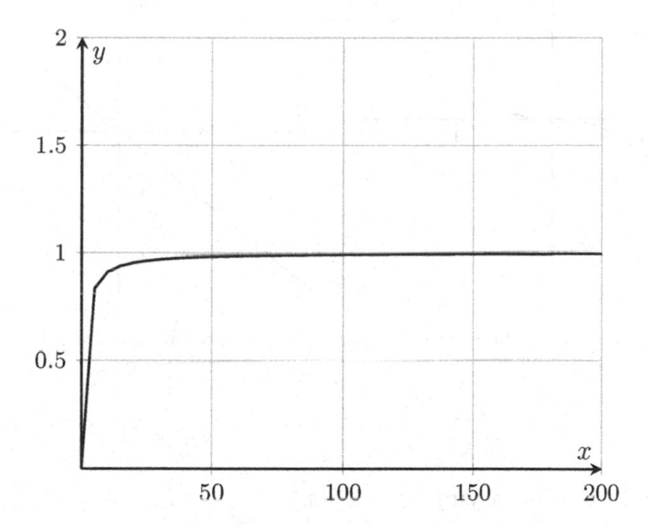

Fig. 2.10 Zooming out on $x/(x+1)$ to show asymptotic behavior.

Asymptotes form a "skeleton" around which the graph of a function is built. Or considered another way, asymptotes are the simple shapes which a function might tend to look like if we zoomed our viewpoint way, way, out. For example, Fig. 2.10 shows the function $f(x) = x/(x+1)$ from $x = 0$ to $x = 100$. Apart from the small area of approach, this function looks a whole lot like $y = 1$ on a large scale.

So far, we've seen horizontal and vertical asymptotes, but there are others kinds. Anytime we identify a large scale trend for a function, we say we've found *asymptotic behavior*. The asymptotes are the "landing strips" for the function, and they don't have to be horizontal or vertical; heck, they don't even have to be straight. Figure 2.11 shows the function $f(x) = x^3/x^2 + 1$ on two different scales — (A) relatively close up, and (B) zoomed far out. In the large scale image (B), the graph of $f(x)$ sure looks an awful lot like the graph $y = x$. This is asymptotic behavior. When an asymptote of a function is a straight line but not horizontal or vertical, we call it a *slant asymptote*. We can discover the equation of the slant asymptote using our technique of factoring out the highest order term from both the numerator and denominator.

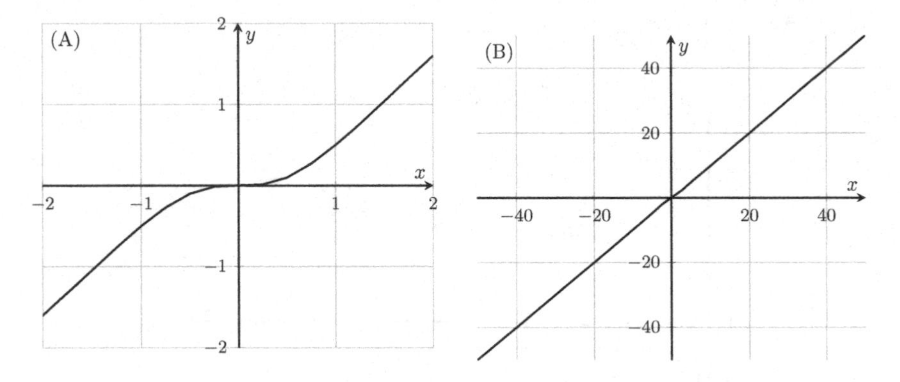

Fig. 2.11 The function $x^3/(x^2+1)$ on two different scales.

EX 6 Find the slant asymptote of $f(x) = \dfrac{x^3}{x^2+1}$.

Let's factor out x^3 from the numerator and x^2 from the denominator of $f(x)$:

$$\frac{x^3}{x^2+1} = \frac{x^3(1)}{x^2(1+1/x^2)} = x \cdot \frac{1}{1+1/x^2}$$

In its new form, we can see that as x gets larger and larger, the fractional part collapses to 1, and the function is controlled by the leading x. This is why $f(x)$ tends to look more and more like $y = x$ as x gets larger and larger. Thus, $y = x$ is a slant asymptote for this function. ■

You Try It

(6) Revisit You Try It 4 and state the slant asymptote for the function given there. Plot that function for yourself using a computer algebra system and zoom out t0 be sure you see the slant asymptote in action.

(7) Find the slant asymptote of $f(x) = \dfrac{x^4 - x^2}{2x^3 + 1}$.

Given the analysis in EX 6, you may have surmised that asymptotes don't even have to be straight lines! Figure 2.12 shows the function $f(x) = x^3/(x+1)$ on two different scales. On the "regular" scale shown in (A), we see the vertical asymptote at $x = -1$ and the growing trend of the function elsewhere. On the zoomed out scale shown in (B), we see that $f(x)$ sure looks an awful lot like the graph $y = x^2$ (the vertical asymptote is still there, but the sampling of points to create the curve misses it). This is also asymptotic behavior, and the asymptote here is $y = x^2$.

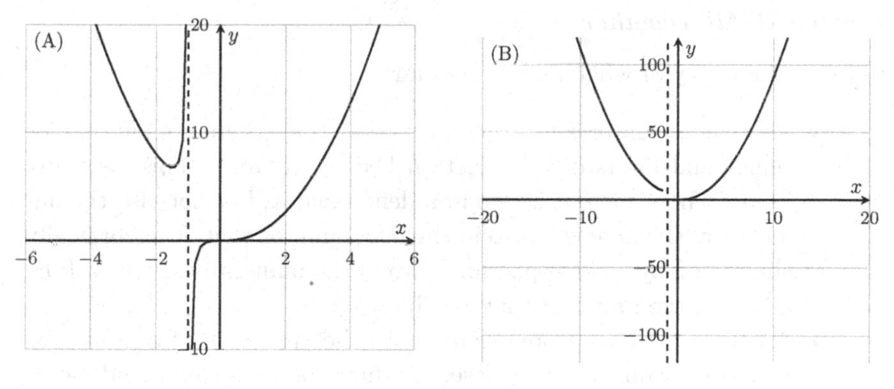

Fig. 2.12 The function $x^3/(x+1)$ on two different scales.

EX 7 Determine the large scale asymptotic behavior of $f(x) = \dfrac{4x^4}{2x^2 + 5}$.

Let's factor out x^4 from the numerator and x^2 from the denominator of $f(x)$:

$$\frac{4x^4}{2x^2+5} = \frac{x^4(4)}{x^2(2+5/x^2)} = x^2 \cdot \frac{4}{2+5/x^2}$$

In its new form, we can see that as x gets larger and larger, the fractional part collapses to 2, and the function is controlled by the leading x^2. Together, this shows that as x gets larger and larger, $f(x)$ will look more and more like $y = 2x^2$. ■

You Try It

(8) Determine the large scale asymptotic behavior of $f(x) = \dfrac{x^4}{9 - x^2}$.

FFT: There is predictability in the presence of slant asymptotes (or other kinds) for rational functions based on a comparison of the degrees of the numerator and denominator. For a rational function $f(x) = P(x)/Q(x)$, can you determine how $deg(P)$ and $deg(Q)$ must relate in order to confirm the presence of a slant asymptote? (The answer to this is given away just below here, but see if you can figure it out beforehand.)

Putting It All Together

Here is a summary of what we know so far:

- Vertical asymptotes form on a function $f(x)$ where the left and/or right hand limits of the function become infinite. This generally occurs where there is a zero in a denominator, but not also the numerator. (If there is a zero in the denominator that is algebraically eliminated by a corresponding zero in the numerator, there will be a hole in the graph at that x value.)
- A horizontal asymptote occurs when the limit of $f(x)$ as $x \to \pm\infty$ is a finite value. In this case, the function gets closer and closer to that finite value as x gets bigger and bigger. This generally occurs for a rational function when the degrees of the numerator and denominators are the same.
- In a rational function where the degree of the numerator is one larger than the degree of the denominator, we will see a slant asymptote.

A function can have mixes of different types of asymptotes. Because functions can be defined only by their graphs, it would be easy to draw a function which has a horizontal asymptote for $x \to -\infty$, but a slant asymptote for $x \to +\infty$. However, when starting with functions that have a single algebraic representation (i.e. are not piecewise functions), then the combinations are usually limited to vertical and horizontal, OR vertical and slant. So it's not like total open season when looking for asymototes.

EX 8 Identify all asymptotes on the graph of $f(x) = \dfrac{x^3}{9 - x^2}$.

This function will have vertical asymptotes at $x = \pm 3$, and a slant asymptote of $y = -x$. The former is evident from the denominator. The latter can be seen through factoring as in examples above:

$$\frac{x^3}{9 - x^2} = \frac{x^3(1)}{x^2(9/x^2 - 1)} = x \cdot \frac{1}{9/x^2 - 1}$$

In this form, we can see $f(x)$ will tend to $x(-1) = -x$ as x trends to $\pm\infty$.

■

You Try It

(9) Identify all asymptotes on the graph of $y = \dfrac{x}{x + 4}$

As you get into the exercises for this section. note that while You Try It problems appear with the variety of wording that was appropriate as they were placed into the stream of content, the Practice Problems and Challenge Problems will all say simply, "Find thc asymptotes for ..." This is because determining the limits in question is the same as finding the asymptotes.

Have You Learned...

- The conceptual difference between a variable approaching $\pm\infty$ and a function approaching $\pm\infty$?
- How to find vertical asymptotes of a rational function?
- How to find horizontal asymptotes of a rational function?
- How to predict asymptotic behavior of a rational function by comparing the degrees of the numerator and denominator?
- How to determine non-linear asymptotic behavior of rational functions which don't have just straight line asymptotes?

Limits Involving Infinity — Problem List

Limits Involving Infinity — You Try It

These appeared above; solutions begin on the next page.

(1) Determine $\lim_{x \to \infty} \dfrac{1}{2x+3}$.

(2) Identify a horizontal asymptote of $f(x) = \dfrac{-2x^2+x+1}{x^2+3}$.

(3) Determine $\lim_{x \to \infty} \dfrac{x^3+5x}{2x^3-x^2+4}$.

(4) Determine $\lim_{x \to \infty} \dfrac{x^4-2x+1}{x^3+3}$.

(5) Determine $\lim_{x \to \infty} \dfrac{\sqrt{9x^6-x}}{x^3+1}$.

(6) Find the slant asymptote of $f(x) = \dfrac{x^4-x^2}{2x^3+1}$.

(7) Determine the large scale asymptotic behavior of $f(x) = \dfrac{x^4}{9-x^2}$.

(8) Identify all asymptotes on the graph of $y = \dfrac{x}{x+4}$.

Limits Involving Infinity — Practice Problems

Try these as you get the hang of the You Try It problems. Solutions to these problems are available in Sec. A.2.3.

Determine all asymptotes of the following functions:

(1) $f(x) = \dfrac{3x+5}{x-4}$.

(2) $g(t) = \dfrac{t^2+2}{t^3+t^2-t-1}$.

(3) $h(x) = \dfrac{x+2}{\sqrt{9x^2+1}}$.

(4) $y = \dfrac{x^2+4}{x^2-1}$.

(5) $y = \dfrac{2x^2-1}{x-2}$.

(6) $y = \tanh(x)$.

Limits Involving Infinity — Challenge Problems

Try these problems to test your skills with the ideas in this section. Solutions to these problems are available in Sec. B.2.3.

(1) Determine all asymptotes of $f(x) = \dfrac{\sqrt{x^2+4}}{(x-1)^2}$.

(2) Determine all asymptotes of $g(x) = \dfrac{x^3-2x+3}{5-2x^2}$.

(3) Determine all asymptotes of $y = \dfrac{x^3+1}{x^3+x}$.

Limits Involving Infinity — You Try It — Solved

(1) Determine $\lim_{x\to\infty} \dfrac{1}{2x+3}$.

☐ The numerator is constant, while the denominator goes to ∞, so: $\lim_{x\to\infty} \dfrac{1}{2x+3} = 0.$ ■

(2) Identify a horizontal asymptote of $f(x) = \dfrac{-2x^2+x+1}{x^2+3}$.

☐ The numerator and denominator have the same degree (2), so there will be a horizontal asymptote. The ratio of coefficients on the terms of highest degree up and down shows that the asymptote should be $y = -2$. But just to be sure, let's do the factoring business — factor x^2 from each of the numerator and denominator:

$$\frac{-2x^2+x+1}{x^2+3} = \frac{x^2(-2+1/x+1/x^2)}{x^2(1+3/x^2)} = \frac{-2+1/x+1/x^2}{1+3/x^2}$$

and so the limit of $f(x)$ as $x \to \infty$ is $(-2+0+0)/(1+0) = -2$ (the same happens for $x \to -\infty$, too). So the horizontal asymptote is at $y = -2$, as predicted. ■

(3) Determine $\lim_{x\to\infty} \dfrac{x^3+5x}{2x^3-x^2+4}$.

☐ Your intuition might tell you that this limit is $1/2$, since the numerator and denominator are polynomials of the same degree, and so the limit is determined by the ratio of the leading coefficients. To confirm this, we can compute the limit directly by factoring out the highest term from the numerator and denominator (which is x^3 in both):

$$\lim_{x\to\infty} \frac{x^3+5x}{2x^3-x^2+4} = \lim_{x\to\infty} \frac{1+5/x^2}{2-1/x+4/x^3} \frac{1+0}{2-0+0} = \frac{1}{2} \quad ■$$

(4) Determine $\lim_{x\to\infty} \dfrac{x^4-2x+1}{x^3+3}$.

☐ Since the degree of the numerator is larger than the degree of the denominator, the numerator "wins", and this limit is ∞. While we should trust our instincts, it's always a good idea to confirm; let's rewrite the given function as:

$$f(x) = \frac{x^4(1-2/x^3+1/x^4)}{x^3(1+3/x^3)} = x \cdot \frac{1-2/x^3+1/x^4}{1+3/x^3}$$

so that

$$\lim_{x\to\infty} f(x) = \lim_{x\to\infty}(x) \cdot \lim_{x\to\infty} \frac{1 - 2/x^3 + 1/x^4}{1 + 3/x^3} = \infty \cdot 1 = \infty \quad \blacksquare$$

---

(5) Determine $\lim\limits_{x\to\infty} \dfrac{\sqrt{9x^6 - x}}{x^3 + 1}$.

□ We can compute the limit by factoring out the highest term from the numerator and denominator:

$$\lim_{x\to\infty} \frac{\sqrt{9x^6 - x}}{x^3 + 1} = \lim_{x\to\infty} \frac{x^3(\sqrt{9 - 1/x^5}}{x^3(1 + 1/x^3)} = \lim_{x\to\infty} \frac{\sqrt{9 - 1/x^5}}{1 + 1/x^3} = 3$$

There is a very subtle thing in the simplification of the function in this problem. Since we're doing the limit as $x \to \infty$, and thus handling *positive* x values, it's true that $\sqrt{x^6} = x^3$. But what if we were taking the limit as $x \to -\infty$, and therefore dealing with *negative* x values? This is something to keep in mind during a Challenge Problem on this topic, and a little hint for those of you who are actually reading these solutions! ■

---

(6) Find the slant asymptote of $f(x) = \dfrac{x^4 - x^2}{2x^3 + 1}$.

□ If you're catching on, you should be able to foresee that the slant asymptote is $y = x/2$, based on comparisons of the leading terms. To confirm this, let's factor out the highest term from the numerator and denominator (x^4, x^3 respectively):

$$\frac{x^4 - x^2}{2x^3 + 1} = \frac{x^4(1 - 1/x^2)}{x^3(2 + 1/x^3)} = x \cdot \frac{1 - 1/x^2}{2 + 1/x^3}$$

From this version of $f(x)$, we can see that as x grows larger, the fractional term collapses towards $1/2$, and so overall the function gets closer to $y = x/2$. ■

---

(7) Determine the large scale asymptotic behavior of $f(x) = \dfrac{x^4}{9 - x^2}$.

□ Let's rewrite $f(x)$ to see the controlling term:

$$f(x) = \frac{x^4}{9 - x^2} = \frac{x^4(1)}{x^2(9/x^2 - 1)} = x^2 \frac{1}{9/x^2 - 1}$$

Now we can see that as x grows larger, the fractional term collapses towards -1, and so overall the function collapses toward $y = -x^2$. This graph will be asymptotic to an inverted parabola. Go ahead and plot it! ■

---

(8) Identify all asymptotes on the graph of $y = \dfrac{x}{x+4}$.

□ The graph has a vertical asymptote at $x = -4$ since the left and right hand limits of the function are infinite there. The graph of y would head to $+\infty$ from the right of $x = -4$ and to $-\infty$ from the left of $x = -4$. There is a horizontal asymptote at $y = 1$ because

$$\lim_{x \to \infty} \frac{x}{x+4} = \lim_{x \to \infty} \frac{1}{1 + 4/x} = 1$$ ■

2.4 Limits and Continuity

Introduction

Continuity is one of the most desirable properties a function can have.[1] Many major concepts from here forward will be prefaced with, "Suppose $f(x)$ is continuous ..." Although an intuitive sense of what continuity means can get us a long way, we need a precise definition — and this definition involves limits.

The Definition of Continuity

Without further ado, here comes the definition of continuity. If you are comfortable with the language of functions, it's easy to interpret. If you're not, then you may have to state at it for a long time.

Definition 2.1. *A function $f(x)$ is continuous at $x = a$ if $\lim_{x \to a} f(x) = f(a)$.*

In other words, a function is continuous at a point $x = a$ if the value the function is aiming at while $x \to a$ (the limit) really does become the value of the function there ($f(a)$). Or, even more informally: if you can draw the graph of $f(x)$ through the point $(a, f(a))$ without lifting your pencil, then the function is continuous there. (Interestingly, the converse of that isn't true; there are some wacko examples of functions which are continuous at a point $(a, f(a))$, but it's still not possible to draw the curve through that point without lifting your pencil — for example, if the function oscillates and infinite number of times before reaching $(a, f(a))$.

The function shown in Fig. 2.13 is continuous at $x = 1$ but not continuous at $x = 3$, as follows:

- At $x = 1$, the left and right hand limits are the same and equal to the value of the function, so

$$\lim_{x \to 1} f(x) = f(1)$$

- At $x = 3$, though, the left and right hand limits are 2, but the value of the function is 1, so

$$\lim_{x \to 3} f(x) \neq f(3)$$

[1] Unless you're a math nerd, because a lot of fun can be had with pathological situations that result from a lack of continuity!

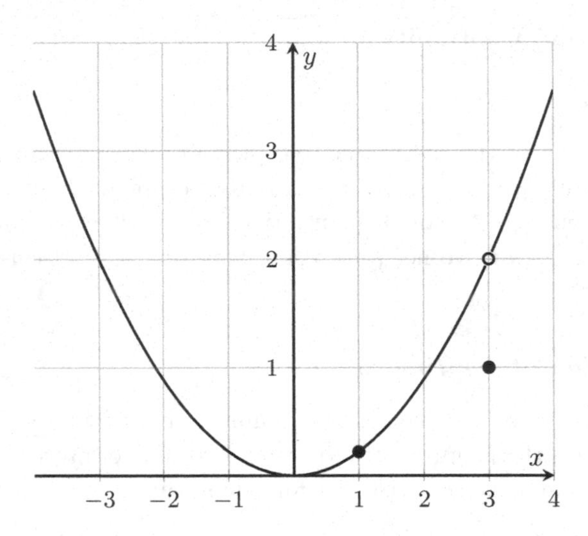

Fig. 2.13 Continuous at $x = 1$ but discontinuous at $x = 3$.

The point $x = 1$ is certainly not the only place where this function is continuous. It is actually continuous at *every* point except $x = 3$. When you collect together into intervals all the places where the function in Fig. 2.13 is continuous, you have *intervals of continuity*. The intervals where this function is continuous are $(-\infty, 3)$ and $(3, \infty)$.

The location where this function is not continuous is called a *discontinuity*. There are three types of discontinuities:

(1) A *removable* discontinuity is a hole in the graph; it can be "repaired" by changing the value of the function (or adding a new value) at a single point. The discontinuity in the above figure is a removable discontinuity.
(2) A *jump* discontinuity is where there is a break in the function, but the endpoits at the jump are finite; this kind of discontinuity cannot be "repaired" without changing an entire piece of the function.
(3) An *infinite* discontinuity is a vertical asymptote.

(Note that talking about "repairing" discontinuities is just for illustration; if you actually change a function to "repair" a discontinuity. you have replaced it with a totally different function.)

The different types of discontinuities become more well defined when use the language of limits:

(1) If the limit of $f(x)$ as $x \to a$ is exists and is finite, then the left and right hand limits are finite and agree with each other. If $f(a)$ itself is not equal to these limits, then we have a hole in the graph — i.e. a removable discontinuity.
(2) If the limit doesn't exist because the left and right hand limits, while finite, do not agree, then we have a jump discontinuity. For example, suppose that as $x \to 1$, $f(x)$ approaches 3 from the left and 5 from the right. Then there is a jump in the graph at $x = 1$.
(3) If the limit doesn't exist because the left and right hand limits do not agree and are infinite, then we have a vertical asymptote — i.e. an infinite discontinuity.
(4) If the left and right hand limits agree but are $\pm\infty$, then we have a vertical asymptote.

Also, because continuity is defined using limits, and we do have left and right hand limits, we also distinguish between continuity *from the left* and *from the right.* Suppose that $f(x)$ is already known to be continuous on some interval (a, b), and we want to decide if it should be considered continuous at the endpoints of that interval, too.

Definition 2.2. *If* $\lim\limits_{x \to a^+} f(x) = f(a)$ *then we say that* $f(x)$ *is continuous from the right at* $x = a$*, we include* $x = a$ *in the interval of continuity, and we report the interval of continuity as* $[a, b)$*. Similarly, if* $\lim\limits_{x \to b^-} f(x) = f(b)$ *then we say that* $f(x)$ *is continuous from the left at* $x = b$*, we include* $x = b$ *in the interval of continuity, and we report the interval of continuity as* $[a, b]$*. Generally, the interval of continuity could be any of* (a, b)*,* $[a, b)$*,* $(a, b]$*, or* $[a, b]$ *depending on whether* $f(x)$ *is continuous from the right at* $x = a$*, from the left at* $x = b$*, or both.*

EX 1 Identify the locations and types of discontinuities in the graph shown in Fig. 2.14, and also state the intervals on which the function is continuous.

There is a jump discontinuity at $x = -1$. The limit as x approaches -1 from the left does not match the value of $f(-1)$, but the limit as x approaches -1 from the right does match $f(-1)$. So, $x = -1$ will be included in the

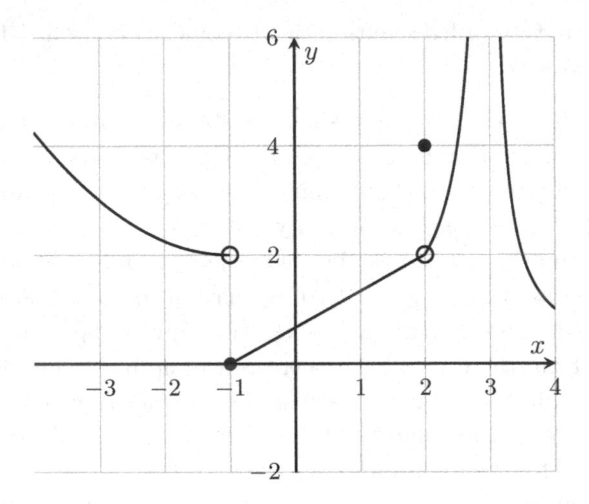

Fig. 2.14 Discontinuous are fun!.

interval of continuity to the *right* of $x = -1$, but not the left. There is a removable discontinuity at $x = 2$, and an infinite discontinuity at $x = 3$. The resulting intervals of continuity are $(-\infty, -1)$, $[-1, 2)$, $(2, 3)$, and $(3, \infty)$.

■

You Try It

(1) Identify the locations and types of discontinuities in the graph shown in Fig. 2.15 (see end of section), and also state the intervals on which the function is continuous.

Assessing Continuity

For most "generic" functions, assessing continuity is often similar to figuring out the domain of the function. For many functions, we know that if there is a defined value for the function at $x = a$, the function will be continuous there. In general,

- A polynomial is continuous at all points in its domain — i.e. polynomials are continuous everywhere.
- A rational function is continuous at all points in its domain — i.e. everywhere except where the denominator is zero.
- As when finding domains of compositions of two functions, the

interval of continuity cannot be determined from the simplified algebraic form of the composition, you must begin with knowledge of the domains / continuity of the individual functions making up the composition, and make any refinements from there.

EX 2 Where is the function $f(x) = 1/(x^2 - 3x + 2)$ continuous?

The function can be written as

$$f(x) = \frac{1}{x^2 - 3x + 2} = \frac{1}{(x-1)(x-2)}$$

So, the points $x = 1$ and $x = 2$ are not in the domain of this function. $f(x)$ is continuous everywhere except $x = 1$ and $x = 2$. ■

EX 3 (a) If $f(x)\sqrt{x}$ and $g(x) = x + 2$, where is the composition $f(g(x))$ continuous? (b) If $f(x)\sqrt{x}$ and $g(x) = x^2 + 2$, where is the composition $g(f(x))$ continuous?

Note that $f(x)$ itself is defined for $x \geq 0$, is certainly continuous for $x > 0$, and is even continuous from the right at $x = 0$ because

$$\lim_{x \to 0^+} f(x) = f(0)$$

So, we report that $f(x)$ itself is continuous on $[0, \infty)$. Now, the final simplified form of $f(g(x))$ is $\sqrt{x+2}$. This is defined for all $x + 2 \geq 0$, is certainly continuous for $x + 2 > 0$, and is continuous from the right at $x = 2$. Therefore, we report that $f(g(x))$ is defined and continuous wherever $x + 2 \geq 0$, i.e. in the interval $[-2, \infty)$.

In (b), the final simplified form of $g(f(x))$ is $x + 2$, and we might want to conclude immediately that $g(f(x))$ will be defined and continuous for all real numbers. However, this would be *incorrect*. The domain of $f(x)$ is $x \geq 0$. The domain of $g(f(x))$ begins with $x \geq 0$, and we must apply any further restrictions required by g. But there are none, and so the domain of $g(f(x))$ is $x \geq 0$. Within this domain, there are no "problems" with continuity except perhaps at $x = 0$ itself; but at $x = 0$, $f(x)$ is continuous from the right and so also is $g(f(x))$. In all, we report that $g(f(x))$ is continuous on $[0, \infty)$. ■

You Try It

(2) Where is the function $F(x) = x/(x^2 + 5x + 6)$ continuous?
(3) If $f(t) = \ln(t)$ and $g(t) = t^4 - 1$, where is the composition $f(g(t))$ continuous?

You may have noticed that all of the functions shown only as graphs in examples and exercises above would have — when put into algebraic form — piecewise expressions. It just so happens that many of the interesting questions about continuity arise with such piecewise functions. Fortunately, things don't really get more complicated. There are only two ways a piecewise function will have discontinuities:

- If one of the pieces of a piecewise function contains discontinuities of its own.
- If any two pieces of the function don't actually meet at their common endpoint.

The latter case is no different than assessing left and right hand limits.

EX 4 Describe everything you know about the continuity of

$$f(x) = \begin{cases} \sin(x) & \text{for} \quad x \leq 0 \\ 3x & \text{for} \quad 0 < x \leq 1 \\ (x-2)^2 & \text{for} \quad x > 1 \end{cases}$$

The three individual pieces of this function have no "breaks". So, the only places this function can have discontinuities are $x = 0$ and $x = 1$, where the pieces try to join. At $x = 0$, we have $\sin(x)$ to the left and $3x$ to the right. Now,

$$\lim_{x \to 0^-} \sin(x) = 0 \quad \text{and} \quad \lim_{x \to 0^+} (3x) = 0$$

So the left and right hand limits of $f(x)$ agree at $x = 0$ and $f(x)$ is continuous at $x = 0$. At $x = 1$, we have $3x$ to the left and $(x-2)^2$ to the right. Now,

$$\lim_{x \to 1^-} 3x = 3 \quad \text{and} \quad \lim_{x \to 1^+} (x-2)^2 = 1$$

So the left and right hand limits of $f(x)$ do NOT agree at $x = 1$ and $f(x)$ is discontinuous there. At $x = 1$, though, at least we have that the function is continuous from the left. The intervals of continuity for $f(x)$ are $(-\infty, 1]$ and $(1, \infty)$. Try graphing this function to see the discontinuity. ■

You Try It

(4) Is this function continuous at $x = 0$?

$$f(x) = \begin{cases} e^x & \text{for } x < 0 \\ x^2 & \text{for } x \geq 0 \end{cases}$$

You Try It

(5) Describe everything you know about the continuity of

$$f(x) = \begin{cases} 1 + x^2 & \text{for } \quad x \leq 0 \\ 2 - x & \text{for } 0 < x \leq 2 \\ (x-2)^2 & \text{for } \quad x > 2 \end{cases}$$

You Try It

(6) What value of c makes the following function continuous?

$$f(x) = \begin{cases} cx + 1 & \text{for } x \leq 3 \\ cx^2 - 1 & \text{for } x > 3 \end{cases}$$

The Intermediate Value Theorem

In a utilitarian Calculus course, you are usually not asked to go into the weeds of mathematical legalese very often, but here is one of the times it's appropriate to do so, given the topic at hand. So roll up your sleeves and get ready to dig in. This is the Intermediate Value Theorem; note it's our first instance of a law starting with "Suppose $f(x)$ is continuous ..."

Theorem 2.1 ***Intermediate Value Theorem.*** *Suppose that $f(x)$ is continuous on the closed interval $[a, b]$ and W is any number between $f(a)$ and $f(b)$. Then there is a number c in $[a, b]$ such that $f(c) = W$.*

Wow, that's a mouth-full! We'll break this down one bit (bite?) at a time. As with all theorems, this has hypotheses ("Suppose that") and a conclusion. If the hypotheses hold, then so must the conclusion. Some necessary information can be given, some can be inferred. To apply this particular theorem, we must be provided with the function $f(x)$ and the interval $[a, b]$. Given that information, we can compute $f(a)$ and $f(b)$. Then, either we must be given a specific target value W (between $f(a)$ and $f(b)$), or the context of a particular question may provide it; that'll make more sense in a little while.

Armed with the data $f(x), [a, b], f(a), f(b), W$, we can move to the conclusion of the theorem (sentence 2). The conclusion tells us about the existence of a magic number c where *something* happens ("...there is a number c ..."). Your understanding of what that event is depends on your ability to read functional notation. How are you doing with that so far? Here are some subtleties:

- The number c is known to live on the x axis, because it is said to sit in the interval $[a, b]$.
- The number W is known to live on the y axis, because it is said to be between $f(a)$ and $f(b)$.
- We are playing target practice with the function $f(x)$; W is the target, and we use this theorem when we're wondering if we can make the function hit that particular target. If so, then there's an x-value called c where $f(c) = W$.

For example, suppose a function starts at the point $(1, 5)$ and then goes uphill to the point $(3, 10)$. If the function is continuous, then there is no way to draw the function from $(1, 5)$ to $(3, 10)$ without passing through every y value between 3 and 10. So if you wanted to know if the function had to cross, say, $y = 6.7125443$, the answer would be, "of course!" That's the essence of the IVT.

Here are some examples where the IVT does (and does not) help out.

EX 5 Consider the function $f(x)$ defined on the closed interval [1,2] by $f(x) = x^5 - x^4 - x^3 + x^2 + 1$. Does this function have to cross the line $y = 5$ somewhere between $x = 1$ and $x = 2$?

We are going to use the IVT to answer this. First, we need the complete endpoints of the graph of $f(x)$ on the given interval $[a, b] = [1, 2]$.

- The left endpoint sits at $a = 1$, and $f(a) = 1^5 - 1^4 - 1^3 + 1^2 + 1$, so $f(a) = 1$.
- The left endpoint sits at $b = 2$, and $f(b) = 2^5 - 2^4 - 2^3 + 2^2 + 1$, so $f(b) = 13$.

We are asking if $f(x)$ has to cross the line $y = 5$ between $x = 1$ and $x = 2$. That is, we are asking if there is a point c in the interval $[1, 2]$ such that $f(c) = 5$. This is a job for the Intermediate Value Theorem!

- *Suppose $f(x)$ is continuous on $[a, b]$...* YES, $f(x)$ is a polynomial, so it's definitely continuous on $[1, 2]$.

- ... *and W is any number between $f(a)$ and $f(b)$* ... YES, 5 is between $f(a) = f(1) = 1$ and $f(b) = f(2) = 13$.
- The IVT then just tells us directly, YES, there will be a number c in between $x = 1$ and $x = 2$ such that $f(c) = 5$.

We might write a concise reply to the question as so: "Since $f(x)$ is a polynomial it is continuous on $[1, 2]$. The value $W = 5$ is between $f(a)$ and $f(b)$ because $f(a) = 1$ and $f(b) = 5$. Therefore by the Intermediate Value Theorem, there is a number c in $[1, 2]$ such that $f(c) = 5$." ■

And that's it! The question in that example asked only "does $f(x)$ have to cross the line $y = 5$ in the given interval?" The answer is YES. We were not asking *where* does the function cross $y = 5$? If we wanted that information, then we'd still have numerical work to do. But the IVT at least guarantees for us that we're not wasting our time looking for such a point. (It turns out that point is about $c = 1.71$, as $f(1.71) \approx 5$.)

The Intermediate Value Theorem is an example of an *existence theorem*. It tells us that under certain circumstances, a magic number c where something specific happens does indeed exist, but it does not give a rule or algorithm for how to find that number . It says only, ".. there is a number c ...," not "and here's how you find c."

EX 6 Prove that the function $f(x) = x^3 - 2x^2 + 5$ has a root somewhere in the interval $[-2, -1]$.

We have an interval $[a, b] = [-2, -1]$ with corresponding y values $f(a) = f(-2) = -11$ and $f(b) = f(-1) = 2$. If the function has a root, the function is equal to 0 at that root, because that's the definition of a root of a function: c is a root of $f(x)$ if $f(c) = 0$. And this gives us our value of W ... we are asking if there is a number c in $[-2, -1]$ such that $f(c) = 0$. Now $W = 0$ is definitely between $f(a) = -11$ and $f(b) = 2$, and since $f(x)$ is continuous, the IVT guarantees that YES, there must indeed be a value $x = c$ between $x = -2$ and $x = -1$ where $f(c) = 0$. ■

You Try It

(7) Prove that the function $f(x) = x^4 + x - 3$ has a root somewhere in the interval $[1, 2]$.

As with continuity in general, using the IVT becomes more fun when we consider piecewise functions.

EX 7 Does the following function have to have a value $g(x) = 0$ somewhere on the interval $[-4, 4]$?

$$h(x) = \begin{cases} -x^2 - 1 & \text{for } -4 \le x \le 1 \\ x^2 + 1 & \text{for } 1 < x \le 4 \end{cases}$$

Our interval of interest is $[a, b] = [-4, 4]$. The values of this function at the endpoints of the interval are $f(-4) = (-4)^2 - 1 = -17$ and $f(4) = 4^2 + 1 = 17$. Now, our target value of $W = 0$ is definitely in between -17 and 17, so things are looking good for h taking on the value of $W = 0$ somewhere in between the endpoints. But, we still have to confirm the most important requirement of the IVT, namely that $h(x)$ is continuous on the interval of interest. Each piece of the function is by itself continuous, and so the function will be continuous overall if the pieces actually connect at the junction point of $x = 0$. However, note that

- The limit of $h(x)$ as x approaches 1 from the left is $-(1)^2 - 1 = -2$.
- The limit of $h(x)$ as x approaches 1 from the right is $(1)^2 + 1 = 2$.

Since the two pieces of the function don't meet up at their junction point (there is a jump discontinuity at $x = 1$), the function is not continuous on the interval of interest! So, the IVT does not apply and we cannot jump to any conclusions one way or the other. We cannot say, based on the IVT alone, whether or not the function takes on the value $h(x) = 0$; it might, or it might not. Further investigation (like a plot!) shows that the function does indeed NOT hit $h(x) = 0$, as it just leaps over the x-axis between the two pieces. But this has nothing to do with the IVT. ■

The wrap up of that example is really important. If the conditions of a theorem are not met, we cannot conclude the opposite. The IVT simply says, IF a bunch of conditions are true, THEN we can make a certain conclusion. We cannot flip that around to say that when the conditions of the theorem are not met, we can conclude that the opposite case will occur. We might say "if you jump in a pool of water you will get wet", but we cannot then flip that around and say, "if you don't jump in a pool, you won't get wet". Maybe it's raining?

You Try It

(8) Does the following function have to have a value $g(x) = 2$ somewhere between $x = -3$ and $x = 2$?

$$g(x) = \begin{cases} x^3 + 3 & \text{for } -3 \leq x \leq 0 \\ 3 - x^2 - x^4 & \text{for } \ 0 < x \leq 2 \end{cases}$$

Here is a slight tweak of the above examples to show a more useful scenario for the IVT.

EX 8 Use the IVT to prove that the equation $\sin(x) = \cos(x)$ must have a solution on the interval $[0, 1]$.

At first, this question does not directly relate to the IVT, because the IVT deals with individual functions and values they take on, not equations and their solutions. But, would you agree that the question, "Does $\sin(x) = \cos(x)$ have a solution on the interval $[0, 1]$?" is the same as the question, "Does the function $\sin(x) - \cos(x)$ have a root on the interval $[0, 1]$?" If you don't agree yet, here's the algebraic step:

$$\begin{aligned} \sin(x) &= \cos(x) \\ \sin(x) - \cos(x) &= 0 \end{aligned}$$

Having agreed that the question posed can be re-interpreted as, "Does the function $\sin(x) - \cos(x)$ have a root on the interval $[0, 1]$?" we can get to work with the IVT:

- The function is continuous on $[0, 1]$ because $\sin(x)$ and $\cos(x)$ are both continuous everywhere.
- The value of the function at the left endpoint is $\sin(0) - \cos(0) = -1$.
- The value of the function at the right endpoint is $\sin(1) - \cos(1) \approx 0.3$.
- The value 0 is definitely between -1 and 0.3.

Since all the conditions of the IVT are met, then we can conclude that YES, $\sin(x) - \cos(x)$ must take on the value $\sin(x) - \cos(x) = 0$ somewhere between $x = 0$ and $x = 1$, therefore $\sin(x) = \cos(x)$ must have a solution on the interval $[0, 1]$. ■

You Try It

(9) Use the IVT to prove that the equation $ln(x) = -x/2$ must have a solution on the interval $[1/2, 2]$.

EX 9 If your oven is at 250 degrees when you turn it off and allow it to cool to room temperature, must there be an instant when the oven temperature is exactly 170.565197092 degrees?

Is this really a math problem? Yes! It seems like the answer would be yes, but can you always trust intuition? The IVT allows us to move intuition into the area of mathematical certainty — or *proof*.

- We are not given a specific function to work with, but suppose we had a function $T(t)$ that gave us the temperature of the oven as a function of time. At time $t = 0$, the oven is turned off. The oven is completely cool at time, oh, say, $t = D$ (D for "done"). Then our interval of interest is $[0, D]$ (it doesn't really matter what D is).
- Is the temperature function $T(t)$ continuous on $[0, D]$? Well, it reflects a physical process that, by its nature, must be continuous. So we will claim that yes, $T(t)$ is continuous on $[0, D]$. (This is still just intuition, but we're going to go with it.)
- The temperature at the first endpoint is $T(0) = 250$. The temperature at the second endpoint is $T(D) = \ldots$ whatever room temperature is. Can we agree that whatever it is, room temperature will be below 170.565197092 degrees? Unless you live on Mercury, we should be able to agree on this. For argument's sake, let's say room temperature is a nice comfortable 72 degrees.
- Given what we've agreed about the temperature T at the endpoints of our interval, $T(0) = 250$ and $T(D) \approx 72$, then our target temperature of $T(t) = 170.565197092$ is definitely between those two extremes.

And so, since all the conditions of the IVT are met, we can conclude that YES, there be an instant when the oven temperature is exactly 170.565197092 degrees. ■

You Try It

(10) If you remove marbles from a bag one at a time, must there always come a time when the bag contains exactly half the number of marbles it began with? You might have an intuitive answer to this, but be sure to put your explanation in the context of the IVT.

Have You Learned...

- The formal definition of continuity?
- An intuitive interpretation of continuity?
- How to assess where a function may have discontinuities?
- How to categorize discontinuities?
- The formal statement of the Intermediate Value Theorem?
- An intuitive interpretation of the Intermediate Value Theorem?
- How the Intermediate Value Theorem can be used to confirm the presence of a root of a function or a solution to an equation within a specified section of the x axis?

Limits and Continuity — Problem List

Limits and Continuity — You Try It

These appeared above; solutions begin on the next page.

(1) Identify the locations and types of discontinuities in the graph shown in Fig. 2.15, and also state the intervals on which the function is continuous.

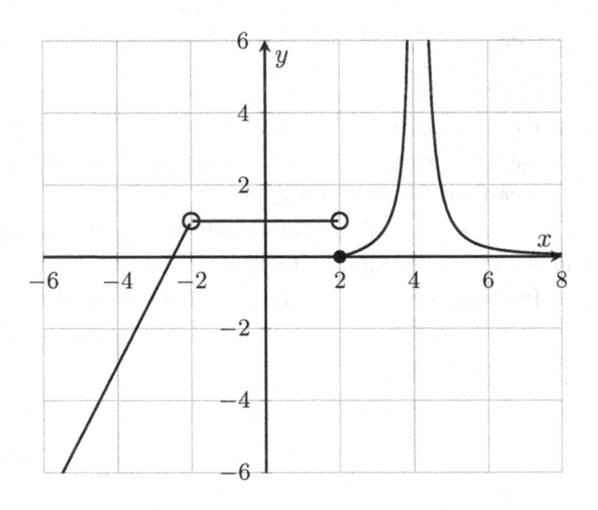

Fig. 2.15 Discontinuous for You Try It 1.

(2) Where is the function $F(x) = x/(x^2 + 5x + 6)$ continuous?

(3) If $f(t) = \ln(t)$ and $g(t) = t^4 - 1$, where is the composition $f(g(t))$ continuous?

(4) Is this function continuous at $x = 0$?

$$f(x) = \begin{cases} e^x & \text{for } x < 0 \\ x^2 & \text{for } x \geq 0 \end{cases}$$

(5) Describe everything you know about the continuity of

$$f(x) = \begin{cases} 1 + x^2 & \text{for } \quad x \leq 0 \\ 2 - x & \text{for } 0 < x \leq 2 \\ (x-2)^2 & \text{for } \quad x > 2 \end{cases}$$

(6) What value of c makes the following function continuous?

$$f(x) = \begin{cases} cx + 1 & \text{for } x \leq 3 \\ cx^2 - 1 & \text{for } x > 3 \end{cases}$$

(7) Prove that the function $f(x) = x^4 + x - 3$ has a root somewhere in the interval $[1, 2]$.

(8) Does the following function have to have a value $g(x) = 2$ somewhere between $x = -3$ and $x = 2$?

$$g(x) = \begin{cases} x^3 + 3 & \text{for } -3 \le x \le 0 \\ 3 - x^2 - x^4 & \text{for } \; 0 < x \le 2 \end{cases}$$

(9) Use the IVT to prove that the equation $\ln(x) = -x/2$ must have a solution on the interval $[1/2, 2]$.

(10) If you remove marbles from a bag one at a time, must there always come a time when the bag contains exactly half the number of marbles it began with? You might have an intuitive answer to this, but be sure to put your explanation in the context of the IVT.

Limits and Continuity — Practice Problems

Try these as you get the hang of the You Try It problems. Solutions to these problems are available in Sec. A.2.4.

(1) Identify the locations and types of discontinuities in the graph shown in Fig. 2.16, and also state the intervals on which the function is continuous.

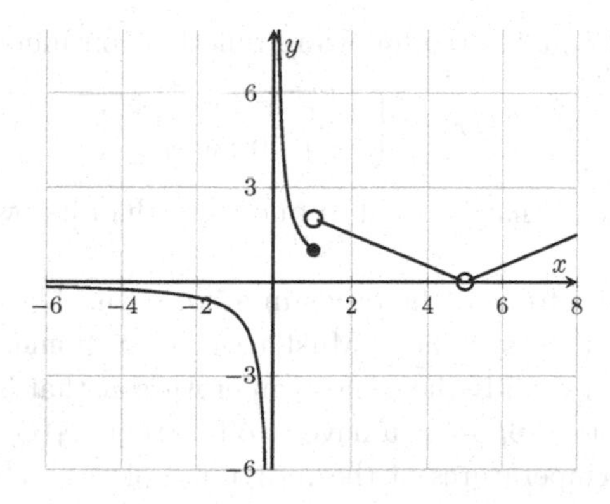

Fig. 2.16 Discontinuities for Practice Problem 1.

(2) If $f(x) = \sqrt[3]{x}$ and $g(x) = 1 + x^3$, where is the composition $f(g(x))$ continuous?

(3) Where is the function $H(x) = \cos(e^{\sqrt{x}})$ continuous?

(4) Is this function continuous at $x = 1$?

$$f(x) = \begin{cases} \frac{x^2-x}{x^2-1} & \text{for } x \neq 1 \\ 1 & \text{for } x = 1 \end{cases}$$

(5) Describe everything you know about the continuity of

$$f(x) = \begin{cases} x+1 & \text{for } \quad x \leq 1 \\ 1/x & \text{for } 1 < x < 3 \\ \sqrt{x-3} & \text{for } \quad x \geq 3 \end{cases}$$

(6) Prove that the equation $\sqrt[3]{x} = 1 - x$ has a solution somewhere in the interval $(0, 1)$.

(7) During the COVID-19 outbreak of 2020, the number of coronavirus cases in the United States on March 01 was reported (due to the spectacular testing program in the country) as 89. By March 12, the number of cases had increased to a reported 1,645. Can we use the IVT to prove that there must have been an instant between March 01 and March 12 when the number of cases was exactly 120?

Limits and Continuity — Challenge Problems

Try these problems to test your skills with the ideas in this section. Solutions to these problems are available in Sec. B.2.4.

(1) What value of c makes the following function continuous?

$$f(x) = \begin{cases} x^2 - c^2 & \text{for } x < 4 \\ cx + 20 & \text{for } x \geq 4 \end{cases}$$

(2) Is there a number that is equal to one more than its own cube? (Hint: Use the IVT.)

(3) One plate is in a freezer, the other in a hot oven. The locations of the two plates are then switched. Must there be a moment in time when the plates are at exactly the same temperature at that instant? Why or why not? (Hint: Suppose you have two functions $f_1(t)$ and $f_2(t)$ which describe the temperatures of the individual plates. The new function $f_1 - f_2$ would be very interesting, indeed!)

Limits and Continuity — You Try It — Solved

(1) Identify the locations and types of discontinuities in the graph shown in Fig. 2.15, and also state the intervals on which the function is continuous.

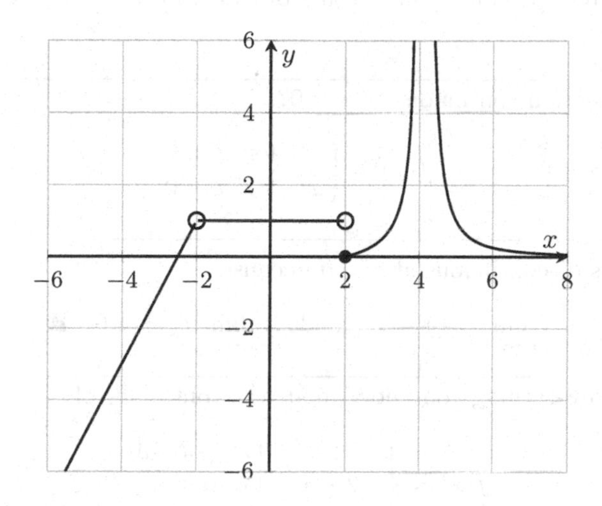

Fig. 2.17 Discontinuous for You Try It 1.

□ $f(x)$ is not continuous at $x = -2$, $x = 2$, and $x = 4$. The discontinuity at $x = -2$ is a removable discontinuity; the discontinuity at $x = 2$ is a jump discontinuity; the discontinuity at $x = 4$ is an infinite discontinuity. The intervals of continuity are $(-\infty, -2)$, $(-2, 2)$, $[2, 4)$, and $(4, \infty)$. The endpoint $x = 2$ is included in the interval $[2, 4)$ because the right hand limit is equal to the value of the function at that point. ■

(2) Where is the function $F(x) = x/(x^2 + 5x + 6)$ continuous?

□ This is a rational function, and a rational function is continuous at all points in its domain. Since $F(x)$ is not defined at $x = -2$ and $x = -3$ since the denominator factors as $x^2 + 5x + 6 = (x + 2)(x + 3)$, these two points are not in its domain. So $F(x)$ is continuous at all reals except $x = -2, -3$. ■

(3) If $f(t) = \ln(t)$ and $g(t) = t^4 - 1$, where is the composition $f(g(t))$ continuous?

□ $g(t) = t^4 - 1$ is continuous everywhere, and $f(t) = \ln(t)$ is continuous everywhere it is defined. Therefore composition $f(g(t)) = \ln(t^4 - 1)$ is defined and continuous is all values of t for which $t^4 - 1 > 0$, i.e. $t > 1$. ■

(4) Is this function continuous at $x = 0$?

$$f(x) = \begin{cases} e^x & \text{for } x < 0 \\ x^2 & \text{for } x \geq 0 \end{cases}$$

□ $f(x)$ is discontinuous at $x = 0$ because

$$\lim_{x \to 0^-} f(x) = 1 \quad \text{and} \quad \lim_{x \to 0^+} f(x) = 0 \quad ■$$

(5) Describe everything you know about the continuity of

$$f(x) = \begin{cases} 1 + x^2 & \text{for} \quad x \leq 0 \\ 2 - x & \text{for} \quad 0 < x \leq 2 \\ (x-2)^2 & \text{for} \quad x > 2 \end{cases}$$

□ Since all three pieces of this function are continuous, the only places that $f(x)$ can be discontinuous are where the pieces connect (or not). So, since

$$\lim_{x \to 0^-} f(x) = 1 \quad \text{and} \quad \lim_{x \to 0^+} f(x) = 2$$

$$\lim_{x \to 2^-} f(x) = 0 \quad \text{and} \quad \lim_{x \to 2^+} f(x) = 0$$

$f(x)$ is discontinuous at $x = 0$ but continuous at $x = 2$. At $x = 0$, $f(x)$ is continuous from the left because the point $x = 0$ is included in the left piece. ■

(6) What value of c makes the following function continuous?

$$f(x) = \begin{cases} cx + 1 & \text{for } x \leq 3 \\ cx^2 - 1 & \text{for } x > 3 \end{cases}$$

☐ Since both pieces of this function are continuous by themselves, the only place this function can be discontinuous is at $x = 3$, where the two pieces may or may not connect. To make them connect, i.e. to make this function continuous at $x = 3$, the pieces need to join there, so we need:

$$c(3) + 1 = c(3)^2 - 1$$
$$3c + 1 = 9c - 1$$
$$c = 3 \quad \blacksquare$$

---

(7) Prove that the function $f(x) = x^4 + x - 3$ has a root somewhere in the interval $[1, 2]$.

☐ The function $f(x) = x^4 + x - 3$ is continuous everywhere because it is a polynomial. Since $f(1) = -1$ and $f(2) = 15$, we see that $f(1) < 0$ and $f(2) > 0$. So by the Intermediate Value Theorem, there is a location $x = c$ somewhere between $x = 1$ and $x = 2$ where $f(x) = 0$, and that value $x = c$ is the root. ■

---

(8) Does the following function have to have a value $g(x) = 2$ somewhere between $x = -3$ and $x = 2$?

$$g(x) = \begin{cases} x^3 + 3 & \text{for } -3 \le x \le 0 \\ 3 - x^2 - x^4 & \text{for } 0 < x \le 2 \end{cases}$$

☐ Our interval of interest is $[a, b] = [-3, 2]$. The values of this function at the endpoints of the interval are $f(-3) = (-3)^2 + 3 = 12$ and $f(2) = 3 - 2^2 - 2^4 = 17$. Now, our target value of $W = 2$ is definitely in between 12 and -17, so things are looking good for g taking on the value of $W = 2$ somewhere in between the endpoints. But, we still have to confirm the most important requirement of the IVT, namely that $f(x)$ is continuous on the interval of interest. Each piece of the function is by itself continuous, and so the function will be continuous overall if the pieces actually connect at the junction point of $x = 0$. So we fall back on the definition of continuity:

- The limit of $g(x)$ as x approaches 0 from the left is $0^3 + 3 = 3$.
- The limit of $g(x)$ as x approaches 0 from the right is $3 - 0^2 - 0^4 = 3$.
- The actual value of $g(0)$ comes from the left piece, and it is $0^3 + 3 = 3$.

Together, the limit of g as x approaches 0 exists and is equal to 3; the value of g at $x = 0$ is 3, and $\lim_{x\to 0} g(x) = g(0)$. The function is continuous everwhere on the interval. And since, now, all the conditions of the IVT are met, we can conclude that YES, g must take on the value $g(x) = 2$ somewhere on the given interval. ■

---

(9) Use the IVT to prove that the equation $ln(x) = -x/2$ must have a solution on the interval $[1/2, 2]$.

□ We translate this to the equivalent, "prove that the function $\ln(x) + x/2$ has a root somewhere in the interval $[1/2, 2]$. Then, we can apply the IVT.

- The function $\ln(x) + x/2$ is continuous for all $x > 0$, and so specifically on the given interval.
- At the first endpoint, $f(1/2) = \ln(1/2) + 1/4 \approx -0.44$
- At the second endpoint, $f(2) = \ln 2 - 1 \approx 1.7$
- The target value of $W = 0$ is in between $f(1/2) \approx -0.44$ and $f(2) \approx 1.7$.

Since all the conditions of the IVT are met, then we can conclude that YES, $\ln(x) + x/2$ must take on the value 0 somewhere between $x = 1/2$ and $x = 2$, therefore $\ln(x) = -x/2$ must have a solution on the interval $[1/2, 2]$. ■

---

(10) If you remove marbles from a bag one at a time, must there always come a time when the bag contains exactly half the number of marbles it began with? You might have an intuitive answer to this, but be sure to put your explanation in the context of the IVT.

□ If we had a function that gave us the number of marbles in the bag with respect to time, say $M(t)$, then this function would not be continuous, as its range would consist of only integers. Therefore, the IVT does not apply, and we can't draw any conclusions.

(Of course, if you zoomed in really close and watched marbles exit the top of the bag, so that you could observe marbles as they left the bag continously, then the answer would be different ... but that's an extreme case.) ■

2.5 Secant Lines and Tangent Lines

Introduction

We've spent a lot of time learning to compute limits. Now, we'll look at how to begin a sequence of calculations which at first don't look related to limits, but will quickly having you realize, "Hey, I'm doing a limit!"

Secant Lines and Tangent Lines

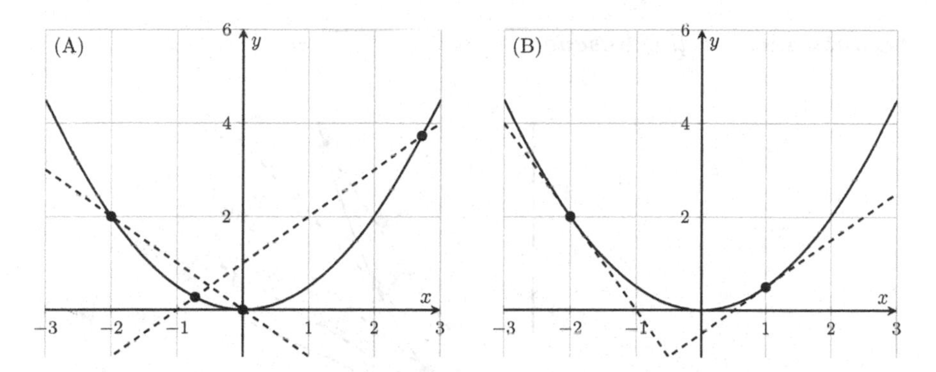

Fig. 2.18 Secant lines (A) and tangent lines (B) on $y = x^2/2$.

Secant lines and tangent lines are lines that are defined in conjunction with a graph of a function $f(x)$. Given such a graph, a *secant line* of $f(x)$ is a line that joins two points on the graph of $f(x)$. A *tangent line* of $f(x)$ is a line that just brushes the graph of $f(x)$ at one point, without actually crossing $f(x)$. Figure 2.18 shows the graph of $f(x) = x^2/2$ along with (A) a pair of secant lines, and (B) a pair of tangent lines. Note that the secant lines each plow through the function at two points, while the tangent lines skim the function at one point.

Believe it or not, many physical or "real world" problems require finding information about secant lines or tangent lines. More often than not, we need to work with tangent lines. For example, if $f(x)$ represents the position of a moving object, the speed of the object at any point is the slope of the tangent line there.

Consider the situation where we are asked to find the equation of one of the secant lines in Fig. 2.18. As long as we know the two points where the secant line hits the function, that's easy: with those two points, we can calculate the slope of the line, and then use the point-slope formula to generate the equation of the secant line. But then consider the situation where we're asked to find the equation of one of the tangent lines in the figure. Even if we know the coordinates of the point where the tangent line touches the function, wouldn't we be stuck? How can we find the slope of a line if we only know one point on the line? That would be magic.

Secant Lines To the Rescue!

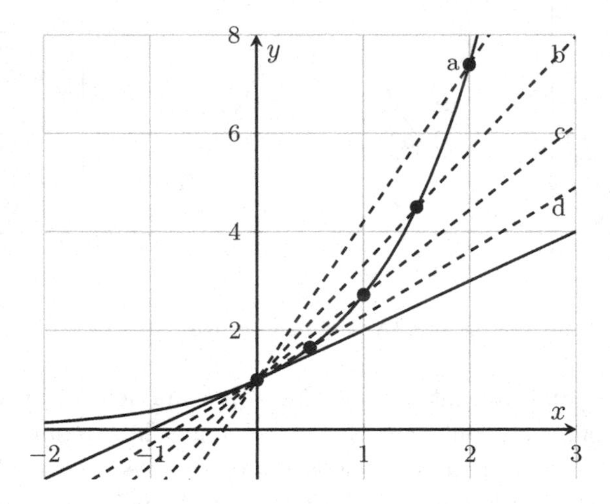

Fig. 2.19 Four secant lines and a tangent line.

Suppose we want the slope of the line tangent to $f(x)$ at some point $(a, f(a))$. The trick to that lies with secant lines — in fact, multiple secant lines originating at the same point. In Fig. 2.19, we can see four secant lines (dashed) on the graph of some function $f(x)$; each of these secant lines connect the point $(0, 1)$ to other points a bit farther away. We can also see the line tangent to $f(x)$ at the point $(0, 1)$. Note that it would be "easy" to compute the slopes of the four secant lines, as long as we knew the second point each one uses. It's currently not possible to find the slope of the tangent line, though, since the only information we have about it is the one point $(0, 1)$. But suppose we found and wrote down the slopes of the four secant lines in order: a, b, c, then d. Then, do you think the trend of those values of the secant lines would point us to what the slope of the tangent line ought to be?

I wonder if we know a word for a process where we compute several values, then use the trend of those values to predict some final value? Hmmm.

It turns out that the function in Fig. 2.19 is e^x. The following table shows, for each of the four secant lines, the coordinates of the second point besides $(0, 1)$, and the slope of the secant line connecting $(0, 1)$ and the second point. The slopes of these lines are laid out in reverse order, d, c, b, a, so we can watch the approach of that second point back to $(0, 1)$:

x :	2.0	1.5	1.0	0.5	0
y :	7.4	4.5	2.72	1.65	1
m :	3.2	2.3	1.72	1.3	?

Can we use the sequence of slopes $3.2, 2.3, 1.72, 1.3$ to predict what slope would fill in the next spot? If we can predict that next spot, then we've found the slope of the tangent line at $(0, 1)$. In this particular case, the trend is not entirely obvious. There are lots of values which could be good guesses for the missing spot. Maybe 1.1? 0.9? 1.15? None of them are crazy guesses. The actual value we're looking for is 1, exactly. That's the slope of the tangent line at $(0, 1)$. So while this picture and chart are helpful for illustrating the process we're going to follow, we need to select more strategically for the secant lines we use in the process. Rather than using, here, x coordinates of $2, 1.5, 1, 0.5$ for our second points on the secant lines, what if we used $1.0, 0.5, 0.2, 0.1, 0.01, 0.001$? That would give a *much* better basis for estimating the slope of the secant line. Even better, maybe we could approach $x = 0$ from both sides!

EX 1 Use a progression of secant lines and their slopes to estimate the slope of the line tangent to $y = 2 - x^2$ at the point $(1, 1)$. Then find the equation of that tangent line.

We'll connect $(1, 1)$ to points on $f(x)$ that have x-coordinates 0.5, 0.9, 0.99, 0.999 on the left of $x = 1$, and 1.5, 1.1, 1.01, and 1.001 on the right of $x = 1$. Then we can examine the list of slopes of the secant lines and see what value they're approaching. There are lots of small computations here that will be summarized in a table. For any point (x, y) on $f(x) = 2 - x^2$, the slope of the secant line connecting $(1, 1)$ to that (x, y) will be $m = (y-1)/(x-1)$. So here is a table summarizing the y-values corresponding to each x-value listed before, and the slopes of all the secant lines connecting those points (x, y) to the point $(1, 1)$. For example, when $x = 0.5$, then $y = f(0.5) = 1.75$, and the slope between the point $(0.5, 1.75)$ and $(1, 1)$ is $(1.75 - 1)/(0.5 - 1) = -1.5$. The rest of the data follows as:

x:	0.5	0.9	0.99	0.999	1	1.001	1.01	1.1	1.5
y:	1.75	1.19	1.02	1.2	1	0.998	0.9799	0.79	−0.25
m:	−1.5	−1.9	−1.99	−1.999	?	−2.001	−2.01	−2.1	−2.5

The spot held by the "?" is the spot where the slope of the tangent line should go. And it's pretty clear from the table that as the x coordinate of our second point gets closer to 1, the value of the slope of the secant line connecting that point to $(1, 1)$ is getting closer to -2. This guess is the same from both sides. So we identify the slope of the tangent line at $(1, 1)$ as -2, and the equation of that line is then $y - 1 = -2(x - 1)$. ■

You Try It

(1) Use a progression of secant lines and their slopes to estimate the slope of the line tangent to $y = \ln(x)$ at the point $(2, \ln 2)$. Then find the equation of that tangent line.

Limits to the Rescue!

We can go through the above procedure every time we need the slope of a tangent line, but it is quite cumbersome. Perhaps there's a way to speed it up a bit. The secret to this speed-up is in the process we just did. You may have noticed that the discussion at the end of Example 1 above says, essentially, "as x gets closer to 1, the slope gets closer to ...." Did you realize that we were doing a limit? Instead of computing specific slopes at several points which lead to the slope of the tangent line, perhaps we can look more generally at this whole situation.

Here are the ingredients for the process:

- We have a function $f(x)$.
- We want a tangent line at the point given by $x = a$, i.e. the point $(a, f(a))$.
- We create several secant lines; one point on each secant line is $(a, f(a))$, the second is at some other point $(x, f(x))$.

- The slope of each secant line, between $(a, f(a))$ and $(x, f(x))$, is then $(f(x) - f(a))/(x - a)$.

So, what we're really doing is asking "as x approaches a, what is the slope approaching?" When we answer that question, we call the result the slope of the tangent line, m_{tan}. In limit notation, we have

$$m_{\text{tan}} = \lim_{x \to a} \frac{f(x) - f(a)}{x - a}$$

And, instead of computing a lot of individual values like we did above, we can build and evaluate that single limit. The only information we need to put in there is $f(x)$, the function, and a, the target value of the function.

Now, that's the good news. The bad news is that we want to adjust the "look" of the limit to set it up in a way that will be more useful later. But it's a pretty simple change. Note that as x approaches a, the *distance* between x and a goes to zero. Let's name that distance h. Then $h = x - a$ and $x = h + a$. And instead of having $x \to a$, we can have $h \to 0$. When we re-do the limit to reflect these changes, we get

$$m_{\text{tan}} = \lim_{h \to 0} \frac{f(a + h) - f(a)}{h}$$

This is more typical of the limit form used to solve these problems.

EX 2 Use limits to find the slope of the tangent line to $f(x) = 2 - x^2$ at (1,1), then find the equation of that line.

Note we're finding the same slope as in Example 1, just in a different way. We want to compute

$$m_{\text{tan}} = \lim_{h \to 0} \frac{f(a + h) - f(a)}{h}$$

To do so, we must build the individual pieces:

- $a = 1$
- $f(a) = f(1) = 2 - (1)^2 = 1$
- $a + h = 1 + h$
- $f(a + h) = 2 - (1 + h)^2 = 2 - (1 + 2h + h^2) = 1 - 2h - h^2$

Putting this all together,

$$\begin{aligned} m_{\text{tan}} &= \lim_{h \to 0} \frac{f(a + h) - f(a)}{h} = \lim_{h \to 0} \frac{(1 - 2h - h^2) - 1}{h} \\ &= \lim_{h \to 0} \frac{-2h - h^2}{h} = \lim_{h \to 0} -2 - h = -2 \end{aligned}$$

So the slope of the tangent line to $f(x) = 2 - x^2$ at $(1, 1)$ is -2, and the equation of that line is then $y - 1 = -2(x - 1)$, the same as in Example 1.

■

There is on important side issue to note in Example 2. When computing $f(a + h)$, it is important to remember that $f(a + h)$ is NOT the same as $f(a) + h$. We spent a lot of time reviewing functional notation precisely for this purpose. Do not confuse $f(a + h)$ with $f(a) + h$, they are not always the same thing! It is a common mistake for students to write something like $f(2 + h)$ as $f(2) + h$. Don't do that.

You Try It

(2) Use limits to find the slope of the line tangent to $f(x) = 2(x + 1)^2$ at (0,2), then find the equation of that line.

(3) Use limits to find the slope of the line tangent to $f(x) = \sqrt{2x + 1}$ at (4,3), then find the equation of that line.

Generalizing the Generalized Process

Suppose we had the graph of a function $f(x)$ and needed the slope / equation of tangent lines at several locations instead of just one. We could repeat the above limit process for each point, one at a time. Or, we could recognize something about that limit process that can be generalized even more.

Definition 2.3. *We define the slope of the line tangent to $f(x)$ at $x = a$ with this limit,*

$$m_{\tan} = \lim_{h \to 0} \frac{f(a + h) - f(a)}{h}$$

If we want slopes of tangent lines for several points, we have two options:

(1) Plug in one $x = a$, run the limit, get the slope, and repeat over and over. *Or*

(2) Run the limit *without* putting in any specific value for a. When we're done, we'll have a little formula in terms of a that represents the slope of the tangent line at *any* $x = a$. Then, we can use that simpler formula for each point.

The choice here seems like a no-brainer to me! When given a choice between evaluting several limits, one at a time, or evaluating a *single* limit and then using the result several times, I'll take the latter. Here's how it works, and yes, we'll use the same old function as the example. (Why stop now?)

EX 3 Use limits to find the form of the slope of the tangent line to $f(x) = 2 - x^2$ at an unspecified point $x = a$. Then use that slope form to get the equations of the lines tangent to $f(x)$ at $x = -1$, $x = 0$, and $x = 1/2$.

We're going to do exactly the same thing as in Example 2, except we're not going to plug anything in for a yet. We still need to build

$$m_{\tan} = \lim_{h \to 0} \frac{f(a+h) - f(a)}{h}$$

so let's create the individual pieces with no particular value for $x = a$ in mind:

- $f(a) = 2 - a^2$
- $f(a+h) = 2 - (a+h)^2 = 2 - (a^2 + 2ah + h^2) = 2 - a^2 - 2ah - h^2$

Putting this all together,

$$\begin{aligned} m_{\tan} = \lim_{h \to 0} \frac{f(a+h) - f(a)}{h} &= \lim_{h \to 0} \frac{(2 - a^2 - 2ah - h^2) - (2 - a^2)}{h} \\ &= \lim_{h \to 0} \frac{-2ah - h^2}{h} = \lim_{h \to 0}(-2a - h) = -2a \end{aligned}$$

What this tells us is that the slope of the tangent line to $f(x) = 2-x^2$ at *any* $x = a$ will be equal to $-2a$. So when $x = -1$, we have $m_{\tan} = -2(-1) = 2$. When $x = 0$, we have $m_{\tan} = -2(0) = 0$. When $x = 1/2$, we have $m_{\tan} = -2(1/2) = -1$. The y-coordinates for these points are as follows: when $x = -1, y = 1$; when $x = 0, y = 2$; when $x = 1/2, y = 7/4$. Therefore:

- At $x = -1$ the tangent line is $y - 1 = 2(x + 1)$
- At $x = 0$ the tangent line is $y = 2$
- At $x = 1/2$ the tangent line is $y - 7/4 = -1(x - 1/2)$

Isn't that easier than doing the limit three times? Just to check that our tangent lines are the correct ones, Fig. 2.20 shows a graph of $f(x)$ with those tangent lines. (The horizontal tangent line is thick instead of dashed to distinguish it from the regular grid lines.) ■

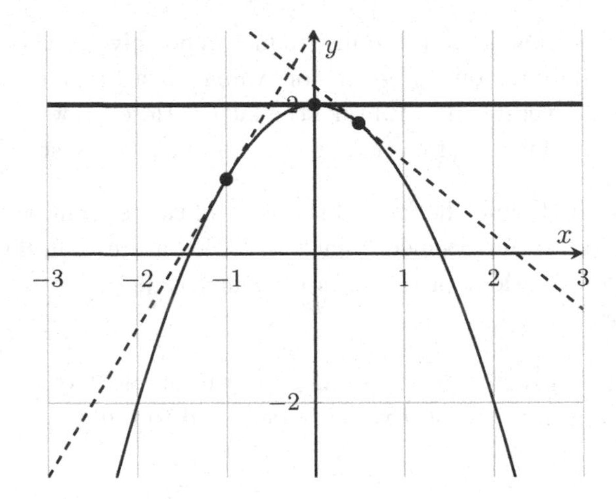

Fig. 2.20 Three tangent lines on $f(x) = 2 - x^2$.

You Try It

(4) Use limits to find the form of the slope of the tangent line to $f(x) = 1+x+x^2$ at an unspecified point $x = a$. Then use that slope form to get the equations of the lines tangent to $f(x)$ at $x = -1$, $x = -1/2$, and $x = 1$.

(5) Use limits to find the form of the slope of the tangent line to $f(x) = \sqrt{x}$ at an unspecified point $x = a$. Then use that slope form to get the equations of the lines tangent to $f(x)$ at $x = 1$ and $x = 4$.

The Difference Quotient

The fraction seen above in two forms,

$$\frac{f(a+h) - f(a)}{h} \quad \text{and} \quad \frac{f(x) - f(a)}{x - a}$$

is called the *difference quotient.* Make sure you understand why both forms of the difference quotient mean the same thing: the slope of a secant line between two points on the graph of $f(x)$. In the first form, the coordinates of these two points are $(a, f(a))$ and $(a+h, f(a+h))$. In the second form, the coordinates of these two points are $(a, f(a))$ and $(x, f(x))$; the link between the two forms is that h represents the distance between the two points, so that $h = x - a$ and so $x = a + h$.

The difference quotient will show up quite a lot now, so you need to recognize it when you see it, and also know how to build it when necessary.

Have You Learned...

- How to form secant lines between two points on the graph of a function?
- How to write the expression describing the slope of a secant line in formal mathematical notation?
- How to form a sequence of secant lines that converge on a tangent line?
- How we can now predict the slope of a line tangent to a function at a given point, even though we only know *one* point on that line?
- How to describe the search for the slope of a tangent line using formal mathematical notation?
- The two common appearances of the difference quotient?

Secant Lines and Tangent Lines — Problem List

Secant Lines and Tangent Lines — You Try It

These appeared above; solutions begin on the next page.

(1) Use a progression of secant lines and their slopes to estimate the slope of the line tangent to $y = \ln(x)$ at the point $(2, \ln 2)$. Then find the equation of that tangent line.
(2) Use limits to find the slope of the tangent line to $f(x) = 2(x+1)^2$ at (0,2), then find the equation of that line.
(3) Use limits to find the slope of the tangent line to $f(x) = \sqrt{2x+1}$ at (4,3), then find the equation of that line.
(4) Use limits to find the form of the slope of the tangent line to $f(x) = 1 + x + x^2$ at an unspecified point $x = a$. Then use that slope form to get the equations of the lines tangent to $f(x)$ at $x = -1$, $x = -1/2$, and $x = 1$.
(5) Use limits to find the form of the slope of the tangent line to $f(x) = \sqrt{x}$ at an unspecified point $x = a$. Then use that slope form to get the equations of the lines tangent to $f(x)$ at $x = 1$ and $x = 4$.

Secant Lines and Tangent Lines — Practice Problems

Try these as you get the hang of the You Try It problems. Solutions to these problems are available in Sec. A.2.5.

(1) Use a progression of secant lines and their slopes to estimate the slope of the line tangent to $y = \dfrac{1}{1+x}$ at the point $(1, 1/2)$. Then find the equation of that tangent line.
(2) Use limits to find the slope of the tangent line to $f(x) = 1 + 2x - x^2$ at (1,2), then find the equation of that line.
(3) Use limits to find the form of the slope of the tangent line to $f(x) = x^3 - 4x + 1$ at an unspecified point $x = a$. Then use that slope form to get the equations of the lines tangent to $f(x)$ at $x = 1$ and $x = 2$.

Secant Lines and Tangent Lines — Challenge Problems

Try these problems to test your skills with the ideas in this section. Solutions to these problems are available in Sec. B.2.5.

(1) Use a progression of secant lines and their slopes to estimate the slope of the line tangent to $y = e^x$ at the point $(1, e)$. Then find the equation of that tangent line.
(2) Use limits to find the slope of the tangent line to $f(x) = 2x/(x+1)^2$ at (0,0), then find the equation of that line.
(3) Use limits to find the form of the slope of the tangent line to $f(x) = 2/(x+3)$ at an unspecified point $x = a$. Then use that slope form to get the equations of the lines tangent to $f(x)$ at $x = -1$, $x = 0$ and $x = 1$. Plot $f(x)$ and these three tangent lines.

Secant Lines & Tangent Lines — You Try It — Solved

(1) Use a progression of secant lines and their slopes to estimate the slope of the line tangent to $y = \ln(x)$ at the point $(2, \ln 2)$. Then find the equation of that tangent line.

☐ We'll form secant lines by connecting $(2, \ln 2)$ to several points to its left and right, and see the trend in slope as we get closer to $x = 2$. For any (x, y), the slope of the line connecting (x, y) to $(2, \ln 2)$ is

$$m = \frac{y - \ln 2}{x - 2}$$

Here is a table of such slope values:

x	1.5	1.9	1.99	1.999	1	2.001	2.01	2.1	2.5
y	0.4054	0.6419	0.6881	0.6926	$\ln 2$	0.6936	0.6981	0.7419	0.9163
m	0.5754	0.5129	0.5013	0.5001	?	0.4999	0.4988	0.4879	0.4463

So it looks like the slope of the line tangent to the curve at $(2, \ln 2)$ should be 0.5. The equation of the tangent line is then:

$$y - \ln 2 = \frac{1}{2}(x - 2) \quad ■$$

(2) Use limits to find the slope of the tangent line to $f(x) = 2(x+1)^2$ at (0,2), then find the equation of that line.

☐ We're looking for the tangent line at $x = a = 0$, so the slope of the tangent line there is:

$$\begin{aligned} m_{tan} &= \lim_{h \to 0} \frac{f(a+h) - f(a)}{h} = \lim_{h \to 0} \frac{f(0+h) - f(0)}{h} \\ &= \lim_{h \to 0} \frac{2(0+h+1)^2 - 2(0+1)^2}{h} = \lim_{h \to 0} \frac{2(h+1)^2 - 2}{h} \\ &= \lim_{h \to 0} \frac{2(h^2 + 2h + 1) - 2}{h} = \lim_{h \to 0} \frac{2h^2 + 2h}{h} \\ &= \lim_{h \to 0} (2h + 2) = 2 \end{aligned}$$

So the equation of the tangent line is $y - 2 = 2(x - 0)$, or $y = 2x + 2$.

■

(3) Use limits to find the slope of the tangent line to $f(x) = \sqrt{2x+1}$ at (4,3), then find the equation of that line.

☐ We're looking for the tangent line at $x = a = 4$, so the slope of the tangent line there is:

$$\begin{aligned} m_{tan} &= \lim_{h\to 0} \frac{f(a+h)-f(a)}{h} = \lim_{h\to 0} \frac{f(4+h)-f(4)}{h} \\ &= \lim_{h\to 0} \frac{\sqrt{2(4+h)+1}-\sqrt{2(4)+1}}{h} = \lim_{h\to 0} \frac{\sqrt{9+2h}-3}{h} \\ &= \lim_{h\to 0} \frac{\sqrt{9+2h}-3}{h} \cdot \frac{\sqrt{9+2h}+3}{\sqrt{9+2h}+3} = \lim_{h\to 0} \frac{9+2h-9}{h(\sqrt{9+2h}+3)} \\ &= \lim_{h\to 0} \frac{2}{\sqrt{9+2h}+3} = \frac{2}{3+3} = \frac{1}{3} \end{aligned}$$

So the equation of the tangent line is $y - 3 = (1/3)(x - 4)$. ■

(4) Use limits to find the form of the slope of the tangent line to $f(x) = 1 + x + x^2$ at an unspecified point $x = a$. Then use that slope form to get the equations of the lines tangent to $f(x)$ at $x = -1$, $x = -1/2$, and $x = 1$.

☐ The slope of the tangent line to $f(x) = 1 + x + x^2$ at $x = a$ comes from:

$$\begin{aligned} m_{tan} &= \lim_{h\to 0} \frac{f(a+h)-f(a)}{h} \\ &= \lim_{h\to 0} \frac{[1+(a+h)+(a+h)^2]-[1+a+a^2]}{h} \\ &= \lim_{h\to 0} \frac{[1+a+h+a^2+2ah+h^2]-[1+a+a^2]}{h} \\ &= \lim_{h\to 0} \frac{h+2ah+h^2}{h} \\ &= \lim_{h\to 0} 1+2a+h = 1+2a \end{aligned}$$

The slopes of the tangent lines at $x = -1$, $x = -1/2$ and $x = 1$ are

$$1 + 2(-1) = -1 \text{ and } 1 + 2(-1/2) = 0 \text{ and } 1 + 2(1) = 3$$

The y-coordinates associated with these x-values are

$$1+(-1)+(-1)^2 = 1 \text{ and } 1+(-1/2)+(1/2)^2 = \frac{3}{4} \text{ and } 1+1+(1)^2 = 3$$

So the tangent lines themselves are

$$y - 1 = -1(x + 1) \text{ and } y - \frac{3}{4} = 0\left(x + \frac{1}{2}\right) \text{ and } y - 3 = 3(x - 1) \quad \blacksquare$$

---

(5) Use limits to find the form of the slope of the tangent line to $f(x) = \sqrt{x}$ at an unspecified point $x = a$. Then use that slope form to get the equations of the lines tangent to $f(x)$ at $x = 1$ and $x = 4$.

□ The slope of the tangent line to $f(x) = \sqrt{x}$ at $x = a$ comes from:

$$\begin{aligned} m_{tan} &= \lim_{h \to 0} \frac{f(a+h) - f(a)}{h} = \lim_{h \to 0} \frac{\sqrt{a+h} - \sqrt{a}}{h} \\ &= \lim_{h \to 0} \frac{\sqrt{a+h} - \sqrt{a}}{h} \cdot \frac{\sqrt{a+h} + \sqrt{a}}{\sqrt{a+h} + \sqrt{a}} = \lim_{h \to 0} \frac{(a+h) - a}{h(\sqrt{a+h} + \sqrt{a})} \\ &= \lim_{h \to 0} \frac{h}{h(\sqrt{a+h} + \sqrt{a})} = \lim_{h \to 0} \frac{1}{\sqrt{a+h} + \sqrt{a}} = \frac{1}{2\sqrt{a}} \end{aligned}$$

So at $x = 1$ we have $m_{\text{tan}} = 1/(2\sqrt{1}) = 1/2$. At $x = 4$ we have $m_{\text{tan}} = 1/(2\sqrt{4}) = 1/4$. For y coordinates, we have: at $x = 1, y = 1$ and at $x = 4, y = 2$. The tangent lines at $(1, 1)$ and $(4, 2)$ are therefore

$$y - 1 = \frac{1}{2}(x - 1) \quad ; \quad y - 2 = \frac{1}{4}(x - 4) \quad \blacksquare$$

The Formal Definitions of Limits

Into the Pit!!

(This entire section is our first visit ... Into the Pit!!)[2]

Introduction

This topical interlude is here only for those of you who want a peek behind the curtain at the more theoretical foundation of math. If you're more about the toolkit aspects of things, then perhaps you can just browse this to see if it piques your interest; if not, no harm done.

With a few sections of problem solving related to the more operational aspect of limits, we can stop and take a look at what technical details really drive limits. Saying things like "this function gets close to this value as its input gets close to that value" is really informal. It's intuitive, but if you had to step into the mathematical court of law and defend it, how would you do it?

Into the Pit! The Actual Definition of a (Finite) Limit

Math can be like chemistry or biology, in that to understand what's really going on, we need to peer through a microscope. Here, it's a numerical microscope rather than a real one. In our numerical microscope, it's traditional to use Greek letters to represent numbers that can get really, really small. So in the following definitions, when you see a δ or an ϵ, just think of them as, "really, really tiny numbers, and in fact, the tinier, the better."

This is the actual formal definition of a limit, in which we quantify all this business about "closeness".

Definition 2.4. $\lim_{x \to a} f(x) = L$ *if, given any* $\epsilon > 0$ *there exists a* $\delta > 0$ *such that* $|f(x) - L| < \epsilon$ *whenever* $0 < |x - a| < \delta$.

That's as clear as mud, right? Let's break it down, item by item.

[2] Read the Preface if you have no idea what that is.

- First, remember that ϵ and δ are numbers that are intended to get tiny ... really, really tiny. We're talking about values like 0.5, 0.1, 0.001, 1×10^{-6}, and so on.
- The expression $|f(x) - L| < \epsilon$ measures the distance between the function $f(x)$ and its supposed limit value L. This is how we quantify $f(x)$ getting "close" to L, because the expression $|f(x) - L| < \epsilon$ means that values of $f(x)$ are falling in the window $L - \epsilon < f(x) < L + \epsilon$. When ϵ is tiny, this is a really small window around L; that is, $f(x)$ is "close" to L.
- Similarly, the expression $0 < |x - a| < \delta$ defines the window $a - \delta < x < a + \delta$. When δ is tiny, this quantifies x getting "close" to a. And, the exclusion of a itself means that we don't really care what happens if x is equal to a, we only care about when x is near a.

A visual representation of this definition is shown in Fig. 2.21, which displays $f(x) = \sqrt{x}$ around $x = 4$. We know $\lim_{x \to 4} \sqrt{x} = 2$. The figure illustrates a value of x selected from within a distance of δ from 4 (on the x-axis) will prodice a value of $f(x)$ within a distance of ϵ from 2 (on the y-axis).

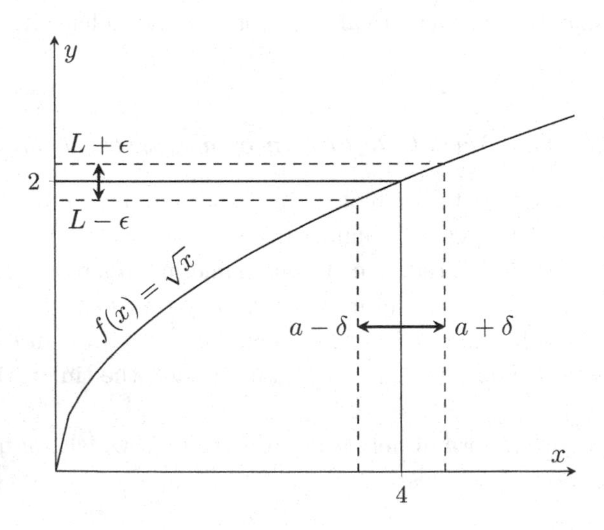

Fig. 2.21 The $\epsilon - \delta$ frame around a limit value.

Definition 2.4 essentially sets up a challenge. We say, "Hey, we think that the limit of $f(x)$ as x approaches a is L!" A skeptical mathematical judge then says, "Oh, really? Can you guarantee that $f(x)$ lands within a distance of 0.1 from L?" And we do a bit of algebra, and reply, "No problem. We can show that as long as x is within a distance of 0.02 from a, then $f(x)$ is guaranteed to be within 0.1 of L!" The judge scowls, but retorts, "Well, smarty pants, now get $f(x)$ within 0.005 of L!" We return to our algebra, and respond, "We can demonstrate that as long as x is within a distance of 0.006 of a, then $f(x)$ will land within 0.005 of L!"

And then, because we're tired of this judge, we say, "Look. We can actually develop an algebraic relationship that gives δ in terms of ϵ. No matter what ϵ you give us, we can immediately provide the corresponding value of δ that guarantees we'll see $|f(x)-L| < \epsilon$ whenever $0 < |x-a| < \delta$." At this point, the skeptical mathematical judge knows that we have him beat, so he tosses his gavel in the air and leaves. We've defeated him by mathematical proof!

And this is the game. Given $f(x)$, a, and a supposed limit L, we should be able to tie δ and ϵ together via Def. 2.4 so that we can always find the right window around a from which we pick x to guarantee that $f(x)$ lands within the chosen window around L. And then, we can demonstrate that as ϵ shrinks, so does *delta* — meaning, that $f(x)$ "gets closer" to L as x "gets closer" to a. Here's a simple example of how this works.

EX 1 Prove that the limit of $f(x) = 2x$ as x approaches 3 is 6. That is, prove $\lim_{x \to 3} x = 3$.

The limit is super obvious, but "super obvious" doesn't fly in the mathematical court of law. So, let's see what $|f(x)-L|$ looks like in this scenario:

$$|f(x) - L| = |2x - 6| = 2 \cdot |x - 3|$$

Now let's see what $|x - a|$ looks like:

$$|x - a| = |x - 3|$$

Then the two phrases in the definition are:

$$|f(x) - L| < \epsilon \quad \text{becomes} \quad 2 \cdot |x - 3| < \epsilon \quad \text{or} \quad |x - 3| < \frac{\epsilon}{2}$$

and

$$0 < |x - a| < \delta \quad \text{becomes} \quad 0 < |x - 3| < \delta$$

In other words, we're trying to show that

$$|x - 3| < \frac{\epsilon}{2} \quad \text{whenever} \quad 0 < |x - 3| < \delta$$

Or, as long as we choose $\delta = \epsilon/2$, then we can demonstrate $|f(x) - L| < \epsilon$ when $0 < |x - a| < \delta$. For example, if we want $f(x)$ to be within 0.04 of L, then we have to be sure we pick x to be within 0.02 of a. Beyond the individual values, though is the relationship: since $\delta = \epsilon/2$, then as ϵ gets smaller and smaller, so does δ — thus, we have quantitative measurement of how $f(x)$ gets "closer" to L as x gets "closer" to a. Our concluding statement is, "Given $\epsilon > 0$, then choosing $\delta = \epsilon/2$ will guarantee $|f(x) - L| < \epsilon$ whenever $0 < |x - a| < \delta$. We rest our case." ■

EX 2 Prove $\lim\limits_{x \to 1}(3x + 5) = 8$.

Again, we are crossing over from "that's obvious" to "oh yeah? Prove it!" To apply Def. 2.4, we have to see how two expressions are related. We start by building $|f(x) - L| < \epsilon$, which in this case is,

$$|(3x + 5) - 8| < \epsilon$$

Now, we manipulate this into the form $|x - 1| < \delta$:

$$|3x - 3| < \epsilon \quad \to \quad |x - 1| < \frac{\epsilon}{3}$$

By comparing $|x - 1| < \epsilon/3$ and $0 < |x - 1| < \delta$, we can see that given any $\epsilon > 0$, if we choose $\delta = \epsilon/3$, we will guarantee that $|f(x) - L| < \epsilon$ whenever $0 < |x - a| < \delta$. By Def. 2.4, this proves $\lim\limits_{x \to 1}(3x + 5) = 8$. FFT: Are we allowed to just blithely update $|x - 1| < \delta$ to $0 < |x - 1| < \delta$? Why can we just exclude the instant when $x = 1$ like that? ■

You Try It

(1) Determine $\lim\limits_{x \to -2}(x - 4)$ and prove your value is correct.

(2) Determine $\lim\limits_{x \to 127}(17x + 44)$ and prove your value is correct.

At this point, we've worked on the formal definition of a finite limit for linear functions only. Unfortunately, we'll have to leave it at that, because using Def. 2.4 for non-linear functions can get complicated. For example, when $f(x)$ is not linear, the δ we find to the left of $x = a$ may be different than the δ to the right of $x = a$. Formally proving a limit as simple as $\lim x \to 2x^2 = 4$ requires the invocation of a rule called the triangle inequality. But you can invent and prove limits for linear functions all day long!

There are other versions of Def. 2.4 that apply to limits with ∞ in them, either as a target for x or as the limit value. In these, some of our target values will be large instead of small; we'll represent large values by capital letters. The first definition is for the case where we have a vertical asymptote at our limit point; we want to formally quantify how $f(x)$ gets large as x gets closer to the location of the asymptote on the x-axis.

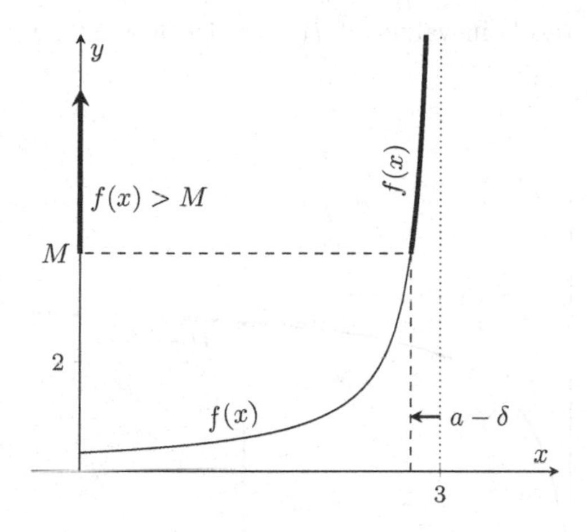

Fig. 2.22 The $M - \delta$ frame for a limit at a vertical asymptote.

Definition 2.5. $\lim_{x \to a} f(x) = \infty$ *if, given any* $M > 0$, *there exists a* $\delta > 0$ *such that* $f(x) > M$ *whenever* $0 < |x - a| < \delta$.

In this definition, we're asserting that we can make $f(x)$ as large as we want to by picking x within a suitably small window around a. This is illustrated in Fig. 2.22, which shows the $M - \delta$ frame for the left-sided limit $\lim_{x \to 3^-} \dfrac{1}{3-x} = +\infty$. The dashed line shows the threshold described in Def. 2.5. Selecting values of x which are closer than δ to a result in values of $f(x)$ that are larger than M.

The second definition is for the case where we have a horizontal asymptote; we want to formally quantify how the values of $f(x)$ will get closer to the value of the asymptote on the y-axis as x gets larger and larger on the x-axis.

Definition 2.6. $\lim_{x\to\infty} f(x) = L$ *if, given any* $\epsilon > 0$*, there exists a number* N *such that* $|f(x) - L| < \epsilon$ *whenever* $x > N$.

In this definition, we're asserting that we can make $f(x)$ as close to a limit value of L as we want to by picking large enough values of x This is illustrated in Fig. 2.23, which shows the $\epsilon - N$ frame for the limit $\lim_{x\to+\infty} \frac{2x}{x+1} = 2$. The dashed line shows the threshold where values of x larger than N result in values of $f(x)$ which are within a distance of ϵ from L.

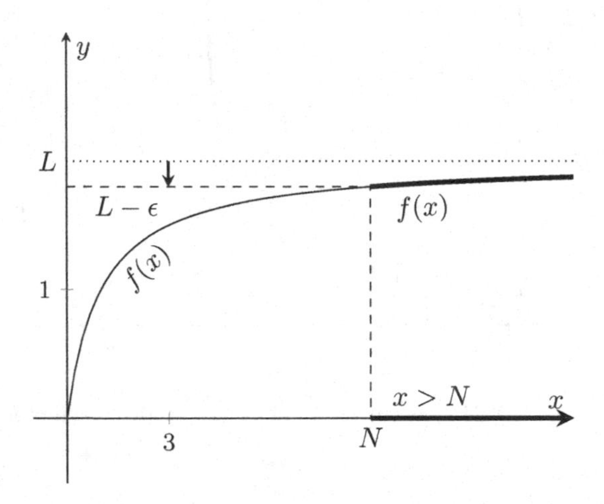

Fig. 2.23 The $\epsilon - N$ frame for a limit at a horizontal asymptote.

We are not going to pursue any full proofs of limits using these definitions, but we can at least watch them operate on a value-by-value basis with algebraic solutions or numerical experimentation.

EX 3 Apply Def. 2.5 to $\lim_{x\to 2} \frac{1}{(x-2)^2}$ by finding a value of δ which corresponds to $M = 500$.

There is a vertical asymptote at $x = 2$ for $f(x) = 1/(x-2)^2$, and so $\lim_{x\to 2} \frac{1}{(x-2)^2} = +\infty$. Our job is to see how close x must get to 2 to ensure we have $f(x) > 500$. We could determine δ by solving $1/(x-2)^2 = 500$ to get $x = 2 \pm 1/(10\sqrt{5})$, i.e. $x = 1.9553$ and $x = 2.0447$. So certainly anything within $\delta = 0.04$ of $x = 2$ is guaranteed to result in $f(x) > 500$.

It's also very satisfying to watch the values of $f(x)$ grow as x gets closer to 2 if these results are laid out in a spreadsheet. Figure 2.24 shows this scenario; the results in bold show where being closer than 0.04 to $x = 2$ results in $f(x) > 500$. Be sure to understand that our results are not a *proof* per Def. 2.5; rather, they are just suggestive numerical evidence. ■

x	200	300	350	400	425	440	**445**	**450**	**500**	**700**	**1000**
f(x)	2.056122	2.037162	2.031792	2.027778	2.026128	2.025229	**2.024943**	**2.024664**	**2.022177**	**2.015805**	**2.011044**

Fig. 2.24 Demonstration of limit behavior near a vertical asymptote, with EX 3.

x	200	300	350	400	425	440	**445**	**450**	**500**	**700**	**1000**
f(x)	2.056122	2.037162	2.031792	2.027778	2.026128	2.025229	**2.024943**	**2.024664**	**2.022177**	**2.015805**	**2.011044**

Fig. 2.25 Demonstration of limit behavior near a vertical asymptote, with EX 4.

EX 4 Apply Def. 2.6 to $\displaystyle\lim_{x\to\infty} \frac{2x+3}{x-4}$ by finding a value of N which corresponds to $\epsilon = 0.025$.

There is a horizontal asymptote at $x = 2$ for $f(x) = (2x+3)/(x-4)$. The graph of $f(x)$ is above the line $y = 2$ for $x > 4$, so we know that $f(x)$ approaches 2 from above. We can spot our magic value of N, then, by solving $(2x+3)/(x-4) = 2.025$ (we would solve $f(x) = 1.975$ if $f(x)$ approached 2 from below). The solution to this equation is $x = 444$. Just to keep things clean, let's round up and say that $f(x)$ will be within 0.025 of $L = 2$ as long as use any x larger than 450. Again, for intuitive purposes, Fig. 2.25 shows this scenario laid out numerically; the results in bold show where choosing values of x that are 450 or larger results in $f(x)$ being within 0.025 of the limit of $L = 2$. ■

You Try It

(3) Illustrate the limit $\displaystyle\lim_{x\to 5} \frac{3}{(x^2-25)^2}$ by finding a value of δ which corresponds to $M = 100$.

(4) Illustrate the limit $\displaystyle\lim_{x\to\infty} (4 - e^{-x})$ by finding a value of N which corresponds to $\epsilon = 10^{-6}$.

Other than the four You Try It problems posed above (solutions follow), there are no exercises in this Interlude. If these sorts of problems intriguing, then you should eventually take an Advanced Calculus or Real Anlysis course where you'll see plenty of similar (and more complicated) problems. These are the courses in which you find out that everything you learn in a regular introductory Calculus sequence is a lie. Okay, that's an exaggeration; it's where you learn that what you see in your first pass through Calculus is not the whole story.

Formal Definition of Limits — You Try It — Solved

(1) Determine $\lim_{x\to-2}(x-4)$ and prove your value is correct.

□ The value of the limit should be $\lim_{x\to-2}(x-4)=-6$. To prove this, we must show — by Def. 2.4 — that given $\epsilon>0$, there is a $\delta>0$ such that $|f(x)-(-6)|<\epsilon$ when $0<|x-(-2)|<\delta$. But when we resolve the individual expressions there, the connection is simple:

- $|f(x)-(-6)|<\epsilon$ becomes $|(x-4)+6|<\epsilon$, or $|x+2|<\epsilon$.
- $0<|x-(-2)|<\delta$ becomes $0<|x+2|<\delta$.

So the phrase, "given $\epsilon>0$, there is a $\delta>0$ such that $|f(x)-(-6)|<\epsilon$ when $0<|x-(-2)|<\delta$" becomes, "given $\epsilon>0$, there is a $\delta>0$ such that $|x+2|<\epsilon$ when $0<|x+2|<\delta$". So, as long as we choose $\delta=\epsilon$, the defining requirement is met: $|f(x)-L|<\epsilon$ when $0<|x-a|<\delta$.

As a numerical illustration, if the challenge is, "guarantee that we get $f(x)$ within 0.05 of the limit -6", then we simply choose $\delta=0.05$ and say, "fine, just pick any x value within 0.05 of $x=-2$ and then $f(x)$ will be guaranteed to be within 0.05 of $L=-6$. Further, and more importantly, ϵ shrinks, so does δ — this formalizes the idea that $f(x)$ gets closer to L as x gets closer to a. ■

(2) Determine $\lim_{x\to127}(17x+44)$ and prove your value is correct.

□ The value of the limit should be $\lim_{x\to127}(17x+44)=2203$. To prove this, we must show — by Def. 2.4 — that given $\epsilon>0$, there is a $\delta>0$ such that $|f(x)-2203|<\epsilon$ when $0<|x-127|<\delta$. The expression $|f(x)-127|<\epsilon$ becomes $|17x-2159|<\epsilon$, or $|17x-2159|<\epsilon/17$. So the defining statement,

- Given $\epsilon>0$ there is a $\delta>0$ such that $|f(x)-L|<\epsilon$ whenever $0<|x-a|<\delta$

becomes

- Given $\epsilon>0$ there is a $\delta>0$ such that $|x-127|<\epsilon/17$ whenever $0<|x-127|<\delta$

And so the connection between δ and ϵ is established. Given $\epsilon>0$, if we choose $\delta=\epsilon/17$, then $|f(x)-L|<\epsilon$ whenever $0<|x-a|<\delta$; the limit is now confirmed. On a numerical basis, the challenge, "guarantee that $f(x)$ will be within 0.034 of the limit 2203", is met with, "we'll choose $\delta = \dfrac{0.034}{17} = 0.02$ and then any x value within 0.02 of $x = 127$ will force $f(x)$ to be within 0.034 of $L = 2203$." ■

x	4.5	4.6	4.7	4.8	**4.9**	**4.95**	**4.97**	**4.98**	**4.99**	**5**	**5.01**	**5.02**	**5.03**	**5.05**	**5.1**	5.2	5.3
f(x)	12.0	18.8	33.3	75.0	**300.0**	**1200.0**	**3333.3**	**7500.0**	**30000.0**	**---**	**30000.0**	**7500.0**	**3333.3**	**1200.0**	**300.0**	75.0	33.3

Fig. 2.26 Demonstration of limit behavior near a vertical asymptote, with YTI 3.

(3) Illustrate the limit $\displaystyle\lim_{x\to 5} \frac{3}{(x^2-25)^2}$ by finding a value of δ which corresponds to $M = 100$.

□ There is a vertical asymptote at $x = 5$ on the graph of $f(x)$ and because $f(x)$ is an even function, the left and right hand limits will be the same:

$$\lim_{x\to 5^-} \frac{1}{(x-2)^2} = +\infty \qquad \text{and} \qquad \lim_{x\to 5^+} \frac{1}{(x-2)^2} = +\infty$$

We want to determine how close x must get to 5 (from either side) to ensure we have $f(x) > 100$. We can determine δ by solving $3/(x-5)^2 = 100$ to get $x = 5 \pm \sqrt{3}/10$, i.e. $x = 4.83$ and $x = 5.15$. So certainly anything within $\delta = 0.1$ of $x = 5$ is guaranteed to result in $f(x) > 100$. Although not necessary, Figure 2.26 shows this scenario; the results in bold show where being within 0.1 of $x = 100$ results in $f(x) > 100$. ■

(4) Illustrate the limit $\displaystyle\lim_{x\to\infty} (4 - e^{-x})$ by finding a value of N which corresponds to $\epsilon = 10^{-6}$.

x	8	10	11	12	13	**14**	**15**	**20**	**50**	**100**
f(x)	3.999665	3.999955	3.999983	3.999994	3.999998	**3.999999**	**4.000000**	**4.000000**	**4.000000**	**4.000000**
|f(x) - L|	3.35E-04	4.54E-05	1.67E-05	6.14E-06	2.26E-06	**8.32E-07**	**3.06E-07**	**2.06E-09**	**0.00E+00**	**0.00E+00**

Fig. 2.27 Demonstration of limit behavior near a horizontal asymptote, with YTI 4.

□ There is a horizontal asymptote at $x = 4$ for $f(x) = 4 - e^{-x}$, and we approach $L = 4$ from below. Using $\epsilon = 10^{-6}$, then an estimate for N comes from the solution to $4 - e^{-N} = 4 - 10^{-6}$; the pure solution is $N = 6\ln 10 \approx 13.8$. Let's just round up to $N = 14$. Then, $f(x)$ will be within 10^{-6} of $L = 4$ as long as use any x that's 14 or larger. Again, for illustration, Fig. 2.27 shows this scenario laid out numerically; the results in bold show where choosing $N \geq 14$ ensures $f(x)$ gets within 10^{-6} of the intended limit. ■

Chapter 3

Embrace the Change

3.1 The Rate of Change of a Function

Introduction

At the end of the last chapter, we looked at a geometry problem: how can we find the slope of a secant line on a graph, and (the harder question) how can we find the slope of a line tangent to that graph? Here we introduce more general terminology for these concepts.

The Average Rate of Change

Suppose you left home at 8am for a destination 180 miles away, and you arrived at 11am. You covered 180 miles in 3 hours, and so this means you drove at a speed of 60 miles per hour, right? Well, no. You probably stopped at a red light. And you probably got slowed down behind a truck, and sped up to pass someone else. You probably stopped to use the restroom. So, no, you didn't drive 60 mph at all times. However, the fact remains that you covered 180 miles in 3 hours — and so your *average* speed was 60 mph.

This is an example of something called the *average rate of change* of a function. Given a function $f(x)$ and two x coordinates of interest, say x_1 and x_2, the average rate of change of the function is a measure of how much the function's value changes as the x coordinate changes from x_1 to x_2:

$$\Delta f_{avg} = \frac{f(x_2) - f(x_1)}{x_2 - x_1} \tag{3.1}$$

This average does not concern itself with the behavior of the function in between the endpoints $(x_1, f(x_1))$ and $(x_2, f(x_2))$. In Fig. 3.1, all three curves have the *same* average rate of change between $x = 0$ and $x = 2$; do you know what that common rate of change is?

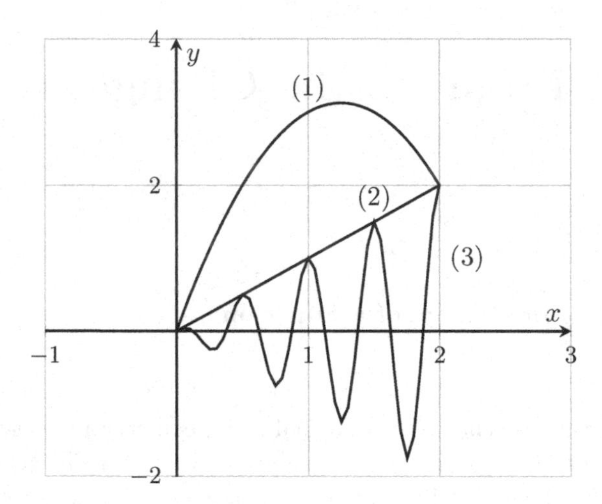

Fig. 3.1 The same average rate of change on $[0, 2]$.

Thinking back to Sec. 2.5, you should recognize that the expression in Eq. (3.1) for the average rate of change of $f(x)$ between x_1 and x_2 is the slope of the secant line through the points $(x_1, f(x_1))$ and $(x_2, f(x_2))$. To be more specific in our language, we should be referring to this average rate of change as the average rate of change of f *with respect to* x. Change in the independent variable x creates change in the function f. Now since the variables in our problems can be all sorts of things, we may not be using x's and f's in all problems, therefore we have to adapt the name of the rate of change accordingly. For example, if we have a function $P(t)$ describing population as a function of time, we would discuss the average rate of change of P with respect to t, and we could even denote it $\Delta P/\Delta t$.

The Instantaneous Rate of Change

The slope of the secant line through the points $(x_1, f(x_1))$ and $(x_2, f(x_2))$ is also known as the average rate of change of the function $f(x)$ between x_1 and x_2. To go with this, the slope of line *tangent* to $f(x)$ at x_1 is also known as the *instantaneous rate of change* of $f(x)$ at x_1.

Revisiting the hypothetical car trip from the start of this section, if we had a graph of your position from home as a function of time over the three hours of your trip, then the slope of the tangent line at any point on the curve would be your speed at that instant, since speed is the rate of change of position.

Using the notation from Sec. 2.5, we have the following definition:

Definition 3.1. *The instantaneous rate of change of a function $f(x)$ at a point $x = a$ is given by*

$$\boxed{\left(\frac{\Delta f}{\Delta x}\right)_{inst} = \lim_{x\to a} \frac{f(x) - f(a)}{x - a}}$$

An equivalent expression uses the parameter h to represent $x - a$,

$$\boxed{\left(\frac{\Delta f}{\Delta x}\right)_{inst} = \lim_{h\to 0} \frac{f(a+h) - f(a)}{h}} \tag{3.2}$$

What we're seeing is that the calculation we already know how to do (finding slopes of secant and tangent lines) are more generally useful in a broader context. You can probably imagine that finding average and/or instantaneous rates of change of functions is more important than the more restrictive sounding "finding slopes of secant lines". The rate of change of position is speed or velocity. The rate of change of the principal in your bank account is related to your interest rate. The rate of change of fluid pressure at a point governs the speed of the movement of fluid past that point. We're in a position to open up tools that are useful for many applications. To start with, we'll just practice the routines we already know using this new language.

Examples

First, let's see how we can compute / estimate average and instantaneous rates of change from tabulated data, for which an algebraic function is not available.

EX 1 The distance D (in miles) from its home nest of the migrating Yellow Headed Duck Duck Goose is given per day in the following table:

t	1	2	3	4	5
D	7	12	19	29	35

(a) Use the data in the table to estimate its average speed (in miles per day) from days 1 to 3, and 3 to 5. (b) Estimate the instantaneous speed on day 3.

(a) Using the data in the table, the average speed is

(i) from day 1 to 3: $\Delta D/\Delta t = (19 - 7)/(3 - 1) = 11/2$ miles per day
(ii) from day 3 to day 5: $\Delta D/\Delta t = (35 - 19)/(5 - 3) = 8$ miles per day

(b) We can only use the data we're given, so an estimate of the instantaneous speed will be pretty lame. The best we can do is borrow information from the data points closest to $t = 3$, in this way: find the average change days 2 to 3, and from 3 to 4, and then average them. The average rate of change from day 2 to 3 is 7 miles per day, and from days 3 to 4, it's 10 miles per day. So the average of 8.5 miles per day is the best estimate of an instantaneous rate of change we have. ■

You Try It

(1) The number of cell phones in use in the small country of Blabbopia is given in this table (N is in thousands):

t	2013	2014	2015	2016	2017
N	304	572	873	1513	2461

(a) Use the data in the table to estimate the average rate of change in the number of phones from 2015 to 2017, and 2015 to 2016. (b) Estimate the instantaneous rate of change in 2015.

Next, let's see how we can estimate average and instantaneous rates of change when an algebraic function is given.

EX 2 The speed of sound c (m/s) in dry air is given by

$$c = 331.3\sqrt{1 + \frac{T}{273.15}}$$

where T is the temperature in degrees Celsius.[1] (a) Find the average rate of change in the speed of sound over intervals starting at $T = 20$ and ending at $T = 22, 21, 20.5, 20.1$. (b) Use those values to estimate the instantaneous rate of change at $T = 20$.

[1]per Wikipedia "speed of sound" page

(a) At $T = 20$, the speed of sound is $c = 343.2146\, m/s$. The average rate of change $\Delta c/\Delta T$ in speed of sound between between any other point (T, c) and this one is $(c - 343.2146)/(T - 20)$. A collection of c values and resulting average rates of change for the given T values are in this table (calculations are easy and omitted):

T	25	21	20.5	20.1	20
c	344.3834	343.7995	343.5072	343.2732	343.2146
$\Delta c/\Delta t$	.584396	.584892	.585141	.585341	?

(b) The numbers in the table are not changing much, and it's not completely clear what value they are tending to. But our guess of the value that should fill in the "?" represents our estimate of the instantaneous rate of change in speed of sound at $T = 20$. A good guess might be 0.5854. (The true value of the instantaneous rate of change is 0.5853908, which is reasonable given the data in the table; the secret to calculating this true value is coming up soon!) ■

You Try It

(2) A ball is tossed in the air at time $t = 0$. The height (position) of the ball as a function of time is given by $y = 40t - 16t^2$. (a) Find the average velocity of the ball over intervals starting at $t = 2$ and ending at $t = 2.5, 2.1, 2.05, 2.01$. (b) Use those values to estimate the instantaneous velocity of the ball at $t = 2$.

Finally, we'll apply limits to estimate instantaneous rates of change of a given function.

EX 3 A spherical balloon is being inflated. Use limits to determine the instantaneous rate of change of the surface area of the balloon at the instant the radius is 10 cm.

For our problem, let's remember that the surface area of a sphere as a function of radius is $S(r) = 4\pi r^2$. Adapting Eq. (3.2) $m_{\tan}$ to these variables, we have

$$\left(\frac{\Delta S}{\Delta r}\right)_{inst} = \lim_{h \to 0} \frac{S(a+h) - S(a)}{h}$$

We are interested in the rate of change at $r = 10$ cm, so $a = 10$:

$$\left(\frac{\Delta S}{\Delta r}\right)_{inst} = \lim_{h\to 0}\frac{S(10+h)-S(10)}{h} = \lim_{h\to 0}\frac{4\pi(10+h)^2-4\pi(10)^2}{h}\cdots$$

$$= \lim_{h\to 0}\frac{4\pi[100+20h+h^2-100]}{h} = \lim_{h\to 0}\frac{4\pi[20h+h^2]}{h}\cdots$$

$$= \lim_{h\to 0} 4\pi(20+h) = 80\pi$$

The instantaneous rate of change of surface area with respect to radius at $r = 10\,cm$ is *exactly* $80\pi\,cm^2$. ■

You Try It

(3) Repeat part (b) of the previous You Try It problem, but use a limit to estimate the instantaneous velocity of the ball at $t = 2$.

Although the reapprearance and use of limits may be distressing, note that using tabulated data will provide — at best — 0 an estimate of an instantaneous rate of change, whereas the limit process can provide the exact value. So... sorry, not sorry!

Have You Learned...

- What an average rate of change is, and what data you need to find it?
- How average rates of change relate to slopes of secant lines?
- What an instantaneous rate of change is, and what data you need to find it?
- How instantaneous rates of change relate to slopes of tangent lines?
- How to write the slope of a tangent line in two different limit forms?

The Rate of Change of a Function — Problem List

The Rate of Change of a Function — You Try It

These appeared above; solutions begin on the next page.

(1) The number of cell phones in use in the small country of Blabbopia is given in this table (N is in thousands):

t	2013	2014	2015	2016	2017
N	304	572	873	1513	2461

(a) Use the data in the table to estimate the average rate of change in the number of phones from 2015 to 2017, and 2015 to 2016. (b) Estimate the instantaneous rate of change in 2015.

(2) A ball is tossed in the air at time $t = 0$. The height (position) of the ball as a function of time is given by $y = 40t - 16t^2$. (a) Find the average velocity of the ball over intervals starting at $t = 2$ and ending at $t = 2.5, 2.1, 2.05, 2.01$. (b) Use those values to estimate the instantaneous velocity of the ball at $t = 2$.

(3) Repeat part (b) of the previous You Try It problem, but use a limit to estimate the instantaneous velocity of the ball at $t = 2$.

The Rate of Change of a Function — Practice Problems

Try these as you get the hang of the You Try It problems. Solutions to these problems are available in Sec. A.3.1.

(1) The average number of hits per day each year on a popular web page about calculus is given in this table:

t	2012	2014	2016	2018
N	10,036	10,109	10,152	10,175

(a) Use the data in the table to estimate the average rate of change in the number of hits from (i) 2012 to 2016, (ii) 2014 to 2016, and (iii) 2016 to 2018. (b) Estimate the instantaneous rate of change in 2016.

(2) If an arrow is shot upwards on the moon with a velocity of 58 m/s, its height in meters after t seconds is $y = 58t - 0.83t^2$. (a) Find the average velocity over time intervals starting at $t = 1$ and ending at (i) 2, (ii) 1.5, (iii) 1.1, (iv) 1.01, (v) 1.001. (b) Use those values to estimate the instantaneous velocity of the arrow at $t = 1$.

(3) Repeat part (b) of the previous Practice Problem, but use a limit to estimate the instantaneous velocity of the arrow at $t = 1$.

The Rate of Change of a Function — Challenge Problems

Try these problems to test your skills with the ideas in this section. Solutions to these problems are available in Sec. B.3.1.

(1) The volume of a spherical balloon is a function of its radius, $V = 4\pi r^3/3$. Use either a succession of slopes of secant lines or a limit to estimate the instantaneous rate of change of the volume with respect to the radius at $r = 1$.
(2) A standard physics formula says that the height in feet of a projectile launched straight upwards with an initial velocity $v_0\, ft/s$ is given by $y = v_0 t - 16t^2$. Use a limit to find the instantaneous velocity of the object at $t = 3$ seconds. (Your answer will be in terms of v_0.)
(3) Suppose the cost of making x gizmos is given by $C(x) = x^3 - 200x^2$. The *marginal cost* of making, say, the 100th gizmo is the instantaneous rate of change of the cost function at $x = 100$. Use a limit to find this marginal cost.

The Rate of Change of a Function — You Try It — Solved

(1) The number of cell phones in use in the small country of Blabbopia is given in this table (N is in thousands):

t	2013	2014	2015	2016	2017
N	304	572	873	1513	2461

(a) Use the data in the table to estimate the average rate of change in the number of phones from 2015 to 2017, and 2015 to 2016. (b) Estimate the instantaneous rate of change in 2015.

□ (a) Using the data in the table, the average rate of growth of change in the number of phones N is

(i) from 2015 to 2017: $\Delta N/\Delta t = (2461 - 873)/2 = 794$ thousand per year

(ii) from 2015 to 2016: $\Delta N/\Delta t = (1513 - 873)/1 = 640$ thousand per year

(b) We could estimate the instantaneous growth in 2015 by averaging the average growth from 2014 to 2015 and 2015 to 2016, which gives approximately $(301 + 640)/2 = 470.5$ thousand per year. ■

(2) A ball is tossed in the air at time $t = 0$. The height (position) of the ball as a function of time is given by $y = 40t - 16t^2$. (a) Find the average velocity of the ball over intervals starting at $t = 2$ and ending at $t = 2.5, 2.1, 2.05, 2.01$. (b) Use those values to estimate the instantaneous velocity of the ball at $t = 2$.

□ (a) If t is time after the toss and y is the height of the ball, then the average velocity between any two points (t_1, y_1) and (t_2, y_2) is

$$v_{avg} = \frac{y_2 - y_1}{t_2 - t_1}$$

That is, v_{avg} is the slope of the secant line connecting the points. Since $y = 40t - 16t^2$, $t_1 = 2$ gives $y_1 = 16$, and to any other point, then, $v_{avg} = (y - 16)/(t - 2)$. So,

t	2.5	2.1	2.05	2.01	2
y	0	13.44	14.76	15.76	16
m	−32	−25.6	−24.8	−24.16	?

(b) The instantaneous velocity at $t = 2$ is the slope of the tangent line to (t, y) at $t = 2$, which looks like it's about -24. ■

(3) Repeat part (b) of the previous You Try It problem, but use a limit to estimate the instantaneous velocity of the ball at $t = 2$.

□ The function describing height of the ball as a function of time is $y = 40t - 16t^2$. The instantaneous rate of change of y with respect to t at $t = 2$ will be:

$$\left(\frac{\Delta y}{\Delta t}\right)_{inst} = \lim_{h \to 0} \frac{y(2+h) - y(2)}{h}$$

Let's find the the pieces in the numerator of this expression:

$$\begin{aligned} y(2) &= 40(2) - 16(2)^2 = 16 \\ y(2+h) &= 40(2+h) - 16(2+h)^2 \\ &= 80 + 40h - 16(4 + 4h + h^2) \\ &= 16 - 24h - 16h^2 \end{aligned}$$

so

$$\begin{aligned} \left(\frac{\Delta y}{\Delta t}\right)_{inst} &= \lim_{h \to 0} \frac{y(2+h) - y(2)}{h} = \lim_{h \to 0} \frac{(16 - 24h - 16h^2) - (16)}{h} \\ &= \lim_{h \to 0} \frac{(-24h - 16h^2}{h} = \lim_{h \to 0}(-24 - 16h) = -24 \end{aligned}$$

The instantaneous rate of change of height with respect to time at $t = 2$ is -24 m/s. ■

3.2 The Derivative at a Point

Introduction

We have developed two expressions and several corresponding interpretations. The expressions

$$\frac{f(x)-f(a)}{x-a} \quad \text{and} \quad \frac{f(a+h)-f(a)}{h}$$

both compute the same thing — the slope of a line between $(a, f(a))$ and $(x, f(x))$. This slope is interpreted as the slope of the secant line of $f(x)$ between those points, and the average rate of change of $f(x)$ between those two points. From these, we developed the expressions

$$\lim_{x \to a} \frac{f(x)-f(a)}{x-a} \quad \text{and} \quad \lim_{h \to 0} \frac{f(a+h)-f(a)}{h}$$

which themselves compute the same thing, the slope of the line tangent to $f(x)$ at $x = a$, and the instantaneous rate of change of $f(x)$ at $x = a$.

We're mostly done with the former expressions, but the latter two limits have one more interpretation.

The Derivative of $f(x)$ at $x = a$

In addition to their earlier interpretations as slopes of tangent lines and instantaneous rates of change, the expressions above are used in the following definition:

Definition 3.2. *The expressions*

$$\boxed{\lim_{x \to a} \frac{f(x)-f(a)}{x-a} \quad \textit{and} \quad \lim_{h \to 0} \frac{f(a+h)-f(a)}{h}}$$

both compute the derivative of $f(x)$ at $x = a$. This value is denoted as $f'(a)$.

The value of $f'(a)$ as given in this definition is the same as the slope of the line tangent to $f(x)$ at $x = a$, and is also the same as the instantaneous rate of change of $f(x)$ at $x = a$. With only this information, we're already able to answer some straightforward questions about derivatives.

EX 1 If $f(x) = \sin(x)$, state whether the following values are positive, negative, or zero: (a) $f'(0)$, (b) $f'(\pi/2)$, (c) $f'(\pi)$, (d) $f'(3\pi/2)$, and (e) $f'(2\pi)$.

Each of these derivative values is equivalent to the slope of the line tangent to $f(x)$ at the given location. If the tangent line goes uphill at $x = a$, then $f'(a)$ is positive. If the tangent line goes downhill at $x = a$, then $f'(a)$ is negative. If the tangent line is horizontal at $x = a$, then $f'(a)$ is zero. Figure 3.2 shows $f(x)$ with tangent lines (dashed) at the given points, we can determine that (a) $f'(0) > 0$, (b) $f'(\pi/2) = 0$, (c) $f'(\pi) < 0$, (d) $f'(3\pi/2) = 0$, and (e) $f'(2\pi) > 0$. ■

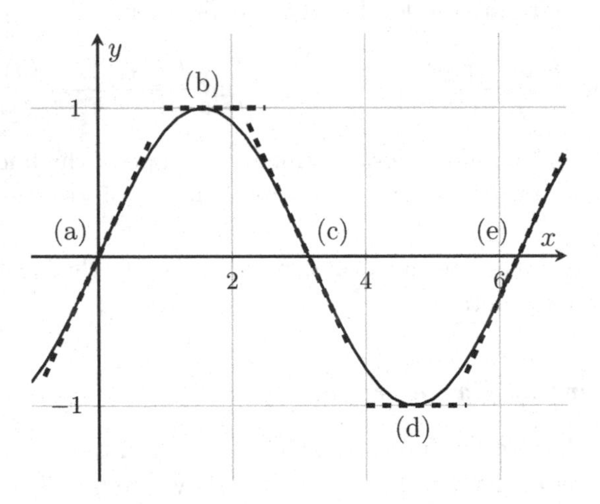

Fig. 3.2 Tangent lines along $f(x) = \sin(x)$.

You Try It

(1) If $f(x) = \cos(x)$, state whether the following values are positive, negative, or zero: (a) $f'(0)$, (b) $f'(\pi/2)$, (c) $f'(\pi)$, (d) $f'(3\pi/2)$, and (e) $f'(2\pi)$.

Now pictures are nice and all, but let's get a little more quantitative.

EX 2 If $f(x) = 1/x$, use a limit to determine $f'(3)$ and provide the equation of the line tangent to $f(x)$ at $x = 3$.

The value of $f'(3)$ is the value of the slope of the the tangent line to

$f(x) = 1/x$ at $(3, 1/3)$. We know how to do this with a limit, via Def. 3.2, which tells us that

$$f'(3) = \lim_{h \to 0} \frac{f(3+h) - f(3)}{h}$$

Note that we are going to need to build $f(3+h) - f(3)$ and then divide that by h. The normal outcome is that the simplified version of $f(3+h) - f(3)$ (the numerator) will have a factor of h in every surviving term, so that it can be factored out and then canceled by the h in the denominator. We have

$$f(3+h) - f(3) = \frac{1}{3+h} - \frac{1}{3} = \frac{3 - (3+h)}{3(3+h)} = \frac{-h}{3(3+h)}$$

so yes, indeed, the division by h will play out nicely

$$\begin{aligned} f'(3) &= \lim_{h \to 0} \frac{f(3+h) - f(3)}{h} = \lim_{h \to 0} \frac{-h}{3(3+h)} \cdot \frac{1}{h} \\ &= \lim_{h \to 0} \frac{-1}{3(3+h)} = -\frac{1}{9} \end{aligned}$$

This value is the slope of the tangent line at $x = 3$, so at $(3, 1/3)$, the equation of the tangent line is $y - \frac{1}{3} = -\frac{1}{9}(x - 3)$. ■

You Try It

(2) If $f(x) = 3x^2 - 5x$, use a limit to determine $f'(2)$ and provide the equation of the line tangent to $f(x)$ at $x = 2$.

Now, finding the derivative at a specific point is fun and all, but wouldn't it be great if we could see a way to write a general expression for a derivative at *any* point? We can determine what a derivative $f'(a)$ looks like without specifying any particular value of a, just by going through the limit process with a in place.

EX 3 If $f(t) = t(t+1)$, use a limit to determine $f'(a)$, where a is unspecified.

With no particular value of a in mind yet, we can still build $f(a+h) - f(a)$ and hope we find that factor of h which is ready for cancellation by the denominator of the quotient posed in Def. 3.2:

$$\begin{aligned} f(a+h) - f(a) &= (a+h)(a+h+1) - a(a+1) \\ &= (a^2 + 2ah + a + h^2 + h) - (a^2 + a) \\ &= 2ah + h^2 + h = h(2a + h + 1) \end{aligned}$$

And now,

$$f'(a) = \lim_{h\to 0} \frac{f(a+h) - f(a)}{h}$$
$$= \lim_{h\to 0} \frac{2ah + h^2 + h}{h} = \lim_{h\to 0}(2a + h + 1) = 2a + 1 \quad \blacksquare$$

You Try It

(3) If $f(x) = 3 - 2x + 4x^2$, use a limit to determine $f'(a)$, where a is unspecified.

Now, being able to use limits to compute derivatives is fun and all. but we should also be able to identify a derivative when we see one!

EX 4 The following expression computes $f'(a)$ for some function $g(x)$ at some $x = a$. What are $g(x)$ and a?

$$\lim_{h\to 0} \frac{\sqrt[3]{-2+h} - \sqrt[3]{-2}}{h}$$

We must match this to the general expression

$$\lim_{h\to 0} \frac{g(a+h) - g(a)}{h}$$

So evidently, $g(a+h) = \sqrt[3]{-2+h}$ and $g(a) = \sqrt[3]{-2}$. This means the given expression is the derivative of $g(x) = \sqrt[3]{x}$ at $a = -2$. $\blacksquare$

You Try It

(4) The following expression computes $f'(a)$ for some function $f(x)$ at some $x = a$. What are $f(x)$ and a?

$$\lim_{h\to 0} \frac{(1+h)^{10} - 1}{h}$$

And, besides computing and recognizing derivatives, we must know how to interpret them in the context of a specific problem. Remember that a derivative is a slope of a tangent line, i.e. a measure of $\Delta f/\Delta x$. So the *units* of a derivative will be the quotient of the units of f and units of x. Often, the phrase "per" is used, such as "miles per hour".

EX 5 Suppose $T(t)$ represents the temperature in degrees Celsius of a cup of coffee as a funcion of time in minutes. Let $t = 0$ represent the time when the cup is poured from the (hot) pot. (a) What is the meaning of $T'(1)$ in the context of the problem, and what are the units of $T'(1)$? (b) Would you expect $T'(1)$ to be larger or smaller in magnitude than $T'(3)$? (c) What is the sign of $T'(a)$ at any point $t = a$ where $a > 0$?

(a) $T'(1)$ is a measure of how fast the temperature T of the cup is changing at $t = 1$ minute. Its units are degrees Celsius per minute. (b) The cup of coffee will cool the fastest when it is the most hot (and so the difference in temperature between the coffee and the room is largest). So as t increases, the rate of change of the temperature T will decrease. So, we should expect $T'(1)$ to be larger in magnitude than $T'(3)$. (c) Since the coffee is cooling, its temperature is always going down, so we would expect $T'(a)$ to be negative at all $t = a$ once the cup is poured from the pot. ■

You Try It

(5) Suppose the number n of bacteria in a Petri dish is a function of time t (in hours), so that $n = f(t)$. (a) What is the meaning of $f'(5)$ in the context of the problem, and what are its units? (b) Would you expect $f'(5)$ to be larger or smaller than $f'(10)$?

Have You Learned...

- The definition of the derivative of a function at a point?
- How to write the derivative of a function at a point in two different limit forms?
- How the derivative of a function at a point relates to the instantaneous rate of change of the function at that point, as well as the slope of the line tangent to that function at that point?
- How to calculate the derivative of a function at a point for some simple functions?
- How to interpret values of a derivative in a given applied context?
- How to recognize derivatives that have already been formed?

The Derivative at a Point — Problem List

The Derivative at a Point — You Try It

These appeared above; solutions begin on the next page.

(1) If $f(x) = \cos(x)$, state whether the following values are positive, negative, or zero: (a) $f'(0)$, (b) $f'(\pi/2)$, (c) $f'(\pi)$, (d) $f'(3\pi/2)$, and (e) $f'(2\pi)$.
(2) If $f(x) = 3x^2 - 5x$, use a limit to determine $f'(2)$ and provide the equation of the line tangent to $f(x)$ at $x = 2$.
(3) If $f(x) = 3 - 2x + 4x^2$, use a limit to determine $f'(a)$, where a is unspecified.
(4) The following expression computes $f'(a)$ for some function $f(x)$ at some $x = a$. What are $f(x)$ and a?

$$\lim_{h \to 0} \frac{(1+h)^{10} - 1}{h}$$

(5) Suppose the number n of bacteria in a Petri dish is a function of time t (in hours), so that $n = f(t)$. (a) What is the meaning of $f'(5)$ in the context of the problem, and what are its units? (b) Would you expect $f'(5)$ to be larger or smaller than $f'(10)$?

The Derivative at a Point — Practice Problems

Try these as you get the hang of the You Try It problems. Solutions to these problems are available in Sec. A.3.2.

(1) If $f(x) = e^{-x}$, is there any point $x = a$ where $f'(a)$ is positive or zero?
(2) If $g(x) = 1 - x^3$, use a limit to determine $g'(0)$ and provide the equation of the line tangent to $g(x)$ at $x = 0$.
(3) If $f(t) = t^4 - 5t$, use a limit to determine $f'(a)$ where a is still unspecified.
(4) The following expression computes $f'(a)$ for some function $f(x)$ at some $x = a$. What are $f(x)$ and a?

$$\lim_{h \to 0} \frac{\sqrt[4]{16+h} - 2}{h}$$

(5) Use the identity $\sin(A+B) = \sin A \cos B + \cos A \sin B$ to find $f'(0)$ for $f(x) = \sin(x)$. Can you use this result to guess the value of $f'(a)$ at other locations, too?

(6) Suppose the fuel consumption (in gallons per hour) c of a car is a function of velocity v (in mph), so that $c = f(v)$. (a) What is the meaning of $f'(20)$ in the context of the problem, and what are its units? (b) What would $f'(20) = -0.05$ mean?

The Derivative at a Point — Challenge Problems

Try these problems to test your skills with the ideas in this section. Solutions to these problems are available in Sec. B.3.2.

(1) If $f(x) = \sqrt{3x+1}$, use a limit to determine $f'(a)$, where a is yet to be specified. Is $f'(a)$ undefined for any value of a at which $f(a)$ is not also undefined?

(2) The following expression computes $f'(a)$ for some function $f(t)$ at some $t = a$. What are $f(t)$ and a?

$$\lim_{t\to 1} \frac{t^4 + t - 2}{t - 1}$$

(3) The quantity (pounds) Q of coffee sold is a function of price p, $Q = f(p)$, where p is in dollars per pound. (a) Interpret the derivative $f'(8)$ in the context of the problem, and state its units. (b) Would you expect $f'(8)$ to be positive or negative? (There's no one right answer to this, but your answer must be consistent with your explanation.)

The Derivative at a Point — You Try It — Solved

(1) If $f(x) = \cos(x)$, state whether the following values are positive, negative, or zero: (a) $f'(0)$, (b) $f'(\pi/2)$, (c) $f'(\pi)$, (d) $f'(3\pi/2)$, and (e) $f'(2\pi)$.

☐ If we visualize the tangent lines to $\cos(x)$ at these locations (shown in Fig. 3.3), we can determine that (a) $f'(0) = 0$, (b) $f'(\pi/2) < 0$, (c) $f'(\pi) = 0$, (d) $f'(3\pi/2) > 0$, and (e) $f'(2\pi) = 0$. ■

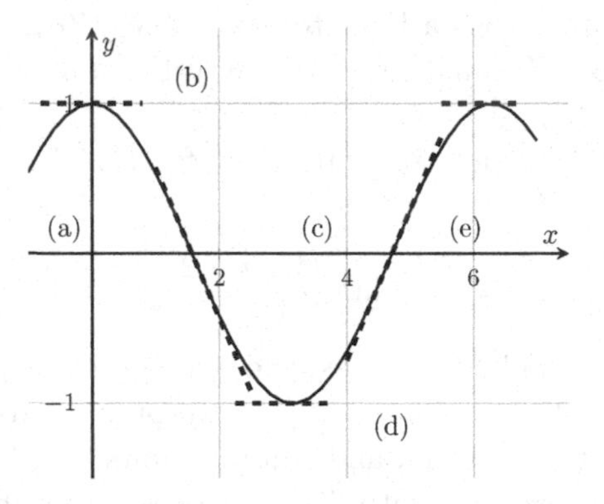

Fig. 3.3 Tangent lines along $f(x) = \cos(x)$.

(2) If $f(x) = 3x^2 - 5x$, use a limit to determine $f'(2)$ and provide the equation of the line tangent to $f(x)$ at $x = 2$.

☐ The value of $f'(2)$ is the value of the slope of the the tangent line to $f(x) = 3x^2 - 5x$ at $(2, 2)$. We know how to do this with a limit, via Def. 3.2. We can prepare for that by building $f(2+h) - f(2)$ and trying to find an h to factor out of each term:

$$\begin{aligned} f(2+h) - f(2) &= [3(2+h)^2 - 5(2+h)] - [3(2)^2 - 5(2)] \\ &= [12 + 12h + 3h^2 - 10 - 5h] - [2] \\ &= 7h + 3h^2 = h(7+3h) \end{aligned}$$

And now,

$$f'(2) = \lim_{h\to 0} \frac{f(2+h) - f(2)}{h} = \lim_{h\to 0} \frac{7h + 3h^2}{h} = \lim_{h\to 0}(7 + 3h) = 7$$

So at (2,2), the equation of the tangent line is $y - 2 = 7(x - 2)$. ■

(3) If $f(x) = 3 - 2x + 4x^2$, use a limit to determine $f'(a)$, where a is unspecified.

□ If $f(x) = 3 - 2x + 4x^2$, then

$$\begin{aligned} f(a+h) - f(a) &= [3 - 2(a+h) + 4(a+h)^2] - [3 - 2a + 4a^2] \\ &= [3 - 2a - 2h + 4a^2 + 8ah + 4h^2] - [3 - 2a + 4a^2] \\ &= -2h + 8ah + 4h^2 = h(-2 + 8a + 4h) \end{aligned}$$

and so

$$\begin{aligned} f'(a) &= \lim_{h\to 0} \frac{f(a+h) - f(a)}{h} = \lim_{h\to 0} \frac{h(-2 + 8a + 4h)}{h} \\ &= \lim_{h\to 0} (-2 + 8a + 4h) = -2 + 8a \quad \blacksquare \end{aligned}$$

(4) The following expression computes $f'(a)$ for some function $f(x)$ at some $x = a$. What are $f(x)$ and a?

$$\lim_{h\to 0} \frac{(1+h)^{10} - 1}{h}$$

□ We must match this to the general expression

$$\lim_{h\to 0} \frac{f(a+h) - f(a)}{h}$$

So evidently, $f(a+h) = (1+h)^{10}$ and $f(a) = 1$. This means the given expression is the derivative of $f(x) = x^{10}$ at $a = 1$. ■

(5) Suppose the number n of bacteria in a Petri dish is a function of time t (in hours), so that $n = f(t)$. (a) What is the meaning of $f'(5)$ in the context of the problem, and what are its units? (b) Would you expect $f'(5)$ to be larger or smaller than $f'(10)$?

□ (a) The derivative $f'(5)$ represents the rate of change of population n at $t = 5$; the units of this would be bacteria per hour. And (b), we should expect $f'(10)$ to be larger than $f'(5)$ because there is a larger population after 10 hours than after only 5 hours, so the population will be growing faster at 10 hours — just like your bank account would gain more interest after 10 months than after 5, assuming you weren't making withdraws! ■

3.3 The Derivative Function

Introduction

By now, we've been asked to build and evaluate several varieties of the expression

$$\lim_{h \to 0} \frac{f(a+h) - f(a)}{h}$$

where a is a yet-to-be-specified point. Here are some examples of what we've found:

- If $f(x) = 1 + x + x^2$, then $f'(a)$ at *any* $x = a$ looks like $1 + 2a$.
- If $f(x) = x^3 - 4x + 1$, then $f'(a)$ at *any* $x = a$ looks like $3a^2 - 4$.
- If $f(x) = 2/(x+3)$, then $f'(a)$ at *any* $x = a$ looks like $-2/(x+3)^2$.
- If $f(x) = 3 - 2x + 4x^2$, then $f'(a)$ at *any* $x = a$ looks like $8a - 2$.
- If $f(t) = t(t+1)$, then $f'(a)$ at *any* $t = a$ looks like $2a + 1$.

Recall that each $f'(a)$ is called the derivative of $f(x)$ at $x = a$. Each of these items provides a relationship between $x = a$ and the value of the derivative. That is, different values of a give different values of the derivative. And since different inputs (a) give different outputs ($f'(a)$), the derivative of $f(x)$ is itself a function! We'll now develop that idea even further.

The Derivative Function

In EX 1 of Sec. 3.2, we saw that if $f(x) = \sin(x)$, then we already know several values of its derivative, such as (a) $f'(0) > 0$, (b) $f'(\pi/2) = 0$, (c) $f'(\pi) < 0$, (d) $f'(3\pi/2) = 0$, and (e) $f'(2\pi) > 0$. We don't have to restrict our assessment of the derivative of $\sin(x)$ to these points alone. In fact, we know that wherever the $\sin(x)$ graph is going uphill, its derivative will be positive. Where $\sin(x)$ is *steeper*, the size of its derivative is larger. Wherever the $\sin(x)$ graph is going downhill, its derivative will be negative. Wherever the $\sin(x)$ graph tops off or bottoms out, its derivative will be zero. When we put all this information together, we get a sense of what the derivative *function* (i.e. the collection of all derivative values) might look like.

EX 1 Prepare a graph of $\sin(x)$ on the interval $[-5\pi/2, 5\pi/2]$. (This interval allows us to see a couple of full cycles of the graph.) Sketch a plausible graph of its derivative on the same axes.

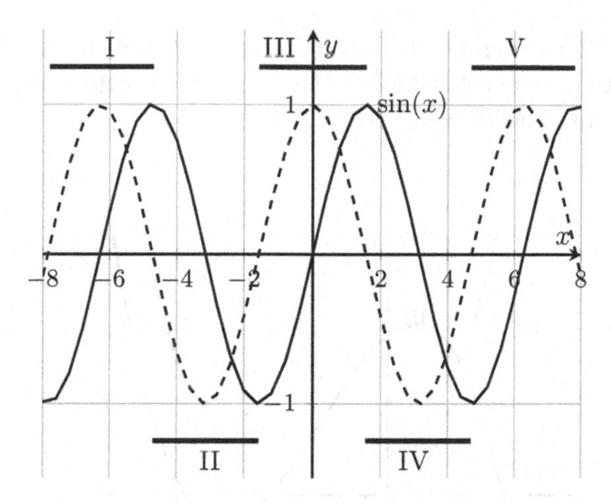

Fig. 3.4 sin(x) and its derivative function (dashed).

Figure 3.4 shows graph of sin(x) together with its derivative function, with a couple of descriptive annotations that show how the graph of the derivative may have been determined. Here are some highlights based on our previous experience:

- By extending EX 1 in Sec. 3.2 we know that the derivative of sin(x) is 0 at $x = -5\pi/2,\ -3\pi/2,\ -\pi/2,\ \pi/2,\ 3\pi/2,\ 5\pi/2$ (because the tangent line is horizontal at these points.
- Similarly, we know $f'(0) > 0$ on intervals $I : (-5\pi/2, -3\pi/2)$, $III : (-\pi/2, \pi/2)$, $V : (3\pi/2, 5\pi/2)$ because the graph of sin(x) is going "uphill" on these intervals — so the tangent lines have positive slope.
- Similarly, we know $f'(0) < 0$ on intervals $II : (-3\pi/2, -\pi/2)$, $IV : (\pi/2, 3\pi/2)$ because the graph of sin(x) is going "downhill" on these intervals — so the tangent lines have negative slope.
- By Practice Problem 5 of Sec. 3.2, we identify the max / min values of the derivative of sin(x) are ± 1, at $x = -5\pi/2,\ -3\pi/2,\ -\pi/2,\ \pi/2,\ 3\pi/2,\ 5\pi/2$.

These trends are used to trace an estimate of the graph of the deritive function for sin(x). If you remember your trigonometric functions, you should recognize that graph! ■

EX 2 Fig. 3.5 shows a function $f(x)$, along with two curves (dashed) and (dotted) — one of which is the derivative function of $f(x)$. Which is it, any how do you know?

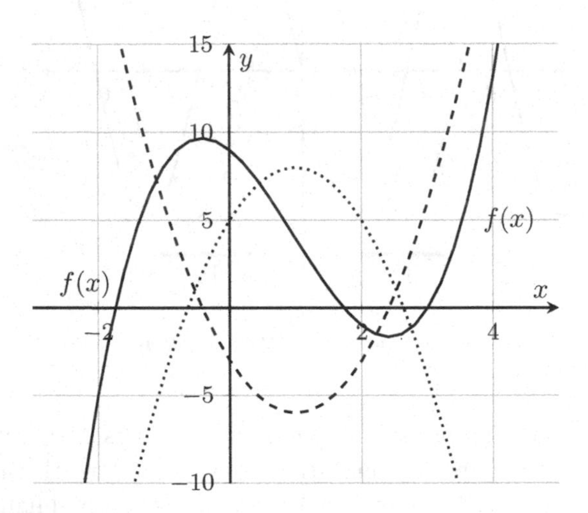

Fig. 3.5 $f(x)$ and two candidate derivatives (dashed, dotted).

The dashed curve must be the derivative of $f(x)$. This curve is zero at the x-values where $f(x)$ tops off or bottoms out (and so has a horizontal tangent line). (This is *almost* true for the dotted curve, but not quite.) This curve is positive where $f(x)$ is going uphill, and negative where $f(x)$ is going downhill. This is not true for the dotted curve. So the dashed curve must be the derivative. ■

You Try It

(1) Draw a graph of $f(x) = 2 - x^2$, and then sketch a plausible representation of its derivative function on the same set of axes.

(2) Figure 3.8 (see the end of this section) shows a function $f(x)$, along with two curves (dashed) and (dotted) — one of which is the derivative function of $f(x)$. Which is it, any how do you know?

We can only get so far by analyzing graphs of functions or deciding what derivative functions sort of look like. To make more progress, we need to compute entire derivative functions for a given function $f(x)$. Fortunately, we already know how to use a limit to find the derivative of $f(x)$ at $x = a$ for an unspecified a. But a is actually disposable, it's just a surrogate for the unknown x. We can just revert to x again. Here is the best, most super-duper, fundamental definition of the derivative that we'll see:

Definition 3.3. *The derivative function of $f(x)$ is defined with the following limit, as long as the limit exists:*

$$\boxed{f'(x) = \lim_{h\to 0} \frac{f(x+h) - f(x)}{h}}$$

Here is an example of the use of that definition, although it's really the same old thing we've been doing for a while now:

EX 3 Use a limit to determine $f'(x)$ for $f(x) = 2x^2 - x$. Use the result to determine $f'(1)$ and $f'(2)$.

Because it gets a little gross, let's build $f(x+h) - f(x)$ ahead of time. And, since the difference quotient in Def. 3.3 will require a division of $f(x+h) - f(x)$ by h, let's hope we find an h to factor out. And we do:

$$\begin{aligned} f(x+h) - f(x) &= [2(x+h)^2 - (x+h)] - [2x^2 - x] \\ &= 2(x^2 + 2xh + h^2) - (x+h) - 2x^2 + 2x \\ &= 4xh + 4h^2 - h = h(4x + 4h - 1) \end{aligned}$$

See? We found an h in all surviving terms of $f(x+h) - f(x)$, and so we're set up nicely for Def. 3.3:

$$\begin{aligned} f'(x) &= \lim_{h\to 0} \frac{f(x+h) - f(x)}{h} \\ &= \lim_{h\to 0} \frac{h(4x + 4h - 1)}{h} = \lim_{h\to 0}(4x + 4h - 1) = 4x - 1 \end{aligned}$$

So, if $f(x) = 2x^2 - x$, then $f'(x) = 4x - 1$, and $f'(1) = 3$, $f'(2) = 7$. ■

You Try It

(3) Use a limit to determine $f'(x)$ for $f(x) = 1 - 3x^2$. Use the result to determine $f'(1)$ and $f'(2)$.
(4) Use a limit to determine $f'(x)$ for $f(x) = \sqrt{1 + 2x}$.

Differentiability

If you are able to determine the derivative of a function $f(x)$ at a point $x = a$ (i.e. if there is a single identifiable value for the slope of the tangent line there), then $f(x)$ is said to be *differentiable* at $x = a$. (Obviously this is a very loose definition.) A function is differentiable in an interval $[a, b]$ if it is differentiable at every point inside the interval.

Here are two things that can stop a function from being differentiable at a point $x = a$:

- If there is a discontinuity at $x = a$, then the function cannot be differentiable at $x = a$.
- If there is a corner or sharp kink in the graph of $f(x)$ at $x = a$, then the function is not differentiable at $x = a$. We usually refer to these corners or kinks as "cusps".

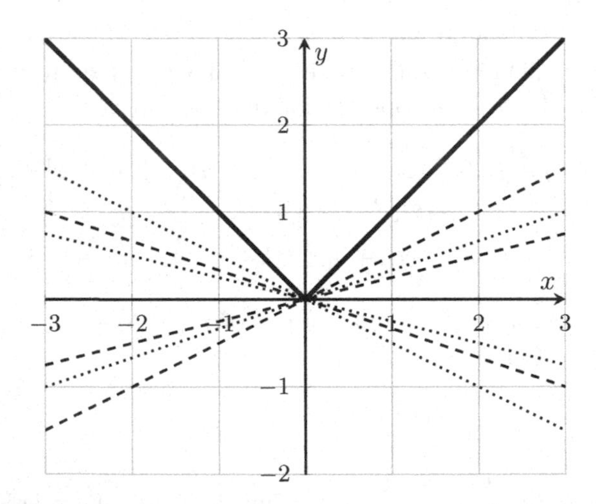

Fig. 3.6 $f(x) = |x|$ and a few of its tangent lines at $(0, 0)$.

We already know how to find discontinuities of functions, so let's take a closer look at the second item. An example of a function with a cusp is $|x|$; there is a cusp at the origin, where the graph comes to a sharp "V". The loose and informal way to see why $f(x) = |x|$ is not differentiable at $x = 0$ is to remember that the derivative is trying to identify the slope of *the* one and only tangent line at any point — but we can draw more than one line that appears to be tangent to $f(x)$ at $x = 0$ ... so how do we know which one is *the* tangent line there? Figure 3.6 shows this conundrum; if we use the most informal idea of a tangent line, i.e. a line which touches the curve at a single point, there are infinitely many "tangent" lines possible at $x = 0$ for $f(x) = |x|$.

The more formal way to consider the failure of the derivative to exist at $x = 0$ is to consider the definition of the derivative. For $f(x) = |x|$, the definition of its derivative at $x = 0$ is (in one of the two possible forms),

$$f'(0) = \lim_{x \to 0} \frac{f(x) - f(0)}{x - 0} = \lim_{x \to 0} \frac{f(x)}{x} = \lim_{x \to 0} \frac{|x|}{x}$$

For this limit to exist, it must be the same from both sides. However, for the function $f(x) = |x|$, we have

$$\lim_{x \to 0^-} \frac{|x|}{x} = -1 \qquad \text{while} \qquad \lim_{x \to 0^+} \frac{|x|}{x} = 1$$

(Make you know why!) Since we get a different value of the limit from the left and right sides, the overall limit does not exist. And so the derivative at $x = 0$ does not exist.

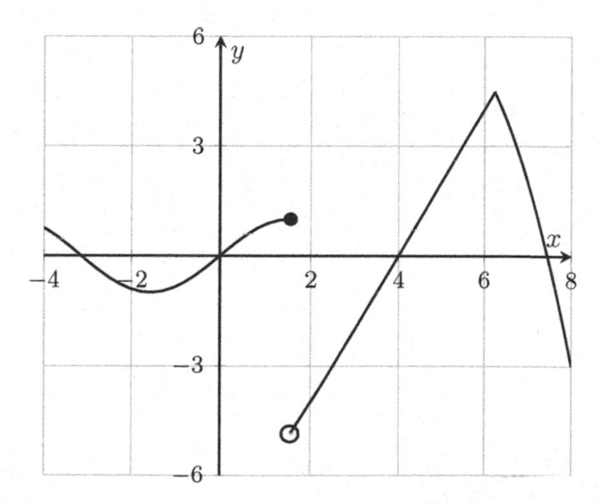

Fig. 3.7 Discontinuities and cusps ruin differentiability (EX 4).

EX 4 Figure 3.7 shows a function $f(x)$. Identify the points where $f(x)$ is not differentiable.

The function is not differentiable at $x = 1$ because of the discontinuity there. It is not differentiable at $x = 5$ because of the cusp there. ■

You Try It

(5) Figure 3.9 (see end of this section) shows a function $f(x)$. Identify the locations where $f(x)$ is not differentiable.

Have You Learned...

- The information carried by the derivative of a function?
- How the behavior of a funtion (increasing, decreasing, steep, etc.) is related to the value / sign of its derivative at any one location?
- How to graph a derivative based on the graph of the original function?
- The proper notation to use when writing derivatives?
- The definition of *differentiable*?
- How to tell where a function is not differentiable?

The Derivative Function — Problem List

The Derivative Function — You Try It

These appeared above; solutions begin on the next page.

(1) Draw a graph of $f(x) = 2 - x^2$, and then sketch a plausible representation of its derivative function on the same set of axes.
(2) Figure 3.8 shows a function $f(x)$, along with two curves (A) and (B) — one of which is the derivative function of $f(x)$. Which is it, any how do you know?

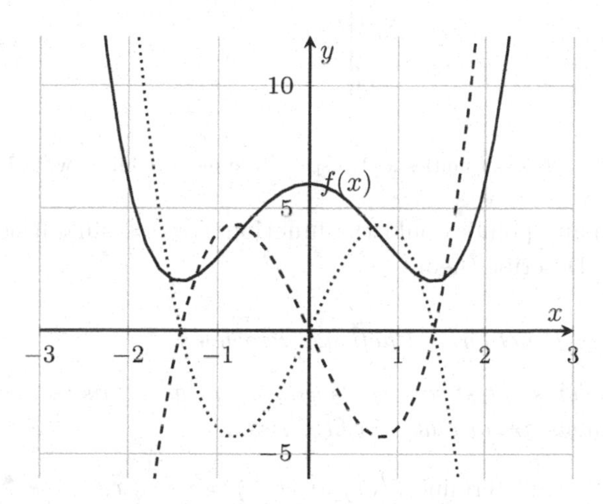

Fig. 3.8 $f(x)$ and two candidate derivatives (dashed, dotted), w/ YTI 2.

(3) Use a limit to determine $f'(x)$ for $f(x) = 1 - 3x^2$.
(4) Use a limit to determine $f'(x)$ for $f(x) = \sqrt{1 + 2x}$.
(5) Figure 3.9 shows a function $f(x)$. Identify the locations where $f(x)$ is not differentiable.

The Derivative Function — Practice Problems

Try these as you get the hang of the You Try It problems. Solutions to these problems are available in Sec. A.3.3.

(1) Use the graph of $f(x) = x^3/3 - x$ to determine where $f'(x)$ is negative.
(2) Use a limit to determine $f'(x)$ for $f(x) = 5x^2 + 3x - 2$.
(3) Use a limit to determine $f'(x)$ for $f(x) = 1/x^2$.

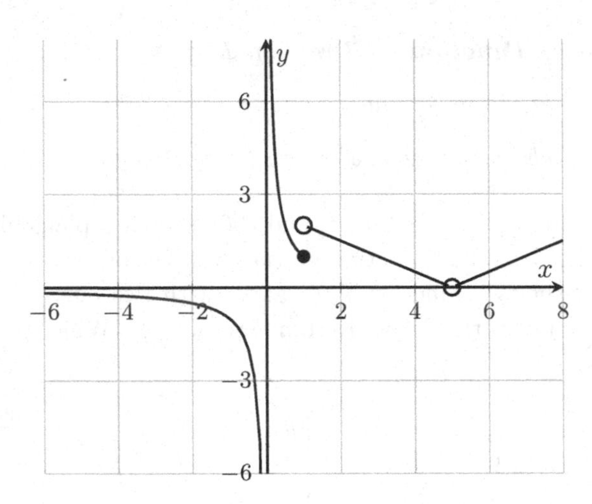

Fig. 3.9 Discontinuities and cusps ruin differentiability, w/ YTI 5.

(4) At how many points would the function $f(x) = |\sin(x)|$ not be differentiable? Describe them.

The Derivative Function — Challenge Problems

Try these problems to test your skills with the ideas in this section. Solutions to these problems are available in Sec. B.3.3.

(1) Use a limit to determine $f'(x)$ for $f(x) = x + \sqrt{x}$.
(2) Use a limit to determine $f'(x)$ for $f(x) = ax^2 + bx + c$ (in which a, b, and c are unspecified constants).
(3) Determine and graph $f'(x)$ for $f(x) = |x - 6|$. You don't necessarily need to use a limit, but you must describe how you come up with $f'(x)$. Are there any points where $f'(x)$ is not defined?

The Derivative Function — You Try It — Solved

---

(1) Draw a graph of $f(x) = 2 - x^2$, and then sketch a plausible representation of its derivative function on the same set of axes.

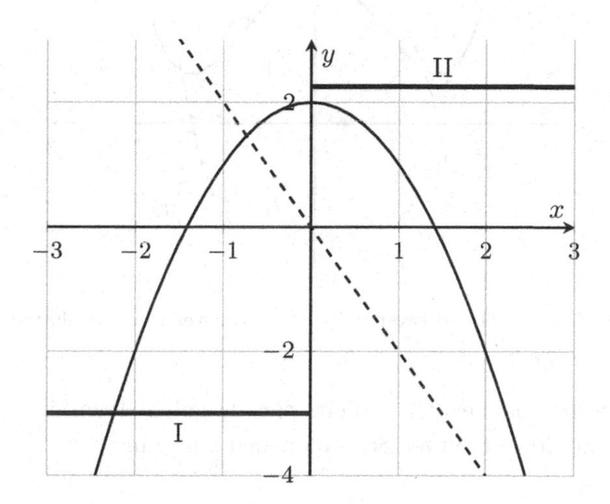

Fig. 3.10 $f(x) = 2 - x^2$ and a plausible sketch of its derivative function (dashed), w/ YTI 1.

□ Note that $f(x) = 2 - x^2$ is an inverted parabola, with a vertex at $(0, 2)$. Here are the trends to look for:

- $f(x)$ is going uphill from left to right until $x = 0$; so, the derivative graph should be positive for $x < 0$; see interval (I) in Fig. 3.10.
- There will be a horizontal tangent line on $f(x)$ at $(0, 2)$, and so the derivative should be zero at $x = 0$
- $f(x)$ is going downhill starting with $x = 0$ and moving right; so, the derivative graph should be negative for $x > 0$; see interval (II) in Fig. 3.10. ■

---

(2) Figure 3.11 shows a function $f(x)$, along with two curves (dashed) and (dotted) — one of which is the derivative function of $f(x)$. Which is it, any how do you know?

□ The dashed curve must be the derivative of $f(x)$. This curve is zero at the x-values where $f(x)$ tops off or bottoms out (and so has a horizontal tangent line). Also, the dashed curve is positive where $f(x)$ is

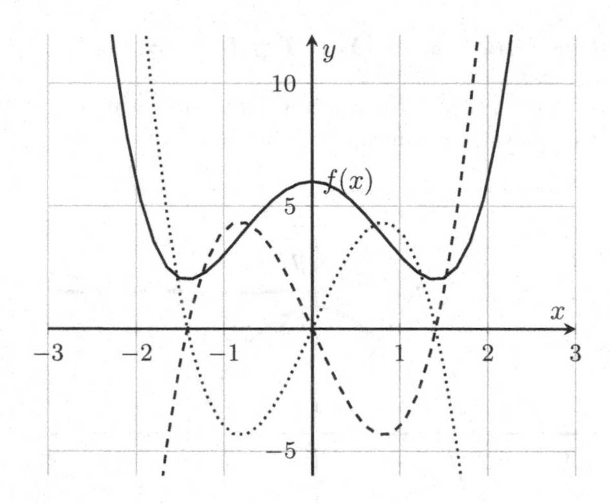

Fig. 3.11 $f(x)$ and two candidate derivatives (dashed, dotted).

going uphill, and negative where $f(x)$ is going downhill. This is not true for the dotted curve. So the dashed curve must be the derivative. ■

(3) Use a limit to determine $f'(x)$ for $f(x) = 1 - 3x^2$.

☐ Let's prepare for Def. 3.3 by building:

$$\begin{aligned} f(x+h) - f(x) &= [1 - 3(x+h)^2] - [1 - 3x^2] \\ &= [1 - 3(x^2 + 2xh + h^2)] - [1 - 3x^2] \\ &= -6xh + 3h^2 = h(-6x + 3h) \end{aligned}$$

And now,

$$\begin{aligned} f'(x) &= \lim_{h\to 0} \frac{f(x+h) - f(x)}{h} \\ &= \lim_{h\to 0} \frac{h(-6x+3h)}{h} \\ &= \lim_{h\to 0} (-6x + 3h) = -6x \end{aligned}$$

So, if $f(x) = 1 - 3x^2$, then $f'(x) = -6x$, and $f'(1) = -6$, $f'(2) = -12$. ■

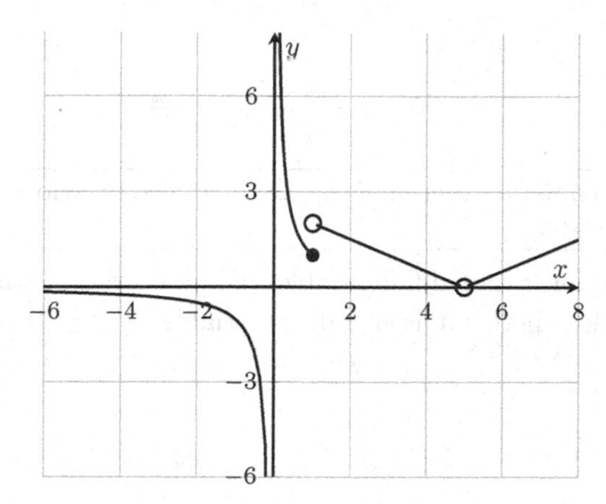

Fig. 3.12 Discontinuities and cusps ruin differentiability, w/ YTI 5.

(4) Use a limit to determine $f'(x)$ for $f(x) = \sqrt{1+2x}$.

□ Let's start with $f(x+h) - f(x)$, then do that thing where we multiply $\sqrt{A} - \sqrt{B}$ up and down by $\sqrt{A} + \sqrt{B}$, because in the long run, it actually helps! Starting with,

$$f(x+h) - f(x) = \sqrt{1+2(x+h)} - \sqrt{1+2x}$$

we have,

$$\begin{aligned}
\left(\sqrt{1+2(x+h)} - \sqrt{1+2x}\right) \cdot \frac{\sqrt{1+2(x+h)} + \sqrt{1+2x}}{\sqrt{1+2(x+h)} + \sqrt{1+2x}} \\
= \frac{1+2(x+h) - (1+2x)}{\sqrt{1+2(x+h)} + \sqrt{1+2x}} \\
\to f(x+h) - f(x) = \frac{2h}{\sqrt{1+2(x+h)} + \sqrt{1+2x}}
\end{aligned}$$

See, it got better! Proceeding,

$$\begin{aligned}
f'(x) &= \lim_{h\to 0} \frac{f(x+h) - f(x)}{h} \\
&= \lim_{h\to 0} \frac{2h}{\sqrt{1+2(x+h)} + \sqrt{1+2x}} \cdot \frac{1}{h} \\
&= \lim_{h\to 0} \frac{2}{\sqrt{1+2(x+h)} + \sqrt{1+2x}} \\
&= \frac{2}{\sqrt{1+2x} + \sqrt{1+2x}} = \frac{2}{2\sqrt{1+2x}} = \frac{1}{\sqrt{1+2x}}
\end{aligned}$$

Altogether,

$$f'(x) = \frac{1}{\sqrt{1+2x}} \quad \blacksquare$$

---

(5) Figure 3.12 shows a function $f(x)$. Identify the locations where $f(x)$ is not differentiable.

□ The function is not differentiable at $x = 0$ or $x = 1$ because of the discontinuities there. It is not differentiable at $x = 5$ because of the cusp there. ■

3.4 Simple Derivatives and Antiderivatives

Introduction

Computing derivatives using the limit definition is fun, and surely almost everyone wishes to continue doing so, but frankly it's just not that efficient. Having used the limit definition to find a bunch of individual derivatives, perhaps we can extend that experience to make some general rules. Here, we will see simple examples of this. In addition, as long as we're asking the question, "*Given* $f(x)$, *what is its derivative* $f'(x)$?", we may as well also consider the reverse question: "Given some derivative $f'(x)$, what function $f(x)$ is it the derivative of?" Bad grammar aside, the reverse question is even more interesting!

New Notation

First, we need to work on our notation. We have referred to the derivative of $y = f(x)$ as both the instantaneous rate of change of $y = f(x)$ (either at a point or as a general expression), as well as the instantaneous rate of change of $f(x)$ (again, either at a point or as a general expression). When we think about rates of change, we think about slopes — in fact, we can refer to the "slope of $f(x)$ at $x = a$" by using the slope of the tangent line there as a surrogate. The language used when discussing slopes is often based on the symbol Δ — such as in $\Delta y/\Delta x$ or $\Delta f/\Delta x$. But the symbol Δ usually implies that there is a notable difference between the points in question — such as, when we ask for the slope of the line between $(1, 4)$ and $(3, 7)$, where $\Delta x = 2$ and $\Delta y = 3$. When finding a derivative, though, we employ a limit process which shrinks that difference to zero. To acknowledge this, we adjust the notation a bit. We write "the derivative with respect to x of $y = f(x)$", we dispense with Δx and Δy and instead write either

$$\frac{d}{dx}(y) \qquad \text{or, shortened,} \qquad \frac{dy}{dx}$$

or

$$\frac{d}{dx}(f) \qquad \text{or, shortened,} \qquad \frac{df}{dx}$$

We can even take out the name of the function and write, for example, "the derivative of $x^2 - 2x$ with respect to x" as

$$\frac{d}{dx}(x^2 - x)$$

The use of d rather than Δ invokes a *differential* — a really, really tiny quantity — in fact, a quantity approaching zero. Thus, functions which have a derivative are said to be "differentiable".

We can still use the prime notation ($f'(x)$), too. The notation around derivatives is actually quite flexible; be prepared to move back and forth between the different versions (differential vs prime). Also be prepared to adjust the notation according to variable names; for example, if we write the volume of a sphere as a function of radius, $V(r) = 4\pi r^3/3$, then we'd write the derivative of volume with respect to radius as

$$\frac{dV}{dr} \quad \text{or} \quad \frac{d}{dr}\left(\frac{4}{3}\pi r^3\right)$$

The Derivative of a Constant

A derivative is a rate of change. But a constant is constant, and does not change — thus, the name "constant". So, we might expect that the rate of change (derivative) of any constant, at any point, is 0. That is, if c is a constant, we'd expect

$$\boxed{\frac{d}{dx}(c) = 0} \tag{3.3}$$

This is true. We can see it (or, dare I say, *prove it*) quickly using the limit definition of the derivative:

$$\frac{d}{dx}(c) = \lim_{h \to 0} \frac{f(x+h) - f(x)}{h} = \lim_{h \to 0} \frac{c - c}{h} = \lim_{h \to 0} (0) = 0$$

Since we now know the derivative of ANY constant is zero, we don't have to use the limit definition to find the derivative of individual constants anymore.

The Derivative of a Linear Function

Any linear function that goes through the origin has the form $y = ax$, for some a. I'll ask you to agree to two ideas:

(1) At a point on a function that's a straight line, the line tangent to the function at any point will have the same slope as the line.
(2) A function which is a straight line has the same slope (and consequently, the same slope of the tangent line) at every point.

If you believe these two things — and you should — then we should certainly agree that the derivative of a linear function should be to be the same everywhere. Even better, we should expect the derivative (rate of change / slope of the tangent line) of $y = ax$ to be the same as the slope of the line itself. Can we confirm our suspicion that $\frac{d}{dx}(ax) = a$ for all x? Sure! We can check this with a quick trip through the limit definition:

$$\frac{d}{dx}(ax) = \lim_{h\to 0} \frac{a(x+h) - ax}{h} = \lim_{h\to 0} \frac{ah}{h} = \lim_{h\to 0}(a) = a$$

So yes,

$$\boxed{\frac{d}{dx}(ax) = a} \tag{3.4}$$

Since we now know the derivative of linear function $y = ax$ is a, we don't have to use the limit definition to find the derivative of individual linear functions anymore. If $y = 3x$, then $\frac{dy}{dx} = 3$; if $y = -x$, then $dy/dx = -1$; if $y = 17x$, then $dy/dx = 17$; and so on.

The Derivative of a Sum of Two Functions

Suppose we have two functions added together, and need the derivative of that sum. Would the derivative of the sum be the sum of the derivatives? In other words, is the following statement true?

$$\frac{d}{dx}(f(x) + g(x)) = \frac{d}{dx}f(x) + \frac{d}{dx}g(x)$$

We don't have to decide this for every single case we come across, we can decide once and for all using the limit definition of the derivative. First, we set it up:

$$\frac{d}{dx}(f(x) + g(x)) = \lim_{h\to 0} \frac{[f(x+h) + g(x+h)] - [f(x) + g(x)]}{h}$$

Then we rearrange some terms:

$$\frac{d}{dx}(f(x) + g(x)) = \lim_{h\to 0} \frac{[f(x+h) - f(x)] + [g(x+h) - g(x)]}{h}$$

Then we split up the right side into two fractions:

$$\frac{d}{dx}(f(x) + g(x)) = \lim_{h\to 0} \left(\frac{f(x+h) - f(x)}{h} + \frac{g(x+h) - g(x)}{h} \right)$$

Then we separate the limit into two:

$$\frac{d}{dx}(f(x) + g(x)) = \lim_{h\to 0} \frac{f(x+h) - f(x)}{h} + \lim_{h\to 0} \frac{g(x+h) - g(x)}{h}$$

But wait... The two limits terms on the right are now the definitions of $f'(x)$ and $g'(x)$ themselves! So yes, indeed, we have found what we expected. Of course, this presumes that $f'(x)$ and $g'(x)$ both exist and can be found. The same procedure works for the difference of two functions as well, and so we can write an overall rule of derivatives:

$$\boxed{\frac{d}{dx}(f(x) \pm g(x)) = \frac{d}{dx}f(x) \pm \frac{d}{dx}g(x)} \tag{3.5}$$

EX 1 Find the derivatives of $y = 3x - 2$, $y = -x + 7$, and $y = \pi x + e$.

Each is the sum of two functions (constant and linear) that we know how to handle through Eqs. (3.5), (3.3), and (3.4):

$$\begin{aligned}
\frac{d}{dx}(3x - 2) &= \frac{d}{dx}(3x) - \frac{d}{dx}(2) = 3 + 0 = 3 \\
\frac{d}{dx}(-x + 7) &= \frac{d}{dx}(-x) + \frac{d}{dx}(7) = -1 + 0 = -1 \\
\frac{d}{dx}(\pi x + e) &= \frac{d}{dx}(\pi x) + \frac{d}{dx}(e) = \pi + 0 = \pi \quad \blacksquare
\end{aligned}$$

From this discussion and this example, we now know the derivative of *any* linear function,

$$\boxed{\frac{d}{dx}(ax + b) = a} \tag{3.6}$$

Note that we can now dispense with Eq. (3.4), since (3.6) is a better, more general, version.

You Try It

(1) Find the derivatives of $y = -2x-1$, $y = -2x+5$, and $y = -2x+10$.

The Derivative of a Constant Multiple of $f(x)$

What if we wanted the derivative of a function multiplied by a constant, i.e. we wanted the derivative of some $cf(x)$? We might expect the constant to factor out of the derivative, so that

$$\boxed{\frac{d}{dx}(c \cdot f(x)) = c \cdot \frac{d}{dx}f(x)} \tag{3.7}$$

Like above, we can learn this once and for all using one pass through the limit definition:

$$\begin{aligned}\frac{d}{dx}(c\cdot f(x)) &= \lim_{h\to 0}\frac{cf(x+h)-cf(x)}{h}\\ &= \lim_{h\to 0} c\cdot\frac{f(x+h)-f(x)}{h}\\ &= c\cdot\lim_{h\to 0}\frac{f(x+h)-f(x)}{h} = c\cdot\frac{d}{dx}f(x)\end{aligned}$$

So yep, we can do that. This might not seem like a big deal until we start putting it together with the rules above, and some other results.

The Derivative of a Quadratic Function

We have seen the derivative of simple quadratic functions in earlier notes and problem sets, but here it is one more time. If $f(x) = x^2$, then

$$f(x+h) - f(x) = (x+h)^2 - x^2 = x^2 + 2xh + h^2 - x^2 = 2xh + h^2$$

so

$$\frac{d}{dx}(x^2) = \lim_{h\to 0}\frac{(x+h)^2 - x^2}{h} = \lim_{h\to 0}\frac{2xh+h^2}{h} = \lim_{h\to 0}(2x+h) = 2x$$

and together,

$$\boxed{\frac{d}{dx}(x^2) = 2x} \tag{3.8}$$

We can now combine this with the new rules listed above to learn the derivative of ANY quadratic function $y = ax^2 + bx + c$. We're going to use a combination of rules regarding the derivatives of a sum, a constant, a linear function, a contant multiple, and a quadratic function:

$$\begin{aligned}\frac{d}{dx}(ax^2+bx+c) &= \frac{d}{dx}(ax^2) + \frac{d}{dx}(bx) + \frac{d}{dx}(c)\\ &= a\cdot\frac{d}{dx}(x^2) + b\cdot\frac{d}{dx}(x) + \frac{d}{dx}(c)\\ &= a(2x) + b(1) + 0\end{aligned}$$

so that together,

$$\boxed{\frac{d}{dx}(ax^2+bx+c) = 2ax + b} \tag{3.9}$$

With this more general expression, we can dispense with Eq. (3.8).

EX 2 Find the derivatives of $y = x^2 - 2x + 1$, $y = 3x^2 + x$, and $r = 5 - 2t^2$.

Each is a quadratic function that we know to handle term by term. For the first two, we have:

$$\frac{d}{dx}(x^2 - 2x + 1) = \frac{d}{dx}(x^2) - \frac{d}{dx}(2x) + \frac{d}{dx}(1) = 2x - 2$$

$$\frac{d}{dx}(3x^2 + x) = \frac{d}{dx}(3x^2) + \frac{d}{dx}(x) = 3(2x) + 1 = 6x + 1$$

In the last one, we see a function $r(t)$, where t is the independent variable. So the derivative is written $r'(t)$, or,

$$\frac{d}{dt}(5 - t^2) = \frac{d}{dt}(5) - \frac{d}{dt}(t^2) = 0 - (2t) = -2t \quad \blacksquare$$

You Try It

(2) Find the derivatives of $y = -x^2 - 1$, $y = x^2 - 3x + 5$, and $w = y^2 - 2y$.

Finding all these derivatives is only good when we can put them to use:

EX 3 Find the equation of the line tangent to $f(x) = x^2 + x - 1$ at $x = 1$.

To find the slope of the tangent line, we need the derivative:

$$\frac{d}{dx}(x^2 + x - 1) = 2x + 1$$

so that when $x = 1$, the slope of the tangent line is $f'(1) = 2(1) + 1 = 3$. The point the tangent line hits is the point it shares with the function; its y-coordinate is $f(1) = (1)^2 + 1 - 1 = 1$. (Pay close attention to where we need to use the function itself, and where we need to use its derivative.) The line through $(1, 1)$ with a slope of 3 has equation $y - 1 = 3(x - 1)$, or if you prefer, $y = 3x - 2$. $\blacksquare$

You Try It

(3) Find the equation of the line tangent to $f(x) = 2x^2 - 4x + 1$ at $x = 0$.

Antiderivatives

With these few derivatives under our belt, we can now start looking at the reverse question. Suppose we have a function $f(x)$ in mind, but rather than asking, "what is the derivative of $f(x)$?" we ask instead, "what is $f(x)$ the derivative of?" This is called finding an *antiderivative*. When finding a derivative, we fill in this blank:

- The derivative of $f(x)$ with respect to x is ______________.

When finding the antiderivative, we fill in this blank:

- The derivative of ______________. with respect to x is $f(x)$.

First, the notation. The notation for an antiderivative must indicate what the known derivative is, and identify the variable the derivative is "with respect to". (Right now, that variable is usually obvious, but later on, it might not be so obvious.) Here is what we write, and we get to use the coolest symbol in math:

$$\int f(x)\,dx$$

This expression represents the antiderivative of $f(x)$ with respect to x — i.e. what is $f(x)$ the derivative of? An an example of a derivative / antiderivative pair,

$$\frac{d}{dx}(x^2 + x) = 2x + 1 \quad \Longleftrightarrow \quad \int (2x + 1)\,dx = x^2 + x$$

But sadly, that antiderivative isn't completely correct, for a reason we'll see now.

The Arbitrary Constant

Let's consider: what is the antiderivative of 2x? Hopefully, an obvious answer is:

$$\int 2x\,dx = x^2 \quad \text{since} \quad \frac{d}{dx}(x^2) = 2x$$

But, here are some other functions whose derivatives are also $2x$:

$$\frac{d}{dx}(x^2 - 1) = 2x \quad ; \quad \frac{d}{dx}(x^2 + 3) = 2x \quad ; \quad \frac{d}{dx}(x^2 + 5) = 2x$$

$$\frac{d}{dx}(x^2 + \sqrt{\pi + 1}) = 2x \quad \cdots$$

This means that $2x$ does not have a unique antiderivative. The possible antiderivatives of $2x$ all start the same, with the x^2, but end with different constants. So, we report our antiderivative like this:

$$\int 2x\,dx = x^2 + C$$

The C is called an *arbitrary constant*, and it is always required. The antiderivative of *ANY* function $f(x)$ will need to include the arbitrary constant. A function has an infinite number of antiderivatives, all distinguished by the possible values of C. Since the inclusion of C takes a single answer and changes it to an infinite number of possible answers, it's sort of important.

Antiderivatives of Constant and Linear Functions

Without any other formulas, we can do antiderivatives of constants and linear functions.

EX 4 Find the antiderivative $\int 3\,dx$.

Hopefully it's clear by now that the derivative of $3x$ is 3, and so the antiderivative of 3 will be $3x$. But don't wrap up so fast, because we need to remember the arbitrary constant. The derivative of $3x$ is indeed 3. But the derivatives of $3x-5$, $3x+6$, $3x+101$, $3x-\sqrt{7}$, and so on, are also ALL 3. Therefore 3 does not have a single antiderivative, but an infinite number of them; we denote the antiderivative of 3 by including the arbitrary constant:

$$\int 3\,dx = 3x + C \quad \blacksquare$$

EX 5 Find the antiderivative $\int 3x\,dx$.

The derivative of x^2 is $2x$. But here, we have a derivative that is $3x$. This means that the original function had not only x^2, but a constant that, in the derivative, cancels the 2 and leaves a 3. If the original function looks like Ax^2, then its derivative is $2Ax$. What value of A makes $2Ax = 3$? It would be $A = 3/2$. So (don't forget the arbitrary constant!),

$$\int 3x\,dx = \frac{3}{2}x^2 + C$$

You can check this result by quickly verifying that the derivatives of the functions

$$\frac{3}{2}x^2 + 1 \quad ; \quad \frac{3}{2}x^2 - 3 \quad ; \quad \frac{3}{2}x^2 - 11 \cdots$$

(as well as anything else of the form $\frac{3}{2}x^2 + C$ are ALL $3x$). $\blacksquare$

In general, if a and b are constants, we have

$$\boxed{\int b\,dx = bx + C} \quad \text{and} \quad \boxed{\int ax\,dx = \frac{1}{2}ax^2 + C}$$

Together,

$$\boxed{\int ax + b\,dx = \frac{1}{2}ax^2 + bx + C}$$

You Try It

(4) Find the following antiderivatives:

$$\int (-2)\, dx \quad ; \quad \int (-x+5)\, dx \quad ; \quad \int (2x-6)\, dx$$

A Quick Note About Derivative Formulas

You are going to see a LOT of derivative (and antiderivative) formulas as we move along. To help you keep track of them in your head for easy reference, I recommend paying close attention to which are *calculational* formulas, and which are *organizational* formulas. So far,

- Equations (3.3), (3.6), (3.9) are *calculational* formulas: each shows you how to find the derivative of a specific type of function.
- Equation (3.5) is an *organizational* formula. It tells you how to break down the derivative of a particularly structured function, but you still need to know how to find the individual derivatives that go into it.

At some point, you may get overwhelmed and scream, "AAAAAH, there are too many derivative formulas!" When that happens, take a deep breath and try to see how many are computational formulas, as opposed to organizational formulas. That might help relieve a bit of your stress. Not all of it, but some!

Have You Learned...

- How to find the derivative of a constant? A linear function? A quadratic function?

$$\frac{d}{dx}(c) = 0 \quad ; \quad \frac{d}{dx}(ax+b) = a \quad ; \quad \frac{d}{dx}(ax^2+bx+c) = 2ax+b$$

- How addition and subtraction affects derivatives?

$$\frac{d}{dx}(f(x) \pm g(x)) = \frac{d}{dx}f(x) \pm \frac{d}{dx}g(x)$$

- How multiplicative constants affect derivatives?

$$\frac{d}{dx}(c \cdot f(x)) = c \cdot \frac{d}{dx}f(x)$$

- The definition of an antiderivative?

- The meaning of an *arbitrary constant*, and why it appears in an antiderivative?
- How to find the antiderivative of a constant? A linear function?

$$\int b\,dx = bx + C \qquad ; \qquad \int ax + b\,dx = \frac{1}{2}ax^2 + bx + C$$

- The proper notation(s) for derivatives and antiderivatives?

Simple Derivatives and Antiderivatives — Problem List

Simple Derivatives and Antiderivatives — You Try It

These appeared above; solutions begin on the next page.

(1) Find the derivatives of $y = -2x - 1$, $y = -2x + 5$, and $y = -2x + 10$.
(2) Find the derivatives of $y = -x^2 - 1$, $y = x^2 - 3x + 5$, and $w = y^2 - 2y$.
(3) Find the equation of the line tangent to $f(x) = 2x^2 - 4x + 1$ at $x = 0$.
(4) Find the following antiderivatives:

$$\int (-2)\, dx \quad ; \quad \int (-x + 5)\, dx \quad ; \quad \int (2x - 6)\, dx$$

Simple Derivatives and Antiderivatives — Practice Problems

Try these as you get the hang of the You Try It problems. Solutions to these problems are available in Sec. A.3.4.

(1) Find the derivatives of $y = x - 1$, $y = 12x - 5$, and $y = \sqrt{5}x$.
(2) Find the derivatives of $y = x^2 - 4x$ and $y = -3x + x^2$.
(3) Find the equation of the line tangent to $f(x) = x^2/2 - 2x - 1$ at $x = 2$.
(4) Find the following antiderivatives:

$$\int (4)\, dx \quad ; \quad \int \frac{2}{3} x\, dx$$

(5) Find the antiderivative $\int 1 - 6t\, dt$.

Simple Derivatives and Antiderivatives — Challenge Problems

Try these problems to test your skills with the ideas in this section. Solutions to these problems are available in Sec. B.3.4.

(1) Find the derivative of $f(x) = (x - 1)(x + 2)$.
(2) Use a derivative to help find the vertex of the parabola $y = -x^2 + 2x - 3$. (Hint: What will the slope of the tangent line be at the vertex of the parabola?)
(3) We know that x has an infinite number of antiderivatives. Find the ONE antiderivative of x that goes through the point $(1, 2)$. (Hint: Get the general antiderivative of x, then discover which value of the arbitrary constant C will cause the antiderivative to hit this point.)

Simple Derivs and Antiderivs — You Try It — Solved

(1) Find the derivatives of $y = -2x - 1$, $y = -2x + 5$, and $y = -2x + 10$.

$$\square\ \frac{d}{dx}(-2x-1) = \frac{d}{dx}(-2x) - \frac{d}{dx}(1) = -2 - 0 = -2$$
$$\frac{d}{dx}(-2x+5) = \frac{d}{dx}(-2x) + \frac{d}{dx}(5) = -2 + 0 = -2$$
$$\frac{d}{dx}(-2x+10) = \frac{d}{dx}(-2x) + \frac{d}{dx}(10) = -2 + 0 = -2 \quad \blacksquare$$

(2) Find the derivatives of $y = -x^2 - 1$, $y = x^2 - 3x + 5$, and $w = y^2 - 2y$.

$$\square\ \frac{d}{dx}(-x^2-1) = \frac{d}{dx}(-x^2) - \frac{d}{dx}(1) = -2x$$
$$\frac{d}{dx}(x^2-3x+5) = \frac{d}{dx}(x^2) - \frac{d}{dx}(3x) + \frac{d}{dx}(5) = 2x - 3$$
$$\frac{d}{dy}(y^2-2y) = \frac{d}{dy}(y^2) - \frac{d}{dy}(2y) = 2y - 2 \quad \blacksquare$$

(3) Find the equation of the line tangent to $f(x) = 2x^2 - 4x + 1$ at $x = 0$.

□ Since $f'(x) = 4x - 4$, then $f'(0) = -4$. This is the slope of the tangent line. The point on the tangent line is $(0, f(0)) = (0, 1)$. So the equation of the tangent line is $y - 1 = -4(x - 0)$, or $y = -4x + 1$. ■

(4) Find the following antiderivatives:

$$\int (-2)\,dx \quad ; \quad \int (-x+5)\,dx \quad ; \quad \int (2x-6)\,dx$$

□ The first ties back to You Try It 1:

$$\int (-2)\,dx = -2x + C$$

The other two are,

$$\int (-x+5)\,dx = -\frac{1}{2}x^2 + 5x + C$$
$$\int (2x-6)\,dx = x^2 - 6x + C \quad \blacksquare$$

3.5 The Power Rule

Introduction

We have done many derivatives of many functions by now, and patterns are emerging. Here, we'll exploit one of them.

The Power Rule for Derivatives

This table shows some derivatives we've created using the formal limit definition, with some "also known as" forms for the derivatives that will help us see a pattern.

$f(x)$	$f'(x)$	AKA
x	1	$1 \cdot x^0$
x^2	$2x$	$2 \cdot x^1$
x^3	$3x^2$	$3 \cdot x^2$
$\sqrt{x}$	$\frac{1}{2\sqrt{x}}$	$\frac{1}{2} \cdot x^{-1/2}$

In each case, what we notice is that the derivative of the function is the same as what we'd get if we took the original function and...

- dropped the original exponent down to the front.
- replaced the original exponent with one less.

This pattern holds true for any function of the form $f(x) = x^p$, and is known as the **power rule for derivatives**:

$$\boxed{\frac{d}{dx}x^p = p\,x^{p-1}} \tag{3.10}$$

This derivative rule can be combined with any of the others we know so far, such as the rules for differentiating sums, constant multiples, etc.

EX 1 Here are several derivatives computed using the power rule and other known rules.

$$\frac{d}{dx}(3x^5 + 3x^3 - 7x) = 3(5x^4) + 3(3x^2) - 7 = 15x^4 + 9x^2 - 7$$

$$\frac{d}{dx}(2x^2 - 4x^{4/5}) = 2(2x) - 4\left(\frac{4}{5}x^{-1/5}\right) = 4x - \frac{16}{5\sqrt[5]{x}}$$

$$\frac{d}{dx}(x^2+1)^2 = \frac{d}{dx}(x^4 + 2x^2 + 1) = 4x^3 + 4x = 4x(x^2+1)$$

Note that in the third one, the derivative of $(x^2+1)^2$ is NOT just $2(x^2+1)^1$. You have to expand the function to see all the cross terms before you can do the derivative. Also note that the power rule can be sneaky when the exponent is negative. For example, since the derivative of $x^2 = 2x^1$, it is tempting to write the derivative of x^{-2} as $-2x^{-1}$. But it's not. No matter whether the exponent is positive or negative, you *subtract 1* to get the derivative. So, the derivative of x^{-2} is $-2x^{-3}$. ■

You Try It

(1) Find the derivatives of $y = x^4 - 3x^{-2}$ and $r = 6s^3 + \sqrt{2}s^5$.

To use the product rule, you must remember how to change roots to exponents (powers), and put proper exponents on reciprocals.

EX 2 Find the derivative of $f(x) = \sqrt[3]{x^2} - \frac{1}{2}x^4 + \frac{1}{x^3}$.

$$\begin{aligned}\frac{d}{dx}\left(\sqrt[3]{x^2} - \frac{1}{2}x^4 + \frac{1}{x^3}\right) &= \frac{d}{dx}\left(x^{2/3} - \frac{1}{2}x^4 + x^{-3}\right)\\ &= \frac{2}{3}x^{-1/3} - \frac{1}{2}(4x^3) + (-3x^{-4})\\ &= \frac{2}{3\sqrt[3]{x}} - 2x^3 - \frac{3}{x^4} \quad ■\end{aligned}$$

You Try It

(2) Find the derivative of $g(x) = \frac{1}{\sqrt{x}} - \frac{1}{2x}$.

Sometimes functions can be power rule functions in disguise. For example:

EX 3 Find the derivative of $f(x) = \frac{x^2 - 2x + 1}{\sqrt{x}}$.

We need to divide the denominator into the terms of the numerator to reveal the true exponent on each term:

$$\begin{aligned}\frac{d}{dx}\frac{x^2 - 2x + 1}{\sqrt{x}} &= \frac{d}{dx}\left(\frac{x^2}{\sqrt{x}} - \frac{2x}{\sqrt{x}} + \frac{1}{\sqrt{x}}\right)\\ &= \frac{d}{dx}\left(x^{3/2} - 2x^{1/2} + x^{-1/2}\right)\\ &= \frac{3}{2}x^{1/2} - 2\cdot\frac{1}{2}x^{-1/2} - \frac{1}{2}x^{-3/2} = \frac{3}{2}\sqrt{x} - \frac{1}{\sqrt{x}} - \frac{1}{2x\sqrt{x}} \quad ■\end{aligned}$$

You Try It

(3) Find the derivative of $f(x) = \dfrac{x+1}{x^2} - \dfrac{1}{x^3}$.

Knowing the power rule, we can answer the usual kinds of questions that involve derivatives:

EX 4 Find the equation of the line tangent to $f(x) = x^4 - 1$ at $x = -1$.

The point the line shares with $f(x)$ is $(-1, 0)$. The slope of the tangent line is $f'(-1)$. Since the power rule gives $f'(x) = 4x^3$, then $f'(-1) = -4$. So the tangent line is $y - (0) = -4(x - (-1))$, or $y = -4x - 4$. ■

You Try It

(4) Find where the line tangent to $f(x) = x - x^3$ is horizontal.

The Power Rule for Antiderivatives

If we want the antiderivative of x^p, we know that it came from a function with an original exponent of $p+1$. But the derivative of x^{p+1} is $(p+1)x^p$, not just x^p. To get only x^p, the original function needs a coefficient of $1/(p+1)$ to cancel the $(p+1)$ which is brought down during the power rule derivative. So,

$$\boxed{\int x^p \, dx = \frac{1}{p+1} x^{p+1} + C} \tag{3.11}$$

(Remember that we put the arbitrary constant $+C$ on all antiderivatives.) This is the **power rule for antiderivatives**. This formula does NOT work for $p = -1$. That is, this doesn't tell us the antiderivative of $1/x$. (Do you see why it can't work in the case that $p = -1$?) We'll get to that specific case later on. Here are examples of the use of Eq. (3.11).

EX 5 What is the antiderivative of $x - x^2 + 3x^3$?

$$\begin{aligned}
\int x - x^2 + 3x^3 \, dx &= \int x \, dx - \int x^2 \, dx + \int 3x^3 \, dx \\
&= \frac{1}{2}x^2 - \frac{1}{3}x^3 + 3\left(\frac{1}{4}x^4\right) + C \\
&= \frac{1}{2}x^2 - \frac{1}{3}x^3 + \frac{3}{4}x^4 + C
\end{aligned}$$

Note that some liberties were taken with the arbitrary constant there; technically, there should be a separate arbitrary constant associated with each individual antiderivatite in the second step; however, even if we inserted three of them, say C_1, C_2, C_3, their sum would still be arbitrary, and so we could replace $C_1 + C_2 + C_3$ with C. Just holding off any putting one arbitrary constant at the end of a multi-term antiderivative is a matter of routine. It will happen in this next example, too! ■

EX 6 What is the antiderivative of $\frac{2}{x^2} + x\sqrt{x}$?

$$\begin{aligned}\int \frac{2}{x^2} + x\sqrt{x}\,dx &= \int 2x^{-2} + x^{3/2}\,dx \\ &= 2\left(\frac{1}{-1}x^{-1}\right) + \frac{1}{5/2}x^{5/2} + C = -\frac{2}{x} + \frac{2}{5}x^{5/2} + C \quad ■\end{aligned}$$

You Try It

(5) Find $\int 1 - x^3 + 5x^5 - 3x^7\,dx$

(6) Find $\int x^2\sqrt[3]{x} + \frac{x}{\sqrt[4]{x}}\,dx$

We can resolve the arbitrary constant if we also supply a point that we want our antiderivative to hit.

EX 7 Find the antiderivative of $f(x) = \sqrt{x} + 1/2$ that goes through the point $(4, 4)$.

The general antiderivative of $f(x)$ is

$$\int \sqrt{x} + \frac{1}{2}\,dx = \int x^{1/2} + \frac{1}{2}\,dx = \frac{2}{3}x^{3/2} + \frac{1}{2}x + C$$

Let's name this antiderivative $F(x)$. We want $F(x)$ to hit the point $(4, 2)$, so we want $F(4) = 2$. This means we want

$$\frac{2}{3}(4)^{3/2} + \frac{1}{2}(4) + C = 2$$

$$\frac{2}{3}(8) + 2 + C = 2$$

so that $C = -16/3$, and the specific antiderivative we're looking for is

$$F(x) = \frac{2}{3}x^{3/2} + \frac{1}{2}x - \frac{16}{3}$$

Another liberty has been taken in this example, as we skipped over the intermediate step where we apply the antiderivative symbol to each term, such as

$$\int x^{1/2} + \frac{1}{2}\,dx = \int x^{1/2}\,dx + \int \frac{1}{2}\,dx$$

But this is just a general application of the rule

$$\int f(x) \pm g(x)\,dx = \int f(x)\,dx + \int g(x)\,dx$$

in the background, and since we're all friends here, we'll agree this is OK. ■

You Try It

(8) Find the antiderivative of $f(x) = 1 - 6x$ that goes through the point $(0, 8)$.

Have You Learned...

- How to find the derivative of a power function using

$$\boxed{\frac{d}{dx}x^p = p\,x^{p-1}}$$

- How to use derivatives of power functions for ancillary information, such as equations of tangent lines?
- How to find the antiderivative of a power function using

$$\boxed{\int x^p\,dx = \frac{1}{p+1}x^{p+1} + C}$$

- How to use antiderivatives of power functions for ancillary information, such as finding the specific antiderivative which uses a particular point?

The Power Rule — Problem List

The Power Rule — You Try It

These appeared above; solutions begin on the next page.

(1) Find the derivatives of $y = x^4 - 3x^{-2}$ and $r = 6s^3 + \sqrt{2}s^5$.
(2) Find the derivative of $g(x) = \dfrac{1}{\sqrt{x}} - \dfrac{1}{2x}$.
(3) Find the derivative of $f(x) = \dfrac{x+1}{x^2} - \dfrac{1}{x^3}$.
(4) Find where the line(s) tangent to $f(x) = x - x^3$ is / are horizontal.
(5) Find $\displaystyle\int 1 - x^3 + 5x^5 - 3x^7\,dx$.
(6) Find $\displaystyle\int x^2\sqrt[3]{x} + \frac{x}{\sqrt[4]{x}}\,dx$.
(7) Find the antiderivative of $f(x) = 1 - 6x$ that goes through the point $(0, 8)$.

The Power Rule — Practice Problems

Try these as you get the hang of the You Try It problems. Solutions to these problems are available in Sec. A.3.5.

(1) Find the derivative of $f(t) = \sqrt{t} - 1/\sqrt{t}$.
(2) Find the derivative of $y = (x^2 - 2\sqrt{x})/x$.
(3) Find the equation of the line tangent to $y = (1 + 2x)^2$ at $x = 1$.
(4) Find the locations where the tangent lines to $y = x^3 + 3x^2 + x + 3$ are horizontal.
(5) Find $\displaystyle\int x^{20} + 4x^{10} + 8\,dx$.
(6) Find $\displaystyle\int \frac{2x^4 - 3\sqrt{x}}{x^2}\,dx$.
(7) Find the antiderivative of $f(x) = 8x^3 + 12x + 3$ that goes through the point $(1, 6)$.

The Power Rule — Challenge Problems

Try these problems to test your skills with the ideas in this section. Solutions to these problems are available in Sec. B.3.5.

(1) (a) Find the locations where the lines tangent to $y = 2(x^2 - 3)^2$ are horizontal. (b) Design a function that has horizontal tangent lines at $x = 1$ and $x = 5$ and provide the formula for this function.
(2) Find the derivative and antiderivative of $y = (x^2 + 4x + 3)/\sqrt{x}$.
(3) Find the value of A for which the following function has a horizontal tangent line at $x = 4$:

$$f(x) = \frac{3}{4}x^4 - 2Ax - \frac{4}{3}x^{3/2}$$

The Power Rule — You Try It — Solved

(1) Find the derivatives of $y = x^4 - 3x^{-2}$ and $r = 6s^3 + \sqrt{2}s^5$.

☐ For the first,

$$\frac{d}{dx}(x^4 - 3x^{-2}) = 4x^3 - 3(-2x^{-3}) = 4x^3 + 6x^{-3}$$

For the second,

$$\frac{d}{ds}(6s^3 + \sqrt{2}s^5) = 6(3s^2) + \sqrt{2}(5s^4) = 18s^2 + 5\sqrt{2}s^4 \quad \blacksquare$$

(2) Find the derivative of $g(x) = \dfrac{1}{\sqrt{x}} - \dfrac{1}{2x}$.

☐ We can rewrite the function as $g(x) = x^{-1/2} - (1/2) \cdot x^{-1}$, so that

$$g'(x) = -\frac{1}{2}x^{-3/2} - \frac{1}{2}\left(-x^{-2}\right) = -\frac{1}{2x\sqrt{x}} + \frac{1}{2x^2} \quad \blacksquare$$

(3) Find the derivative of $f(x) = \dfrac{x+1}{x^2} - \dfrac{1}{x^3}$.

☐ Rewrite the function as

$$f(x) = \frac{1}{x} + \frac{1}{x^2} - \frac{1}{x^3} = x^{-1} + x^{-2} - x^{-3}$$

so that

$$f'(x) = -x^{-2} - 2x^{-3} + 3x^{-4} = -\frac{1}{x^2} - \frac{2}{x^3} + \frac{3}{x^4} \quad \blacksquare$$

(4) Find where the line(s) tangent to $f(x) = x - x^3$ is / are horizontal.

☐ The tangent line is horizontal where the derivative is zero. Since $f'(x) = 1 - 3x^2$, we are looking for where $1 - 3x^2 = 0$, which means we have found two such locations: $x = \pm 1/\sqrt{3}$. These are the two x-coordinates where the tangent lines are horizontal. $\blacksquare$

(5) Find $\displaystyle\int 1 - x^3 + 5x^5 - 3x^7 \, dx$.

☐ Using the power rule (don't forget the arbitrary constant!) we get

$$\begin{aligned}\int 1 - x^3 + 5x^5 - 3x^7 \, dx &= x - \frac{1}{4}x^4 + 5 \cdot \frac{1}{6}x^6 - 3 \cdot \frac{1}{8}x^8 + C \\ &= x - \frac{1}{4}x^4 + \frac{5}{6}x^6 - \frac{3}{8}x^8 + C \quad \blacksquare\end{aligned}$$

(6) Find $\int x^2 \sqrt[3]{x} + \frac{x}{\sqrt[4]{x}}\, dx$.

□

$$\int x^2 \sqrt[3]{x} + \frac{x}{\sqrt[4]{x}}\, dx = \int x^2 \cdot x^1 3 + x \cdot x^{-1/4}\, dx$$
$$= \int x^7 3 + x^{3/4}\, dx = \frac{3}{10} x^{10} 3 + \frac{4}{7} x^{7/4} + C \quad ■$$

(7) Find the antiderivative of $f(x) = 1 - 6x$ that goes through the point $(0, 8)$.

□ We have

$$\int (1 - 6x)\, dx = x - 3x^2 + C$$

Let's name this antiderivative $F(x)$. To make $F(0) = 8$, we need $0 - 3(0) + C = 8$, so that $C = 8$ and $F(x) = x - 3x^2 + 8$. ■

3.6 The Power Rule on Steroids

Introduction

In Sec. 3.5, we saw how to find derivatives (and antiderivatives) of functions in the form x^p, that is,

$$\frac{d}{dx}x^p = px^{p-1} \quad \text{and} \quad \int x^p\,dx = \frac{1}{p+1}x^{p+1} + C$$

for $p \neq -1$. During that conversation, we noted that these power rules do not immediately extend to similar, but slightly more complicated structures; for example, an overly simplistic guess as to the derivative of $(2x+1)^3$ might be $3(2x+1)^2$, but that's incorrect. Similarly,

$$\int (2x+1)^3\,dx \neq \frac{1}{4}(2x+1)^4 + C$$

It still stands to reason, though, that since $(2x+1)^3$ has the overall structure X^3, its derivative and antiderivative should at least be based on $3X^2$ and $X^4/4$, respectively. And that is the case. So like we did at the start of Sec. 3.5, let's build a set of individual results that we can examine for a pattern. We'll start on the derivative end of things.

The Power Rule Extended to Compositions

In Sec. 3.5, our strategy for finding the derivative of $(2x+1)^3$ was to multiply out $(2x+1)^3$ and examine the derivative term by term:

$$\begin{aligned}\frac{d}{dx}(2x+1)^3 &= \frac{d}{dx}(8x^3 + 12x^2 + 6x + 1) = 24x^2 + 24x + 6\\ &= 6(4x^2 + 4x + 1) = 6(2x+1)^2\end{aligned}$$

Now (because I know the secret), I'm going to write $6(2x+1)^2$ as $3(2x+1)^2 \cdot 2$. Let's put this aside for a bit and see more examples.

The derivative of a more general function, $(ax+b)^3$, can be generated like this:

$$\begin{aligned}\frac{d}{dx}(ax+b)^3 &= \frac{d}{dx}(a^3x^3 + 3a^2bx^2 + 3ab^2x + b^3) = 3a^3x^2 + 6a^2bx + 6ab^2\\ &= 3a(a^2x^2 + 2abx + b^2) = 3a(ax+b)^2\end{aligned}$$

Again, since I know the secret, I'll write the result as $\frac{d}{dx}(ax+b)^3 = 3(ax+b)^2 \cdot a$.

Ready to make it a bit worse?

$$
\begin{aligned}
\frac{d}{dx}(ax^2+bx+c)^2 &= \frac{d}{dx}(a^2x^4+2abx^3+2acx^2+b^2x^2+2bcx+c^2)\\
&= 4a^2x^3+6abx^2+4acx+2b^2x+2bc\\
&= \text{(trust me on the factoring)}\\
&= 2(2ax+b)(ax^2+bx+c)
\end{aligned}
$$

which I'll summarize as $\frac{d}{dx}(ax^2+bx+c)^2 = 2(ax^2+bx+c)^1 \cdot (2ax+b)$.

Finally, I'll present the start and end of another gruesome expansion / differentiation / refactoring and I encourage you to try it out on your own either by hand or using tech such as Wolfram Alpha. It will add nicely to our set of examples:

$$\frac{d}{dx}(ax^3+bx^2+c)^3 = \ldots = 3(ax^3+bx^2+c)^2 \cdot (3ax^2+2bx)$$

And, just to get away from polynomials (and also the letter x), let's try the derivative of $(\sqrt{t}+2t)^4$:

$$
\begin{aligned}
\frac{d}{dt}(\sqrt{t}+2t)^4 &= t^2+8t^{(}5/2)+24t^3+32t^{(}7/2)+16t^4\\
&= 2t+20t^{(}3/2)+72t^2+112t^{(}5/2)+64t^3\\
&= \text{(trust me on the factoring again)}\\
&= 4(\sqrt{t}+2t)^3 \cdot \left(\frac{1}{2\sqrt{t}}+2\right)
\end{aligned}
$$

Let's take stock of what we have so far, because there is indeed a pattern to be noticed. First, remember that we're looking at functions which are built as *compositions*. For example, the function $(ax+b)^3$ is the composition $f(g(x))$ for $f(x) = x^3$ and $g(x) = ax+b$. Similarly, $(\sqrt{t}+2t)^4$ is the composition $f(g(t))$ for $f(t) = t^4$ and $g(t) = \sqrt{t}+2t$. The "outer" function is a power function, and the inner function is a polynomial or polynomial-ish. Here is a chart that shows the full function, the component functions, and the derivative (reminder: the notation $f \circ g$ means $f(g(x))$ or $f(g(t))$):

$f \circ g$	f	g	$(f \circ g)'$
$(ax+b)^3$	x^3	$ax+b$	$3(ax+b)^2 \cdot a$
$(ax^2+bx+c)^2$	x^2	ax^2+bx+c	$2(ax^2+bx+c)^1 \cdot (2ax+b)$
$(ax^3+bx^2+c)^3$	x^3	ax^3+bx^2+c	$3(ax^3+bx^2+c)^2 \cdot (3ax^2+2bx)$
$(\sqrt{t}+2t)^4$	t^4	$\sqrt{t}+2t$	$4(\sqrt{t}+2t)^3 \cdot \left(\frac{1}{2\sqrt{t}}+2\right)$

The pattern should be very evident. The derivative of these power function compositions proceed as follows: The derivative of the outside power function is performed as usual, on the "large scale" (while the inside function stays intact and goes along for the ride). Then, this result is multiplied by the derivative of the inside function. Or, succinctly,

$$\boxed{\frac{d}{dx}(g(x))^p = p(g(x))^{p-1} \cdot g'(x)} \tag{3.12}$$

Note that the original, regular, boring power rule follows this rule, too — so we don't actually have two rules. Observe:

$$\frac{d}{dx}(x)^5 = 5(x)^4 \cdot \frac{d}{dx}(x) = 5x^4 \cdot 1 = 5x^4$$

With this new and improved power rule, we've significantly expanded the suite of functions we can attack with the derivative operation. The mathematical lawyers among you will raise objections, saying, "Look, building a table with a small number of examples and trying to claim you've discovered a global rule is not how this works!" And you're right. So for those of you who are interested, this section will end with an actual derivation of this expanded power rule.

EX 1 Find the derivatives of $(x^2+1)^2$, $(x^3-4)^3$, and $3\sqrt[3]{x^2+1}+x^5$.

The first function is in the form $(g(x))^p$ where $g(x) = x^2+1$. Using (3.12), we deal with the power of 2 as usual, and then multiply by the derivative of the inner function g:

$$\frac{d}{dx}(x^2+1)^2 = 2(x^2+1)^1 \cdot \frac{d}{dx}(x^2+1) = 2(x^2+1)(2x) = 4x(x^2+1)$$

For the second function, we have (a bit more quickly):

$$\frac{d}{dx}(x^3-4)^3 = 3(x^3-4)^2 \cdot \frac{d}{dx}(x^3-4) = 3(x^3-4)^2(3x^2) = 9x^2(x^3-4)$$

For the third, we have two terms to worry about — but the second term is just a regular power rule:

$$\begin{aligned}\frac{d}{dx}(3\sqrt[3]{x^2+1}+x^5) &= 3 \cdot \frac{d}{dx}(x^2+1)^{1/3} + \frac{d}{dx}(x^5)\\ &= 3 \cdot \frac{1}{3}(x^2+1)^{-2/3} \cdot \frac{d}{dx}(x^2+1) + 5x^4\\ &= (x^2+1)^{-2/3}(2x) + 5x^4 = \frac{2x}{(x^2+1)^{2/3}} + 5x^4 \quad \blacksquare\end{aligned}$$

You Try It

(1) Find the derivative of $\sqrt{3+x^3}$.
(2) Find ds/dt for $s = 1/(t^2+1)$.
(3) Find dy/dx for $y = \sqrt[3]{x^2+2} + (3-5x)^4$.

Antiderivatives Involving Compositions and the Power Rule

It's time to introduce a new word, because we'll use it a lot. At some point in the distant future, we will start referring to antiderivatives as "integrals". As a consequence, the function at hand in an antiderivative is the *integrand.* For example, in the antiderivative $\int 9x^2(x^3-4)^2\,dx$, we call $9x^2(x^3-4)^2$ the *integrand.* We want to use this term now because it's easier than calling it the antiderivigrand (which is not a word), or even worse, "that function whose antiderivative we are seeking."

Anyway, we know how derivatives work when we have a power rule enhanced by a composition:

$$\frac{d}{dx}(g(x))^p = p(g(x))^{p-1} \cdot g'(x)$$

At the top level, this derivative works just like the regular power rule. If we hide the $g(x)$ for a second by just masking it as X, the derivative of X^p has the structure pX^{p-1}. When we do a derivative, we pull the old exponent down in front, and put in a new exponent that's one less. So to do an antiderivative, it looks like we should add one back to the exponent. But, is this next expression correct?

$$\int X^p\,dx = X^{p+1} + C$$

No, that's not correct, it's slightly off. Just like in Sec. 3.5, we know that since the derivative of X^{p+1} looks like $(p+1)X^p$, we need to have a constant waiting in front to cancel that $p+1$; the correct adjusted relationship is

$$\frac{d}{dx}\frac{1}{p+1}X^{p+1} = X^p$$

and so now we have a correct antiderivative relationship, too:

$$\int X^p\,dx = \frac{1}{p+1}X^{p+1} + C$$

Now we have to remember that the X is masking the inner function of the composition. So is this next statement correct?

$$\int (g(x))^p \, dx = \frac{1}{p+1}(g(x))^{p+1} + C$$

Nope. We know that if we take the derivative of the right hand side, we'd pop out a $g'(x)$ as part of the result. So to correct this relation, we need to say:

$$\int g'(x)(g(x))^p \, dx = \frac{1}{p+1}(g(x))^{p+1} + C$$

And now we can upgrade out previous, simpler, pair of equivalent statements from Sec. 3.5:

$$\frac{d}{dx}x^p = px^{p-1} \quad \Leftrightarrow \quad \int x^p \, dx = \frac{1}{p+1}x^{p+1} + C$$

with this new, more impressive pair of equivalent statements:

$$\boxed{\frac{d}{dx}(g(x))^p = p(g(x))^{p-1} \cdot g'(x)} \Leftrightarrow$$

$$\boxed{\int g'(x)(g(x))^p \, dx = \frac{1}{p+1}(g(x))^{p+1} + C} \quad (3.13)$$

The sad truth is that in most integrands of this type, we won't see an expression that looks *exactly* like $g'(x)(g(x))^p$, but rather something that is quite close. But, if we don't see $g'(x)$ exactly, we need to see something only different by a multiplicative constant — otherwise, we won't be able to solve it yet. Just remember the fundamental rule of math. "If you need something, put it there!" We are in charge of these expressions, and if we need to see, say, a 3 in front of the integrand, then dang it, we can put a 3 there. We just have to remember to also multiply by $1/3$ so that we're multiplying in net by 1.

One other thing to look out for is the ordering of the expressions in each rule. When looking at the derivative, we order the power function part first, and then the derivative of the inside function; note the position of $g'(x)$ in

$$\frac{d}{dx}(g(x))^p = p(g(x))^{p-1} \cdot \mathbf{g'(x)}$$

But when looking at the antiderivative rule, the $g'(x)$ is out front:

$$\int \mathbf{g'(x)}(g(x))^p \, dx = \frac{1}{p+1}(g(x))^{p+1} + C$$

The flip-flopped arrangement in the derivative vs antiderivative rule is just a book-keeping convention, don't let it get you mixed up.

EX 2 Find $\int 9x^2(x^3 - 4)^2\,dx$.

Let's think through the solution in two ways. One, just using what we know about derivatives to piece together an antiderivative. And then second, by setting up for use of Eq. (3.13).

Method 1: Using our smarts! The *integrand* (see, we really needed that word) contains the composition $f(g(x)) = (x^3 - 4)^2$, where $f(x) = x^2$ and $g(x) = x^3 - 4$. So then,

- Since the integrand is supposedly a derivative, with exponent 2, the original parent function must have been based on $(x^3 - 4)^3$.
- The derivative of $(x^3-4)^3$ would be $3(x^3-4)^2\cdot(3x^2) = 9x^2(x^3-4)^2$
- And that wraps it up; since we know the derivative of $(x^3 - 4)^3$ is $9x^2(x^3 - 4)^2$, then we also have the equivalent antiderivative:

$$\int 9x^2(x^3 - 4)^2\,dx = (x^3 - 4)^3 + C$$

(Don't forget the arbitrary constant $+C$!) That worked really well, but more often, the expressions won't settle out that nicely. In those cases, we need to be more systematic and use Eq. (3.13).

Method 2: Setting up for Eq. (3.13). In matching $9x^2(x^3 - 4)^2$ to $g'(x)(g(x))^p$ we see that $g(x) = x^3 - 4$. Now, with that, we would like to see $g'(x) = 3x^2$ at the front of the integrand. There is a 3 available, but it's buried in the 9. So, let's yank it out:

$$\int 9x^2(x^3 - 4)^2\,dx = \int \mathbf{3}\cdot\mathbf{3}x^2(x^3 - 4)^2\,dx = \mathbf{3}\int (\mathbf{3}x^2)(x^3 - 4)^2\,dx$$

Now we've built an exact match to $g'(x)(g(x))^2$, and we can apply Eq. (3.13):

$$\begin{aligned}\int \mathbf{9}x^2(x^3 - 4)^2\,dx &= \mathbf{3}\int (\mathbf{3}x^2)(x^3 - 4)^2\,dx \\ &= \mathbf{3}\left(\frac{1}{3}(x^3 - 4)^3 + C\right) \\ &= (x^3 - 4)^3 + 3C = (x^3 - 4)^3 + C\end{aligned}$$

(Note the game with the arbitrary constant: when we multiply the 3 through the parenthetical expression in the second to last step, we actually produce a $3C$. But if C is arbitrary, so is $3C$ ... we might as well keep calling it C. We often say that the arbitrary constant C *absorbed* the constant 3. In most cases, we won't see the intermediate step, we'll just keep C called C no matter how many constants it absorbs.) ■

Here's another variation on the antiderivative in EX 2. Let's go straight toward the use of Eq. (3.13).

EX 3 Find $\int x^2(x^3-4)^2\,dx$.

In matching $x^2(x^3-4)^2$ to $g'(x)(g(x))^p$ we see that $g(x)=x^3-4$. Now, with that, is the x^2 sitting out front equal to $g'(x)$ exactly? No. If $g(x)=x^3-4$, we want to see $g'(x)=3x^2$ out front. We're missing the 3, and we don't have a larger constant we can borrow it from as in EX 2. So we need to introduce the 3 ourselves. As said above, we are in charge of the integrand, and if we want a 3 in front of the x^2, then dang it, we can put a 3 in front of it! We just have to do it smartly, and include a multiplication by 1/3 as well. So let's get to it:

$$\int x^2(x^3-4)^2\,dx = \int \mathbf{\frac{1}{3}}\cdot\mathbf{3}x^2(x^3-4)^2\,dx = \mathbf{\frac{1}{3}}\int 3x^2(x^3-4)^2\,dx$$

Then we apply Eq. (3.13):

$$\begin{aligned}\int x^2(x^3-4)^2\,dx &= \mathbf{\frac{1}{3}}\int \mathbf{3}x^2(x^3-4)^2\,dx\\ &= \mathbf{\frac{1}{3}}\left(\frac{1}{3}(x^3-4)^3+C\right) = \frac{1}{9}(x^3-4)^3+C\end{aligned}$$

Again we see the game with the arbitrary constant: in the last step, we technically should see $C/3$ ... but C *absorbed* the 1/3. Together, joining the beginning to the end:

$$\int x^2(x^3-4)^2\,dx = \frac{1}{9}(x^3-4)^3+C \quad ■$$

You Try It

(4) Find the antiderivatives

(a) $\int 24x^2(2x^3-5)^4\,dx$, (b) $\int 11x^2(2x^3-5)^4\,dx$,

(c) $\int x^2(2x^3-5)^4\,dx$

Don't forget that the power rule works with fractional exponents, too!

EX 4 Find $\int 3x\sqrt{x^2+1}\,dx$.

We need to see exponents explicitly, so let's write the integrand as $3x(x^2+1)^{1/2}$. In preparation for matching to Eq. (3.13), we identify $g(x) = x^2+1$, which requires $g'(x) = 2x$. Now, we don't see that $2x$ exactly, so let's make the proper adjustments:

$$\int 3x\sqrt{x^2+1}\,dx = \int 3\cdot\frac{1}{2}\cdot 2x(x^2+1)^{1/2}\,dx = \frac{3}{2}\int 2x(x^2+1)^{1/2}\,dx$$

And now we have an exact match to Eq. (3.13) in the integrand, and can proceed by bumping the given power up by 1 to restore the derivative to its parent function:

$$\begin{aligned}\int 3x\sqrt{x^2+1}\,dx &= \frac{3}{2}\int 2x(x^2+1)^{1/2}\,dx = \frac{3}{2}\left(\frac{1}{1/2+1}(x^2+1)^{1/2+1}+C\right)\\ &= \frac{3}{2}\left(\frac{2}{3}(x^2+1)^{3/2}+C\right) = (x^2+1)^{3/2}+C\end{aligned}$$

Wrapping up,

$$\int 3x\sqrt{x^2+1}\,dx = (x^2+1)^{3/2}+C \quad \blacksquare$$

Let's try that again, with a slightly different antiderivative:

EX 5 Find $\int x\sqrt{x^2+1}\,dx$.

In the integrand, we are working with $(g(x))^{1/2}$ for $g(x) = x^2+1$. Since $g'(x) = 2x$, it would be just great to have $2x$ leading the integrand. But it's not, and so we have to play the game of constants to put it there and apply Eq. (3.13):

$$\begin{aligned}\int x\sqrt{x^2+1}\,dx &= \frac{1}{2}\int 2x(x^2+1)^{1/2}\,dx = \frac{1}{2}\left(\frac{1}{3/2}(x^2+1)^{3/2}+C\right)\\ &= \frac{1}{2}\left(\frac{2}{3}(x^2+1)^{3/2}+C\right) = \frac{1}{3}(x^2+1)^{3/2}+C\end{aligned}$$

Together,

$$\int x\sqrt{x^2+1}\,dx = \frac{1}{3}(x^2+1)^{3/2}+C \quad \blacksquare$$

See if you can track the thinking in this rapid result, using a variation on EX 4 and EX 5:

EX 6 Find $\int 4x\sqrt{x^2+1}\,dx$.

$$\int 4x\sqrt{x^2+1}\,dx = 4\cdot\frac{1}{2}\int 2x(x^2+1)^{1/2}\,dx$$
$$= 2\left(\frac{2}{3}(x^2+1)^{3/2}+C\right) = \frac{4}{3}(x^2+1)^{3/2}+C \quad \blacksquare$$

You Try It

(5) Find the antiderivatives

$$(a)\ \frac{x}{\sqrt{x^2+1}}\,dx \ ; \quad (b)\ \int \frac{3x}{\sqrt{x^2+1}}\,dx \ ; \quad (c)\ \int \frac{x}{3\sqrt{x^2+1}}\,dx$$

The more you practice with this antiderivative process, the quicker you can fight through the solutions. Eventually, you should find that you solve these by thinking backwards through the derivative process, rather than relying on rote application of Eq. (3.13). Understanding the process is always much better than reliance on formulas, but we have to start somewhere.

Additionally, practicing with antiderivatives of more complicated power functions gets us ready for antiderivatives of more intimidating function types. Here are a few last examples for you to wrap up with.

You Try It

(6) Find all of the following antiderivatives:

$$(a)\ \int x^3(x^4+3)^3\,dx \ , \quad (b)\ \int 3x^3(x^4+3)^3$$
$$(c)\ \int 10x^3(x^4+3)^3\,dx \ , \quad (d)\ \int -4x^3(x^4+3)^3\,dx$$

You Try It

(7) One of these two antiderivatives can be solved with the current technique, one cannot. Identify which can be solved, and solve it!

$$(a)\ \int \frac{x^2}{(x^3+1)^{3/4}}\,dx \ , \quad (b)\ \int \frac{x}{(x^3+1)^{3/4}}\,dx$$

Into the Pit!!

Derivation of the Power Rule for Compositions

Are you the type of person who says, "I don't want to know how my car works, I just want to drive it"? Or are you the type of person who lifts the hood to take a look at the engine? If you are one of the latter, this section is for you! There are no additional exercises here, just information.

The goal here is to derivative formula (3.12). First, we have to set a couple of restrictions. The function $g(x)$ involved in the composition $g(x)^p$ must be differentiable; this should not be surprising — since $g'(x)$ shows up in the formula for the derivative of $g(x)^p$, it's logical to expect that $g'(x)$ always exists. The second restriction is not ultimately necessary, but it leads to a simpler derivation of the formula. In the the following sequence of events leading to the derivative formula we want, we must require that the function $g(x)$ is one-to-one. There is a place in this argument where the possibility of having g repeat a value at two or more locations could lead to bad news.

Here we go. First, you should already know how to form the limit definition of the derivative of $(g(x))^p$:

$$\frac{d}{dx}(g(x))^p = \lim_{h\to 0} \frac{(g(x+h))^p - (g(x))^p}{h}$$

This looks pretty horrible, right? Well, get ready, because we're going to exploit a rule of math which says: sometimes you have to make something look a lot worse before you can make it look better. We will multiply the limit expression up and down by something that will make it look worse than it already does:

$$\frac{d}{dx}(g(x))^p = \lim_{h\to 0} \left(\frac{(g(x+h))^p - (g(x))^p}{h} \cdot \frac{g(x+h) - g(x)}{g(x+h) - g(x)} \right)$$

For reasons which will soon become clear, we're going to exchange the denominators:

$$\frac{d}{dx}(g(x))^p = \left(\lim_{h\to 0} \frac{(g(x+h))^p - (g(x))^p}{g(x+h) - g(x)} \cdot \frac{g(x+h) - g(x)}{h} \right)$$

One of our known limit laws is that under proper circumstances, the limit of a product is the product of the limits, so we can write this as:

$$\frac{d}{dx}(g(x))^p = \lim_{h\to 0} \frac{(g(x+h))^p - (g(x))^p}{g(x+h) - g(x)} \cdot \lim_{h\to 0} \frac{g(x+h) - g(x)}{h} \qquad (3.14)$$

Do you recognize the limit on the right? You should! It's simply $g'(x)$, since

$$g'(x) = \lim_{h \to 0} \frac{g(x+h) - g(x)}{h}$$

But, what is that limit on the left? To figure this one out, we need a bit of smoke and mirrors. We're going to change variables using the following scheme / facts:

- Let's fix a value of x in mind, and then rename the term $g(x+h)$ to be called y.
- Let's also now rename the term $g(x)$ to be a.
- In the original limit, we have $h \to 0$; with the renaming $g(x+h) = y$, then $h \to 0$ means $y \to g(x)$. And since we've renamed $g(x)$ to be "a", we have the following fact:

$$h \to 0 \quad \text{is the same as } y \to a$$

Putting all this information together, we can reconfigure the limit on the left:

$$\lim_{h \to 0} \frac{(g(x+h))^p - (g(x))^p}{g(x+h) - g(x)} = \lim_{y \to a} \frac{y^p - a^p}{y - a}$$

Hopefully, you remember that there are actually *two* ways to write the limit definition of a derivative, one using $h \to 0$ and the other using $x \to a$ or, in this case, $y \to a$. The limit we have now is just the limit definition of the derivative of plain old y^p at $y = a$, which we know to be py^{p-1} at $y = a$, or pa^{p-1}. Since "a" was just the temporary name for $g(x)$, we can complete this sequence of events:

$$\lim_{h \to 0} \frac{(g(x+h))^p - (g(x))^p}{g(x+h) - g(x)} = \lim_{y \to a} \frac{y^p - a^p}{y - a} = pa^{p-1} = p(g(x))^{p-1}$$

Now we have both limits in (3.14) reduced to something much simpler. We've taken an ugly expression, made it look a lot uglier, but eventually whittled it down to something simple. The final chain of events is:

$$\begin{aligned} \frac{d}{dx}(g(x))^p &= \lim_{h \to 0} \frac{(g(x+h))^p - (g(x))^p}{g(x+h) - g(x)} \cdot \lim_{h \to 0} \frac{g(x+h) - g(x)}{h} \\ &= p(g(x))^{p-1} \cdot g'(x) \end{aligned}$$

And that is where we get the power rule for derivatives of compositions in the form $(g(x))^p$!

This discussion started with the warning that the steps only work when a differentiable function $g(x)$ is also one-to-one. (IOI FFT: Do you see where the argument might break down if we allow $g(x)$ to have the same value at two or more different locations? IOI) Ultimately this restriction is not necessary, but at this point it's reasonable to postulate that the same formula would hold for any differentiable $g(x)$, not just those that are one-to-one. If you want to see the full explanation of this formula that's good for ALL differentiable functions $g(x)$, then make a note in your calendar to try out a course in Real Analysis (sometimes called Advanced Calculus) in a couple of years. Heck, you should do that anyway!

Have You Learned...

- How to find derivatives of power functions which are constructed via composition as $(g(x))^p$?

$$\frac{d}{dx}(g(x))^p = p(g(x))^{p-1} \cdot g'(x)$$

- How to find antiderivatives of these enhanced power functions?

$$\int g'(x)(g(x))^p\, dx = \frac{1}{p+1}(g(x))^{p+1} + C$$

The Power Rule on Steroids — Problem List

The Power Rule on Steroids — You Try It

These appeared above; solutions begin on the next page.

(1) Find the derivative of $\sqrt{3+x^3}$.
(2) Find ds/dt if $s = 1/(t^2+1)$.
(3) Find dy/dx for $y = \sqrt[3]{x^2+2} + (3-5x)^4$.
(4) Find the antiderivatives

$$(a) \int 24x^2(2x^3-5)^4\,dx \quad , \quad (b) \quad \int 11x^2(2x^3-5)^4\,dx \quad ,$$
$$(c) \int x^2(2x^3-5)^4\,dx$$

(5) Find the antiderivatives

$$(a) \frac{x}{\sqrt{x^2+1}}\,dx \quad , \quad (b) \int \frac{3x}{\sqrt{x^2+1}}\,dx \quad , \quad (c) \int \frac{x}{3\sqrt{x^2+1}}\,dx$$

(6) Find all of the following antiderivatives:

$$(a) \int x^3(x^4+3)^3\,dx \quad , \quad (b) \int 3x^3(x^4+3)^3$$
$$(c) \int 10x^3(x^4+3)^3\,dx \quad , \quad (d) \int -4x^3(x^4+3)^3\,dx$$

(7) One of these two antiderivatives can be solved with the current technique, one cannot. Identify which can be solved, and solve it!

$$(a) \int \frac{x^2}{(x^3+1)^{3/4}}\,dx \quad , \quad (b) \int \frac{x}{(x^3+1)^{3/4}}\,dx$$

The Power Rule — Practice Problems

Try these as you get the hang of the You Try It problems. Solutions to these problems are available in Sec. A.3.6.

(1) Find dy/dx for $y = (2x^2-5)^6$.
(2) Find $h'(t)$ if $h(t) = \sqrt[4]{t^2+2}$.
(3) Find the instantaneous rate of change of $f(x) = \sqrt[5]{(2x^4+x+1)^3}$ at $x = 0$.
(4) Find the equation of the line tangent to $y = \dfrac{3}{(x^2-4)^2+16}$ at $x = 2$.

(5) Analyze the heck out of the graph of $f(x) = 2/(x^3+1)$. ("Analyze the heck out of" is a very technical phrase meaning "find all horizontal and vertical asymptotes, roots, and locations of horizontal tangent lines".) Produce a sketch of the curve based on your work.

(6) Find the antiderivatives

$$(a) \int x^3(1-x^4)^3\,dx \quad ; \quad (b) \int 11x^3(1-x^4)^3\,dx$$

(7) Find the antiderivatives

$$(a) \int \frac{s^2}{\sqrt{s^3+1}}\,ds \quad ; \quad (b) \int \frac{s^2}{3\sqrt{s^3+1}}\,ds$$

(8) Find $\int (2x^3+1)(x^4+2x+3)^3\,dx$.

The Power Rule — Challenge Problems

Try these problems to test your skills with the ideas in this section. Solutions to these problems are available in Sec. B.3.6.

(1) Analyze the heck out of the graph of $y = 1/((x^2-1)^2+3)$ (see Practice Problem 2 for terminology). Produce a sketch of the curve based on your work — I hereby declare this as the "cowboy hat function".

(2) Find the antiderivative of $f(x) = 2x/(x^2-1)^2$ that goes through the point $(3, 7/8)$.

(3) Find the derivative and antiderivative of $H(x) = \sqrt{2x-5} + \dfrac{1}{3x+7}$.

The Power Rule on Steroids — You Try It — Solved

(1) Find the derivative of $\sqrt{3+x^3}$.

$$\square \frac{d}{dx}(3+x^3)^{1/2} = \frac{1}{2}(3+x^3)^{-1/2}\frac{d}{dx}(3+x^3)$$
$$= \frac{1}{2}(3+x^3)^{-1/2}(3x^2) = \frac{3x^2}{2\sqrt{3+x^3}} \quad \blacksquare$$

(2) Find ds/dt if $s = 1/(t^2+1)$.

$$\square \frac{ds}{dt} = \frac{d}{dt}(t^2+1)^{-1} = -(t^2+1)^{-2}\cdot\frac{d}{dt}(t^2) = -\frac{2t}{(t^2+1)^2} \quad \blacksquare$$

(3) Find dy/dx for $y = \sqrt[3]{x^2+2} + (3-5x)^4$.

☐ This function contains two terms, which much each be tackled with the expanded power rule.

$$\frac{d}{dx}\sqrt[3]{x^2+2} = \frac{d}{dx}(x^2+2)^{1/3} = \frac{1}{3}(x^2+2)^{-2/3}\cdot\frac{d}{dx}(x^2+2)$$
$$= \frac{1}{3}(x^2+2)^{-2/3}(2x) = \frac{2x}{3\sqrt[3]{(x^2+2)^2}} \quad \blacksquare$$

(4) Find the antiderivatives

(*a*) $\int 24x^2(2x^3-5)^4\,dx$, (*b*) $\int 11x^2(2x^3-5)^4\,dx$,

(*c*) $\int x^2(2x^3-5)^4\,dx$

☐ All of the integrands are variations of $g'(x)(g(x))^4$ in which $g(x) = 2x^3-5$. For this $g(x)$, we have $g'(x) = 6x^2$. In each antidervative, we ask: can we spot that exact $g'(x)$ alongside the power function itself, or do we see an expression which is different from $g'(x)$ by some multiplicative constant? If the latter, we can introduce (properly) whatever constant is needed to force in an exact match to $g'(x)$, and then use Eq. (3.13).

In (a), we can create an exact match to $g'(x)$ by splitting up the coefficient of 24 into $4\cdot 6$,

$$\int 24x^2(2x^3-5)^4\,dx = 4\left(\int 6x^2(2x^3-5)^4\,dx\right)$$

Inside the big parentheses, we now have an exact match to the form $\int g'(x)(g(x))^p\,dx$, and we can complete the solution using Eq. (3.13):

$$\begin{aligned}\int 24x^2(2x^3-5)^4\,dx &= 4\left(\int 6x^2(2x^3-5)^4\,dx\right)\\ &= 4\left(\frac{1}{5}(2x^3-5)+C\right) = \frac{4}{5}(2x^3-5)+C\end{aligned}$$

For (b), we can create an exact match to $g'(x)$ by multiplying with both 6 and $\frac{1}{6}$; let's also make the solution more streamlined:

$$\begin{aligned}\int 11x^2(2x^3-5)^4\,dx &= \frac{11}{6}\left(\int 6x^2(2x^3-5)^4\,dx\right)\\ &= \frac{11}{6}\left(\frac{1}{5}(2x^3-5)+C\right) = \frac{11}{30}(2x^3-5)+C\end{aligned}$$

Finally for (c), let's just get to it:

$$\begin{aligned}\int x^2(2x^3-5)^4\,dx &= \frac{1}{6}\left(\int 6x^2(2x^3-5)^4\,dx\right)\\ &= \frac{1}{6}\left(\frac{1}{5}(2x^3-5)+C\right) = \frac{1}{30}(2x^3-5)+C \quad \blacksquare\end{aligned}$$

---

(5) Find the antiderivatives

$$(a)\ \frac{x}{\sqrt{x^2+1}}\,dx \quad , \quad (b)\int \frac{3x}{\sqrt{x^2+1}}\,dx \quad , \quad (c)\int \frac{x}{3\sqrt{x^2+1}}\,dx$$

□ All of the integrands are variations of $g'(x)(g(x))^{-1/2}$ in which $g(x) = x^2+1$, thus requiring $g(x)' = 2x$ for an exact match to Eq. (3.13). Let's play with the constants in each case to force a match to $g(x)'$. (I'm presuming you've examined the solution to YTI 4, so that the solutions here can now be more concise.)

$$\begin{aligned}(a)\int \frac{x}{\sqrt{x^2+1}}\,dx &= \frac{1}{2}\int (2x)(x^2+1)^{-1/2}\,dx\\ &= \frac{1}{2}\left(\frac{1}{1/2}(x^2+1)^{1/2}+C\right)\\ &= \frac{1}{2}\left(2\sqrt{x^2+1}+C\right) = \sqrt{x^2+1}+C\end{aligned}$$

$$(b)\ \int \frac{3x}{\sqrt{x^2+1}}\,dx = 3\cdot\frac{1}{2}\int (2x)(x^2+1)^{-1/2}\,dx$$
$$= \frac{3}{2}\left(\frac{1}{1/2}(x^2+1)^{1/2}+C\right)$$
$$= \frac{3}{2}\left(2\sqrt{x^2+1}+C\right) = 3\sqrt{x^2+1}+C$$

$$(c)\ \int \frac{x}{3\sqrt{x^2+1}}\,dx = \frac{1}{3}\cdot\frac{1}{2}\int (2x)(x^2+1)^{-1/2}\,dx$$
$$= \frac{1}{6}\left(\frac{1}{1/2}(x^2+1)^{1/2}+C\right)$$
$$= \frac{1}{6}\left(2\sqrt{x^2+1}+C\right) = \frac{1}{3}\sqrt{x^2+1}+C \quad \blacksquare$$

---

(6) Find all of the following antiderivatives:

$$(a)\ \int x^3(x^4+3)^3\,dx \quad , \quad (b)\ \int 3x^3(x^4+3)^3\,dx$$
$$(c)\ \int 10x^3(x^4+3)^3\,dx \quad , \quad (d)\ \int -4x^3(x^4+3)^3\,dx$$

□ What?! We're supposed to solve four?? Well, yes. But also, no. Since they are all variations of the same antiderivative, let's combine them all into one representative problem, using a constant a:

$$\int ax^3(x^4+3)^3\,dx$$

Now, with our inner function $g(x) = x^3+3$, then we'd like to see exactly $4x^3$ out in front as $g'(x)$. So let's make that happen by multiplying in both 4 and $\frac{1}{4}$:

$$\frac{1}{4}\int a(4x^3)(x^4+3)^3\,dx = \frac{a}{4}\int (4x^3)(x^4+3)^3\,dx$$
$$= \frac{a}{4}\left(\frac{1}{4}(x^4+3)^4+C\right) = \frac{a}{16}(x^4+3)^4+C$$

So, now we can go back and insert the proper value of a from the four variants of this antiderivative: (a) $a=1$; (b) $a=3$; (c) $a=10$, and

(d) $a = -4$:

$$(a) \int x^3(x^4+3)^3\,dx = \frac{1}{16}(x^4+3)^4 + C$$

$$(b) \int 3x^3(x^4+3)^3\,dx = \frac{3}{16}(x^4+3)^4 + C$$

$$(c) \int 10x^3(x^4+3)^3\,dx = \frac{10}{16}(x^4+3)^4 + C = \frac{5}{8}(x^4+3)^4 + C$$

$$(d) \int -4x^3(x^4+3)^3\,dx = \frac{-4}{16}(x^4+3)^4 + C = -\frac{1}{4}(x^4+3)^4 + C \quad \blacksquare$$

---

(7) One of these two antiderivatives can be solved with the current technique, one cannot. Identify which can be solved, and solve it!

$$(a) \int \frac{x^2}{(x^3+1)^{3/4}}\,dx \quad , \quad (b) \int \frac{x}{(x^3+1)^{3/4}}\,dx$$

□ Both of these are based on the structure $(x^3+1)^{3/4}$, in which $g(x) = x^3+1$. So we'd like to see $g'(x) = 3x^2$ as part of the integrand, too. In (a) we have x^2, which is only different by a multiplicative constant. In (b), we have x, which is a fundamentally different function than x^2, and there is no way to account for that. (FFT: Why can't we just multiply in another x to bump x to x^2 and also multiply in by $\frac{1}{x}$ to account for that?) So (a) is the keeper:

$$\begin{aligned}
\int \frac{x^2}{(x^3+1)^{3/4}}\,dx &= \frac{1}{3}\int (3x^2)(x^3+1)^{-3/4}\,dx \\
&= \frac{1}{3}\left(\frac{1}{-3/4+1}(x^3+1)^{-3/4+1} + C\right) \\
&= \frac{1}{3}\left(4(x^3+1)^{1/4} + C\right) = \frac{4}{3}(x^3+1)^{1/4} + C
\end{aligned}$$

Concluding,

$$\int \frac{x^2}{(x^3+1)^{3/4}}\,dx = \frac{4}{3}(x^3+1)^{1/4} + C \quad \blacksquare$$

Chapter 4

Abandon Hope All Ye Who Enter Here

4.1 How to Make Exponential and Logarithmic Functions Worse

Introduction

Here we see derivatives and antiderivatives involving the exponential function e^x and the natural logarithm function $\ln(x)$. As a bonus, once we know how to find derivatives of exponential functions, we can also play with derivatives of hyperbolic functions! First, we'll see the operational aspects of these derivatives. At the end of the section, we'll take a step *Into the Pit!* and see some theoretical underpinnings of the derivative rules.

The Derivative and Antiderivative of e^x and $e^{g(x)}$

The number e, called Euler's number, has several definitions. One is that

$$e = \lim_{x\to\infty}\left(1+\frac{1}{x}\right)^x$$

Another is that e is defined to be the number for which

$$\lim_{h\to 0}\frac{e^h-1}{h} = 1$$

This latter definition comes in handy when determining the derivative of the function $y = e^x$:

$$\frac{d}{dx}(e^x) = \lim_{h\to 0}\frac{e^{x+h}-e^x}{h} = \lim_{h\to 0}\frac{e^x(e^h-1)}{h} = e^x\cdot\lim_{h\to 0}\frac{e^h-1}{h} = e^x\cdot(1) = e^x$$

So e^x is its own derivative! In fact, it is the only function with this property. In summary, both the derivative and antiderivative of e^x are as easy as it gets:

$$\boxed{\frac{d}{dx}e^x = e^x} \quad\Leftrightarrow\quad \boxed{\int e^x\,dx = e^x + C} \tag{4.1}$$

The only way to make these even mildly interesting is to combine them with other functions, whose derivatives and antiderivatives are also already known.

EX 1 Here is an assortment of examples of derivatives and antiderivatives involving e^x:

$$\frac{d}{dx}(e^x + 2x - 5) = e^x + 2$$

$$\int e^x + 2x - 5\,dx = e^x + x^2 - 5x + C$$

$$\frac{d}{dx}\left(-e^x + \sqrt{x}\right) = \frac{d}{dx}\left(-e^x + x^{1}2\right) = -e^x + \frac{1}{2\sqrt{x}}$$

$$\int \left(-e^x + \sqrt{x}\right)\,dx = -e^x + \frac{2}{3}x^{3/2} + C = -e^x + \frac{2}{3}x\sqrt{x} + C$$

$$\frac{d}{dx}(x^3 - 3e^x) = 3x^2 - 3e^x$$

$$\int x^3 - 3e^x\,dx = \frac{1}{4}x^4 - 3e^x + C \quad \blacksquare$$

You Try It

(1) Find the equation of the line tangent to $y = x^4 + 2e^x$ at (0,2).

The way to make these more interesting is to find the derivatives of things like e^{-x}, e^{x^2}, e^{2x-1}, $e^{\cos(x)}$, and so on. Note that each of those is an instance of a general composition of the form $e^{g(x)}$. Since the derivative of e^x is merely e^x, it's tempting to conjecture that the derivative of $e^{g(x)}$ is just $e^{g(x)}$. But, sadly, it isn't; there's more to it than that. We'll go through this story in reverse. First, we'll see the method for doing these derivatives, some examples, and how to start the reverse process of antiderivatives, and then we'll see why this new formula works. The payoff of this story is that

$$\boxed{\frac{d}{dx}e^{g(x)} = e^{g(x)} \cdot g'(x)} \tag{4.2}$$

In other words, you do get $e^{g(x)}$ back again as part of the derivative, but you must also multiply by the derivative of $g(x)$, the "inside" function of the composition $e^{g(x)}$. (Does that sound familiar? If not, go have another look at Eq. (3.13).)

EX 2 Here are some examples of derivatives of functions of the form $e^{g(x)}$:

$$\frac{d}{dx}e^{-x} = e^{-x} \cdot \frac{d}{dx}(-x) = e^{-x} \cdot (-1) = -e^{-x}$$
$$\frac{d}{dx}e^{x^2} = e^{x^2} \cdot \frac{d}{dx}(x^2) = e^{x^2} \cdot (2x) = 2xe^{x^2}$$
$$\frac{d}{dx}e^{2x-1} = e^{2x-1} \cdot \frac{d}{dx}(2x-1) = e^{2x-1} \cdot (2) = 2e^{2x-1}$$
$$\frac{d}{dx}e^{\sqrt{x}} = e^{\sqrt{x}} \cdot \frac{d}{dx}(\sqrt{x}) = e^{\sqrt{x}} \cdot \frac{1}{2\sqrt{x}} = \frac{e^{\sqrt{x}}}{2\sqrt{x}} \quad \blacksquare$$

Can you combine derivatives like this with others we already know?

You Try It

(2) Find the following derivatives:

$$\frac{d}{dx}(x^2 - e^{-2x}) \quad ; \quad \frac{d}{dx}e^{\sqrt[4]{x}} \quad ; \quad \frac{d}{dt}\left(e^{1/t} + \frac{1}{t^2}\right)$$

Next, let's try this backwards and think about antiderivatives, starting with one specific example. Since we now know that

$$\frac{d}{dx}e^{-x^2} = -2xe^{-x^2}$$

then we also know an antiderivative:

$$\int -2xe^{-x^2}\,dx = e^{-x^2} + C$$

But, what about this antiderivative:

$$\int 4xe^{-x^2}\,dx$$

This is close to the known antiderivative right above it, but not quite the same. How can we adjust? The secret is in this match:

$$\boxed{\frac{d}{dx}e^{g(x)} = e^{g(x)} \cdot g'(x)} \quad \Leftrightarrow \quad \int \boxed{g'(x)e^{g(x)}\,dx = e^{g(x)} + C} \tag{4.3}$$

In most such antiderivatives, we will not have an integrand that looks exactly like $g'(x)e^{g(x)}$, but rather something close. We can still get it done if the actual integrand vs the ideal integrand $g'(x)e^{g(x)}$ only differ by a multiplicative constant. If that's the case, we can identify $g(x)$ and note what $g'(x)$ *should* look like, and then do cosmetic surgery on the integrand to introduce the right constants in the right places. As an example, lets compare an antiderivative with the template from (4.3):

$$\int 4xe^{-x^2}dx \qquad \Leftrightarrow \qquad \int g'(x)e^{g(x)}dx$$

We identify $g(x) = -x^2$, and so hope to also see $g'(x) = -2x$ as a multiple in the integrand. However, we don't see $-2x$, we see $4x$ instead. But the difference is just a matter of a multiplicative constant, and we can adjust things easily by breaking up the 4 into $(-2)(-2)$:

$$\int 4xe^{-x^2}\,dx = \mathbf{-2}\int(\mathbf{-2}x)e^{-x^2}\,dx$$

and the new integrand now displays exactly the form $g'(x)e^{g(x)}$. So, the antiderivative can and solved:

$$\int 4xe^{-x^2}\,dx = -2\int(-2x)e^{-x^2}\,dx = -2\left(e^{-x^2}+C\right) = -2e^{-x^2}+C$$

(Do you remember why we can write the constant term as just $+C$ rather than $-2C$?)

As a quick note, now that we have Eq. (4.3), we can dispense with (4.1). When $g(x) = x$, then the two parts of (4.3) collapse to:

$$\frac{d}{dx}e^{(x)} = e^x \cdot \frac{d}{dx}(x) = e^x(1) = e^x$$

$$\int(1)e^{(x)}\,dx = e^x + C$$

i.e. we have now stated what we originally knew as (4.1).

Here are more examples of antiderivatives of functions close to the form $\int g'(x)e^{g(x)}dx$:

EX 3 Evaluate the following:

$$A)\int e^{5x}\,dx \quad ; \quad B)\int x^2e^{x^3}\,dx \quad ; \quad C)\int \frac{e^{\sqrt{y}}}{\sqrt{y}}\,dy$$

In each case, we will manipulate the integrand so that, having identified $g(x)$, we find $g'(x)$ precisely and use the antiderivative formula in (4.3). For (A), we try to match e^{5x} to $g'(x)e^{g(x)}$. We identify $g(x) = 5x$, and so we look for $g'(x) = 5$ out front. But we don't see any 5's. So, let's adjust things; remember that we're in charge, and if we need a multiple of 5 on the integrand, then dang it, we'll put in a multiple of 5 on the integrand ... we just have to be sure to also put in a balancing multiple of $1/5$. Once we do that, we're ready to use (4.3):

$$\int e^{5x}\,dx = \frac{1}{5}\int 5e^{5x}\,dx = \frac{1}{5}\left(e^{5x}+C\right) = \frac{1}{5}e^{5x}+C$$

For (B), we identify $g(x) = x^3$, and so we look for $g'(x) = 3x^2$. But we don't see $3x^2$, we see x^2. So, let's adjust things:

$$\int x^2 e^{x^3}\,dx = \frac{1}{3}\int 3x^2 e^{x^3}\,dx = \frac{1}{3}e^{x^3}+C$$

(Again, why can we write $+C$ rather than $+C/3$ as the constant term?) For (C), we see $g(y) = \sqrt{y}$, and so we look for $g'(y) = 1/(2\sqrt{y})$. But we don't see $1/(2\sqrt{y}$, we only see $1/\sqrt{y}$. So, let's adjust things:

$$\int \frac{e^{\sqrt{y}}}{\sqrt{y}}\,dy = 2\int \frac{1}{2\sqrt{y}}e^{\sqrt{y}}\,dy = 2e^{\sqrt{y}}+C \quad \blacksquare$$

You Try It

(3) Find the following antiderivatives:

$$\int (4x-2)e^{x^2-x}\,dx \quad ; \quad \int 5t^3 e^{-t^4}\,dt \quad ; \quad \int \frac{e^{\sqrt{x}+1}}{2\sqrt{x}}\,dx$$

Now that we've seen the formula for the derivative of $e^{g(x)}$ and have used it to help find derivatives and antiderivatives, you should be rabidly eager to see how that formula gets developed. Rather than disrupt the flow of things here, I have set that conversation at the end of this section.

Derivatives and Antiderivatives Involving ln(x) *and* ln($g(x)$)

The derivative of the basic natural logarithm function is:

$$\boxed{\frac{d}{dx}\ln(x) = \frac{1}{x}} \tag{4.4}$$

(the derivation is held off until the end of this section). IT seems like we should also reverse this to give:

$$\int \frac{1}{x}\,dx = \ln(x)+C \tag{4.5}$$

But, there's a catch. Consider the domains of the two functions $1/x$ and $\ln(x)$. The domain of $1/x$ is all numbers $x \neq 0$ (including negative values), while the domain of $\ln(x)$ is only $x > 0$. Therefore, the expression on the left can't really be equal to the expression on the right. Instead, we fix that as:

$$\boxed{\int \frac{1}{x}\,dx = \ln|x| + C} \tag{4.6}$$

The inclusion of the absolute value ensures that any x value which is "OK" for the left side is also "OK" for the right side. In this way, the left half of the hyperbola $y = 1/x$ is not left out of the antiderivative process.

Equation (4.6) patches a hole in the power rule. You may recall that the power rule for antiderivatives from Eq. (3.11),

$$\int x^p\,dx = \frac{1}{p+1}x^{p+1} + C$$

does not work when $p = -1$.[1] We now have the information to close that gap. The derivative formula for $\ln x$ now provides the antiderivative of $1/x$.

If you're thinking ahead, you may be wondering: What the heck is the antiderivative of $\ln x$ itself? Well, it turns out that

$$\int \ln(x)\,dx = x\ln x - x + C$$

But we won't see that "officially" until we learn something called integration by parts, and we won't see that for several chapters in this book, which could mean you'll wait for an entirely different semester of a traditional Calculus sequence. So we won't use this formula, but I wanted you to know it's there so you would not lose sleep over it.

We can easily combine these new formulas for the derivative of $\ln x$ and the antiderivative of $1/x$ with others that are known:

EX 4 Here are some examples of derivatives and antiderivatives using e^x and $\ln x$:

[1] Even if you don't recall that, just try it for $p = -1$. See?

$$\frac{d}{dx}(e^x + \ln(x)) = e^x + \frac{1}{x}$$
$$\int x^2 - \frac{2}{x}\,dx = \frac{1}{3}x^3 - 2\ln|x| + C$$
$$\frac{d}{dx}(x^{3/2} - 3\ln x) = \frac{3}{2}\sqrt{x} - \frac{3}{x}$$
$$\int \frac{2}{x^4} - \frac{5}{x}\,dx = -\frac{2}{3x^3} - 5\ln|x| + C \quad \blacksquare$$

You Try It

(4) Find the derivative of $f(x) = \ln x + \sqrt{x-1}$.

(5) Find the derivative and antiderivative of $f(x) = \sqrt{x} - 2e^x - \dfrac{1}{x}$.

(6) Evaluate $\displaystyle\int \frac{2}{x^2} + \frac{1}{x}\,dx$.

Much like with the exponential function, it's much more fun to differentiate functions like $\ln(x^2-1)$ or $\ln((x+2)^2)$ or $\ln(1-\sqrt{x})$. In order to do that, we need the following formula that expands (4.4):

$$\frac{d}{dx}\ln g(x) = \frac{1}{g(x)}\cdot g'(x) = \frac{g'(x)}{g(x)} \tag{4.7}$$

We see the same "action" as in other extensions of derivative formulas. We still start with the derivative of ln(*whatever*) as 1 over that *whatever*, but then we also have multiply by the derivative of *whatever*. Here are the three derivatives mentioned just above:

EX 5 Examples of derivatives of functions of the form $\ln g(x)$:

$$\frac{d}{dx}\ln(x^2-1) = \frac{1}{x^2-1}\cdot\frac{d}{dx}(x^2-1) = \frac{2x}{x^2-1}$$
$$\frac{d}{dx}\ln((x+2)^2) = \frac{1}{(x+2)^2}\cdot\frac{d}{dx}(x+2)^2 = \frac{2x+4}{(x+2)^2} = \frac{2}{x+2}$$
$$\frac{d}{dx}\ln(1-\sqrt{x}) = \frac{1}{1-\sqrt{x}}\cdot\frac{d}{dx}(1-\sqrt{x}) = \frac{1}{1-\sqrt{x}}\cdot\frac{-1}{2\sqrt{x}}$$
$$= -\frac{1}{2\sqrt{x}(1-\sqrt{x})} \quad \blacksquare$$

Note that the middle derivative could also have been found in a more clever way, with a bit of preparation using properties of logarithms:

$$\frac{d}{dx}\ln((x+2)^2) = \frac{d}{dx}2\ln((x+2)) = 2\cdot\frac{1}{x+2}\cdot\frac{d}{dx}(x+2) = \frac{2}{x+2}$$

You Try It

(7) Find the following derivatives:

$$\frac{d}{dx}(\ln(x^3-1)+e^{x-1}) \quad ; \quad \frac{d}{dx}\ln(e^{2x}+x^{1.7}) \quad ; \quad \frac{d}{dt}\ln(\ln t-2)$$

Next, let's try this backwards and think about antiderivatives. In general, we depend on the matching pair of formulas

$$\boxed{\frac{d}{dx}\ln g(x)=\frac{g'(x)}{g(x)}} \quad \Leftrightarrow \quad \boxed{\int \frac{g'(x)}{g(x)}\,dx=\ln|g(x)|+C} \qquad (4.8)$$

Note the use of absolute values, just as in the simpler Eq. (4.6). In problems of this type, we will be posed an antiderivative which has an integrand that's close to the form $g'(x)/g(x)$. Like before, we can shuffle multiplicative constants tn ensure an explicit match, so that we can use Eq. (4.8).

The more extended expressions involving $\ln(g(x))$ found in (4.8) allows us to dispose of the previous simpler versions in (4.4) and (4.6).

EX 6 Evaluate the following antiderivatives:

$$A)\int \frac{x-2}{x^2-4x+1}\,dx\,;\; B)\int \frac{e^x}{e^x-\pi}\,dx\,;\; C)\int \frac{1}{2\sqrt{y}(\sqrt{y}+1)}\,dy$$

In each case, we'll compare the integrand to the form $g'(x)/g(x)$. The goal is to identify $g(x)$ and, if we don't also see $g'(x)$ exactly, manipulate the expression so that we do see $g'(x)$ exactly. Then we can use the antiderivative formula in (4.8). For (A), we identify $g(x)=x^2-4x+1$, and so we look for $g'(x)=2x-4$. But we don't see $2x-4$, we see $x-2$. Recognizing that $x-2$ is just half of $2x-4$, though,we can write

$$\int \frac{x-2}{x^2-4x+1}\,dx=\frac{1}{2}\int \frac{2x-4}{x^2-4x+1}\,dx$$

The latter integrand now displays the precise form we need, $g'(x)/g(x)$, and we can invoke Eq. (4.8):

$$\begin{aligned}\int \frac{x-2}{x^2-4x+1}\,dx&=\frac{1}{2}\int \frac{2x-4}{x^2-4x+1}\,dx\\&=\frac{1}{2}\left(\ln|x^2-4x+1|+C\right)=\frac{1}{2}\ln|x^2-4x+1|+C\end{aligned}$$

For (B), we identify $g(x) = e^x - \pi$, and so we look for $g'(x) = e^x$. That is already what we see in the numerator, so we can use (4.8) right away:

$$\int \frac{e^x}{e^x - \pi}\, dx = \ln|e^x - \pi| + C$$

The antiderivative (C) is a bit trickier, since the numerator is just 1 and the denominator has multiple terms. But consider what happens if we try to set $g(y) = \sqrt{y} + 1$. We'd expect to see $g'(y) = 1/(2\sqrt{y})$, which we *do* in fact see there:

$$\begin{aligned}\int \frac{1}{2\sqrt{y}(\sqrt{y}+1)}\, dy &= \int \frac{1}{2\sqrt{y}} \frac{1}{\sqrt{y}+1}\, dy \\ &= \int g'(y)\frac{1}{g(y)}\, dy = \ln|g(y)| + C = \ln|\sqrt{y}+1| + C \quad \blacksquare\end{aligned}$$

You Try It

(8) Find the following antiderivatives:

$$\int \frac{x^3}{x^4 - 10}\, dx \quad ; \quad \int \frac{1}{2t+1}\, dt \quad ; \quad \int \frac{2xe^{x^2} + 1}{e^{x^2} + x}\, dx$$

Derivatives and Antiderivatives Involving Some Hyperbolic Functions

In Sec. 1.6, we used exponential functions as building blocks for *hyperbolic functions.* We're now in a position to look at derivatives of these functions as well ... or, at least two of them. We'll save the rest for later.

Definition 1.3 presented the following constructions:

$$\sinh(x) = \frac{e^x - e^{-x}}{2} \qquad \text{and} \qquad \cosh(x) = \frac{e^x + e^{-x}}{2}$$

So now we're ready do determine the following derivatives:

$$\begin{aligned}\frac{d}{dx}\sinh(x) &= \frac{d}{dx}\frac{e^x - e^{-x}}{2} = \frac{1}{2}\left(\frac{d}{dx}e^x - \frac{d}{dx}e^{-x}\right) = \frac{1}{2}(e^x - (-e^x)) \\ &= \frac{e^x + e^{-x}}{2} \\ \frac{d}{dx}\cosh(x) &= \frac{d}{dx}\frac{e^x + e^{-x}}{2} = \frac{1}{2}\left(\frac{d}{dx}e^x + \frac{d}{dx}e^{-x}\right) = \frac{1}{2}(e^x + (-e^x)) \\ &= \frac{e^x - e^{-x}}{2}\end{aligned}$$

and a careful look at the results reveals these relationships:

$$\boxed{\frac{d}{dx}\sinh(x) = \cosh(x)} \quad \text{and} \quad \boxed{\frac{d}{dx}\cosh(x) = \sinh(x)} \tag{4.9}$$

We then also have the immediate counterparts,

$$\boxed{\int \sinh(x)\,dx = \cosh(x) + C} \quad \text{and} \quad \boxed{\int \cosh(x)\,dx = \sinh(x) + C} \tag{4.10}$$

These results provide more insight into why these functions, built out of e^x and e^{-x}, have names of trigonometric functions built into their own names.

The use of these fundamental derivative and antiderivative formulas also extends to compositions. One matching pair is:

$$\boxed{\frac{d}{dx}\sinh(g(x)) = \cosh(g(x)) \cdot g'(x)} \quad \text{and}$$

$$\boxed{\int g'(x)\cosh(g(x))\,dx = \sinh(g(x)) + C} \tag{4.11}$$

Another is:

$$\boxed{\frac{d}{dx}\cosh(g(x)) = \sinh(g(x)) \cdot g'(x)} \quad \text{and}$$

$$\boxed{\int g'(x)\sinh(g(x))\,dx = \cosh(g(x)) + C} \tag{4.12}$$

Here are some examples of their uses:

EX 7 Examples of derivatives of functions of the form $\sinh(g(x))$ and $\cosh(g(x))$:

$$\frac{d}{dx}\cosh(2x^3+5) = \sinh(2x^3+5) \cdot \frac{d}{dx}(2x^3+5) = 6x^2\sinh(2x^3+5)$$

$$\frac{d}{dx}\sinh(\sqrt{x}+1) = \cosh(\sqrt{x}+1) \cdot \frac{d}{dx}(\sqrt{x}+1) = \frac{\cosh(\sqrt{x}+1)}{2\sqrt{x}}$$

$$\frac{d}{dx}\sinh(\cosh(x)) = \cosh(\cosh(x)) \cdot \frac{d}{dx}(\cosh(x)) = \sinh(x)\cosh(\cosh(x)) \quad \blacksquare$$

And in the reverse direction, we have:

EX 8 Evaluate the following antiderivatives:

$$A)\int x\sinh(2-x^2)\,dx\,;\; B)\int e^x\cosh(e^x+1)\,dx\,;\; C)\int \frac{\cosh(t)}{\sinh(t)}\,dt$$

For (A), if we tinker with the coefficient and write

$$\int x \sinh(2 - x^2)\, dx = -\frac{1}{2} \int (-2x) \sinh(2 - x^2)\, dx$$

then in the latter, we have an exact match to 4.12 with $g(x) = 2 - x^2$ and $g'(x) = -2x$, and so

$$\int x \sinh(2 - x^2)\, dx = -\frac{1}{2} \cosh(2 - x^2) + C$$

For (B), we have an immediate match to 4.11 with $g(x) = e^x + 1$ and $g'(x) = e^x$, and so

$$\int e^x \cosh(e^x + 1)\, dx = \sinh(e^x + 1) + C$$

(C) is a bit less obvious, but if we rewrite the antiderivative posed as

$$\int \frac{\cosh(t)}{\sinh(t)}\, dt = \int \frac{1}{\sinh(t)} \cdot \cosh(t)$$

then in the latter, we have an exact match to 4.12 with $g(t) = \sinh(t)$ and $g'(t) = \cosh(t)$, and so

$$\int \frac{\cosh(t)}{\sinh(t)}\, dt = \ln \sinh(t) + C \quad \blacksquare$$

You Try It

(9) Find the following derivatives:

(a) $\dfrac{d}{dx} \sinh(\ln x)$, (b) $\dfrac{d}{dx} \cosh(x^2 + 4x + 4)$,

(c) $\dfrac{d}{dt} \left(\sinh(e^{2t}) + \cosh(2t^{3/2} + t) \right)$

You Try It

(10) Find the following antiderivatives:

(a) $\displaystyle\int e^{\sinh(x)} \cosh(x)\, dx$, (b) $\displaystyle\int 4 \sinh(x) \cosh^3(x)\, dx$,

(c) $\displaystyle\int \frac{\sinh(y) + 2y}{\cosh(y) + y^2}\, dy$

At this point, we've had a lot of fun, but we should take a few moments (or a lot of moments?) to see how the derivative formulas for $e^{g(x)}$ and $\ln(g(x))$ get developed.

Into the Pit!!

Derivation of the Derivative of $e^{g(x)}$

This will be set up and attacked just like in Sec. 3.6. In fact, having seen a very similar derivation there for the derivative of $g(x)^p$, and having seen it here, and then seeing it again a few paragraphs down from here for functions like $\ln(g(x))$, you will have seen it enough that you should be able to carry it out on your own for any new species of functions.

We have to preface this derivation, again, with some warnings. First: the functions $g(x)$ involved here must be differentiable (which is not surprising, since we already know that $g'(x)$ is part of the resulting formula). But we're also going to require that $g(x)$ be one-to-one, so that it never repeats any values. Ultimately this restriction is not necessary, but there is a step in this particular derivation where it helps. But, once we're done, it's reasonable to postulate that the same formula would hold for any differentiable $g(x)$, not just ones that are one-to-one.

Second, we need a reminder that there are two ways to indicate the derivative of a function at a point. The derivative of any function of x at a point $x = a$ can be written as both

$$\lim_{h\to 0}\frac{f(a+h)-f(a)}{h} \qquad \text{AND} \qquad \lim_{x\to a}\frac{f(x)-f(a)}{x-a}$$

In fact, let's make sure to recognize that if the exponential function happened to be written as a function of y, then the derivative of e^y at the point $y = a$ would be written as

$$\lim_{y\to a}\frac{e^y - e^a}{y-a}$$

Further, since we already know that the derivative of e^y is e^y itself, then the derivative of e^y at the point $y = a$ is e^a and so

$$\lim_{y\to a}\frac{e^y - e^a}{y-a} = e^a \tag{4.13}$$

I'm pointing this out just in case, you know, we happen to see this expression soon or something.

Now we're ready! By the fundamental limit definition of the derivative, we know that

$$\frac{d}{dx}e^{g(x)} = \lim_{h\to 0}\frac{e^{g(x+h)} - e^{g(x)}}{h}$$

We're going to exploit a rule of math which says: sometimes you have to make something look a lot worse before you can make it look better. Here, we'll multiply the limit expression up and down by something to make it look at lot worse:

$$\frac{d}{dx}e^{g(x)} = \lim_{h\to 0}\frac{e^{g(x+h)} - e^{g(x)}}{h} = \lim_{h\to 0}\left(\frac{e^{g(x+h)} - e^{g(x)}}{h}\cdot\frac{g(x+h)-g(x)}{g(x+h)-g(x)}\right)$$

Now we can strategically move a few things around:

$$\frac{d}{dx}e^{g(x)} = \left(\lim_{h\to 0}\frac{e^{g(x+h)} - e^{g(x)}}{g(x+h)-g(x)}\cdot\frac{g(x+h)-g(x)}{h}\right)$$

Recall that one of our old limit laws is that under proper circumstances, the limit of a product is the product of the limits, so we can write this as:

$$\frac{d}{dx}e^{g(x)} = \lim_{h\to 0}\frac{e^{g(x+h)} - e^{g(x)}}{g(x+h)-g(x)}\cdot\lim_{h\to 0}\frac{g(x+h)-g(x)}{h}$$

At this point, you should recognize the second of those two limits and say, "A-ha! That's just $g'(x)$!" The first limit takes a bit of work to tidy up. For a fixed value of x, let's name $g(x) = a$. Then while h changes, the term $x + h$ varies, as does $g(x + h)$ — so let's name this as a new variable, y. Under these circumstances, we know a few things:

- The difference quotient in the limit becomes:

$$\frac{e^{g(x+h)} - e^{g(x)}}{g(x+h)-g(x)} = \frac{e^y - e^a}{y-a}$$

- The limit process $h \to 0$ is the same as the limit process $y \to a$. This is because $y = g(x + h)$, so $h \to 0$ means $y \to g(x)$, but we named $g(x)$ to be a, so $h \to 0$ means $y \to a$.
- Because of the above two items, the whole limit can be rewritten:

$$\lim_{h\to 0}\frac{e^{g(x+h)} - e^{g(x)}}{g(x+h)-g(x)} = \lim_{y\to a}\frac{e^y - e^a}{y-a}$$

Here, you should recall (4.13) and realize, "A-ha! That limit represents the limit of e^y at $y = a$, which we know to be e^a!" And since $a = g(x)$, this first limit is simply equal to $e^{g(x)}$. Put it all together, and see that:

$$\frac{d}{dx}e^{g(x)} = \lim_{h\to 0}\frac{e^{g(x+h)} - e^{g(x)}}{h}$$

$$= \lim_{h\to 0}\frac{e^{g(x+h)} - e^{g(x)}}{g(x+h) - g(x)} \cdot \lim_{h\to 0}\frac{g(x+h) - g(x)}{h} = e^{g(x)} \cdot g'(x)$$

Voila! That's why we have the formula $\frac{d}{dx}e^{g(x)} = e^{g(x)} \cdot g'(x)$.

And now to almost identically repeat the concluding paragraph of Sec. 3.6: This derivation was restricted to the case that $g(x)$ is one-to-one. FFT: Do you see where the argument might break down if we allow $g(x)$ to have the same value at two different locations? If you want to see the full explanation of this formula that's good for ALL differentiable functions $g(x)$, then nicely ask your instructor (if you're reading this while taking a class) to take at least an entire day of the class to give a different derivation. Or better yet, sign up for a course in Real Analysis (sometimes called Advanced Calculus) in a couple of years! Heck, you should do that anyway.

Still in the Pit! Derivation of the Derivative of $\ln(g(x))$

Now let's look at the origin of the derivative formula in Eq. (4.7). The same restrictions as above apply here, too: this particular derivation works when $g(x)$ is both differentiable and one-to-one. We also must recognize, like before, that the expression

$$\lim_{y\to a}\frac{\ln(y) - \ln(a)}{y - a}$$

represents the derivative of $\ln y$ at the point $y = a$, and so is equal to $1/a$. With that in mind, here we go. We'll go through the steps more quickly than for the exponential function. The derivative of $\ln g(x)$ is defined as:

$$\frac{d}{dx}\ln g(x) = \lim_{h\to 0}\frac{\ln g(x+h) - \ln g(x)}{h}$$

$$= \lim_{h\to 0}\frac{\ln g(x+h) - \ln g(x)}{h} \cdot \frac{g(x+h) - g(x)}{g(x+h) - g(x)}$$

With regrouping and application of the limit law regarding products, this becomes

$$\frac{d}{dx}\ln g(x) = \lim_{h\to 0}\frac{\ln g(x+h) - \ln g(x)}{g(x+h) - g(x)} \cdot \lim_{h\to 0}\frac{g(x+h) - g(x)}{h}$$

The latter limit is just $g'(x)$. The first limit can be redesigned by setting $a = g(x)$, $y = g(x+h)$, and remembering that the limit process $h \to 0$ is now equivalent to $y \to g(x)$, i.e. $y \to a$. Therefore,

$$\lim_{h\to 0} \frac{\ln g(x+h) - \ln g(x)}{g(x+h) - g(x)} = \lim_{y\to a} \frac{\ln y - \ln a}{y - a}$$

which defines the derivative of $\ln y$ at the point $y = a$, and this was noted above to be equal to $1/a$. But since we have $a = g(x)$, we have, finally, that:

$$\lim_{h\to 0} \frac{\ln g(x+h) - \ln g(x)}{g(x+h) - g(x)} = \frac{1}{g(x)}$$

Together then,

$$\frac{d}{dx} \ln g(x) = \lim_{h\to 0} \frac{\ln g(x+h) - \ln g(x)}{g(x+h) - g(x)} \cdot \lim_{h\to 0} \frac{g(x+h) - g(x)}{g(x+h) - g(x)} = \frac{1}{g(x)} \cdot g'(x)$$

and this is where the derivative formula (4.7) comes from. Remember once again that this particular argument only works when $g(x)$ is not allowed to repeat values (do you see why?), and a fuller derivation that accounts for ALL differentiable functions $g(x)$ requires more advanced techniques.

Have You Learned...

- How to find the derivative and antiderivative of e^x?
- How to find the derivative of a function that is a composition of the form $e^{g(x)}$?

$$\frac{d}{dx} e^{g(x)} = e^{g(x)} \cdot g'(x)$$

- How to find the antiderivative of a function of the form $g'(x)e^{g(x)}$?

$$\int g'(x)e^{g(x)}\, dx = e^{g(x)} + C$$

- How to find the derivative of $\ln(x)$?
- How to find the derivative of a function that is a composition of the form $\ln(g(x))$?

$$\frac{d}{dx} \ln g(x) = \frac{g'(x)}{g(x)}$$

- How to find the antiderivative of a function of the form $g'(x)/g(x)$?

$$\int \frac{g'(x)}{g(x)}\, dx = \ln |g(x)| + C$$

- Is there an antiderivative of $\ln(x)$ itself?
- How the function $\ln x$ fills in a "gap" in the product rule for antiderivatives?
- How to find the derivative and antiderivative of $\sinh(x)$?
- How to find the derivative of a function that is a composition of the form $\sinh(g(x))$ or $\cosh(g(x))$?

$$\frac{d}{dx}\sinh(g(x)) = \cosh(g(x))\cdot g'(x)\,;\ \frac{d}{dx}\cosh(g(x)) = \sinh(g(x))\cdot g'(x)$$

- How to find the antiderivative of a function of the form $g'(x)\sinh(g(x))$ or $g'(x)\cosh(g(x))$?

$$\int g'(x)\cosh(g(x))\,dx = \sinh(g(x)) + C\,;$$

$$\int g'(x)\sinh(g(x))\,dx = \cosh(g(x)) + C$$

- A little bit about how formulas for derivatives of more complicated functions can be generated?

Exponential and Logarithmic Functions — Problem List

Exponential and Logarithmic Functions — You Try It

These appeared above; solutions begin on the next page.

(1) Find the equation of the line tangent to $y = x^4 + 2e^x$ at (0,2).

(2) Find the following derivatives:

$$\frac{d}{dx}(x^2 - e^{-2x}) \quad ; \quad \frac{d}{dx}e^{\sqrt[4]{x}} \quad ; \quad \frac{d}{dt}\left(e^{1/t} + \frac{1}{t^2}\right)$$

(3) Find the following antiderivatives:

$$\int (4x-2)e^{x^2-x}\,dx \quad ; \quad \int 5t^3 e^{-t^4}\,dt \quad ; \quad \int \frac{e^{\sqrt{x}+1}}{2\sqrt{x}}\,dx$$

(4) Find the derivative of $f(x) = \ln x + \sqrt{x-1}$.

(5) Find the derivative and antiderivative of $f(x) = \sqrt{x} - 2e^x - \dfrac{1}{x}$.

(6) Evaluate $\displaystyle\int \frac{2}{x^2} + \frac{1}{x}\,dx$.

(7) Find the following derivatives:

$$\frac{d}{dx}(\ln(x^3 - 1) + e^{x-1}) \quad ; \quad \frac{d}{dx}\ln(e^{2x} + x^{1.7}) \quad ; \quad \frac{d}{dt}\ln(\ln t - 2)$$

(8) Find the following antiderivatives:

$$\int \frac{x^3}{x^4 - 10}\,dx \quad ; \quad \int \frac{1}{2t+1}\,dt \quad ; \quad \int \frac{2xe^{x^2} + 1}{e^{x^2} + x}\,dx$$

(9) Find the following derivatives:

$$\frac{d}{dx}\sinh(\ln x) \quad , \quad \frac{d}{dx}\cosh(x^2 + 4x + 4) \quad ,$$

$$\frac{d}{dt}\left(\sinh(e^{2t}) + \cosh(2t^{3/2} + t)\right)$$

(10) Find the following antiderivatives:

$$\int e^{\sinh(x)}\cosh(x)\,dx \quad , \quad \int 4\sinh(x)\cosh^3(x)\,dx \quad ,$$

$$\int \frac{\sinh(y) + 2y}{\cosh(y) + y^2}\,dy$$

Exponential and Logarithmic Functions — Practice Problems

Try these as you get the hang of the You Try It problems. Solutions to these problems are available in Sec. A.4.1.

(1) Find the derivative and antiderivative of $y = 5e^x + \dfrac{3}{x}$.
(2) Find the derivative and antiderivative of $y = 7\sqrt{x} - e^x - \dfrac{2}{x}$.
(3) Find the equation of the line tangent to $y = \ln(x) + x$ at $x = 1$.
(4) Find the derivative of $f(x) = e^{5x - \ln x^2}$.
(5) Find $h'(t)$ if $h(t) = 2t^2 - e^{-t^3}$.
(6) Find $\dfrac{dy}{dx}$ if $y = \ln\left(\dfrac{1}{x^2} - 2\right)$.
(7) Find the derivative of $r(s) = \sqrt[4]{s} + \ln(5 - s^5)$.
(8) Find the antiderivative $\displaystyle\int x^{-3} e^{x^{-2}}\, dx$.
(9) Find the antiderivative $\displaystyle\int \frac{e^{3-\ln x}}{x}\, dx$.
(10) Find the antiderivative $\displaystyle\int \frac{5}{2 - 3t}\, dt$.
(11) Find the antiderivative $\displaystyle\int \frac{2 - x^2}{6x - x^3}\, dx$.
(12) Find the derivative and antiderivative of $\sinh(6x)$.
(13) Find the derivative and antiderivative of $\dfrac{1}{3}\sinh(3x) + x^2$.
(14) If $g(x) = e^{-3\cosh(x)}$, what is $g'(x)$?
(15) Evaluate $\displaystyle\int e^{2x} \cosh(e^{2x})\, dx$.

Exponential and Logarithmic Functions — Challenge Problems

Try these problems to test your skills with the ideas in this section. Solutions to these problems are available in Sec. B.4.1.

(1) Find the derivative of $y = \ln(x^2 + e^{x-1}) + \sinh^2(x)$.
(2) Find the antiderivative $\displaystyle\int \frac{1}{x^{2/3}(\sqrt[3]{x} - 1)}\, dx$.
(3) Find the antiderivative of $f(x)$ that goes through the point $(1, 0)$ for the function

$$f(x) = \frac{1 - x}{x^2 - 2x + 2} + x^2 e^{-x^3}$$

Exp & Log Functions Derivs — You Try It — Solved

(1) Find the equation of the line tangent to $y = x^4 + 2e^x$ at (0,2).

□ The slope of the line tangent to $y = x^4 + 2e^x$ at (0,2) is the value of the derivative at $x = 0$. Since

$$\frac{dy}{dx} = \frac{d}{dx}x^4 + 2\frac{d}{dx}e^x = 4x^3 + 2e^x$$

then $y'(0) = 4(0) + 2e^0 = 2$. The equation of the tangent line is therefore $y - 2 = 2(x - 0)$, or $y = 2x + 2$. ■

(2) Find the following derivatives:

$$\frac{d}{dx}(x^2 - e^{-2x}) \quad ; \quad \frac{d}{dx}e^{\sqrt[4]{x}} \quad ; \quad \frac{d}{dt}\left(e^{1/t} + \frac{1}{t^2}\right)$$

□ $\frac{d}{dx}(x^2 - e^{-2x}) = 2x - e^{-2x} \cdot \frac{d}{dx}(-2x) = 2x + 2e^{-2x}$

$$\frac{d}{dx}e^{\sqrt[4]{x}} = e^{\sqrt[4]{x}} \cdot \frac{d}{dx}(\sqrt[4]{x}) = e^{\sqrt[4]{x}} \cdot \frac{d}{dx}(x^{1/4}) = \frac{1}{4}x^{-3/4}e^{\sqrt[4]{x}}$$

$$\frac{d}{dt}\left(e^{1/t} + \frac{1}{t^2}\right) = e^{1/t} \cdot \frac{d}{dt}\frac{1}{t} + \frac{d}{dt}t^{-2} = -\frac{e^{1/t}}{t^2} - 2t^{-3} = -\frac{e^{1/t}}{t^2} - \frac{2}{t^3}$$ ■

(3) Find the following antiderivatives:

$$\int (4x - 2)e^{x^2 - x}\,dx \quad ; \quad \int 5t^3 e^{-t^4}\,dt \quad ; \quad \int \frac{e^{\sqrt{x}+1}}{2\sqrt{x}}\,dx$$

□ We plan for the use of Eq. (4.3) by finding or creating a match of each integrand to $g'(x)e^{g(x)}$ (adjusted for a different variable if needed).

In the first, we have $g(x) = x^2 - x$, we must arrange to introduce $g'(x) = 2x - 1$:

$$\int (4x - 2)e^{x^2 - x}\,dx = 2\int (2x - 1)e^{x^2 - x}\,dx = 2e^{x^2 - x} + C$$

In the second, we have $g(t) = -t^4$, and thus look for $g'(t) = -4t^3$:

$$\int 5t^3 e^{-t^4}\,dt = -\frac{5}{4}\int -4t^3 e^{-t^4}\,dt = -\frac{5}{4}e^{-t^4} + C$$

In the third, we have $g(x) = \sqrt{x} + 1$, and must arrange to find $g'(x) = 1/(2\sqrt{x})$:

$$\int \frac{e^{\sqrt{x}+1}}{2\sqrt{x}}\,dx = e^{\sqrt{x}+1} + C$$ ■

(4) Find the derivative of $f(x) = \ln x + \dfrac{1}{\sqrt{x}}$.

☐ Since we can write $(x) = \ln x + x^{-1/2}$, the power rule applies to the second term:

$$f'(x) = \frac{d}{dx}(\ln x + x^{-1/2}) = \frac{1}{x} - \frac{1}{2}x^{-3/2} \quad \blacksquare$$

(5) Find the derivative and antiderivative of $f(x) = \sqrt{x} - 2e^x - \dfrac{1}{x}$.

☐ Since we can write $f(x) = x^{1/2} - 2e^x - x^{-1}$, then the power rule applies to both the derivative and antiderivative;

$$f'(x) = \frac{1}{2\sqrt{x}} - 2e^x + \frac{1}{x^2}$$

$$\int f(x)\,dx = \frac{2}{3}x^{3/2} - 2e^x - \ln|x| + C \quad \blacksquare$$

(6) Evaluate $\displaystyle\int \frac{2}{x^2} + \frac{1}{x}\,dx$.

☐ We can rewrite and solve,

$$\int \frac{2}{x^2}+\frac{1}{x}\,dx = \int 2x^{-2}+\frac{1}{x}\,dx = \frac{2}{-1}x^{-1}+\ln|x|+C = -\frac{2}{x}+\ln|x|+C \quad \blacksquare$$

(7) Find the following derivatives:

$$\frac{d}{dx}(\ln(x^3-1)+e^{x-1}) \quad ; \quad \frac{d}{dx}\ln(e^{2x}+x^{1.7}) \quad ; \quad \frac{d}{dt}\ln(\ln t - 2)$$

☐
$$\frac{d}{dx}(\ln(x^3-1)+e^{x-1}) = \frac{1}{x^3-1}\cdot\frac{d}{dx}(x^3-1) + e^{x-1}\cdot\frac{d}{dx}(x-1)$$
$$= \frac{3x^2}{x^3-1} + e^{x-1}$$

$$\frac{d}{dx}\ln(e^{2x}+x^{1.7}) = \frac{1}{e^{2x}+x^{1.7}}\cdot\frac{d}{dx}(e^{2x}+x^{1.7})$$
$$= \frac{1}{e^{2x}+x^{1.7}}\left(e^{2x}\cdot\frac{d}{dx}(2x) + 1.7x^{0.7}\right)$$
$$= \frac{2e^{2x}+1.7x^{0.7}}{e^{2x}+x^{1.7}}$$

$$\frac{d}{dt}\ln(\ln t - 2) = \frac{1}{\ln t - 2}\cdot\frac{d}{dt}(\ln t - 2)$$
$$= \frac{1}{\ln t - 2}\cdot\frac{1}{t} = \frac{1}{t(\ln t - 2)} \quad \blacksquare$$

(8) Find the following antiderivatives:

$$\int \frac{x^3}{x^4 - 10}\, dx \quad ; \quad \int \frac{1}{2t+1}\, dt \quad ; \quad \frac{2xe^{x^2}+1}{e^{x^2}+x}\, dx \quad .$$

☐ In each, we hope to use Eq. (4.8) by matching the integrand to $g'(x)/g(x)$.

In the first, we identifying $g(x) = x^4 - 10$, so that we must arrange to find precisely $g'(x) = 4x^3$:

$$\int \frac{x^3}{x^4 - 10}\, dx = \frac{1}{4}\int \frac{4x^3}{x^4 - 10}\, dx = \frac{1}{4}\ln|x^4 - 10| + C$$

For the second, we identify $g(t) = 2t + 1$ and seek $g'(t) = 2$:

$$\int \frac{1}{2t+1}\, dt = \frac{1}{2}\int \frac{2}{2t+1}\, dt = \frac{1}{2}\ln|2t+1| + C$$

For the last, we have $g(x) = e^{x^2} + x$ and so $g'(x) = 2xe^{x^2} + 1$:

$$\frac{2xe^{x^2}+1}{e^{x^2}+x}\, dx = \frac{1}{4}\ln|e^{x^2} + x| + C \quad ■$$

(9) Find the following derivatives:

$$\frac{d}{dx}\sinh(\ln x) \quad , \quad \frac{d}{dx}\cosh(x^2 + 4x + 4) \quad ,$$

$$\frac{d}{dt}\left(\sinh(e^{2t}) + \cosh(2t^{3/2} + t)\right)$$

☐ $$\frac{d}{dx}\sinh(\ln x) = \cosh(\ln x)\cdot\frac{d}{dx}(\ln x) = \frac{\cosh(\ln(x))}{x}$$

$$\begin{aligned}\frac{d}{dx}\cosh(x^2 + 4x + 4) &= \sinh(x^2 + 4x + 4)\cdot\frac{d}{dx}(x^2 + 4x + 4)\\ &= (2x + 4)\sinh(x^2 + 4x + 4)\end{aligned}$$

$$\begin{aligned}&\frac{d}{dt}\left(\sinh(e^{2t}) + \cosh(2t^{3/2} + t)\right)\\ &\qquad = \cosh(e^{2t})\cdot\frac{d}{dt}(e^{2t}) + \sinh(2t^{3/2} + t)\cdot\frac{d}{dt}(2t^{3/2} + t)\\ &\qquad = \cosh(e^{2t})(2e^{2t}) + \sinh(2t^{3/2} + t)(3t^{1/2} + 1) \quad ■\end{aligned}$$

---

(10) Find the following antiderivatives:

$$\int e^{\sinh(x)} \cosh(x)\, dx \quad , \quad \int 4\sinh(x)\cosh^3(x)\, dx \quad ,$$

$$\int \frac{\sinh(y) + 2y}{\cosh(y) + y^2}\, dy$$

□ The first antiderivative is a direct application of (4.12) with $g(x) = \sinh(x)$, so

$$\int e^{\sinh(x)} \cosh(x)\, dx = e^{\sinh(x)} + C$$

The second antiderivative has a power rule structure, so we can tidy it to match (3.13) with $g(x) = \cosh(x)$:

$$\int 4\sinh(x)\cosh^3(x)\, dx = \int \sinh(x) \cdot 4(\cosh(x))^3\, dx = (\cosh(x))^4 + C$$

The only quotients we've seen are the result of a natural log's derivative, so let's tailor the third to match (4.8) with $g(y) = \cosh(y) + y^2$:

$$\int \frac{\sinh(y) + 2y}{\cosh(y) + y^2}\, dy = \int (\sinh(y) + 2y) \cdot \frac{1}{\cosh(y) + y^2}\, dy$$
$$= \ln(\cosh(y) + y^2) + C \quad ■$$

4.2 How to Make Trigonometric Functions Worse

Introduction

We're adding to our list of functions whose derivatives we know. Here, we look at the functions $\sin x$ and $\cos x$, and even use what we learn about them to find information about other trig functions.

Some Preliminary Information

Here are two limits that will come in handy later on. The first is:

$$\lim_{x \to 0} \frac{\sin x}{x} = 1 \tag{4.14}$$

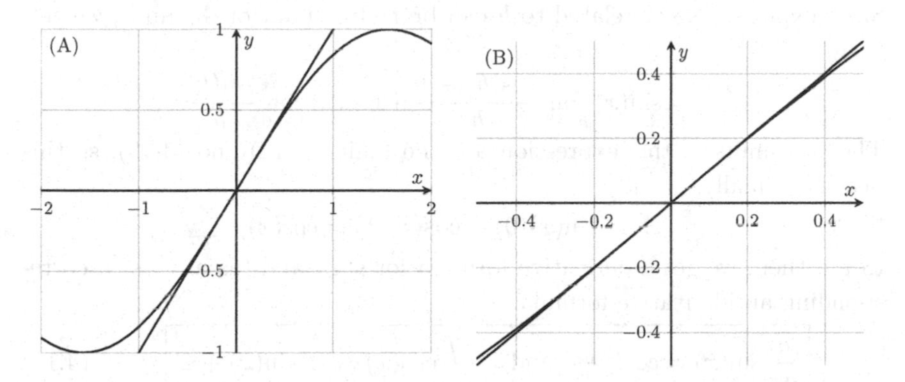

Fig. 4.1 $y = \sin(x)$ vs. $y = x$; both curves are present in both pictures.

There are several ways to demonstrate this limit. But, informally, consider Fig. 4.1, which shows $y = \sin x$ and $y = x$ on two scales. Note that the closer we get to $x = 0$, the harder it becomes to tell $y = \sin x$ apart from $y = x$. In other words, the closer we are to $x = 0$, the closer $y = \sin x$ and $y = x$ are to being the same thing; so, the limit of their ratio approaches 1 as x approaches 0. (In fact, the tangent line to $\sin x$ at $x = 0$ is exactly $y = x$.)

Also, we have this other limit:

$$\lim_{h \to 0} \frac{\cos h - 1}{h} = 0 \tag{4.15}$$

Each of these limits will be useful below.

***The Derivative of* $\sin x$**

To find the derivative of $\sin x$, we can apply the usual limit definition:

$$\frac{d}{dx}\sin x = \lim_{h\to 0}\frac{\sin(x+h)-\sin x}{h}$$

First, let's use our old friend, the trig identity $\sin(A+B) = \sin A\cos B + \cos A \sin B$, to get things started (we are using the rule that "we have to make something look worse before we can make it look better"):

$$\begin{aligned}\frac{d}{dx}\sin x &= \lim_{h\to 0}\frac{\sin(x)\cos(h)+\cos(x)\sin(h)-\sin x}{h}\\ &= \lim_{h\to 0}\frac{\sin(x)(\cos(h)-1)+\cos(x)\sin(h)}{h}\\ &= \lim_{h\to 0}\frac{\sin(x)(\cos(h)-1)}{h}+\lim_{h\to 0}\frac{\cos(x)\sin(h)}{h}\end{aligned}$$

Since expressions not related to h can be factored out of the limit, we get, then,

$$\ldots = \sin(x)\lim_{h\to 0}\frac{\cos(h)-1}{h}+\cos(x)\lim_{h\to 0}\frac{\sin(h)}{h}$$

The two limits in this expression are the limits (4.14) and (4.15), so this becomes, finally,

$$\ldots = \sin(x)(0)+\cos(x)(1) = \cos(x)$$

In all, then, we get a derivative formula for $\sin x$ and (as always) a corresponding antiderivative formula:

$$\boxed{\frac{d}{dx}\sin(x) = \cos(x) \quad \text{and} \quad \int \cos(x)\,dx = \sin(x)+C} \tag{4.16}$$

***The Derivative of* $\cos x$**

This proceeds very much like the above; application of a trig identity, factoring, and known limits gives:

$$\begin{aligned}\frac{d}{dx}\cos x &= \lim_{h\to 0}\frac{\cos(x+h)-\cos x}{h} = \lim_{h\to 0}\frac{\cos(x)\cos(h)-\sin(x)\sin(h)-\cos x}{h}\\ &= \lim_{h\to 0}\frac{\cos(x)(\cos(h)-1)-\sin(x)\sin(h)}{h}\\ &= \lim_{h\to 0}\frac{\cos(x)(\cos(h)-1)}{h}-\lim_{h\to 0}\frac{\sin(x)\sin(h)}{h}\\ &= \cos(x)\lim_{h\to 0}\frac{\cos(h)-1}{h}-\sin(x)\lim_{h\to 0}\frac{\sin(h)}{h}\\ &= \cos(x)(0)-\sin(x)(1) = -\sin(x)\end{aligned}$$

In all, then, we get a derivative formula for $\cos x$ and (as always) a corresponding antiderivative formula,

$$\boxed{\frac{d}{dx}\cos(x) = -\sin(x) \quad \text{and} \quad \int \sin(x)\,dx = -\cos(x) + C} \tag{4.17}$$

Now that we know the simplest derivatives and antiderivatives (4.16) and (4.17) involving $\sin x$ and $\cos x$, we can toss these into the mix when finding derivatives. Here are several examples, in which the only new thing is the presence of $\sin x$ and $\cos x$:

EX 1 Here are some derivatives computed using these new rules, with other simple functions thrown in to make them more interesting.

$$\frac{d}{dx}(2\sin x + e^x - \ln x) = 2\cos x + e^x - \frac{1}{x}$$

$$\frac{d}{dx}(2x^{3/5} - \cos(x)) = \frac{6}{5}x^{-2/5} - (-\sin x) = \frac{6}{5x^{2/5}} + \sin(x)$$

$$h(r) = e^{-2r} - \sin(r) + \sqrt{2}\cos(r) \to h'(r) = -2e^{-2r} - \cos(r) - \sqrt{2}\sin(r)$$

$$p(t) = \frac{2}{3}\sin x + \ln(e^x + 3) \to p'(t) = \frac{2}{3}\cos x + \frac{e^x}{e^x + 3} \quad \blacksquare$$

You Try It

(1) Find and properly name the derivatives of the following:

$(A)\ y = -2\sin(x) - 3e^x$ $\quad (C)\ y(w) = (w^2 - 1)^3 - \sqrt{3}\sin(w)$

$(B)\ s = \dfrac{2}{t} + 2\cos(t)$ $\quad (D)\ y = (\pi - 1)\cos(x) - \ln(x^4 + 2x^2 + 2)$

Don't forget that derivatives carry graphical information:

You Try It

(2) Sketch a graph of $y = x + \sin x$. State at how many points this function has a horizontal tangent line, and find three of those points.

(3) Find the equation of the line tangent to $y = \ln x + \cos x$ at $x = e$. All constants must be in exact form.

In the problems above, the trig functions were kept separate from the other kinds of functions. Now let's scramble them up a bit more, by mixing these trig functions in with others via composition, so that we can use (3.13), (4.3), and (4.8).

EX 2 Find the following derivatives:

(A) $\dfrac{d}{dt}e^{\sin t}$

(B) $\dfrac{d}{dx}\cos^2 x$

(C) $h'(y)$ when $h(y) = \ln(\cos y)$ for $0 \leq y < \dfrac{\pi}{2}$

For (A), we use (4.3):

$$\frac{d}{dt}e^{\sin t} = e^{\sin t} \cdot \frac{d}{dt}\sin t = \cos t\, e^{\sin t}$$

For (B), we can write $\cos^2 x$ as $(\cos x)^2$ and then use the extended power rule, 3.13):

$$\frac{d}{dx}\cos^2 x = \frac{d}{dx}(\cos x)^2 = 2(\cos x)^1 \cdot \frac{d}{dx}(\cos x) = -2\sin x \cos x$$

Finally for (C), if $h(y) = \ln(\cos y)$ then using (4.8),

$$h'(y) = \frac{1}{\cos y} \cdot \frac{d}{dy}\cos y = -\frac{\sin y}{\cos y} = -\tan y$$

FFT: In (C), why is the restriction $0 \leq y < \pi/2$ necessary? ■

You Try It

(4) Find and properly name the derivatives of the following:

(A) $g(x) = \sin^3 x$

(B) $h = \dfrac{1}{\cos^2 y} + e^{-2\cos y}$

(C) $y = \ln(\sin x - 4\cos x)$

(D) $z(t) = \sin^2 t + \cos^2 t$

Did you notice that part (C) of EX 2 gave us (indirectly) the antiderivative of $\tan x$? By EX 2(C), the derivative of $h(y) = \ln(\cos y)$ (over an appropriate domain) is $-\tan y$, and so the antiderivative of $\tan y$ is $-\ln(\cos(y)) + C$. We can extend this result to a larger domain by introducing absolute values, as in (4.8); we can also rename y to x so this formula matches others, and also use a property of logarithms to apply the negative sign elsewhere:

$$\int \tan x\, dx = -\ln|\cos x| + C = \ln|(\cos x)^{-1}| + C = \ln|\sec x| + C$$

Together,

$$\boxed{\int \tan(x)\,dx = \ln|\sec(x)| + C} \tag{4.18}$$

and we have antiderivatives for the full set of $\sin x$, $\cos x$ and $\tan x$. It's odd that we've discovered an antiderivative for $\tan(x)$ before the derivative, but we'll take any useful information when we can get it. There is a fairly direct way to get the derivative of $\tan(x)$, but we don't have the tools for it quite yet.

FFT: There are still individual points where (4.18) is not defined. What are they?

So, we've discovered the antiderivative of $\tan x$ by remembering its definition as $\sin x/\cos x$. Can you use the same ideas to derive

$$\int \cot x\,dx = \ln|\sin x| + C$$

If you are feeling up to this, it is presented as a challenge problem!

EX 3 Here are some antiderivatives computed using the new rules in (4.16), (4.17) and (4.18), plus some older simple rules.

$$\begin{aligned}\int 2\sin x + e^x - \frac{5}{x}\,dx &= 2(-\cos x) + e^x - 5\ln|x| + C\\ &= -2\cos x + e^x - 5\ln|x| + C\end{aligned}$$

$$\int 2x^{3/5} - \cos(x)\,dx = 2\cdot\frac{5}{8}x^{8/5} - (\sin x) + C = \frac{5}{4}x^{8/5} - \sin x + C$$

$$\int \left(4\tan t + \frac{1}{\sqrt{t}}\right)dt = 4\ln|\sec t| + 2\sqrt{t} + C \quad \blacksquare$$

You Try It

(5) Find the following antiderivatives.

(A) $\int -2\sin(x) - 3e^x\,dx$ (C) $\int e^{-t} + \frac{1}{3}\sin t\,dt$

(B) $\int \frac{2}{x} + 2\cos(x)\,dx$ (D) $\int 2x(x^2-1)^2 - \sqrt{7}\tan x\,dx$

You Try It

(6) Find the antiderivative of $y = 2\sin x + 3\cos x$ that goes through the point $(\pi, 0)$.

Now let's practice antiderivatives when trig functions are mixed up a bit more with the other usual suspects. Remember that these antiderivatives can involve extended antiderivative formulas for power, expontial, and logarithmic functions.

EX 4 Find the following antiderivatives:

$$(A)\ \int \sin x e^{\cos x}\, dx \qquad (C)\ \int \sin t(\cos^2 t - 3\cos t)\, dt$$

$$(B)\ \int \frac{\cos x}{1+\sin x}\, dx \qquad (D)\ \int \frac{\cos(w)}{\sqrt{\sin w + 2}}\, dw$$

For (A), we use (4.3) by matching the integrand to the form $g'(x)e^{g(x)}dx$. Identifying $g(x) = \cos x$, we'd like to arrange to see $g'(x) = -\sin(x)$:

$$\int \sin x e^{\cos x}\, dx = -\int (-\sin x)e^{\cos x}\, dx - e^{\cos x} + C$$

For (B), we use (4.8) by matching the integrand to the form $g'(x)/g(x)$. Identifying $g(x) = 1 + \sin x$, we need to see $g'(x) = \cos x$, and it's already right there:

$$\int \frac{\cos x}{1+\sin x}\, dx = \ln|1 + \sin x| + C$$

For (C), we can separate the expression into two terms, and in each use $g(t) = \cos t$ in conjunction with the extended power rule (3.13):

$$\begin{aligned}\int \sin t(\cos^2 t - 3\cos t)\, dt &= \int \sin t \cos^2 t\, dt - 3\int \sin t \cos t\, dt \\ &= -\frac{1}{3}\cos^3 t + \frac{3}{2}\cos^2 t + C\end{aligned}$$

For (D), we also rewrite the integrand to present a clear link to the extended power rule,

$$\int \frac{\cos(w)}{\sqrt{\sin w + 2}}\, dw = \int \cos(w)(\sin w + 2)^{-1/2}\, dw$$

Matching the integrand to $g'(w)(g(w))^p$, we identify $g(w) = \sin w + 2$. This requires us to spot $g'(w) = \cos(w)$, and that term is waiting for us without any adjustments:

$$\begin{aligned}\int \cos(w)(\sin w + 2)^{-1/2}\,dw &= \frac{1}{-1/2+1}(\sin w + 2)^{-1/2+1} + C \\ &= 2\sqrt{\sin w + 2} + C \quad \blacksquare\end{aligned}$$

You Try It

(7) Find the following antiderivatives. Remember that they can now involve antiderivative formulas for the power rule, exponential functions, and logarithmic functions as well as trigonometric functions:

(A) $\int \cos t\, e^{\sin t}\,dt$ (C) $\int \frac{\cos t}{2\sin t + 3}\,dt$

(B) $\int \sin(x)\cos^2 x\,dx$ (D) $\int \tan x - \sin x\sqrt[3]{2 - \cos x}\,dx$

In EX 2, we unlocked an antiderivative achievement by remembering how to write $\tan x$ in terms of $\sin x$ and $\cos x$. (And similarly, the antiderivative for $\cot x$ is posed as a Challenge Problem.) Well, we are not done unlocking derivatives and / or antiderivatives of the other trig functions by relating them back to $\sin x$ and $\cos x$, via the power rule.

EX 5 What is the derivative of $y = \sec x$?

If we remember that $\sec x = (\cos x)^{-1}$, then we can employ the power rule as posed in (3.13):

$$\frac{d}{dx}\sec x = \frac{d}{dx}(\cos x)^{-1} = -(\cos x)^{-2}\frac{d}{dx}\cos x = \sin x(\cos x)^{-2}$$

Now this final result is not the best way to leave things, so let's do some rearrangement:

$$\frac{d}{dx}\sec x = \sin x(\cos x)^{-2} = \frac{\sin x}{\cos^2 x} = \frac{\sin x}{\cos x}\cdot\frac{1}{\cos x} = \tan x \sec x \quad \blacksquare$$

This example gives us a new derivative formula, along with a corresponding antiderivative formula:

$$\boxed{\frac{d}{dx}\sec x = \sec x \tan x \quad \text{and} \quad \int \sec x \tan x\,dx = \sec x + C} \tag{4.19}$$

You should be able to do the same for $\csc x$, and also use these derivative formulas (and their corresponding antiderivative formulas) right away.

You Try It

(8) Use a procedure similar to that of EX 5 to find the derivative of $\csc x$, and state the corresponding antiderivative formula that arises.

You Try It

(9) Find and properly name the derivatives of the following:

$(A)\ y = 2\sec x - 5\csc x$ $\quad (C)\ z = \sec^3 x$

$(B)\ h(r) = e^{\csc r} - \dfrac{1}{\sec r}$ $\quad (D)\ y = \sqrt{\csc x + 3}$

You Try It

(10) Find the following antiderivatives

$(A)\ \displaystyle\int \sec x \tan x e^{\sec x}\, dx$ $\quad (C)\ \displaystyle\int \frac{\csc x \cot x}{\csc x + 1}\, dx$

$(B)\ \displaystyle\int \csc x \cot x \csc^2 x - \cot x\, dx$ $\quad (D)\ \displaystyle\int \frac{\sec x \tan x}{(\sec x + 1)^2}\, dx$

Here is a quick recap of what we know so far that's related to derivatives and antiderivatives of trig functions:

$$\frac{d}{dx}\sin x = \cos x \qquad \frac{d}{dx}\cos x = -\sin x \tag{4.20}$$

$$\frac{d}{dx}\sec x = \sec x \tan x \qquad \frac{d}{dx}\csc x = -\csc x \cot x \tag{4.21}$$

$$\frac{d}{dx}\tan x = ?? \qquad \frac{d}{dx}\cot x = ?? \tag{4.22}$$

$$\int \sin x\, dx = -\cos x + C \qquad \int \cos x\, dx = \sin x + C \tag{4.23}$$

$$\int \tan x\, dx = \ln|\sec x| + C \qquad \int \cot x\, dx = \ln|\sin x| + C \tag{4.24}$$

$$\int \sec x \tan x\, dx = \sec x + C \qquad \int \csc x \cot x\, dx = -\csc x + C \tag{4.25}$$

Make sure you understand why we are unable to find the derivatives of $\tan x$ and $\cot x$ using the methods we have available to this point. (What's different about these two functions and how they're built from $\sin x$ and $\cos x$, as opposed to $\sec x$ and $\csc x$?) Now, just to fill this gap, here's a sneak peak of Sec. 4.3:

$$\boxed{\frac{d}{dx}\tan(x) = \sec^2(x) \quad \text{and so} \quad \int \sec^2(x)\,dx = \tan(x) + C} \tag{4.26}$$

and

$$\boxed{\frac{d}{dx}\cot(x) = -\csc^2(x) \quad \text{and so} \quad \int \csc^2(x)\,dx = -\cot(x) + C} \tag{4.27}$$

Odds are pretty good these will be needed in the exercises at the end of the section.

Derivatives of $\sin(g(x))$ *and* $\cos(g(x))$

We can now find the derivatives of many functions in which a trigonometric function is the inside function, like $\sin^p(x)$ or $e^{\cos x}$. But, what if we have a composition in the form $\sin(g(x))$ or $\cos(g(x))$?

First, let's make a conjecture; so far, in previous sections, we've seen the following relationships:

$$\begin{aligned}
\frac{d}{dx}(g(x))^p &= p(g(x))^{p-1} \cdot g'(x) \\
\frac{d}{dx}e^{g(x)} &= e^{g(x)} \cdot g'(x) \\
\frac{d}{dx}\ln(g(x)) &= \frac{1}{g(x)} \cdot g'(x)
\end{aligned}$$

In each case, we proceed in exactly the same way to find a derivative: we act on the outside function (while leaving the inside part idle), and then multiply by the derivative of the inside function. So it's reasonable to expect similar behavior for trig functions, such as:

$$\boxed{\frac{d}{dx}\sin(g(x)) = \cos(g(x)) \cdot g'(x) \quad \text{and} \quad \frac{d}{dx}\cos(g(x)) = -\sin(g(x)) \cdot g'(x)} \tag{4.28}$$

And it just so happens that these are indeed correct derivative formulas. It would also then be reasonable to propose the following antiderivative formulas:

$$\boxed{\int g'(x)\cos(g(x))\,dx = \sin(g(x)) + C} \quad \text{and}$$

$$\boxed{\int g'(x)\sin(g(x))\,dx = -\cos(g(x)) + C} \qquad (4.29)$$

Let's use these derivative and antiderivative formulas, then look at how one of the derivative formulas in (4.28) is found. The second will be left to you in a Challenge Problem.

EX 6 Here are some miscellaneous derivatives using the extended derivative rules for $\sin(g(x))$ and $\cos(g(x))$ seen in (4.28).

$$\begin{aligned}
\frac{d}{dx}\sin(x^2) &= \cos(x^2)\cdot\frac{d}{dx}(x^2) = 2x\cos(x^2)\\
\frac{d}{dx}\cos(\sin x) &= -\sin(\sin x)\cdot\frac{d}{dx}(\sin x) = -\sin x\sin(\sin x)\\
\frac{d}{dr}\sin(r^2-2r+2) &= \cos(r^2-2r+2)\cdot\frac{d}{dr}(r^2-2r+2)\\
&= (2r-2)\cos(r^2-2r+2) \quad \blacksquare
\end{aligned}$$

You Try It

(11) Find and properly name the derivatives of the following:

$(A)\ y = \cos(x^4)$ $\qquad (C)\ p(t) = \sin(e^t + \tan t)$

$(B)\ z = \sin(\sec x)$ $\qquad (D)\ y = e^{\cos(x^2)}$

EX 7 Here are some miscellaneous antiderivatives that can be solved using (4.29). See if you can figure out how each one works before reading the follow-up description.

$$\text{A)} \int e^x \sin(e^x)\,dx = -\cos(e^x) + C$$

Here in (A), the integrand $e^x\sin(e^x)$ is already a direct match to $g'(x)\sin(g(x))$, with $g(x) = e^x$

$$\text{B)} \int \sec^2 x\cos(\tan x)\,dx = \sin(\tan x) + C$$

In (B), the integrand $\sec^2 x \cos(\tan x)$ is already a direct match to $g'(x)\cos(g(x))$ with $g(x) = \tan x$.

$$\text{C)} \quad \int 3x^4 \sin(x^5)\, dx = \frac{3}{5}\int 5x^4 \sin(x^5)\, dx = -\frac{3}{5}\cos(x^5) + C$$

In (C), we adjust the integrand $3x^4 \sin(x^5)$ so that we had a match to $g'(x)\sin(g(x))$ with $g(x) = x^5$.

$$\text{D)} \quad \int \frac{\cos(\ln x)}{2x}\, dx = \frac{1}{2}\int \frac{\cos(\ln x)}{x}\, dx = \frac{1}{2}\sin(\ln x) + C$$

In (D), after factoring out the 1/2, the integrand presented a direct match to the form $g'(x)\cos(g(x))$ with $g(x) = \ln x$. ■

You Try It

(12) Find the following antiderivatives:

(A) $\displaystyle\int (6t^2 - 4)\sin(t^3 - 2t)\, dt$ $\qquad$ (C) $\displaystyle\int \frac{\sin(\ln(v+6))}{v+6}\, dv$

(B) $\displaystyle\int \sec^2 x \cos(\tan x)\, dx$ $\qquad$ (D) $\displaystyle\int 2xe^{x^2}\cos(e^{x^2})\, dx$

There are still more trig functions to deal with, let's not forget about them! Since we already know the derivatives of plain old $\sec x$ and $\csc x$,

$$\frac{d}{dx}\sec(x) = \sec(x)\tan(x) \quad \text{and} \quad \frac{d}{dx}\csc(x) = -\csc(x)\cot(x)$$

we can also suggest these (correct) derivative formulas:

$$\boxed{\frac{d}{dx}\sec(g(x)) = \sec(g(x))\tan(g(x)) \cdot g'(x)} \quad \text{and}$$

$$\boxed{\frac{d}{dx}\csc(g(x)) = -\csc(g(x))\cot(g(x)) \cdot g'(x)} \qquad (4.30)$$

You Try It

(13) Write down the antiderivative formulas that are a consequence of the derivative formulas presented in (4.30).

(14) Find the derivatives

$$\frac{d}{dx}\sec(\sin x + \cos x) \quad \text{and} \quad \frac{d}{dx}\csc(x^3 + 2x + 1)$$

(15) Find the antiderivatives

$$\int \sin x \sec(\cos x)\tan(\cos x)\,dx \quad \text{and} \quad \int x^2 \csc(x^3)\cot(x^3)\,dx$$

A summary of all of the extended derivatives and antiderivatives related to trigonometric functions we've collected is at the end of the section. Note that because we have the extended formulas, we no longer have to separately keep track of the basic expressions as seen in (4.20). Simply using $g(x) = x$ in any of the extended formulas will collapse the formula back to the basic version. Try it!

Into the Pit!!

Derivation of the Extended Derivative Rule for $\sin(g(x))$

If you have been following the derivations of the extended derivative rules for $(g(x))^p$, $e^{g(x)}$ and $\ln(g(x))$, then this should be very familiar. Here, we'll look at the rule for the derivative of $\sin(g(x))$. If you're feeling brave, you can attempt your own derivation for the derivative of $\cos(g(x))$ in the Challenge Problems.

As before, some restrictions apply: this particular derivation works when $g(x)$ is both differentiable and one-to-one. We also need to recognize that the expression

$$\lim_{y\to a} \frac{\sin(y) - \sin(a)}{y - a}$$

represents the derivative of $\sin(y)$ at the point $y = a$, and so is equal to $\cos(a)$. With that in mind, here we go. The derivative of $\sin(g(x))$ is defined as:

$$\begin{aligned}\frac{d}{dx}\sin(g(x)) &\stackrel{\bullet}{=} \lim_{h\to 0}\frac{\sin(g(x+h))-\sin(g(x))}{h}\\ &= \lim_{h\to 0}\frac{\sin(g(x+h))-\sin(g(x))}{h}\cdot\frac{g(x+h)-g(x)}{g(x+h)-g(x)}\end{aligned}$$

With regrouping and application of the limit law regarding products, this becomes

$$\frac{d}{dx}\sin(g(x)) = \lim_{h\to 0}\frac{\sin(g(x+h))-\sin(g(x))}{g(x+h)-g(x)}\cdot\lim_{h\to 0}\frac{g(x+h)-g(x)}{h}$$

The latter limit is just $g'(x)$. The first limit can be redesigned by setting $a = g(x)$, $y = g(x+h)$, and remembering that the limit process $h \to 0$ is now equivalent to $y \to g(x)$, i.e. $y \to a$. Therefore,

$$\lim_{h\to 0}\frac{\sin(g(x+h))-\sin(g(x))}{g(x+h)-g(x)} = \lim_{y\to a}\frac{\sin(y)-\sin(a)}{y-a}$$

which defines the derivative of $\sin(y)$ at the point $y = a$, and this was noted above to be equal to $\cos(a)$. But since we have $a = g(x)$, we have, finally, that:

$$\lim_{h\to 0}\frac{\sin(g(x+h))-\sin(g(x))}{g(x+h)-g(x)} = \cos(g(x))$$

Together then,

$$\begin{aligned}\frac{d}{dx}\sin(g(x)) &= \lim_{h\to 0}\frac{\sin(g(x+h))-\sin(g(x))}{g(x+h)-g(x)}\cdot\lim_{h\to 0}\frac{g(x+h)-g(x)}{h}\\ &= \cos(g(x))\cdot g'(x)\end{aligned}$$

and this is where the derivative formula for $\sin(g(x))$ in (4.28) comes from, at least when $g(x)$ is differentiable and one-to-one.

Have You Learned...

- How to find the derivative of simple functions involving trigonometric functions?

- How to find the derivative of functions involving compositions of trigonometric functions?

$$\frac{d}{dx}\sin(g(x)) = g'(x)\cos(g(x))$$

$$\frac{d}{dx}\cos(g(x)) = -g'(x)\sin(g(x))$$

$$\frac{d}{dx}\sec(g(x)) = g'(x)\sec(g(x))\tan(g(x))$$

$$\frac{d}{dx}\csc(g(x)) = -g'(x)\csc(g(x))\cot(g(x))$$

$$\frac{d}{dx}\tan(g(x)) = g'(x)\sec^2(g(x))$$

$$\frac{d}{dx}\cot(g(x)) = -g'(x)\csc^2(g(x))$$

- That all those formulas for derivatives of $f(g(x))$, where f is a trigonometric function, aren't really needed as long as you remember the fundamental derivative formulas for trig functions?
- How to find the antiderivative of functions involving $\sin x$, $\cos x$, $\sec x \tan x$ and $\csc x \cot x$?
- How to find the antiderivative of functions of the form $g'(x)\sin(g(x))$, $g'(x)\cos(g(x))$, $g'(x)\sec(g(x))\tan(g(x))$ and $g'(x)\csc(g(x))\cot(g(x))$?

$$\int g'(x)\sin(g(x))\,dx = -\cos(g(x)) + C$$

$$\int g'(x)\cos(g(x))\,dx = \sin(g(x)) + C$$

$$\int g'(x)\ln|\sec(g(x))| = \tan(g(x)) + C$$

$$\int g'(x)\ln|\sin(g(x))| = \cot(g(x)) + C$$

$$\int g'(x)\sec(g(x))\tan(g(x))\,dx = \sec(g(x)) + C$$

$$\int g'(x)\csc(g(x))\cot(g(x))\,dx = -\csc(g(x)) + C$$

- That all those formulas for antiderivatives of $g'(x)F(g(x))$, where F is a derivative of a trigonometric function, aren't really needed as long as you remember the fundamental antiderivative formulas for trig functions?
- That the antiderivatives of $\tan x$ and $\cot x$ are the cases in which antiderivatives of trig functions produce other functions that are not trigonometric functions?

Trigonometric Functions — Problem List

Trigonometric Functions — You Try It

These appeared above; solutions begin on the next page.

(1) Find and properly name the derivatives of the following:

$(A)\, y = -2\sin(x) - 3e^x$ $\quad (C)\, y(w) = (w^2-1)^3 - \sqrt{3}\sin(w)$

$(B)\, s = \dfrac{2}{t} + 2\cos(t)$ $\quad (D)\, y = (\pi - 1)\cos(x) - \ln(x^4 + 2x^2 + 2)$

(2) Sketch a graph of $y = x + \sin x$. State at how many points this function has a horizontal tangent line, and find three of those points.

(3) Find the equation of the line tangent to $y = \ln x + \cos x$ at $x = e$. All constants must be in exact form.

(4) Find and properly name the derivatives of the following:

$(A)\, g(x) = \sin^3 x$ $\quad (C)\, y = \ln(\sin x - 4\cos x)$

$(B)\, h = \dfrac{1}{\cos^2 y} + e^{-2\cos y}$ $\quad (D)\, z(t) = \sin^2 t + \cos^2 t$

(5) Find the following antiderivatives.

$(A) \displaystyle\int -2\sin(x) - 3e^x\,dx$ $\quad (C) \displaystyle\int e^{-t} + \frac{1}{3}\sin t\,dt$

$(B) \displaystyle\int \frac{2}{x} + 2\cos(x)\,dx$ $\quad (D) \displaystyle\int 2x(x^2-1)^2 - \sqrt{7}\tan x\,dx$

(6) Find the antiderivative of $y = 2\sin x + 3\cos x$ that goes through the point $(\pi, 0)$.

(7) Find the following antiderivatives. Remember that they can now involve antiderivative formulas for the power rule, exponential functions, and logarithmic functions as well as trigonometric functions:

$(A) \displaystyle\int \cos t\, e^{\sin t}\,dt$ $\quad (C) \displaystyle\int \frac{\cos t}{2\sin t + 3}\,dt$

$(B) \displaystyle\int \sin(x)\cos^2 x\,dx$ $\quad (D) \displaystyle\int \tan x - \sin x\sqrt[3]{2 - \cos x}\,dx$

(8) Use a procedure similar to that of EX 5 to find the derivative of $\csc x$, and state the corresponding antiderivative formula that arises.

(9) Find and properly name the derivatives of the following:

$(A)\, y = 2\sec x - 5\csc x$ $\quad (C)\, z = \sec^3 x$

$(B)\, h(r) = e^{\csc r} - \dfrac{1}{\sec r}$ $\quad (D)\, y = \sqrt{\csc x + 3}$

(10) Find the following antiderivatives:

$$(A)\ \int \sec x \tan x e^{\sec x}\, dx \qquad (C)\ \int \frac{\csc x \cot x}{\csc x + 1}\, dx$$

$$(B)\ \int \csc x \cot x \csc^2 x - \cot x\, dx \qquad (D)\ \int \frac{\sec x \tan x}{(\sec x + 1)^2}\, dx$$

(11) Find and properly name the derivatives of the following:

$$(A)\ y = \cos(x^4) \qquad (C)\ p(t) = \sin(e^t + \tan t)$$

$$(B)\ z = \sin(\sec x) \qquad (D)\ y = e^{\cos(x^2)}$$

(12) Find the following antiderivatives:

$$(A)\ \int (6t^2 - 4)\sin(t^3 - 2t)\, dt \qquad (C)\ \int \frac{\sin(\ln(v+6))}{v+6}\, dv$$

$$(B)\ \int \sec^2 x \cos(\tan x)\, dx \qquad (D)\ \int 2xe^{x^2}\cos(e^{x^2})\, dx$$

(13) Write down the antiderivative formulas that are a consequence of the derivative formulas presented for $\sec(g(x))$ and $\csc(g(x))$.

(14) Find the derivatives

$$\frac{d}{dx}\sec(\sin x + \cos x) \quad \text{and} \quad \frac{d}{dx}\csc(x^3 + 2x + 1)$$

(15) Find the antiderivatives

$$\int \sin x \sec(\cos x)\tan(\cos x)\, dx \quad \text{and} \quad \int x^2 \csc(x^3)\cot(x^3)\, dx$$

Trigonometric Functions — Practice Problems

Try these as you get the hang of the You Try It problems. Solutions to these problems are available in Sec. A.4.2.

(1) Find the derivative and antiderivative of $y = 3\cos(x) - \sqrt{2}\sin(x)$.

(2) At how many points will the graph of $y = \sin(x^2)$ have a horizontal tangent line? Find the three of those points that are closest to or at the origin.

(3) Find the equation of the line tangent to $y = \cos(\ln x)$ at $x = e$. All constants must be in exact form.

(4) Find the antiderivative of $y = 2\sin x \cos^2(x) + 3\cos x$ that goes through the point $(\pi, 0)$.

(5) Find the derivative of $g(t) = e^{\sec t} + \sec(e^t)$.

(6) Find the derivative of $y(x) = \ln(\csc x) + \csc(\ln x)$.

(7) One of these antiderivatives cannot be solved using the formulas discussed in this section. Identify it, then solve the other two.

$$\text{A)} \int \sin x \cos^2 x\, dx \quad \text{B)} \int \sin^2 x \cos x\, dx \quad \text{C)} \int \sin^2 x \cos^2 x\, dx$$

(8) One of these antiderivatives cannot be solved using the formulas discussed in this section. Identify it, then solve the other two.

$$\text{A)} \int \tan^2 x\, dx \qquad \text{B)} \int \sec^2 x \tan^2 x\, dx \qquad \text{C)} \int \sec^2 x \tan x\, dx$$

(9) Find the antiderivative $\int \sec^2 x(\tan^2 x + \tan x + 1)\, dx$.

(10) Find the derivative of $y = \ln(\sin^2 x + \cos^2 x) + e^{\sec^2 x - \tan^2 x}$. (Hint: Don't just plunge in without looking closely at the function.)

Trigonometric Functions — Challenge Problems

Try these problems to test your skills with the ideas in this section. Solutions to these problems are available in Sec. B.4.2.

(1) Find dy/dx for $y = \sin(\cos(\tan x))$. (Hint: This requires more than one use of an extended derivative formula.)

(2) Follow the procedure in the last part of Example 2 to show that

$$\int \cot x\, dx = \ln|\sin x| + C$$

(3) Follow the procedure shown in this section and the two before it to show why

$$\frac{d}{dx}\cos(g(x)) = -g'(x)\sin(g(x))$$

Trig Functions Derivs — You Try It — Solved

---

(1) Find and properly name the derivatives of the following:

$$(A)\, y = -2\sin(x) - 3e^x \qquad (C)\, y(w) = (w^2-1)^3 - \sqrt{3}\sin(w)$$

$$(B)\, s = \frac{2}{t} + 2\cos(t) \qquad (D)\, y = (\pi - 1)\cos(x) - \ln(x^4 + 2x^2 + 2)$$

☐ The derivatives and their proper names are:

$$(A)\, y = -2\sin(x) - 3e^x \rightarrow \frac{dy}{dx} = -2\cos x - 3e^x$$

$$(B)\, s = 2t^{-1} + 2\cos(t) \rightarrow \frac{ds}{dt} = -\frac{2}{t^2} - 2\sin(t)$$

$$(C)\, y(w) = (w^2-1)^3 - \sqrt{3}\sin(w) \rightarrow y'(w) = 6w(w^2-1)^2 - \sqrt{3}\cos(w)$$

$$(D)\, y = (\pi - 1)\cos(x) - \ln(x^4 + 2x^2 + 2)$$

$$\frac{dy}{dx} = -(\pi - 1)\sin(x) - \frac{4x^3 + 4x}{x^4 + 2x^2 + 2} \quad ■$$

---

(2) Sketch a graph of $y = x + \sin x$. State at how many points this function has a horizontal tangent line, and find three of those points.

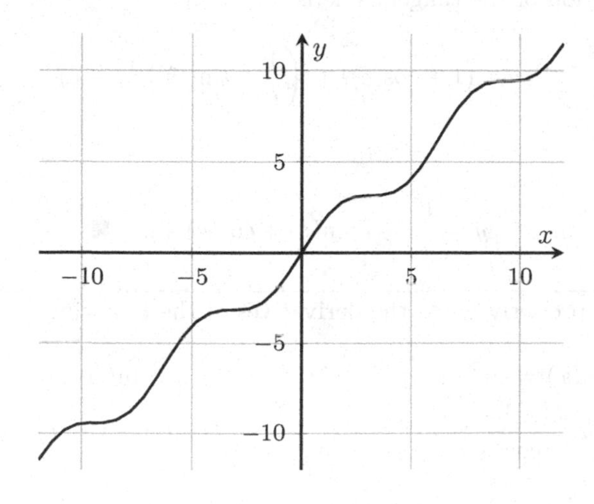

Fig. 4.2 $y = x + \sin(x)$.

□ Figure 4.2 shows a graph of $y = x + \sin(x)$. This function has horizontal tangent lines at an infinite number of points. Specifically, there is a horizontal tangent line whenever $dy/dx = 0$, that is, where:

$$\frac{d}{dx}(x + \sin x) = 0$$
$$1 + \cos x = 0$$

This occurs where $\cos x = -1$, or $x = \pi \pm 2n\pi$. ■

---

(3) Find the equation of the line tangent to $y = \ln x + \cos x$ at $x = e$. All constants must be in exact form.

□ At $x = e$, this line shares the point $(e, f(e))$ with the function; since $f(e) = \ln e + \cos(e) = 1 + \cos(e)$, the tangent line has the point $(e, 1+\cos(e))$. The slope of this tangent line comes from the derivative,

$$\frac{dy}{dx} = \frac{1}{x} - \sin(x)$$

so the slope of the tangent line at $x = e$ is

$$m_{\tan} = \frac{1}{e} - \sin(e)$$

The equation of the tangent line is

$$y - (1 + \cos(e)) = \left(\frac{1}{e} - \sin(e)\right)(x - e)$$

or also,

$$y = \frac{1}{e}x + e\sin(e) + \cos(e) + 1 \quad ■$$

---

(4) Find and properly name the derivatives of the following:

$(A)\ g(x) = \sin^3 x$

$(B)\ h = \dfrac{1}{\cos^2 y} + e^{-2\cos y}$

$(C)\ y = \ln(\sin x - 4\cos x)$

$(D)\ z(t) = \sin^2 t + \cos^2 t$

□ (A) $g(x) = \sin^3 x = (\sin x)^3$

$$g'(x) = 3\sin^2(x)\frac{d}{dx}(\sin x) = 3\cos x \sin^2 x$$

(B) $h = \dfrac{1}{\cos^2 y} + e^{-2\cos y} = (\cos y)^{-2} + e^{-2\cos y}$

$$\begin{aligned}\frac{dh}{dy} &= -2(\cos y)^{-3}\frac{d}{dy}(\cos y) + e^{-2\cos y}\frac{d}{dy}(-2\cos y)\\ &= 2\sin y(\cos y)^{-3} + 2\sin y\, e^{-2\cos y}\\ &= 2\sin y\left(\frac{1}{\cos^3 y} + e^{-2\cos y}\right)\end{aligned}$$

(C) $y = \ln(\sin x - 4\cos x)$

$$\frac{dy}{dx} = \frac{1}{\sin x - 4\cos} \cdot \frac{d}{dx}(\sin x - 4\cos x) = \frac{\cos x + 4\sin x}{\sin x - 4\cos}$$

The clever way to do (D) is to apply the trig identity $\sin^2 t + \cos^2 t = 1$ to write $z(t) = 1$ and so $z'(t) = 0$. But to practice with the extended power rule, we have

$$\begin{aligned}z'(t) &= 2\sin t\frac{d}{dt}(\sin t) + 2\cos t\frac{d}{dt}(\cos t)\\ &= 2\sin t(\cos t) + 2\cos t(-\sin t)\\ &= 2\sin t\cos t - 2\sin t\cos t = 0 \quad ■\end{aligned}$$

---

(5) Find the following antiderivatives.

(A) $\displaystyle\int -2\sin(x) - 3e^x\,dx$ *(C)* $\displaystyle\int e^{-t} + \frac{1}{3}\sin t\,dt$

(B) $\displaystyle\int \frac{2}{x} + 2\cos(x)\,dx$ *(D)* $\displaystyle\int 2x(x^2-1)^2 - \sqrt{7}\tan x\,dx$

□ (A) $\displaystyle\int -2\sin(x) - 3e^x\,dx = 2\cos x - 3e^x + C$

(B) $\displaystyle\int \frac{2}{x} + 2\cos(x)\,dx = 2\ln|x| + 2\sin x + C$

(C) $\displaystyle\int e^{-t} + \frac{1}{3}\sin t\,dt = -e^{-t} - \frac{1}{3}\cos t + C$

(D) $\displaystyle\int 2x(x^2-1)^2 - \sqrt{7}\tan x\,dx = \frac{1}{3}(x^2-1)^3 - \sqrt{7}\ln|\sec x| + C$ ■

(6) Find the antiderivative of $y = 2\sin x + 3\cos x$ that goes through the point $(\pi, 0)$.

☐ The general antiderivative is

$$\int 2\sin x + 3\cos x\, dx = -2\cos x + 3\sin x + C$$

To find the antiderivative that goes through $(\pi, 0)$ (let's name it $F(x)$), we need:

$$\begin{aligned} F(\pi) &= 0 \\ -2\cos\pi + 3\sin\pi + C &= 0 \\ -2(-1) + 3(0) + C &= 0 \\ C &= -2 \end{aligned}$$

So the specific antiderivative we need is

$$F(x) = -2\cos x + 3\sin x - 2 \quad \blacksquare$$

(7) Find the following antiderivatives. Remember that they can now involve antiderivative formulas for the power rule, exponential functions, and logarithmic functions as well as trigonometric functions:

$(A)\ \int \cos t\, e^{\sin t}\, dt$

$(B)\ \int \sin(x)\cos^2 x\, dx$

$(C)\ \int \frac{\cos t}{2\sin t + 3}\, dt$

$(D)\ \int \tan x - \sin x \sqrt[3]{2 - \cos x}\, dx$

☐ For (A), we directly match the integrand to $g'(t)e^{g(t)}$ with $g(t) = \sin t$, so Eq. (4.3) gives

$$\int \cos t\, e^{\sin t}\, dt = e^{\sin t} + C$$

For (B), we can use the extended power rule for integrand $g'(x)(g(x))^p$ with $g(x) = \cos x$:

$$\int \sin(x)\cos^2 x\, dx = -\frac{1}{3}\cos^3 x + C$$

For (C), we can maneuver constants and match the integrand to $g'(t)/g(t)$ with $g(t) = 2\sin t + 3$ and use Eq. (4.8):

$$\int \frac{\cos t}{2\sin t + 3}\, dt = \frac{1}{2}\int \frac{2\cos t}{2\sin t + 3}\, dt = \frac{1}{2}\ln|2\sin t + 3| + C$$

In (D), we have a combination of terms. The first term's antiderivative is given directly by Eq. (4.18). For the second, we must tinker with constants to have a match to the extended power rule: with $g(x) = 2-\cos x$ and thus $g'(x) = \sin x$, we can link to $g'(x)(g(x))^p$ for $p = 1/3$:

$$\begin{aligned}\int \tan x - \sin x\sqrt[3]{2-\cos x}\,dx &= \int \tan x - \int \sin x(2-\cos x)^{1/3}\,dx \\ &= \ln|\sec x| - \frac{1}{4/3}(2-\cos x)^{4/3} + C \\ &= \ln|\sec x| - \frac{3}{4}(2-\cos x)^{4/3} + C \quad \blacksquare\end{aligned}$$

(8) Use a procedure similar to that of Example 5 to find the derivative of $\csc x$, and state the corresponding antiderivative formula that arises.

□ Since $\csc x = (\sin x)^{-1}$, we have

$$\begin{aligned}\frac{d}{dx}\csc x &= -(\sin x)^{-2}\frac{d}{dx}(\sin x) = -\frac{1}{\sin^2 x}(\cos x) \\ &= -\frac{1}{\sin x}\frac{\cos x}{\sin x} = -\csc x\cot x\end{aligned}$$

and the corresponding antiderivative formula is

$$\int \csc x\cot x\,dx = -\csc x + C \quad \blacksquare$$

(9) Find and properly name the derivatives of the following:

$(A)\; y = 2\sec x - 5\csc x$ $\qquad$ $(C)\; z = \sec^3 x$

$(B)\; h(r) = e^{\csc r} - \dfrac{1}{\sec r}$ $\qquad$ $(D)\; y = \sqrt{\csc x + 3}$

□ (A) $y = 2\sec x - 5\csc x$

$$\frac{dy}{dx} = 2\sec x\tan x + 5\csc x\cot x$$

(B) $h(r) = e^{\csc r} - \dfrac{1}{\sec r} = e^{\csc r} - \cos r$

$$h'(r) = -\csc r\cot r e^{\csc r} + \sin r$$

(C) $z = \sec^3 x = (\sec x)^3$

$$\frac{dz}{dx} = 3(\sec x)^2\frac{d}{dx}(\sec x) = 3\sec^2 x\sec x\tan x = 3\sec^3 x\tan x$$

(D) $y = \sqrt{\csc x + 3}$

$$\frac{dy}{dx} = \frac{1}{2}(\csc x + 3)^{-1/2}\frac{d}{dx}(\csc x + 3) = -\frac{\csc x\cot x}{2\sqrt{\csc x + 3}} \quad \blacksquare$$

(10) Find the following antiderivatives:

$$(A)\ \int \sec x \tan x e^{\sec x}\, dx \qquad (C)\ \int \frac{\csc x \cot x}{\csc x + 1}\, dx$$

$$(B)\ \int \csc x \cot x \csc^2 x - \cot x\, dx \qquad (D)\ \int \frac{\sec x \tan x}{(\sec x + 1)^2}\, dx$$

$$\square\,\text{(A)}\ \int \sec x \tan x e^{\sec x}\, dx = e^{\sec x} + C$$

(We matched the integrand to $g'(x)e^{g(x)}$ with $g(x) = \sec x$.)

$$\text{(B)}\ \int \csc x \cot x \csc^2 x - \cot x\, dx = \frac{1}{3}\csc^3 x - \ln|\sin x| + C$$

(For the first term, we matched the integrand as $g'(x)(g(x))^p$ for the extended power rule with $g(x) = \csc x$ amd $p = 2$. The second term is a direct formula.)

$$\text{(C)}\ \int \frac{\csc x \cot x}{\csc x + 1}\, dx = \ln|\csc x + 1| + C$$

(We matched the integrand to $g'(x)/g(x)$ with $g(x) = \csc x + 1$.)

$$\text{(D)}\ \int \frac{\sec x \tan x}{(\sec x + 1)^2}\, dx = \frac{1}{3}(\sec x + 1)^3 + C$$

(We matched the integrand as $g'(x)(g(x))^p$ for the extended power rule with $g(x) = \sec x + 1$.) ■

(11) Find and properly name the derivatives of the following:

$$(A)\ y = \cos(x^4) \qquad (C)\ p(t) = \sin(e^t + \tan t)$$

$$(B)\ z = \sin(\sec x) \qquad (D)\ y = e^{\cos(x^2)}$$

□ For (A), with $y = \cos(x^4)$, we have

$$\frac{dy}{dx} = -\sin(x^4)\frac{d}{dx}(x^4) = -4x^3\sin(x^4)$$

For (B), with $z = \sin(\sec x)$, we have

$$\frac{dz}{dx} = \cos(\sec x)\frac{d}{dx}(\sec x) = \sec x \tan x \cos(\sec x)$$

For (C), with $p(t) = \sin(e^t + \tan t)$, we have

$$p'(t) = \cos(e^t + \tan t)\frac{d}{dt}(e^t + \tan t) = (e^t + \sec^2 t)\cos(e^t + \tan t)$$

For (D), with $y = e^{\cos(x^2)}$, we have

$$\begin{aligned}\frac{dy}{dx} &= e^{\cos(x^2)}\frac{d}{dx}\cos(x^2) = e^{\cos(x^2)}(-\sin(x^2))\frac{d}{dx}(x^2)\\ &= e^{\cos(x^2)}(-\sin(x^2))(2x) = -2x\sin(x^2)e^{\cos(x^2)}\end{aligned}$$ ■

---

(12) Find the following antiderivatives:

$$(A)\ \int (6t^2 - 4)\sin(t^3 - 2t)\,dt \qquad (C)\ \int \frac{\sin(\ln(v+6))}{v+6}\,dv$$

$$(B)\ \int \sec^2 x\cos(\tan x)\,dx \qquad (D)\ \int 2xe^{x^2}\cos(e^{x^2})\,dx$$

□ Each of these presents an antiderivative of an extended sine or cosine function, and thus will require one of the formulas in Eq. (4.29):

$$\begin{aligned}\text{(A)}\quad \int (6t^2 - 4)\sin(t^3 - 2t)\,dt &= 2\int (3t^2 - 2)\sin(t^3 - 2t)\\ &= -2\cos(t^3 - 2t) + C\end{aligned}$$

(We matched the integrand to $g'(t)\sin(g(t))$ with $g(t) = t^3 - 2t$.)

$$\text{(B)}\quad \int \sec^2 x\cos(\tan x)\,dx = \sin(\tan x) + C$$

(We matched the integrand to $g'(x)\cos(g(x))$ with $g(x) = \tan x$.)

$$\text{(C)}\quad \int \frac{\sin(\ln(v+6))}{v+6}\,dv = -\cos(\ln(v+6)) + C$$

(We matched the integrand to $g'(v)\sin(g(v))$ with $g(v) = \ln(v+6)$.)

$$\text{(D)}\quad \int 2xe^{x^2}\cos(e^{x^2})\,dx = \sin(e^{x^2}) + C$$

(We matched the integrand to $g'(x)\cos(g(x))$ with $g(x) = e^{x^2}$.) ■

---

(13) Write down the antiderivative formulas that are a consequence of the derivative formulas presented for $\sec(g(x))$ and $\csc(g(x))$.

□ Since we know

$$\begin{aligned}\frac{d}{dx}\sec(g(x)) &= g'(x)\sec(g(x))\tan(g(x))\\ \frac{d}{dx}\csc(g(x)) &= -g'(x)\csc(g(x))\cot(g(x))\end{aligned}$$

then we also know

$$\int g'(x)\sec(g(x))\tan(g(x))\,dx = \sec(g(x)) + C$$

$$\int g'(x)\csc(g(x))\cot(g(x))\,dx = -\csc(g(x)) + C \quad \blacksquare$$

---

(14) Find the derivatives:

$$\frac{d}{dx}\sec(\sin x + \cos x) \quad \text{and} \quad \frac{d}{dx}\csc(x^3 + 2x + 1)$$

$\square$ $\dfrac{d}{dx}\sec(\sin x + \cos x)$

$$= \sec(\sin x + \cos x)\tan(\sin x + \cos x)\frac{d}{dx}(\sin x + \cos x)$$

$$= \sec(\sin x + \cos x)\tan(\sin x + \cos x)(\cos x - \sin x)$$

$\dfrac{d}{dx}\csc(x^3 + 2x + 1)$

$$= -\csc(x^3 + 2x + 1)\cot(x^3 + 2x + 1)\frac{d}{dx}(x^3 + 2x + 1)$$

$$= -(3x^2 + 2)\csc(x^3 + 2x + 1)\cot(x^3 + 2x + 1) \quad \blacksquare$$

---

(15) Find the antiderivatives

$$\int \sin x \sec(\cos x)\tan(\cos x)\,dx \quad \text{and} \quad \int x^2\csc(x^3)\cot(x^3)\,dx$$

$\square$ For these, we should plan to use the formulas demonstrated in YTI 13, which reverse the derivatives originally given as Eq. (4.30). For the first, we match the integrand to $g'(x)\sec(g(x))\tan(g(x))$, with $g(x) = \cos x$, for which we'll need to adjust with a negative sign, as $g'(x)$ should be *sinx*:

$$\int \sin x \sec(\cos x)\tan(\cos x)\,dx$$

$$= -\int(-\sin x)\sec(\cos x)\tan(\cos x)\,dx = -\sec(\cos x) + C$$

For the second, it looks like we can present $g'(x)\csc(g(x))\cot(g(x))$ with $g(x) = x^3$ and $g'(x) = 3x^2$, with minimal tinkering needed:

$$\int x^2\csc(x^3)\cot(x^3)\,dx = \frac{1}{3}\int(3x^2)\csc(x^3)\cot(x^3)\,dx$$

$$= -\frac{1}{3}\csc(x^3) + C \quad \blacksquare$$

4.3 The Product and Quotient Rules

Introduction

So far, we have learned the derivatives of several simple functions, as well as more complicated variations built as compositions. However, there are still other constructions we can make with these individual functions. For example, in the last section, our basic knowledge of the derivative of $\sin x$ and our knowledge of the power rule allowed us to find the derivative of $\csc x$, since $\csc x = (\sin x)^{-1}$. But we have not yet seen how we can come up with the derivative of $\tan x$. While we can rewrite $\tan x = \sin x / \cos x$, we can't yet use what we know about the derivatives of $\sin x$ and $\cos x$ themselves here because the latter is a *quotient* of the two, not a composition — and we do not know how to find the derivative of something in the form $f(x)/g(x)$. That is what this new section is about: finding derivatives of functions that are constructed as products or quotients of simpler functions with known derivatives.

The Product Rule

We have been lucky so far in that many things have worked out the way we hoped they'd work out. For example, with limits, we know that limit of the sum of two functions is just the sum of the limits,

$$\lim_{x \to a} (f(x) + g(x)) = \lim_{x \to a} f(x) + \lim_{x \to a} g(x)$$

and we know that the limit of a product is the product of the limits,

$$\lim_{x \to a} (f(x) \cdot g(x)) = \lim_{x \to a} f(x) \cdot \lim_{x \to a} g(x)$$

And when we move to derivatives, the derivative of a sum is the sum of derivatives,

$$\frac{d}{dx}(f(x) + g(x)) = \frac{d}{dx} f(x) + \frac{d}{dx} g(x)$$

.... and, well, that's as far as it goes. It would be really nice if the derivative of a product was the product of the derivatives; in other words, it would be great if

$$\frac{d}{dx}(f(x) \cdot g(x)) = \frac{d}{dx} f(x) \cdot \frac{d}{dx} g(x)$$

But the bad news is that's not true, as a simple example will show. Let's take the functions $f(x) = x$ and $g(x) = 1/x$ and test it out. The derivative of the product is

$$\frac{d}{dx}(f(x) \cdot g(x)) = \frac{d}{dx}\left(x \cdot \frac{1}{x}\right) = \frac{d}{dx}(1) = 0$$

while the product of the derivatives is

$$\frac{d}{dx}f(x) \cdot \frac{d}{dx}g(x) = \frac{d}{dx}x \cdot \frac{d}{dx}\frac{1}{x} = (1) \cdot \frac{-1}{x^2} = -\frac{1}{x^2}$$

So at least in this case, the derivative of the product is not equal to the product of the derivatives. But this spoils the whole thing, because if we can find a counterexample to a rule, then the rule isn't true. We need an alternative formula for the derivative of a product, and here it is (its development is at the end of this section).

Useful Fact 4.1. *As long as the derivatives of $f(x)$ and $g(x)$ themselves exist, then*

$$\boxed{\frac{d}{dx}(f(x) \cdot g(x)) = f'(x)g(x) + f(x)g'(x)} \tag{4.31}$$

This is the ***product rule.***

The product rule tells is that if we want the derivative of a product of two functions, we take turns: each function gets its derivative found, while the other sits idle; having done both, we add the results. Note that this is an *organizational* rule for derivatives: it does not tell us what the derivatives $f'(x)$ and $g'(x)$ are, it just tells us what to do with them if we know we can find them already.

Here are several examples.

EX 1 Find the derivative of $f(x) = \sin x \cos x$.

In the product rule, we just take turns. There will be two terms in the derivative; in each term one function sits idle while the other's derivative is found:

$$\begin{aligned} f'(x) = \sin x \frac{d}{dx} \cos x + \cos x \frac{d}{dx} \sin x &= \sin x(-\sin x) + \cos x(\cos x) \\ &= -\sin^2 x + \cos^2 x \quad \blacksquare \end{aligned}$$

EX 2 Find the derivative of $y = e^x\sqrt{x}$.

By the product rule, we just take turns. There will be two terms in the derivative; in each term one function sits idle while the other's derivative is found:

$$\frac{dy}{dx} = e^x \frac{d}{dx}\sqrt{x} + \sqrt{x}\frac{d}{dx}e^x = e^x \cdot \frac{1}{2\sqrt{x}} + \sqrt{x}(e^x) = e^x\left(\frac{1}{2\sqrt{x}} + \sqrt{x}\right) \quad \blacksquare$$

You Try It

(1) Find the derivative of $f(x) = x^2 e^x$.
(2) Find the derivative of $g(t) = t^3 \cos(t^2)$.

You Try It

(3) Find the derivative of $h(\theta) = e^{2\theta} \cot 3\theta$.
(4) Find the derivative of $y = (r^2 - 2r)e^r$.

EX 3 Find the equation of the tangent line to $y = e^x \cos x$ at (0,1).

This requires the slope of that tangent line, i.e. the value of $\frac{dy}{dx}$ at $x = 0$. By the product rule,

$$\frac{dy}{dx} = e^x \frac{d}{dx} \cos x + \frac{d}{dx} e^x \cdot \cos x = e^x(-\sin x) + e^x \cos x = e^x(\cos x - \sin x)$$

So at $x = 0$, $\frac{dy}{dx} = 1$. The equation of the tangent line is then $y - 1 = 1(x - 0)$, i.e. $y = x + 1$. ■

You Try It

(5) Find the equation of the line tangent to $y = 2xe^x$ at (0,0).

The Quotient Rule

It is a sad fact that this statement is FALSE:

$$\frac{d}{dx} \frac{f(x)}{g(x)} = \frac{f'(x)}{g'(x)}$$

It would be great if that was true, since the derivative of a quotient would be straightforward. But it's not true. Like with products above, we need a *correct* way to handle derivatives of quotients. Here it is:

Useful Fact 4.2. *As long as the derivatives of $f(x)$ and $g(x)$ themselves exist, then*

$$\boxed{\frac{d}{dx} \frac{f(x)}{g(x)} = \frac{g(x)f'(x) - f(x)g'(x)}{[g(x)]^2}} \tag{4.32}$$

This is the ***quotient rule****.*

The derivation of the quotient rule is below. Given a function in the form of a quotient, we identify the numerator of our function as $f(x)$, the denominator as $g(x)$, then then fill in the formula with $f(x)$, $g(x)$, and their derivatives. It's like the product rule, but worse.

One of the first uses of the quotient rule can be to (finally!) develop the derivative of $\tan x$.

EX 4 Find the derivative of $y = \tan x$.

We write the function as

$$y = \tan x = \frac{\sin x}{\cos x}$$

and then apply the quotient rule:

$$\begin{aligned}\frac{dy}{dx} &= \frac{(\cos x)\frac{d}{dx}(\sin x) - (\sin x)\frac{d}{dx}(\cos x)}{(\cos x)^2} \\ &= \frac{(\cos x)(\cos x) - (\sin x)(-\sin x)}{(\cos x)^2} = \frac{\cos^2 x + \sin^2 x}{(\cos x)^2} \\ &= \frac{1}{(\cos x)^2} = \sec^2 x \quad \blacksquare\end{aligned}$$

EX 5 Find the derivative of $w = \dfrac{\tan z}{e^{2z-1}}$.

Hey, new variables! No sweat, we can just see that the independent variable is z, and the name of the derivative we want is $\dfrac{dz}{dw}$. We need to use the quotient rule on this, so

$$\begin{aligned}\frac{dw}{dz} &= \frac{(e^{2z-1})\frac{d}{dz}(\tan z) - (\tan z)\frac{d}{dz}e^{2z-1}}{(e^{2z-1})^2} \\ &= \frac{(e^{2z-1})(\sec^2 z) - (\tan z)(2e^{2z-1})}{(e^{2z-1})^2} = \frac{\sec^2 z - 2\tan z}{e^{2z-1}} \quad \blacksquare\end{aligned}$$

EX 6 Find the derivative of $y = \dfrac{t^3 + t}{t^4 - 2}$.

Using the quotient rule,

$$\begin{aligned}\frac{dy}{dt} &= \frac{(t^4 - 2)\frac{d}{dt}(t^3 + t) - (t^3 + t)\frac{d}{dt}(t^4 - 2)}{(t^4 - 2)^2} \\ &= \frac{(t^4 - 2)(3t^2 + 1) - (t^3 + t)(4t^3)}{(t^4 - 2)^2} = \frac{-t^6 - 3t^4 - 6t^2 - 2}{(t^4 - 2)^2} \quad \blacksquare\end{aligned}$$

EX 7 Find the derivative of $y = \dfrac{\sqrt{x}}{x^2 - 1}$.

Using the quotient rule,

$$\begin{aligned}
\frac{dy}{dx} &= \frac{(x^2-1)\frac{d}{dx}(\sqrt{x}) - \sqrt{x}\frac{d}{dx}(x^2-1)}{(x^2-1)^2} = \frac{(x^2-1)\frac{1}{2\sqrt{x}} - \sqrt{x}(2x)}{(x^2-1)^2} \\
&= \frac{(x^2-1) - 2\sqrt{x}(2x\sqrt{x})}{2\sqrt{x}(x^2-1)^2} = \frac{x^2 - 1 - 4x^2}{2\sqrt{x}(x^2-1)^2} \\
&= \frac{-3x^2 - 1}{2\sqrt{x}(x^2-1)^2} \quad \blacksquare
\end{aligned}$$

So clearly, simplification and clean-up after the quotient rule is going to be needed in many cases.

You Try It

(6) Find the derivative of $f(x) = \dfrac{3x-1}{2x+1}$.

(7) Find the derivative of $f(\theta) = \dfrac{\sec\theta}{1+\sec\theta}$.

You Try It

(8) Find the equation of the line tangent to $y = \dfrac{\cos(2x)}{e^x + x}$ at $x = 0$.

(9) Find the derivative of $y = \dfrac{\sin^2 x}{\cos^2 x + 1}$.

You Try It

(10) Where is the line tangent to $y = \dfrac{\ln x}{x}$ horizontal?

With the quotient rule in hand, we are ready to provide the derivatives of the four hyperbolic functions besides $\sinh(x)$ and $\cosh(x)$. We already have those derivatives in Eq. (4.9). You may recall (or you can look up) that the hyperbolic tangent function is defined as $\tanh(x) = \dfrac{\sinh(x)}{\cosh(x)}$. And so ...

EX 8 Find the derivative of $y = \tanh(x)$.

This derivative uses the quotient rule with no surprises,

$$\begin{aligned}\frac{dy}{dx} = \frac{d}{dx}\frac{\sinh(x)}{\cosh(x)} &= \frac{\cosh(x)\frac{d}{dx}(\sinh x) - \sinh x\frac{d}{dx}(\cosh(x))}{(\cosh(x))^2} \\ &= \frac{\cosh(x)(\cosh x) - \sinh x(\sinh(x))}{\cosh^2(x)} \\ &= \frac{\cosh^2(x) - \sinh^2(x)}{\cosh^2(x)} = \frac{1}{\cosh^2(x)} = \text{sech}^2(x) \quad \blacksquare\end{aligned}$$

Let's set this off as a new result:

$$\boxed{\frac{d}{dx}\tanh(x) = \text{sech}^2(x)} \quad \Leftrightarrow \quad \boxed{\int \text{sech}^2(x)\,dx = \tanh(x) + C} \tag{4.33}$$

You Try It

(11) Find the derivative of $f(x) = \text{sech}(x)$.

You may have noticed we did not tackle any antiderivatives in this section. In previous sections, most derivative rules automatically came with corresponding antiderivative rules. In this section, though, it's more complicated. The natural antiderivative partner to the product rule is called *integration by parts*; while it might be suitable to place it here, it usually waits until the content normally associated with a second semester in Calculus. Regardless of ordering of topics in individual sections, I'm not brave enough to start juggling topics across semesters. So a discussion of integration by parts awaits you in Chapter 7 (in Volume 2).

Into the Pit!!

Derivation of The Product Rule

The development of this formula requires some mathematical shenanigans via the strategy of "make it look worse before we make it look better". We'll apply the usual limit definition of derivative, and add / subtract some clever terms. Here is the limit definition at work:

$$\frac{d}{dx}(f(x)\cdot g(x)) = \lim_{h\to 0}\frac{f(x+h)g(x+h) - f(x)g(x)}{h}$$

Now we add and subtract a clever term; we get a numerator that is equal to the old one, but looks much worse:

$$\cdots = \lim_{h\to 0} \frac{f(x+h)g(x+h) - f(x)g(x+h) + f(x)g(x+h) - f(x)g(x)}{h}$$

Now we group and factor,

$$\cdots = \lim_{h\to 0} \frac{g(x+h)(f(x+h) - f(x)) + f(x)(g(x+h) - g(x))}{h}$$

Next, we can split this one big limit into two smaller ones,

$$\cdots = \lim_{h\to 0} \frac{g(x+h)(f(x+h) - f(x))}{h} + \lim_{h\to 0} \frac{f(x)(g(x+h) - g(x))}{h}$$

and then split these two limits into four (we *can* split limits of products):

$$\cdots = \lim_{h\to 0} g(x+h) \cdot \lim_{h\to 0} \frac{f(x+h) - f(x)}{h} + \lim_{h\to 0} f(x) \cdot \lim_{h\to 0} \frac{g(x+h) - g(x)}{h}$$

Of those four limits, two are easy — the first and third are just $g(x)$ and $f(x)$ themselves. The other two should be recognizable as the individual derivatives $f'(x)$ and $g'(x)$. So, together, with a bit of rearranging,

$$\frac{d}{dx}(f(x) \cdot g(x)) = f'(x)g(x) + f(x)g'(x)$$

and there's the product rule!

Derivation of The Quotient Rule

Here is what you've been waiting for! The derivation of the quotient rule will be very similar to that of the product rule; we'll set up the appropriate limit definition, add / subtract a helpful term, and shake things out.

Here is the limit definition of the derivative of our quotient:

$$\frac{d}{dx}\frac{f(x)}{g(x)} = \lim_{h\to 0} \frac{\frac{f(x+h)}{g(x+h)} - \frac{f(x)}{g(x)}}{h} \cdots$$

Those two terms in the numerator really ought to be combined with a least common denominator,

$$\begin{aligned} \cdots &= \lim_{h\to 0} \frac{\frac{f(x+h)g(x)}{g(x+h)g(x)} - \frac{g(x+h)f(x)}{g(x+h)g(x)}}{h} \\ &= \lim_{h\to 0} \frac{f(x+h)g(x) - g(x+h)f(x)}{h \cdot g(x+h)g(x)} \cdots \end{aligned}$$

And now it's time for "make things look worse before we can make them look better"; we add and subtract $f(x)g(x)$ into the numerator for reasons to be determined:

$$\ldots = \lim_{h\to 0} \frac{f(x+h)g(x) - f(x)g(x) - g(x+h)f(x) + f(x)g(x)}{h \cdot g(x+h)g(x)} \ldots$$

and now do some factoring that has been made possible with these new terms:

$$\ldots = \lim_{h\to 0} \frac{g(x)[f(x+h) - f(x)] - f(x)[g(x+h) - g(x)]}{h \cdot g(x+h)g(x)} \ldots$$

Can you see what is starting to develop? If we now pull the h from the denominator up into both terms of the numerator,

$$\ldots = \lim_{h\to 0} \frac{g(x) \cdot \frac{f(x+h)-f(x)}{h} - f(x) \cdot \frac{g(x+h)-g(x)}{h}}{g(x+h)g(x)} \ldots$$

The limit operation $h \to 0$ can be passed along to both terms of the numerator and the denominator, and the pieces are ready to fall into place. If we consider the big quotient as having the form $\frac{A-B}{C}$, then applying the limit gives us

$$\lim_{h\to 0} \frac{A-B}{C} = \frac{\lim_{h\to 0} A - \lim_{h\to 0} B}{\lim_{h\to 0} C} \tag{4.34}$$

(as long as $\lim_{h\to 0} C \neq 0$). Taking each piece one at a time,

$$\begin{aligned}
\lim_{h\to 0} A &= \lim_{h\to 0} g(x) \cdot \frac{f(x+h) - f(x)}{h} = \lim_{h\to 0} g(x) \cdot \lim_{h\to 0} \frac{f(x+h) - f(x)}{h} \\
&= g(x)f'(x) \\
\lim_{h\to 0} B &= \lim_{h\to 0} f(x) \cdot \frac{g(x+h) - g(x)}{h} = \lim_{h\to 0} f(x) \cdot \lim_{h\to 0} \frac{g(x+h) - g(x)}{h} \\
&= f(x)g'(x) \\
\lim_{h\to 0} C &= \lim_{h\to 0} g(x+h)g(x) = g(x) \cdot g(x) = [g(x)]^2
\end{aligned}$$

Putting it all back together from the beginning,

$$\begin{aligned}
\frac{d}{dx}\frac{f(x)}{g(x)} &= \lim_{h\to 0} \frac{g(x) \cdot \frac{f(x+h)-f(x)}{h} - f(x) \cdot \frac{g(x+h)-g(x)}{h}}{g(x+h)g(x)} \\
&= \frac{g(x)f'(x) - f(x)g'(x)}{[g(x)]^2}
\end{aligned}$$

And this is the quotient rule we've been using! FFT: The distribution of the limit operation to the numerator and denominator of expression (4.34) required that $\lim_{h\to 0} C \neq 0$. How are we sure we've met that condition?

Have You Learned...

- ... how to use the product rule?

$$\frac{d}{dx}(f(x) \cdot g(x)) = f'(x)g(x) + f(x)g'(x)$$

- ... how to use the quotient rule?

$$\frac{d}{dx}\frac{f(x)}{g(x)} = \frac{g(x)f'(x) - f(x)g'(x)}{[g(x)]^2}$$

- ... how to use the product and quotient rules in combination with other rules such as the power rule?
- ... some new derivatives related to hyperbolic functions?

The Product and Quotient Rules — Problem List

The Product and Quotient Rules — You Try It

These appeared above; solutions begin on the next page.

(1) Find the derivative of $f(x) = x^2 e^x$.
(2) Find the derivative of $g(t) = t^3 \cos(t^2)$.
(3) Find the derivative of $h(\theta) = e^{2\theta} \cot 3\theta$.
(4) Find the derivative of $y = (r^2 - 2r)e^r$.
(5) Find the equation of the line tangent to $y = 2xe^x$ at (0,0).
(6) Find the derivative of $f(x) = \dfrac{3x-1}{2x+1}$.
(7) Find the derivative of $f(\theta) = \dfrac{\sec\theta}{1+\sec\theta}$.
(8) Find the equation of the line tangent to $y = \dfrac{\cos(2x)}{e^x + x}$ at $x = 0$.
(9) Find the derivative of $y = \dfrac{\sin^2 x}{\cos^2 x + 1}$.
(10) Where is the line tangent to $y = \dfrac{\ln x}{x}$ horizontal?
(11) Find the derivative of $f(x) = \text{sech}(x)$.

The Product and Quotient Rules — Practice Problems

Try these as you get the hang of the You Try It problems. Solutions to these problems are available in Sec. A.4.3.

(1) Find the derivative of $f(x) = \sqrt{x}e^x$.
(2) Find the derivative of $y = e^t(\cos t + 5t)$.
(3) Find the derivative of $f(t) = \dfrac{2t}{4+t^2}$.
(4) Find the derivative of $y = \dfrac{\tan x - 1}{\sec x}$.
(5) Find the equation of the line tangent to $y = \dfrac{e^x}{x}$ at $(1, e)$.
(6) Find the equation of the line tangent to $y = \tanh(x)$ where $x = 1$.
(7) At what locations are the lines tangent to $y = xe^{-x^2}$ horizontal?
(8) Recall that velocity is the derivative of position. If an object's position as a function of time is given by $s(t) = \sin(t^2)/(t^2+1)$, find the first positive t value at which the object has a velocity of zero.
(9) Find the derivative of $f(x) = \tan\left(\dfrac{x}{x+1}\right)$.
(10) Find the derivatives of $f(x) = \text{csch}(x)$ and $g(x) = \coth(x)$.

The Product and Quotient Rules — Challenge Problems

Try these problems to test your skills with the ideas in this section. Solutions to these problems are available in Sec. B.4.3.

(1) Find and properly name the derivative of $z = w^{3/2}\sin(2w + e^{-w})$.
(2) Find the equations of all lines tangent to $y = (x-1)/(x+1)$ which are parallel to $x - 2y = 2$.
(3) At which point will the line tangent to $y = e^{-\sin x}/x - 1/2$ at $x = 1$ cross the x-axis? (If you provide a decimal approximation to the answer, it must have at least 3 decimal place accuracy.)

Product and Quotient Rules — You Try It — Solved

(1) Find the derivative of $f(x) = x^2 e^x$.

☐ Using the product rule,

$$f'(x) = x^2 \frac{d}{dx} e^x + \frac{d}{dx} x^2 \cdot e^x = x^2 e^x + 2xe^x = xe^x(x+2) \quad \blacksquare$$

(2) Find the derivative of $g(t) = t^3 \cos(t^2)$.

☐ By the product rule,

$$\begin{aligned} g'(t) = t^3 \frac{d}{dt} \cos(t^2) + \frac{d}{dt} t^3 \cdot \cos(t^2) &= t^3(-2t \sin(t^2)) + 3t^2 \cos(t^2) \\ &= -2t^4 \sin(t^2) + 3t^2 \cos(t^2) \quad \blacksquare \end{aligned}$$

(3) Find the derivative of $h(\theta) = e^{2\theta} \cot 3\theta$.

☐ By the product rule,

$$\begin{aligned} h'(\theta) &= e^{2\theta} \frac{d}{dt} \cot 3\theta + \frac{d}{dt} e^{2\theta} \cdot 3 \cot \theta \\ &= e^{2\theta}(-3 \csc^2 3\theta) + 2e^{2\theta} \cot 3\theta = e^{2\theta}(-3 \csc^2 3\theta + 2 \cot 3\theta) \quad \blacksquare \end{aligned}$$

(4) Find the derivative of $y = (r^2 - 2r)e^r$.

☐ The independent variable is r, so the derivative we want is dy/dr. You can multiply this out first or not before needing the product rule. I will not multiply it out:

$$\begin{aligned} \frac{dy}{dr} &= (r^2 - 2r) \frac{d}{dr} e^r + \frac{d}{dr} (r^2 - 2r) \cdot e^r \\ &= (r^2 - 2r)e^r + (2r - 2)e^r = e^r(r^2 - 2) \quad \blacksquare \end{aligned}$$

(5) Find the equation of the line tangent to $y = 2xe^x$ at (0,0).

☐ This requires the slope of that tangent line, i.e. the value of dy/dx at $x = 0$. So since

$$\frac{dy}{dx} = 2x \frac{d}{dx} e^x + \frac{d}{dx} 2x \cdot e^x = 2xe^x + 2e^x = 2e^x(x+2)$$

we see that at $x = 0$, $dy/dx = 2$. So the equation of the tangent line is $y - 0 = 2(x - 0)$, i.e. $y = 2x$. ■

(6) Find the derivative of $f(x) = \dfrac{3x-1}{2x+1}$.

☐ Using the quotient rule,

$$\begin{aligned} f'(x) &= \frac{(2x+1)\frac{d}{dx}(3x-1) - (3x-1)\frac{d}{dx}(3x+1)}{(2x+1)^2} \\ &= \frac{(2x+1)(3) - (3x-1)(2)}{(2x+1)^2} = \frac{6x+3-(6x-2))}{(2x+1)^2} \\ &= \frac{5}{(2x+1)^2} \quad \blacksquare \end{aligned}$$

(7) Find the derivative of $f(\theta) = \dfrac{\sec\theta}{1+\sec\theta}$.

☐ Using the quotient rule,

$$\begin{aligned} f'(\theta) &= \frac{(1+\sec\theta)\frac{d}{d\theta}(\sec\theta) - (\sec\theta)\frac{d}{dt}(1+\sec\theta)}{(1+\sec\theta)^2} \\ &= \frac{(1+\sec\theta)(\sec\theta\tan\theta) - \sec\theta(\sec\theta\tan\theta)}{(1+\sec\theta)^2} \\ &= \frac{(1+\sec\theta)(\sec\theta\tan\theta) - \sec^2\theta\tan\theta}{(1+\sec\theta)^2} = \frac{\sec\theta\tan\theta}{(1+\sec\theta)^2} \end{aligned}$$

Alternately, we could have multiplied the original equation up and down by $\cos\theta$ to get

$$y = \frac{1}{\cos\theta + 1}$$

and performed that derivative by the quotient rule. ■

(8) Find the equation of the line tangent to $y = \dfrac{\cos(2x)}{e^x + x}$ at $x = 0$.

☐ The point the tangent line shares with the function is $(0, f(0)) = \left(0, \dfrac{1}{e}\right)$. The slope of the tangent line comes from the derivative:

$$\begin{aligned} \frac{dy}{dx} &= \frac{(e^x+x)\frac{d}{dx}\cos(2x) - \cos(2x)\frac{d}{dx}(e^x+x)}{(e^x+x)^2} \\ &= \frac{(e^x+x)(-2\sin(2x)) - \cos(2x)(e^x+1)}{(e^x+x)^2} \end{aligned}$$

Now to use the specific point:

$$\left.\frac{dy}{dx}\right|_{x=0} = \frac{(e^0+0)(-2\sin(0)) - \cos(0)(e^0+1)}{(e^0+0)^2}$$
$$= \frac{0-(1)(2)}{(1)^2} = -2$$

So the equation of the tangent line is

$$y - \frac{1}{e} = -2(x-0) \qquad \text{i.e.} \qquad y = -2x + \frac{1}{e} \quad \blacksquare$$

---

(9) Find the derivative of $y = \dfrac{\sin^2 x}{\cos^2 x + 1}$.

☐ This will be a combination of the quotient rule and the extended power rule. Just take the derivatives in the order they appear.

$$\frac{dy}{dx} = \frac{(\cos^2 x + 1)\frac{d}{dx}(\sin^2 x) - (\sin^2 x)\frac{d}{dx}(\cos^2 x + 1)}{(\cos^2 x + 1)^2}$$
$$= \frac{(\cos^2 x + 1)(2\sin x\cos x) - (\sin^2 x)(-2\cos x \sin x)}{(\cos^2 x + 1)^2}$$
$$= \frac{(2\sin x\cos x)(\cos^2 x + 1 + \sin^2 x)}{(\cos^2 x + 1)^2} = \frac{4\sin x\cos x}{(\cos^2 x + 1)^2}$$

(Did you see where the identity $\sin^2 x + \cos^2 x = 1$ was used?) ■

---

(10) Where is the line tangent to $y = \dfrac{\ln x}{x}$ horizontal?

☐ We are looking for where the derivative is zero.

$$\frac{dy}{dx} = \frac{(x)\frac{d}{dx}(\ln x) - (\ln x)\frac{d}{dx}(x)}{(x)^2} = \frac{x\cdot\frac{1}{x} - (\ln x)(1)}{x^2} = \frac{1-\ln x}{x^2}$$

Then $\dfrac{dy}{dx}$ can only be zero where $\ln x = 1$, and this is at $x = e$. ■

---

(11) Find the derivative of $f(x) = \operatorname{sech}(x)$.

☐ Since $\operatorname{sech}(x) = \frac{1}{\cosh(x)}$, we can use the quotient rule to get:

$$f'(x) = \frac{d}{dx}\frac{1}{\cosh(x)} = \frac{\cosh(x)\cdot(0) - (1)\cdot\sinh(x)}{(\cosh(x))^2}$$
$$= \frac{-\sinh(x)}{\cosh^2(x)} = -\frac{\sinh(x)}{\cosh(x)}\cdot\frac{1}{\cosh(x)}$$
$$= -\tanh(x)\operatorname{sech}(x) \quad \blacksquare$$

4.4 The Chain Rule

Introduction

In a few previous sections, we have learned how to find derivatives of some individual functions, as well as more complicated variations built as compositions. Those basic and extended derivative formulas have included:

$$\begin{aligned}
\frac{d}{dx}x^p = px^{p-1} \quad &\to \quad \frac{d}{dx}(g(x))^p = p(g(x))^{p-1} \cdot g'(x) \\
\frac{d}{dx}e^x = e^x \quad &\to \quad \frac{d}{dx}e^{g(x)} = e^{g(x)} \cdot g'(x) \\
\frac{d}{dx}\ln(x) = \frac{1}{x} \quad &\to \quad \frac{d}{dx}\ln(g(x)) = \frac{1}{g(x)} \cdot g'(x) \\
\frac{d}{dx}\sin(x) = \cos(x) \quad &\to \quad \frac{d}{dx}\sin(g(x)) = \cos(g(x)) \cdot g'(x) \\
\frac{d}{dx}\cos(x) = -\sin(x) \quad &\to \quad \frac{d}{dx}\cos(g(x)) = -\sin(g(x)) \cdot g'(x)
\end{aligned}$$

There is certainly a distinct pattern to the extended derivative formulas. In this section, we'll look at that pattern.

The Chain Rule

Each of the functions whose derivatives are shown in the five extended derivative formulas given above is a function of the form $f(g(x))$. And in words, the resulting derivative of each displays the derivative of the "outside function" $f(x)$ done as usual, based on its basic derivative formula, while leaving the "inside" function $g(x)$ intact; then, we've multiplied that result by the derivative $g'(x)$ of the inside function. We seem to have discovered this:

Useful Fact 4.3. *As long as we can find the individual derivatites $f'(x)$ and $g'(x)$, then*

$$\boxed{\frac{d}{dx}f(g(x)) = f'(g(x)) \cdot g'(x)} \tag{4.35}$$

*This is called the **Chain Rule**.*

The actual derivation of the Chain Rule is given at the end of this section. If we know and understand the Chain Rule, then we don't need all of those separate extended derivative formulas listed above. We have actually been doing the Chain Rule all along in the previous few sections. To illustrate its use again, let's introduce a *new* derivative formula and then see how to use it in conjunction with the Chain Rule.

You will recall (with some dismay, I'm sure) that we introduced inverse trigonometric functions back in Chapter 1. In the next section, we will learn how to discover the derivative of the inverse sine and cosine functions — but a sneak preview of those results is:

$$\boxed{\frac{d}{dx}\sin^{-1}(x) = \frac{1}{\sqrt{1-x^2}} \quad \text{and} \quad \frac{d}{dx}\cos^{-1}(x) = -\frac{1}{\sqrt{1-x^2}}} \tag{4.36}$$

With these basic formulas as well as the Chain Rule, we now know how to find the derivatives of all sorts of more complicated functions involving the inverse sine and cosine.

EX 1 Find the derivative of $y = 3\sin^{-1}(x) - 2\cos^{-1}(x)$.

This is nothing more than direct application of the derivatives in (4.36), and is not a chain rule problem:

$$\begin{aligned}\frac{d}{dx}(3\sin^{-1}(x) - 2\cos^{-1}(x)) &= 3\left(\frac{1}{\sqrt{1-x^2}}\right) - 2\left(-\frac{1}{\sqrt{1-x^2}}\right)\\ &= \frac{3}{\sqrt{1-x^2}} + \frac{2}{\sqrt{1-x^2}} \quad \blacksquare\end{aligned}$$

That's no fun. But what about this one?

EX 2 Find the derivative of $y = \sin^{-1}(4x)$.

This is a bit more interesting. We now have a function in the form $f(g(x))$ where $f(x) = \sin^{-1}(x)$ and $g(x) = 4x$. The Chain Rule (4.35) says that we just do the derivative of the outer part according to the basic formula in (4.36), carrying the inner part along:

$$f'(g(x)) = \frac{1}{\sqrt{1-(4x)^2}}$$

and then multiply by the derivative of the inner part $g(x) = 4x$:

$$g'(x) = 4$$

Putting it together, for $f(g(x)) = \sin^{-1}(4x)$,

$$\frac{d}{dx}\sin^{-1}(4x) = f'(g(x)) \cdot g'(x) = \frac{1}{\sqrt{1-(4x)^2}} \cdot (4) = \frac{4}{\sqrt{1-16x^2}} \quad \blacksquare$$

EX 3 Find the derivative of $y = 5\cos^{-1}(e^x)$.

We have a function in the form $f(g(x))$ where $f(x) = \cos^{-1}(x)$ and $g(x) = e^x$. By the Chain Rule,

$$\frac{d}{dx}5\cos^{-1}(e^x) = 5\left(-\frac{1}{\sqrt{1-(e^x)^2}}\right)\cdot\frac{d}{dx}(e^x) = -\frac{5e^x}{\sqrt{1-e^{2x}}} \quad ■$$

You should now be able to operate with derivatives on any function which is introduced to you along with its derivative. For example, let me introduce you to a function named *Fred*. We have no idea what *Fred* himself looks like, but the derivative of *Fred* is $e^{x^2} + \sin(x^2)$. (And no, you can't figure out what *Fred* looks like, because we can't find the antiderivative of either term.) Given that information about *Fred*, though, we can certainly build a composition $Fred(2x+1)$ and determine the derivative of that. In fact,

$$\begin{aligned}\frac{d}{dx}Fred(2x+1) &= Fred'(2x+1)\cdot\frac{d}{dx}(2x+1)\\ &= \left(e^{(2x+1)^2} + \sin((2x+1)^2)\right)\cdot(2)\\ &= 2(e^{(2x+1)^2} + \sin((2x+1)^2)\end{aligned}$$

You Try It

(1) Find the derivative of $y = \sin^{-1}(\cos x)$.
(2) Find the derivative of $y = -2\cos^{-1}(\sqrt{x})$.

You Try It

(3) We will see later that the derivative of the inverse tangent function is:

$$\frac{d}{dx}\tan^{-1}(x) = \frac{1}{1+x^2}$$

Use this formula and the Chain Rule to write the derivative of $y = \tan^{-1}(2x^2+3)$.

Further practice with the Chain Rule at this point involves either revisiting derivatives like those in the last few sections and / or presenting more new individual functions and their derivative formulas. The Practice Problems will contain many problems similar to those we've seen before, as practice with the Chain Rule.

Interlude: Substitution and Derivatives

This is going to start off looking like a completely new topic, but it really isn't.

Substitution is what we do to give a complicated mathematical expression a makeover so that it appears simpler. It involves giving a new, simpler, name to part of an expression. In fancy talk, substitution is a *change of variable*. For example, consider the expression $\sin(x^2 - 3)$. Now, we're not afraid of a simple sine function, but this is worse than a simple sine function because of the $x^2 - 3$. So let's "hide" the $x^2 - 3$ by renaming it something simpler, like u. (This is the typical letter of substitution for some reason.) With the identification $u = x^2 - 3$, the expression $\sin(x^2 - 3)$ becomes $\sin(u)$. This new expression looks a lot less scary, but we have to remember that it's a disguise and the u is hiding some unpleasantness.

When doing substitution, we must replace *all* instances of the original variable with the new variable, according to the rule given by the substitution.

EX 4 Rewrite the expression $x^2 \sin(x^2 - 3)$ with the substitution $u = x^2 - 3$.

If $u = x^2 - 3$, then $\sin(x^2 - 3)$ becomes $\sin u$. But what about the leading x^2? We have to convert that, too, using the same substitution. Well, if $u = x^2 - 3$, then $x^2 = u + 3$, and so the full conversion becomes

$$x^2 \sin(x^2 - 3) = (u + 3)\sin(u) \qquad \text{with} \qquad u = x^2 - 3 \quad \blacksquare$$

EX 5 Rewrite the expression xe^{x^2} with the substitution $u = x^2$.

If $u = x^2$, then e^{x^2} becomes e^u. But what about the x in front? If $u = x^2$, then $x = \sqrt{u}$, and so the full conversion becomes

$$xe^{x^2} = \sqrt{u}e^u \qquad \text{with} \qquad u = x^2 \quad \blacksquare$$

You Try It

(4) Rewrite the expression $\sin(x)/(x-1)$ with the substitution $u = x - 1$.
(5) Rewrite the expression $x\ln(1/x)$ with the substitution $u = 1/x$.
(6) Rewrite the expression $(x-2)(x+2)$ with the substitution $u = x+2$.
(7) Rewrite the expression $\cos(x)\sqrt{1-\cos^2 x}$ with the substitution $u = \cos x$.

Our immediate concern is: how does a substitution propagate down into derivatives? To work on this, you really have to be on your toes with knowing how to name derivatives properly. Suppose we want the derivative of $y = \sin(x^2 - 3)$ — that is, we are seeking a derivative whose name is dy/dx. If we make the substitution $u = x^2 - 3$ to rewrite the function as $y = \sin(u)$, then the derivative we get from that new, simpler expression is named dy/du, and specifically we know $dy/du = \cos(u)$. How does dy/du relate to the derivative of the original function, dy/dx?

The secret is in the substitution itself. When we create the substitution $u = x^2 - 3$, we generate another derivative by default: $du/dx = 2x$. So when a substitution is in play,

$$\text{We need } \frac{dy}{dx} \quad \text{but we have} \quad \frac{dy}{du} \quad \text{and} \quad \frac{du}{dx}$$

Whatever will we do? Take a close look at those derivative terms right above this sentence. If we pretend the derivative symbols are fractions (which they aren't, really, but I won't tell if you don't), it's easy to see that:

$$\boxed{\frac{dy}{dx} = \frac{dy}{du} \cdot \frac{du}{dx}} \tag{4.37}$$

So with $y = \sin(x^2 - 3)$ (with derivative dy/dx) becoming $y = \sin(u)$ (with derivative dy/du) via the substitution $u = x^2 - 3$ (with derivative du/dx), we know that:

$$\frac{dy}{du} = \cos(u) \quad \text{and} \quad \frac{du}{dx} = 2x$$

and so the derivative we really want is

$$\frac{dy}{dx} = \frac{dy}{du} \cdot \frac{du}{dx} = \cos(u) \cdot (2x) = 2x\cos(u)$$

This is an awkward way to leave it. The name of the derivative dy/dx frames the output for us: it tells us we are looking at a function named y with independent variable x, and so the variable in the derivative should also be x. The variable u was just a place-holder, and we don't want it in the final answer. In a case like this, it's easy to return u to x, since we declared at the start that $u = x^2 - 3$:

$$\frac{dy}{dx} = 2x\cos(u) = 2x\cos(x^2 - 3)$$

In some cases, the intermediate variable u is so snarled up in the final expression that we can't extract it without making a huge mess — in those cases, it's OK to leave it in place. But generally, when it's possible to cleanly revert the variable of substitution (here, u) back to the original independent variable (here, x), that should be done.

You Try It

(8) Find the derivative of $y = 3\ln(\cos x)$ using the substitution $u = \cos(x)$ and the Chain Rule (4.37).
(9) Find the derivative of $y = 1/(x^2 - 1)$ using the substitution $u = x^2 - 1$ and the Chain Rule (4.37).

A Better Formulation of the Chain Rule

The relation between derivatives given above in (4.37) is just the Chain Rule in disguise! The original Chain Rule (4.35) says that if $y = f(g(x))$ then

$$\frac{dy}{dx} = f'(g(x)) \cdot g'(x)$$

But above, we talked about how we can take a messy function $y = f(g(x))$ and simplify it a bit through a substitution. For $y = f(g(x))$ we'd set $u = g(x)$ to rewrite the function as $y = f(u)$. Now just play the name game with derivatives.

- If we set $u = g(x)$, then the derivative of this expression can be named both du/dx and $g'(x)$. That is,
$$g'(x) = \frac{du}{dx}$$
- If $u = g(x)$, then the original function becomes $y = f(u)$ and the derivative of this is $dy/du = f'(u)$. But when $u = g(x)$, then $f'(u)$ becomes $f'(g(x))$. Together,
$$f'(g(x)) = f'(u) = \frac{dy}{du}$$

and now we can write the Chain Rule in a different way. If $y = f(g(x))$, then:

$$\begin{aligned}\frac{dy}{dx} &= f'(g(x)) \cdot g'(x) \\ &= \frac{dy}{du} \cdot \frac{du}{dx}\end{aligned}$$

which is the derivative expression in (4.37).

This other version of the Chain Rule is very powerful, as we'll see. First, note that the formula given in (4.37) is only an *example* of what the Chain Rule can look like. If we have different variables, then the appearance of the Chain Rule will change. You have to know how to assemble a Chain Rule for any given derivative with any given variables. For example, if we need a derivative named dr/dt but we have the derivatives dr/dx and dx/dt, do you see the proper "chain" of derivatives that link these? It would be:

$$\frac{dr}{dt} = \frac{dr}{dx} \cdot \frac{dx}{dt}$$

To verify its correctness, just pretend the derivatives are fractions and simplify the right side.

EX 6 Suppose the independent variable y depends on the variable w, but w itself depends on x. Ultimately, then, y depends on x, and changes in x will cause changes in y. The derivative dy/dx will provide information about how changes in x cause changes in y. Write a Chain Rule that could be used to display dy/dx.

We do not have a function that directly relates y to x. But, we do have a function that relates y to w; the name of that function is $y = y(w)$, and its derivative is dy/dw. We also have a function that relates w to x; the name of that function is $w = w(x)$, and its derivative is dw/dx. Therefore, we need dy/dx but have dy/dw and dw/dx. The Chain Rule for these derivatives is:

$$\frac{dy}{dx} = \frac{dy}{dw} \cdot \frac{dw}{dx} \quad \blacksquare$$

You Try It

(10) Suppose the independent variable s depends on the variable y, so that we have a function $s = s(y)$, but y itself depends on t, so that we have another function $y = y(t)$. Ultimately, then, s depends on t. Write a Chain Rule for the derivative ds/dt.

The reason this Chain Rule is powerful is that it allows us to find the rate of change of one variable with respect to another, even though we don't have a formula directly linking those two variables! As long as there is a *chain* of equations leading from one variable to the next, we can also link the derivatives together. Why would this be important? Because the mathematics behind physics, engineering, meteorology, economics, and all other math-dependent disciplines would fail to work without it. In many real world problems, our independent variable is time t. Many quantities of interest in all kinds of disciplines are rates of change of something with respect to time, but time t itself may not be a variable in the formula at hand. Here's an example:

EX 7 The radius of a circle is increasing at a known rate. Thus, so is the area of the circle. Write a Chain Rule which tells us how fast the area is changing based on how fast the radius is changing.

First, let's assign obvious variable names: A for area, and r for radius. When we ask something like "How fast is the area changing?" we are asking for the rate of change of area with respect to time t, i.e. we are asking for dA/dt. When we are told that we know the rate of increase of the radius of a circle, we are told we know the derivative dr/dt. If we had a function for the area of a circle that had time t in it, we could compute dA/dt directly. But when is the last time you saw a formula for the area of a circle that involved t? There isn't one! However, we DO know the formula $A = \pi r^2$, which has a derivative named dA/dr. So to summarize,

$$\text{We need } \frac{dA}{dt} \quad \text{but we have} \quad \frac{dA}{dr} \quad \text{and} \quad \frac{dr}{dt}$$

The Chain Rule puts this together:

$$\frac{dA}{dt} = \frac{dA}{dr} \cdot \frac{dr}{dt}$$

Knowing dr/dt and the formula connecting radius to area allows us to find dA/dt even though there is no formula for area that depends on time t. That's pretty nifty. ■

You Try It

(11) The radius of a sphere is decreasing at a known rate. Thus, so is the volume of the sphere. Write a Chain Rule which tells us how fast the volume is changing based on how fast the radius is changing.

In an upcoming section, we'll start giving actual data to solve problems of this sort to completion.

The Onion Principle

Regardless of which version of the Chain Rule you like to see,

$$\frac{d}{dx}f(g(x)) = f'(g(x)) \cdot g'(x) \qquad \text{or} \qquad \frac{dy}{dx} = \frac{dy}{du} \cdot \frac{du}{dx}$$

it all boils down to this: doing derivatives with the Chain Rule is like peeling an onion. You do both from the outside in, and both may make you cry. To find the derivative of a composition, we take the derivative from the outside first, working inward.

If there are 3 or more functions involved in the composition, apply the same principle: work from the outside in, like peeling the layers of an onion.

EX 8 Find the derivative of $y = \sin(e^{x^2})$.

We'll do the derivative from the outside in, according to the procedure given by the Chain Rule. First, we handle the outside function $\sin(\cdot)$, while multiplying by the derivative of inner function:

$$\frac{d}{dx}\sin(e^{x^2}) = \cos(e^{x^2}) \cdot \frac{d}{dx}(e^{x^2})$$

And now we want to handle the derivative of e^{x^2}. But that itself is a composition, which also requires the Chain Rule!

$$\frac{d}{dx}e^{x^2} = e^{x^2} \cdot \frac{d}{dx}(x^2) = e^{x^2} \cdot (2x) = 2xe^{x^2}$$

Putting it all together,

$$\frac{d}{dx}\sin(e^{x^2}) = \cos(e^{x^2}) \cdot \frac{d}{dx}(e^{x^2}) = 2xe^{x^2}\cos(e^{x^2})$$

This is a great example of why it's called the *Chain* Rule. ■

You Try It

(12) Find the derivative of $y = \sqrt[4]{3 + 2\sin(x^2)}$.
(13) Find the derivative of $y = \sqrt{x + \sqrt{x+1}}$.

Combining The Chain Rule With Other Rules

As a final note, (a) let's stop capitalizing Chain Rule, and (b) be aware that the chain rule can be required in combination with any other derivative process. A derivative which starts with a chain rule might also need the product or quotient rule. Or, a derivative that starts off as a product or quotient rule might eventually need the chain rule.

EX 9 Find the derivative of $y = 2xe^{\sqrt{x}}$.

To do this derivative, we first see the need for the product rule:

$$\frac{d}{dx}2xe^{\sqrt{x}} = \frac{d}{dx}(2x) \cdot e^{\sqrt{x}} + (2x) \cdot \frac{d}{dx}e^{\sqrt{x}}$$

But in the second term, the derivative of $e^{\sqrt{x}}$ requires the chain rule! So, continuing,

$$\begin{aligned}\frac{d}{dx}2xe^{\sqrt{x}} = \frac{d}{dx}(2x) \cdot e^{\sqrt{x}} + (2x) \cdot \frac{d}{dx}e^{\sqrt{x}} &= (2) \cdot e^{\sqrt{x}} + (2x) \cdot e^{\sqrt{x}} \cdot \frac{d}{dx}(\sqrt{x}) \\ &= 2e^{\sqrt{x}} + 2xe^{\sqrt{x}} \cdot \frac{1}{2\sqrt{x}} = e^{\sqrt{x}}(2 + \sqrt{x}) \quad \blacksquare\end{aligned}$$

EX 10 Find the derivative of $y = \ln(x/e^x)$.

At first, we see a composition $f(g(x))$, with $f(x) = \ln(x)$ and $g(x) = x/e^x$, so the chain rule gives

$$\frac{d}{dx}\ln\left(\frac{x}{e^x}\right) = \frac{1}{x/e^x} \cdot \frac{d}{dx}\left(\frac{x}{e^x}\right)$$

and then this derivative on the right, which popped out of the chain rule, needs a quotient rule. The complete solution is:

$$\begin{aligned}\frac{d}{dx}\ln\left(\frac{x}{e^x}\right) = \frac{1}{x/e^x} \cdot \frac{d}{dx}\left(\frac{x}{e^x}\right) = \frac{e^x}{x} \cdot \frac{e^x(1) - x(e^x)}{(e^x)^2} &= \frac{e^x}{x} \cdot \frac{e^x(1-x)}{(e^x)^2} \\ &= \frac{1-x}{x} = \frac{1}{x} - 1\end{aligned}$$

Gosh, that's a pretty simple final result for all that work! Is there a mistake? Well, no. Consider that we could have just applied laws of logarithms at the start to write:

$$y = \ln\left(\frac{x}{e^x}\right) = \ln x - \ln e^x = \ln x - x$$

and from that simplified version of the function,

$$\frac{dy}{dx} = \frac{1}{x} - 1 \quad \blacksquare$$

> **You Try It**
>
> (14) Find the derivative of $y = xe^{-x^2}$.
> (15) Find the derivative of $y = x/\sqrt{x^2+1}$.
> (16) Find the derivative of $y = \cos(x\sin(2x))$.

Into the Pit!!

Derivation of the Chain Rule

If you have been following the derivations of the extended derivative rules for $(g(x))^p$, $e^{g(x)}$, $\ln(g(x))$, and $\sin(g(x))$ then this should be very familiar. In fact, this derivation supercedes all of those, since it covers them all at once. As in each case before, some restrictions apply: this particular derivation works when $g(x)$ is both differentiable and one-to-one. In addition, since we're talking about a generic outer function $f(x)$, we must require that $f(x)$ is differentiable over its domain. (In our previous specific cases where $f(x)$ was x^p, e^x, $\ln x$ and $\sin x$, this was always true.) We also need to recognize that the expression

$$\lim_{y \to a} \frac{f(y) - f(a)}{y - a}$$

represents the derivative of $f(y)$ at the point $y = a$, and so is equal to $f'(a)$. With that in mind, here we go. The derivative of $f(g(x))$ is defined as:

$$\begin{aligned}\frac{d}{dx}f(g(x)) &= \lim_{h \to 0} \frac{f(g(x+h)) - f(g(x))}{h} \\ &= \lim_{h \to 0} \frac{f(g(x+h)) - f(g(x))}{h} \cdot \frac{g(x+h) - g(x)}{g(x+h) - g(x)}\end{aligned}$$

With regrouping and application of the limit law regarding products, this becomes

$$\frac{d}{dx}f(g(x)) = \lim_{h\to 0}\frac{f(g(x+h))-f(g(x))}{g(x+h)-g(x)}\cdot\lim_{h\to 0}\frac{g(x+h)-g(x)}{h}$$

The latter limit is just $g'(x)$. The first limit can be redesigned by setting $a = g(x)$, $y = g(x+h)$, and remembering that the limit process $h \to 0$ is now equivalent to $y \to g(x)$, i.e. $y \to a$. Therefore,

$$\lim_{h\to 0}\frac{f(g(x+h))-f(g(x))}{g(x+h)-g(x)} = \lim_{y\to a}\frac{f(y)-f(a)}{y-a}$$

which defines the derivative of $f(y)$ at the point $y = a$, and this was noted above to be equal to $f'(a)$. But remembering that $a = g(x)$ and so $f'(a) = f'(g(x))$:

$$\lim_{h\to 0}\frac{f(g(x+h))-f(g(x))}{g(x+h)-g(x)} = f'(g(x))$$

Together then,

$$\begin{aligned}\frac{d}{dx}f(g(x)) &= \lim_{h\to 0}\frac{f(g(x+h))-f(g(x))}{g(x+h)-g(x)}\cdot\lim_{h\to 0}\frac{g(x+h)-g(x)}{g(x+h)-g(x)}\\ &= \cos(g(x))\cdot g'(x)\end{aligned}$$

and this is the Chain Rule for the derivative of $f(g(x))$ where $f(x)$ is differentiable over its domain, and $g(x)$ is differentiable and one-to-one.

Have You Learned...

- Both of the formulations of the Chain Rule?
$$\frac{dy}{dx} = f'(g(x))\cdot g'(x) \quad ; \quad \frac{dy}{dx} = \frac{dy}{du}\cdot\frac{du}{dx}$$
- How the two formulations of the Chain Rule really are the same?
- That the Chain Rule really is more of a process than a formula?
- How to carry out a substitution?
- How to name and juggle derivatives involving multiple variables?

The Chain Rule — Problem List

The Chain Rule — You Try It

These appeared above; solutions begin on the next page.

(1) Find the derivative of $y = \sin^{-1}(\cos x)$.
(2) Find the derivative of $y = -2\cos^{-1}(\sqrt{x})$.
(3) We will see later that the derivative of the inverse tangent function is:
$$\frac{d}{dx}\tan^{-1}(x) = \frac{1}{1+x^2}$$
Use this formula and the Chain Rule to write the derivative of $y = \tan^{-1}(2x^2+3)$.
(4) Rewrite the expression $\sin(x)/(x-1)$ with the substitution $u = x-1$.
(5) Rewrite the expression $x\ln(1/x)$ with the substitution $u = 1/x$.
(6) Rewrite the expression $(x-2)(x+2)$ with the substitution $u = x+2$.
(7) Rewrite the expression $\cos(x)\sqrt{1-\cos^2 x}$ with the substitution $u = \cos x$.
(8) Find the derivative of $y = 3\ln(\cos x)$ using the substitution $u = \cos(x)$ and the Chain Rule (4.37).
(9) Find the derivative of $y = 1/(x^2-1)$ using the substitution $u = x^2-1$ and the Chain Rule (4.37).
(10) Suppose the independent variable s depends on the variable y, so that we have a function $s = s(y)$, but y itself depends on t, so that we have another function $y = y(t)$. Ultimately, then, s depends on t. Write a Chain Rule for the derivative ds/dt.
(11) The radius of a sphere is decreasing at a known rate. Thus, so is the volume of the sphere. Write a Chain Rule which tells us how fast the volume is changing based on how fast the radius is changing.
(12) Find the derivative of $y = xe^{-x^2}$.
(13) Find the derivative of $y = x/\sqrt{x^2+1}$.
(14) Find the derivative of $y = \cos(x\sin(2x))$.

The Chain Rule — Practice Problems

Try these as you get the hang of the You Try It problems. Solutions to these problems are available in Sec. A.4.4.

(1) Rewrite the expression $2x/(3-5x)$ with the substitution $u = 3-5x$.
(2) Rewrite the expression $e^{2x}\tan(2e^x)$ with the substitution $u = 2e^x$.
(3) Rewrite the expression $x/(1-\sqrt{x})$ with the substitution $u = 1-\sqrt{x}$.

(4) Rewrite the expression $x^{3/2}$ with the substitution $u = \sqrt{x}$.
(5) Find the derivatives of $y = (1 + x^4)^{2/3}$ and $z = \sqrt[4]{1 + 2x + x^3}$.
(6) Suppose y depends on x via $y = \sqrt{\tanh(x) + 1}$, and x depends on t via $x = t^2 - 2t + 3$, so that ultimately y depends on t. Use the Chain Rule to develop the derivative dy/dt (it's OK to leave a mix of x and t in the result).
(7) Find the derivatives of $y = \tan(\sin^{-1}(x))$ and $y = \tanh(\sin^{-1}(x))$.
(8) Find the derivative of $y = e^{-5x} \cos^{-1}(3x)$.
(9) Find the derivative of $y = \sin^2 x / \cos x$.
(10) Find the derivative of $y = \sin(\sin(\sin x))$.
(11) Find the equation of the line tangent to $y = \sinh(x) + \sin^2 x$ at (0,0).
(12) Find the velocity function for the position function $s(t) = A\cos(\omega t + \delta)$.

The Chain Rule — Challenge Problems

Try these problems to test your skills with the ideas in this section. Solutions to these problems are available in Sec. B.4.4.

(1) Find the derivative of $y = x \sin^{-1}(1/x)$.
(2) Find the derivatives of $y = e^{\sin(x^2)}$ and $y = e^{\sinh(x^2)}$.
(3) Find the equation of the line tangent to $y = \cos 2x/(x^2+1)$ at $x = \pi/4$.

The Chain Rule — You Try It — Solved

(1) Find the derivative of $y = \sin^{-1}(\cos x)$.

☐ This is a composition of the form $f(g(x))$ with $f(x) = \sin^{-1}(x)$ and $g(x) = \cos x$. Combining the basic derivative of $\sin^{-1}(x)$ with the Chain Rule, we get:

$$\frac{d}{dx}\sin^{-1}(\cos x) = \frac{1}{\sqrt{1+(\cos x)^2}} \cdot \frac{d}{dx} cosx = -\frac{\sin x}{\sqrt{1+\cos^2 x}} \quad \blacksquare$$

(2) Find the derivative of $y = -2\cos^{-1}(\sqrt{x})$.

☐ This is a composition of the form $f(g(x))$ with $f(x) = -2\cos^{-1}(x)$ and $g(x) = \sqrt{x}$. Combining the basic derivative of $\cos^{-1}(x)$ with the Chain Rule, we get:

$$\begin{aligned}\frac{d}{dx}\left(-2\cos^{-1}(\sqrt{x})\right) &= -2 \cdot \frac{1}{\sqrt{1+(\sqrt{x})^2}} \cdot \frac{d}{dx}\sqrt{x} \\ &= -\frac{2}{\sqrt{1+x}} \cdot \frac{1}{2\sqrt{x}} = -\frac{1}{\sqrt{x+x^2}} \quad \blacksquare\end{aligned}$$

(3) We will see later that the derivative of the inverse tangent function is:

$$\frac{d}{dx}\tan^{-1}(x) = \frac{1}{1+x^2}$$

Use this formula and the Chain Rule to write the derivative of $y = \tan^{-1}(2x^2+3)$.

$$\square\ \frac{dy}{dx} = \frac{1}{1+(2x^2+3)^2} \cdot \frac{d}{dx}(2x^2+3) = \frac{4x}{1+(2x^2+3)^2} \quad \blacksquare$$

(4) Rewrite the expression $\sin(x)/(x-1)$ with the substitution $u = x-1$.

☐ If $u = x - 1$ then $x = u + 1$ and

$$\frac{\sin x}{x-1} = \frac{\sin(u+1)}{u} \qquad \text{with} \qquad u = x-1 \quad \blacksquare$$

(5) Rewrite the expression $x\ln(1/x)$ with the substitution $u = 1/x$.

☐ If $u = 1/x$ then $x = 1/u$ and

$$x\ln\frac{1}{x} = \frac{1}{u}\ln u \qquad \text{with} \qquad u = \frac{1}{x} \quad \blacksquare$$

(6) Rewrite the expression $(x-2)(x+2)$ with the substitution $u = x+2$.

☐ If $u = x+2$ then $x = u-2$ and $x-2 = (u-2)-2 = u-4$, so

$$(x-2)(x+2) = (u-4)(u) \qquad \text{with} \qquad u = x+2 \quad \blacksquare$$

(7) Rewrite the expression $\cos(x)\sqrt{1-\cos^2 x}$ with the substitution $u = \cos x$.

$$\square \; \cos(x)\sqrt{1-\cos^2 x} = u\sqrt{1-u^2} \qquad \text{with} \qquad u = \cos x \quad \blacksquare$$

(8) Find the derivative of $y = 3\ln(\cos x)$ using the substitution $u = \cos(x)$ and the Chain Rule (4.37).

☐ This is a composition $f(g(x))$ where $f(x) = 3\ln x$ and $g(x) = \cos x$. With the substitution $u = \cos x$ we can write the function as $y = 3\ln u$. With these expressions we get

$$\frac{du}{dx} = -\sin x \qquad \text{and} \qquad \frac{dy}{du} = \frac{3}{u}$$

Then from the chain rule, we get

$$\frac{dy}{dx} = \frac{dy}{du} \cdot \frac{du}{dx} = \frac{3}{u} \cdot (-\sin x)$$

and by returning u to $\cos x$, this is

$$\frac{dy}{dx} = -\frac{3}{\cos x} \cdot (\sin x) = -3\tan x \quad \blacksquare$$

(9) Find the derivative of $y = 1/(x^2-1)$ using the substitution $u = x^2-1$ and the Chain Rule (4.37).

☐ This is a composition $f(g(x))$ where $f(x) = 1/x$ and $g(x) = x^2-1$. With the substitution $u = x^2-1$ we can write the function as $y = 1/u$. With these expressions we get

$$\frac{du}{dx} = 2x \qquad \text{and} \qquad \frac{dy}{du} = -\frac{1}{u^2}$$

Then from the chain rule, we get

$$\frac{dy}{dx} = \frac{dy}{du} \cdot \frac{du}{dx} = -\frac{1}{u^2} \cdot (2x)$$

and by returning u to x^2-1, this is

$$\frac{dy}{dx} = -\frac{1}{(x^2-1)^2} \cdot (2x) = -\frac{2x}{(x^2-1)^2} \quad \blacksquare$$

(10) Suppose the independent variable s depends on the variable y, so that we have a function $s = s(y)$, but y itself depends on t, so that we have another function $y = y(t)$. Ultimately, then, s depends on t. Write a Chain Rule for the derivative ds/dt.

□ Since s ultimately depends on t, we want the derivative ds/dt. From the function $s(y)$ we'd get ds/dy. From the function $y(t)$ we'd get dy/dt. Putting these derivatives together,

$$\frac{ds}{dt} = \frac{ds}{dy} \cdot \frac{dy}{dt} \quad \blacksquare$$

(11) The radius of a sphere is decreasing at a known rate. Thus, so is the volume of the sphere. Write a Chain Rule which tells us how fast the volume is changing based on how fast the radius is changing.

□ We are seeking the derivative dV/dt. The known rate of decrease of the radius would be dr/dt. So we need to fill the blank:

$$\frac{dV}{dt} = \left(\frac{d?}{d?}\right) \cdot \frac{dr}{dt}$$

To make a proper Chain Rule, the missing derivative should be dV/dr so that we'd have:

$$\frac{dV}{dt} = \frac{dV}{dr} \cdot \frac{dr}{dt}$$

(Where would we get that derivative dV/dr? Easy! From the standard formula formula for the volume of a sphere, $V = 4\pi r^3/3$.) ■

(12) Find the derivative of $y = xe^{-x^2}$.

□ We use a combination of the product rule and the chain rule:

$$\frac{dy}{dx} = x\frac{d}{dx}e^{-x^2} + e^{-x^2} \cdot \frac{d}{dx}x = x\left(e^{-x^2}\frac{d}{dx}(-x^2)\right) + e^{-x^2}$$

$$= x\left(e^{-x^2}(-2x)\right) + e^{-x^2} = e^{-x^2}(-2x^2+1) = e^{-x^2}(1-2x^2) \quad \blacksquare$$

(13) Find the derivative of $y = x/\sqrt{x^2+1}$.

□ We start with a quotient rule to get

$$\frac{dy}{dx} = \frac{\sqrt{x^2+1}\frac{d}{dx}(x) - x\frac{d}{dx}\sqrt{x^2+1}}{(\sqrt{x^2+1})^2}$$

and then see that we need the chain rule to evaluate the second derivative in the numerator. By itself, we have

$$\begin{aligned}\frac{d}{dx}\sqrt{x^2+1} &= \frac{d}{dx}(x^2+1)^{1/2} = \frac{1}{2}(x^2+1)^{-1/2}\frac{d}{dx}(x^2+1)\\ &= \frac{1}{2\sqrt{x^2+1}}\cdot(2x) = \frac{x}{\sqrt{x^2+1}}\end{aligned}$$

and so the entire derivative becomes

$$\begin{aligned}\frac{dy}{dx} &= \frac{\sqrt{x^2+1}\frac{d}{dx}(x) - x\frac{d}{dx}\sqrt{x^2+1}}{(\sqrt{x^2+1})^2} = \frac{\sqrt{x^2+1}(1) - x\cdot\frac{x}{\sqrt{x^2+1}}}{x^2+1}\\ &= \frac{\sqrt{x^2+1} - \frac{x^2}{\sqrt{x^2+1}}}{x^2+1}\cdot\frac{\sqrt{x^2+1}}{\sqrt{x^2+1}} = \frac{x^2+1-x^2}{(x^2+1)^{3/2}} = \frac{1}{(x^2+1)^{3/2}}\end{aligned}$$ ■

(14) Find the derivative of $y = \cos(x\sin(2x))$.

□ We start with the chain rule and write

$$\frac{dy}{dx} = -\sin(x\sin(2x))\cdot\frac{d}{dx}x\sin(2x)$$

and then realize that for the new derivative that popped out in the chain rule, we need a product rule ... and a little mini chain rule inside that! Doing this new derivative by itself, we have

$$\frac{d}{dx}x\sin(2x) = \sin(2x)\frac{d}{dx}(x) + x\frac{d}{dx}\sin(2x) = \sin(2x) + 2x\cos(2x)$$

(Do you know where the extra 2 at the front of the right-most term came from?) Plugging this back into the original,

$$\begin{aligned}\frac{dy}{dx} &= -\sin(x\sin(2x))\cdot\frac{d}{dx}x\sin(2x)\\ &= -\sin(x\sin(2x))\cdot(\sin(2x) + 2x\cos(2x))\end{aligned}$$

and there's really no further simplification to do. ■

4.5 Implicit Differentiation

Introduction

Have you noticed that every expression we've handled so far for the purpose of finding a derivative has been pre-solved for y or $f(x)$? Every one has been delivered in the form $y = f(x)$. This has been very convenient. It's important to know that when we're finding the derivative of a function $y = f(x)$, we're not just acting on the $f(x)$ part; we're applying a derivative *operator* to both sides of the expression, like so:

$$y = x^2 + 1$$
$$\frac{d}{dx}(y) = \frac{d}{dx}(x^2 + 1)$$
$$\frac{dy}{dx} = 2x$$

The y doesn't just magically turn into dy/dx; rather, it gets acted on by the operator d/dx. Since the function is already solved explicitly for y, then when both sides of the equation are acted on by the operator d/dx, we receive our derivative dy/dx without any other ado.

(By the way, don't stress over the term *operator.* An operator is just a mathematical process that has a job to do, and which doesn't spring into action until you attach it to something. The square root $\sqrt{}$ is an operator, too; it has a job to do, and it is never seen alone — it doesn't activate until you hand it something to act on, like $\sqrt{3}$. The same thing goes for $\dfrac{d}{dx}$; it's meaningless unless it's attached to something.)

While all our functions so far have been pre-solved for y, i.e. written *explicitly* for y, there are also perfectly good expressions not already solved for y which need derivatives, too! This is not news to you, even though it may seem so. Consider the unit circle $x^2 + y^2 = 1$. Every point on the circle can have a tangent line attached to it, and so we should be able to find the slope of any tangent line at any point on the circle (well, except for the two points at which the slope is undefined). We have two choices how to proceed.

(1) Solve this equation for y first, leading to an unpleasant expression with a square root, and apply the derivative operation to this now explicit expression. This would require us to decouple the

expression into two, $y = \pm\sqrt{1 - x^2}$ and find the derivative of each, explicitly. Or,

(2) We can learn how to find dy/dx for an expression that is NOT already explicitly solved for y.

Since an expression pre-solved for y is said to be written explicitly for y, I'll bet you can guess what we call an expression like $x^2 + y^2 = 1$, which is NOT pre-solved for y. This expression is said to be written *implicitly* for y, and the act of using the derivative operator to dig out dy/dx is called *implicit differentiation.*

Sometimes functions are written implicitly because solving for y would create a big mess; in this case, the use of implicit differentiation is a choice. In other instances, where a function remains written implicitly because solving for y is just not possible, then the use of implicit differentiation is a necessity.

Yes, It's The Chain Rule Again

Implicit differentiation does not require any new formulas or procedures. The only thing you need to remember is a mathematical mantra; presuming we have the usual variables of x (independent) and y (dependent), then since y depends on x, we say: y is a function of x, and must be treated accordingly. To repeat:

- **y is a function of x, and must be treated accordingly**

What does this mean? Well, it means that every time y appears in an expression, you have to remember that

- **y is a function of x, and must be treated accordingly**

The derivatives of the following expressions would require the product rule:

$$xe^x \qquad x^2 \sin x \qquad \frac{1}{x} \cdot \ln(x) \qquad x^2 y$$

because each is a product of two functions of x — even the last one! The expression x^2y is a product of two functions of x (one is x^2, and the other is y). To generate the derivative of x^2y, you have to use the product rule, because

- **y is a function of x, and it must be treated accordingly**

The derivative of that term is properly determined like this:

$$\frac{d}{dx}(x^2y) = x^2 \cdot \frac{d}{dx}(y) + y \cdot \frac{d}{dx}(x^2) = x^2 \cdot \frac{dy}{dx} + 2xy$$

Note also that the derivatives of each of these functions would require the chain rule:

$$(\sin x)^2 \qquad (\ln x)^2 \qquad (3x - 4x^3 + x^5)^2 \qquad y^2$$

because each has one function of x composed with (plugged in to) the function x^2. So if you need the derivative of the term y^2, you also have to use the chain rule just like we know to do with the other three, since

- **y is a function of x, and must be treated accordingly**

Since y is a function of x and therefore y^2 is a function of x, too, its derivative goes like this (using the chain rule):

$$\frac{d}{dx}(y^2) = 2y \cdot \frac{d}{dx}(y) = 2y\,\frac{dy}{dx}$$

Some other sample individual derivative terms for expressions involving y are:

$$\text{(A)} \quad \frac{d}{dx}\left(\frac{x}{y}\right) = \frac{y\frac{d}{dx}(x) - x\frac{d}{dx}(y)}{y^2} = \frac{y - x\frac{dy}{dx}}{y^2}$$

$$\text{(B)} \quad \frac{d}{dx}\sin(x+2y) = \cos(x+2y) \cdot \frac{d}{dx}(x+2y) = \cos(x+2y)\left(1 + 2\frac{dy}{dx}\right)$$

$$\begin{aligned}
\text{(C)} \quad \frac{d}{dx}(xy^3 + \sqrt{y}) &= \frac{d}{dx}(xy^3) + \frac{d}{dx}y^{1/2} \\
&= x \cdot \frac{d}{dx}(y^3) + y^3 \cdot \frac{d}{dx}(x) + \frac{1}{2} \cdot y^{-1/2} \cdot \frac{d}{dx}(y) \\
&= x \cdot (3y^2)\frac{dy}{dx} + y^3(1) + \frac{1}{2\sqrt{y}} \cdot \frac{dy}{dx} \\
&= \left(3xy^2 + y^3 + \frac{1}{2\sqrt{y}}\right)\frac{dy}{dx}
\end{aligned}$$

To summarize, if you remember that:

- **y is a function of x and must be treated accordingly**

then you can do implicit differentiation. Here is the procedure:

(1) Given your function, apply the derivative operator d/dx to both sides.

(2) Treat each term as you normally would, remembering that y is a function of x.

(3) If you've done things properly, one or more instances of dy/dx will appear.
(4) Solve for dy/dx using normal algebraic tools.
(5) If there are any y's left over in your solution, that's OK!

EX 1 Find dy/dx for $xy + e^x = 1$.

We could solve this for y right now, but that would lead to an expression needing the quotient rule for its derivative. Ick. So let's do it implicitly. First, take the derivative with respect to x of both sides and distribute the derivative:

$$\frac{d}{dx}(xy) + \frac{d}{dx}(e^x) = \frac{d}{dx}(1)$$

The second and third derivatives in this line are easy. The first derivative will require a product rule, since (all together now) y is a function of x, and must be treated accordingly.

$$\left(x\frac{d}{dx}(y) + y\frac{d}{dx}(x)\right) + \frac{d}{dx}(e^x) = \frac{d}{dx}(1)$$

$$x\frac{dy}{dx} + y + e^x = 0$$

The term dy/dx has now appeared. But since the function was not pre-solved for y, the derivative expression is not pre-solved for dy/dx. So now, we just have to solve for it:

$$\begin{aligned} x\frac{dy}{dx} + y + e^x &= 0 \\ x\frac{dy}{dx} &= -e^x - y \\ \frac{dy}{dx} &= -\frac{e^x + y}{x} \end{aligned}$$

Now there is a y left over in the derivative, but that's OK. The only consequence is that if we wanted the slope of a tangent line at a particular point, we'd need to supply BOTH the x and y coordinates of that point. ■

EX 2 Find dy/dx for $x^2 + y^2 = 1$.

Let's not solve this for y, since that would be gross. Rather, we'll just tackle it as-is:

$$\begin{aligned} \frac{d}{dx}(x^2 + y^2) &= \frac{d}{dx}(1) \\ \frac{d}{dx}(x^2) + \frac{d}{dx}(y)^2 &= \frac{d}{dx}(1) \end{aligned}$$

$$
\begin{aligned}
2x + 2y \cdot \frac{dy}{dx} &= 0 \\
2y \cdot \frac{dy}{dx} &= -2x \\
\frac{dy}{dx} &= -\frac{x}{y} \quad \blacksquare
\end{aligned}
$$

EX 3 Find dy/dx for $\sin(xy) = x + y$.

This is a function we cannot solve for y, even if we wanted to. So we have no choice but to find the derivative implicitly. In the first step, we see a chain rule needed on the left side, and then a product rule after that!

$$
\begin{aligned}
\frac{d}{dx}\sin(xy) &= \frac{d}{dx}(x+y) \\
\cos(xy)\frac{d}{dx}(xy) &= \frac{d}{dx}(x) + \frac{d}{dx}(y) \\
\cos(xy)\left(y\frac{d}{dx}(x) + x\frac{d}{dx}(y)\right) &= \frac{d}{dx}(x) + \frac{d}{dx}(y) \\
\cos(xy)\left(y(1) + x\frac{dy}{dx}\right) &= 1 + \frac{dy}{dx}
\end{aligned}
$$

Here we see that dy/dx has shown up twice! So, we just heave a big sigh, accept the inevitable, and solve for it. We begin by multiplying out the left side, then collecting all terms containing dy/dx on one side, and terms without it on the other:

$$
\begin{aligned}
y\cos(xy) + x\cos(xy)\frac{dy}{dx} &= 1 + \frac{dy}{dx} \\
x\cos(xy)\frac{dy}{dx} - \frac{dy}{dx} &= 1 - y\cos(xy)
\end{aligned}
$$

Then we factor and solve for dy/dx:

$$
\begin{aligned}
(x\cos(xy) - 1)\frac{dy}{dx} &= 1 - y\cos(xy) \\
\frac{dy}{dx} &= \frac{1 - y\cos(xy)}{(x\cos(xy) - 1)}
\end{aligned}
$$

That's a pretty impressive derivative! $\blacksquare$

You Try It

(1) Find dy/dx for $x^2 + y^2 = 1$.
(2) Find dy/dx for $x^2y^2 + x\sin y = 4$.
(3) Find dy/dx for $e^{x^2y} = x + y$.

We can use implicit differentiation to answer questions that need, but aren't completely about, derivatives:

EX 4 Find the equation of the line tangent to $x^{2/3} + y^{2/3} = 1$ at the point $(1/8, 3\sqrt{3}/8)$.

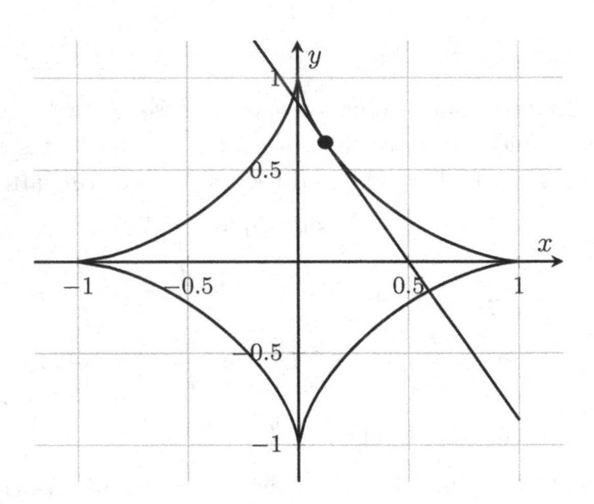

Fig. 4.3 The graph of $x^{2/3} + y^{2/2} = 1$ and a tangent line.

First of all, this implicitly written expression has a pretty cool graph; Fig. 4.3 shows this graph as well as the tangent line we're after. To find the equation of this tangent line, we need its slope, which means we need the value of dy/dx at the given point. Here is that process; be sure to observe and understand the treatment of y:

$$\begin{aligned}
\frac{d}{dx}(x^{2/3} + y^{2/3}) &= \frac{d}{dx}(1) \\
\frac{d}{dx}x^{2/3} + \frac{d}{dx}y^{2/3} &= \frac{d}{dx}(1) \\
\frac{2}{3}x^{-1/3} + \frac{2}{3}y^{-1/3}\frac{d}{dx}(y) &= 0 \\
\frac{2}{3}y^{-1/3}\frac{dy}{dx} &= -\frac{2}{3}x^{-1/3} \\
\frac{dy}{dx} &= -y^{1/3}x^{-1/3} \\
\frac{dy}{dx} &= -\left(\frac{y}{x}\right)^{1/3}
\end{aligned}$$

So at the point $(x, y) = (1/8, 3\sqrt{3}/8)$, the slope of the tangent line is given by:

$$\frac{dy}{dx} = -\left(\frac{3\sqrt{3}/8}{1/8}\right)^{1/3} = -(3\sqrt{(3)})^{1/3} = -(3^{3/2})^{1/3} = -3^{1/2} = -\sqrt{3}$$

and the equation of the tangent line is:

$$y - \frac{3\sqrt{3}}{8} = -\sqrt{3}\left(x - \frac{1}{8}\right)$$

which tidies up (try it!) to

$$y = -\sqrt{3}x + \frac{\sqrt{3}}{2} \quad \blacksquare$$

You Try It

(4) Find the equation of the line tangent to $x^2 + y^2 = (2x^2 + 2y^2 - x)^2$ at $(0, 1/2)$.

Implicit Differentiation as a Tool

Implicit differentiation is very useful as a tool for finding derivatives that would otherwise be difficult or impossible. For example, let's find the derivatives of some inverse trigonometric functions. That sound like fun, right?

EX 5 Find dy/dx for $y = \sin^{-1}(x)$.

At first, this does not look like an implicit differentiation problem since the expression is solved for y already. But we don't know how to find the derivative of the expression when it's written this way, that's the whole problem. While we don't know about derivatives of inverse trig functions, we're familiar with how to find derivatives of the basic trig functions. So let's remember that

$$y = \sin^{-1}(x) \qquad \text{means} \qquad \sin y = x$$

and work with the implicit version of the relationship rather than the explicit version:

$$\begin{aligned} \frac{d}{dx}\sin(y) &= \frac{d}{dx}(x) \\ \cos(y)\frac{dy}{dx} &= 1 \\ \frac{dy}{dx} &= \frac{1}{\cos y} \end{aligned}$$

When it's possible to do so cleanly, we'd like to put a derivative dy/dx in terms of the independent variable x. Doing so here requires some trigonometric shenanigans. The relationship $\sin y = x$ automatically comes with a right triangle that's loaded with good information. The expression $\sin y$ demands that y is an acute angle in a right triangle; further, by the definition of sine, the opposite side and hypotenuse have lengths of x and 1, respectively. Figure 4.4 shows this information encoded into a right triangle (on the left). By the Pythagorean Theorem, though, we know the length of the missing side in the triangle is $\sqrt{1-x^2}$; this is shown in the updated triangle (on the right). Now we have all three sides labelled and can complete any of the six trigonometric functions. Specifically,

$$\cos(y) = \frac{\text{adjacent}}{\text{hypotenuse}} = \frac{\sqrt{1-x^2}}{1} = \sqrt{1-x^2}$$

Altogether then, for $y = \sin^{-1}(x)$ we have

$$\frac{d}{dx}\sin^{-1}(x) = \frac{1}{\sqrt{1-x^2}} \quad \blacksquare$$

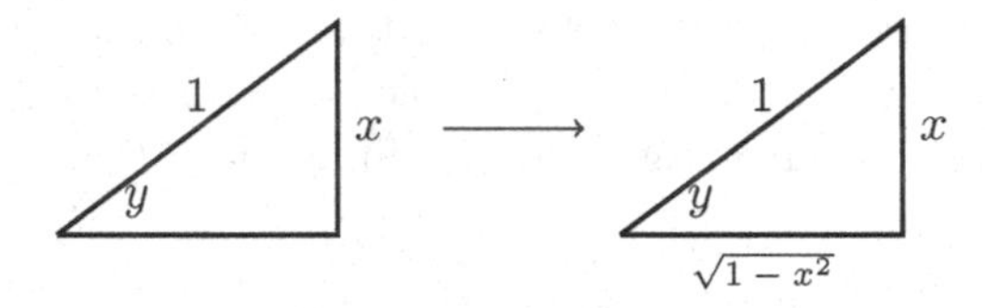

Fig. 4.4 Encoding $\sin y = x$ into a right triangle.

You Try It

(5) Find the derivative dy/dx for the inverse cosine function $y = \cos^{-1}(x)$.

Another tool that involves a combination of implicit differentiation with properties of logarithms is called *logarithmic differentiation*. Suppose you were asked to find the derivative of this function:

$$y = \frac{\sqrt{x+1}(x^2+4)(x^3-1)}{(2x^2+1)^2}$$

At first, this looks like a term project! It would take multiple product rules embedded in a quotient rule, and a chain rule for a couple of terms, just to spice things up. Fortunately, a combination of (1) properties of logarithms and (2) implicit differentiation turns this beast into a teddy bear. First, apply the natural logarithm to both sides (another example of an *operator*!):

$$\ln y = \ln\left(\frac{\sqrt{x+1}(x^2+4)(x^3-1)}{(2x^2+1)^2}\right)$$

Then we use rules of logarithms such as :

$$\ln(a^b) = b\ln a \quad ; \quad \ln(ab) = \ln a + \ln b \quad ; \quad \ln\frac{a}{b} = \ln a - \ln b$$

to systematically disassemble the function:

$$\begin{aligned}\ln y &= \ln\left((x+1)^{1/2}(x^2+4)(x^3-1)\right) - \ln\left((2x^2+1)^2\right)\\ &= \ln(x+1)^{1/2} + \ln(x^2+4) + \ln(x^3-1) - 2\ln(2x^2+1)\\ &= \frac{1}{2}\ln(x+1) + \ln(x^2+4) + \ln(x^3-1) - 2\ln(2x^2+1)\end{aligned}$$

Now, we can take the derivative of each side with respect to x. The left side requires implicit differentiation, and the right side is multiple *simple* derivatives:

$$\begin{aligned}\frac{d}{dx}\ln y &= \frac{d}{dx}\left(\frac{1}{2}\ln(x+1) + \ln(x^2+4) + \ln(x^3-1) - 2\ln(2x^2+1)\right)\\ \frac{1}{y}\frac{dy}{dx} &= \frac{1}{2}\cdot\frac{1}{x+1} + \frac{2x}{x^2+4} + \frac{3x^2}{x^3-1} - \frac{2(4x)}{2x^2+1}\end{aligned}$$

Now we can solve for dy/dx:

$$\begin{aligned}\frac{dy}{dx} &= y\left(\frac{1}{2(x+1)} + \frac{2x}{x^2+4} + \frac{3x^2}{x^3-1} - \frac{8x}{2x^2+1}\right)\\ &= \frac{\sqrt{x+1}(x^2+4)(x^3-1)}{(2x^2+1)^2}\left(\frac{1}{2(x+1)} + \frac{2x}{x^2+4} + \frac{3x^2}{x^3-1} - \frac{8x}{2x^2+1}\right)\end{aligned}$$

Not bad for a day's work! Who would have thought that logarithms and implicit differentiation would work together to make a hard problem simpler!

EX 6 Find the derivative of $y = \dfrac{e^{-x}\sin(x)}{x^2+4}$.

First, we use logarithms to break this down:

$$\begin{aligned}y &= \frac{e^{-x}\sin(x)}{x^2+4}\\ \ln y &= \ln\left(\frac{e^{-x}\sin(x)}{x^2+4}\right)\\ \ln y &= \ln e^{-x} + \ln(\sin x) - \ln(x^2+4)\end{aligned}$$

Having disassembled one humongous function into several smaller pieces, we can apply the derivative operation to both sides and then seek and solve for dy/dx:

$$\begin{aligned}
\frac{d}{dx}\ln y &= \frac{d}{dx}\left(\ln e^{-x} + \ln(\sin x) - \ln(x^2+4)\right) \\
&= \frac{d}{dx}\left(-x + \ln(\sin x) - \ln(x^2+4)\right) \\
\frac{1}{y}\frac{dy}{dx} &= -1 + \frac{1}{\sin x}\cdot\cos x - \frac{2x}{x^2+4} \\
\frac{dy}{dx} &= y\cdot\left(-1+\cot x - \frac{2x}{x^2+4}\right) \\
&= \frac{e^{-x}\sin(x)}{x^2+4}\cdot\left(-1+\cot x - \frac{2x}{x^2+4}\right) \quad \blacksquare
\end{aligned}$$

You Try It

(6) Find the derivative of $y = \dfrac{(x^3+2x^2-1)^2\sin(x)}{\sqrt{5+x^4}}$.

(7) Find the derivative of $z = 2^x$.

Relating Rates of Change

Another use for implicit differentiation (which is really the chain rule in disguise) is that it allows us to take an expression written using one variable, and find its derivative with respect to another variable — assuming there is dependence between the two. To make sense of that, let's look at a "regular" derivative with every single possible step explicitly shown according to the chain rule:

$$\frac{d}{dx}(x^2) = 2x\cdot\frac{d}{dx}(x) = 2x\cdot(1) = 2x$$

When the variable in the expression matches the variable of differentiation, the tail end of the chain rule always returns a 1; here, that's happening in the term $\frac{d}{dx}(x)$). Normally, we don't even see this; rather, we just go straight from $f(x) = x^2$ to $f'(x) = 2x$.

But what if the variable of differentiation is *not* the same as the variable in the expression? We've seen that happen with implicit differentiation. Let's recap what we've seen for the derivative with respect to x of y^2.

- If y does NOT depend on x then

$$\frac{d}{dx}(y^2) = 0$$

- If y does depend on x, then

$$\frac{d}{dx}(y^2) = 2y \cdot \frac{d}{dx}(y) = 2y\,\frac{dy}{dx}$$

Let's suppose we have a different mix of variables. Say a variable x depends on a variable t, then

$$\frac{d}{dt}(x^2) = 2x \cdot \frac{d}{dt}(x) = 2x\,\frac{dx}{dt}$$

Or, if y depends on s,

$$\frac{d}{ds}(y^2) = 2y \cdot \frac{d}{ds}(y) = 2y\,\frac{dy}{ds}$$

Altogether, we can take any expression written with any variable, and determine its rate of change (derivative) with respect to another variable. If one variable, say y, does NOT depend on another variable, say t, then the derivative of *any* function of y with respect to t is zero. But if y *does* depend on t, then we now know how to find an expression that gives the derivative of some $f(y)$ with respect to t.

EX 7 Find the derivative of πr^2 with respect to x if r does not depend on x. What if r does depend on x?

If r does not depend on x, then

$$\frac{d}{dx}(\pi r^2) = 0$$

If r does depend on x, then

$$\frac{d}{dx}(\pi r^2) = 2\pi r \cdot \frac{d}{dx}(r) = 2\pi r\,\frac{dr}{dx} \quad \blacksquare$$

EX 8 Find the derivative of $x^2 + y^2 = 1$ with respect to t if (a) x and y do not depend on t; (b) only x depends on t; (c) both x and y depend on t.

(a) If neither x nor y depend on t, then

$$\frac{d}{dt}(x^2 + y^2) = 0$$

(b) If only x depends on t, then

$$\frac{d}{dx}(x^2 + y^2) = 2x\frac{d}{dt}(x) + 0 = 2x\frac{dx}{dt}$$

(c) If both x and y depend on t, then

$$\frac{d}{dx}(x^2+y^2) = 2x\frac{d}{dt}(x) + 2y\frac{d}{dt}(y) = 2x\frac{dx}{dt} + 2y\frac{dy}{dt} \quad \blacksquare$$

You Try It

(8) Find the derivative of $\sqrt{1-x^2}$ with respect to r if (a) x does not depend on r, and (b) if x does depend on r.

(9) Find the derivative of $\pi r^2 h$ with respect to t if (a) neither h nor r depends on t, and (b) if only h depends on t.

What does this do for us? A lot! Whether you call it "using implicit differentiation" or "using the chain rule", we now have the ability to take static equations from geometry and "power them up" by applying a derivative with respect to time. Geometry itself is very static; we ask questions like, "If the radius of this circle is r, what is the circumference?" Now we can ask questions like, "If the radius of this circle is increasing at a certain rate, how fast is the circumference increasing?" In the language of derivatives, this is the question: "If we know dr/dt, what is the resulting dC/dt?" Any phrase that includes "*how fast*" is, by default, referencing a derivative with respect to time. Any unit that is something-per-second or something-per-hour is, by default, referencing a derivative with respect to time. So, when we are given a geometry like a rectangle, circle, triangle, cylinder, sphere, and so on, we can now examine how the rate of change of one dimension in that geometry affects the rate of change of another, like length, circumference, surface area, volume, etc.

EX 9 Given a right triangle with legs and hypotenuse of length x, y and h respectively, what is the relationship between the rates of change of x, y and h (with respect to time)?

The Pythagorean Theorem $h^2 = x^2 + y^2$ gives us the direct relationship between x, y and h themselves, so their rates of change are related through the derivative of this expression with respect to time t. Assuming all three depend on time,

$$\begin{aligned} \frac{d}{dt}h^2 &= \frac{d}{dt}(x^2+y^2) \\ 2h\frac{dh}{dt} &= 2x\frac{dx}{dt} + 2y\frac{dy}{dt} \\ h\frac{dh}{dt} &= x\frac{dx}{dt} + y\frac{dy}{dt} \end{aligned}$$

This gives us a relationship between the rates of change of the lengths of the sides. ■

You Try It

(10) Given a circle with radius r and circumference C, what is the relationship between the rates of change of r and C (with respect to time)?

(11) Given a sphere with radius r and surface area S, what is the relationship between the rates of change of r and S (with respect to time)?

The ability to generate the relations between rates of change will be explored in the next chapter.

Have You Learned...

- How to find the derivative dy/dx of a function not explicitly solved in advance for y?
- That the process of implicit differentiation is really just the Chain Rule all over again?
- How previously unknown derivatives can be found by rewriting a function implicitly, so as to involve familiar functions with known derivatives?
- How to combine properties of logarithms with implicit differentiation to make seemingly complicated derivatives quite a lot simpler?
- How to take an expression written using one variable and find its derivative with respect to a second variable if there is no dependence between the two variables?
- How to take an expression written using one variable and find its derivative with respect to a second variable if there is dependence between the two variables?
- How we can take an expression that directly relates certain variables, and generate a new expression that relates the rates of change of those variables?

Implicit Differentiation — Problem List

Implicit Differentiation — You Try It

These appeared above; solutions begin on the next page.

(1) Find dy/dx for $x^2 + y^2 = 1$.
(2) Find dy/dx for $x^2y^2 + x\sin y = 4$.
(3) Find dy/dx for $e^{x^2y} = x + y$.
(4) Find the equation of the line tangent to $x^2 + y^2 = (2x^2 + 2y^2 - x)^2$ at $\left(0, \frac{1}{2}\right)$.
(5) Find the derivative dy/dx for the inverse cosine function $y = \cos^{-1}(x)$.
(6) Find the derivative of $y = \dfrac{(x^3 + 2x^2 - 1)^2 \sin(x)}{\sqrt{5 + x^4}}$.
(7) Find the derivative of $z = 2^x$.
(8) Find the derivative of $5/x$ with respect to z if (a) x does not depend on z, and (b) if x does depend on z.
(9) Find the derivative of $\pi r^2 h$ with respect to t if (a) neither h nor r depends on t, and (b) if only h depends on t.
(10) Given a circle with radius r and circumference C, what is the relationship between the rates of change of r and C (with respect to time)?
(11) Given a sphere with radius r and surface area S, what is the relationship between the rates of change of r and S (with respect to time)?

Implicit Differentiation — Practice Problems

Try these as you get the hang of the You Try It problems. Solutions to these problems are available in Sec. A.4.5.

(1) Find dy/dx for $x^2 - y^2 = 1$.
(2) Find dy/dx for $1 + x = \sin(xy^2)$.
(3) Find dy/dx for $\sqrt{x+y} = 1 + x^2y^2$.
(4) Find the equation of the line tangent to $x^{2/3} + y^{2/3} = 4$ at $(-3\sqrt{3}, 1)$.
(5) Find dy/dx for the inverse tangent function $y = \tan^{-1}(x)$.
(6) Find dy/dx for $y = \dfrac{2(x^4 + 3x^2 + x)(x^2 + 1)^5}{(x^3 + 1)^4}$.
(7) Find the slope of the line tangent to $f(x)$ at $x = 0$ where

$$f(x) = \frac{(x + \cos x)\sqrt[3]{(2 - x)^2}}{5 + \sin^2 x}$$

(8) Find the derivative of $\sqrt{1-x^2}$ with respect to r if (a) x does not depend on r, and (b) if x does depend on r.
(9) Find the derivative of $\sin^{-1} y$ with respect to t if (a) y does not depend on t, and (b) if y does depend on t.
(10) Given a cylinder with radius r, height h, and surface area S, what is the relationship between the rates of change of r and S (with respect to time)? (Assume h is constant.)
(11) Given a sphere with radius r and volume V, what is the relationship between the rates of change of r and V (with respect to time)?
(12) Find dy/dx for $y\sin(x^2) = x\sin(y^2)$.
(13) Find the equation of the line tangent to $2(x^2+y^2)^2 = 25(x^2-y^2)$ at $(3,1)$.

Implicit Differentiation — Challenge Problems

Try these problems to test your skills with the ideas in this section. Solutions to these problems are available in Sec. B.4.5.

(1) Find dy/dx for the inverse secant function $y = \sec^{-1}(x)$.
(2) We know that if $y = e^x$ then $dy/dx = e^x$. Also, if $y = \ln x$, then $dy/dx = 1/x$. But when we have a different base, say b, the derivative of $y = b^x$ is not simply $dy/dx = b^x$ and the derivative of $y = \log_b(x)$ is not just $1/x$. Use implicit differentiation to find what the derivatives of b^x and $\log_b(x)$ really are. (Hint: For the first, use implicit differentiation directly. For the second, write $y = \log_b(x)$ in exponential form, *then* use implicit differentiation.)
(3) Given a cube with side length x and full cross diagonal (from one corner to the opposite corner) D, what is the relationship between the rates of change of x and D (with respect to time)? (Hint: If one corner of the cube is set at the origin $(0,0,0)$, then the far opposite corner sits at (x,x,x).)

Implicit Differentiation — You Try It — Solved

(1) Find dy/dx for $x^2 + y^2 = 1$.

☐ We treat y as a function of x. So when we take the derivative of both sides, we'll need the chain rule. Once we make dy/dx pop up somewhere in the mix, we have to solve for it.

$$\frac{d}{dx}x^2 + \frac{d}{dx}y^2 = \frac{d}{dx}(1)$$
$$2x + 2y\frac{d}{dx}(y) = 0$$
$$2x + 2y\frac{dy}{dx} = 0$$
$$\frac{dy}{dx} = -\frac{x}{y} \quad \blacksquare$$

(2) Find dy/dx for $x^2y^2 + x\sin y = 4$.

☐ We treat y as a function of x. and start by applying the derivative operation to both sides:

$$\frac{d}{dx}\left(x^2y^2 + x\sin y\right) = \frac{d}{dx}(4)$$
$$\frac{d}{dx}x^2y^2 + \frac{d}{dx}x\sin y = \frac{d}{dx}(4)$$

Each term on the left requires the product rule:

$$\left(x^2\frac{d}{dx}y^2 + y^2\frac{d}{dx}x^2\right) + \left(x\frac{d}{dx}\sin y + \sin y \cdot \frac{d}{dx}x\right) = \frac{d}{dx}(4)$$

And now we can start tackling each individual derivative:

$$\left(x^2 2y\frac{d}{dx}(y) + 2xy^2\right) + \left(x\cos y\frac{d}{dx}(y) + \sin y\right) = 0$$
$$\left(x^2 2y\frac{dy}{dx} + 2xy^2\right) + \left(x\cos y\frac{dy}{dx} + \sin y\right) = 0$$

Next, we collect together the instances of dy/dx that have bubbled up, and solve:

$$\frac{dy}{dx}\left(2x^2y + x\cos y\right) + 2xy^2 + \sin y = 0$$
$$\frac{dy}{dx} = \frac{2xy^2 + \sin y}{2x^2y + x\cos y} \quad \blacksquare$$

(3) Find dy/dx for $e^{x^2y} = x + y$.

☐ We treat y as a function of x. When we apply the derivative operator to both sides, the derivative on the left side follows via the chain rule on $e^{g(x)}$.

$$\frac{d}{dx}e^{x^2y} = \frac{d}{dx}(x+y)$$
$$e^{x^2y}\frac{d}{dx}(x^2y) = \frac{d}{dx}(x) + \frac{d}{dx}(y)$$

The derivative remaining on the left side requires the product rule because ... sing along with me ... y is a function of x, and must be treated accordingly:

$$e^{x^2y}\left(x^2\frac{d}{dx}(y) + y\cdot\frac{d}{dx}x^2\right) = 1 + \frac{dy}{dx}$$
$$e^{x^2y}\left(x^2\frac{dy}{dx} + 2xy\right) = 1 + \frac{dy}{dx}$$

And now we clean things up: collect / factor the instances of dy/dx and solve:

$$\frac{dy}{dx}\left(x^2e^{x^2y} - 1\right) = 1 - 2xye^{x^2y}$$
$$\frac{dy}{dx} = \frac{1 - 2xye^{x^2y}}{x^2e^{x^2y} - 1} \quad \blacksquare$$

(4) Find the equation of the line tangent to $x^2 + y^2 = (2x^2 + 2y^2 - x)^2$ at $(0, 1/2)$.

☐ The main ingredient of the equation of the tangent line is the slope of that tangent line, i.e. the value of dy/dx at $(0, 1/2)$. We find dy/dx by taking the derivative of both sides. Having seen some steps in painful detail in the above 3 problems, I will quicken the process a bit:

$$\frac{d}{dx}(x^2 + y^2) = \frac{d}{dx}(2x^2 + 2y^2 - x)^2$$
$$2x + 2y\frac{dy}{dx} = 2(2x^2 + 2y^2 - x)\left(4x + 4y\frac{dy}{dx} - 1\right)$$
$$y\frac{dy}{dx} - 4y(2x^2 + 2y^2 - x)\frac{dy}{dx} = (2x^2 + 2y^2 - x)(4x - 1) - x$$
$$\frac{dy}{dx} = \frac{(2x^2 + 2y^2 - x)(4x - 1) - x}{y - 4y(2x^2 + 2y^2 - x)}$$

With $x = 0$ and $y = 1/2$, we get $dy/dx = 1$ (tedious arithmetic not shown). So the equation of the tangent line is $y - 1/2 = 1(x - 0)$ or $y = x + 1/2$. ■

---

(5) Find the derivative dy/dx for the inverse cosine function $y = \cos^{-1}(x)$.

□ Rewrite the function as $\cos(y) = x$ then use implicit differentiation:

$$\frac{d}{dx}\cos(y) = \frac{d}{dx}(x)$$
$$-\sin(y)\frac{dy}{dx} = 1$$
$$\frac{dy}{dx} = -\frac{1}{\sin y}$$

We can encode $\cos y = x$ into a right triangle by putting y in as an acute angle, with x and 1 being the lengths of the adjacent side and hypotenuse, respectively. Then the Pythagorean Theorem lets us write the length of the missing opposite side to be $\sqrt{1-x^2}$. (See Fig. 4.5.) Then we have the lengths of all three sides, and can immediately find that $\sin y = \sqrt{1-x^2}$. So,

$$\frac{dy}{dx} = -\frac{1}{\sin y} = -\frac{1}{\sqrt{1-x^2}} \quad ■$$

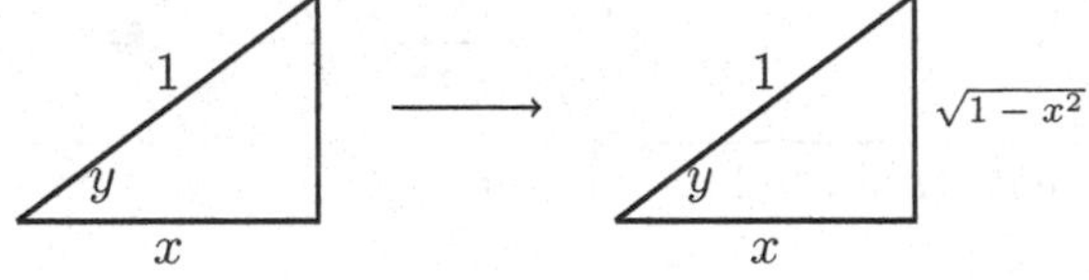

Fig. 4.5 Encoding $\cos y = x$ into a right triangle.

---

(6) Find the derivative of $y = \dfrac{(x^3+2x^2-1)^2\sin(x)}{\sqrt{5+x^4}}$.

□ We can use the natural logarithm to disassemble the one big blob of a function into multiple smaller pieces; we can also show the root as an exponent in anticipation of using the general rule $\ln a^b = b\ln a$.

$$\ln\, y = \ln, \frac{(x^3+2x^2-1)^2\sin(x)}{(5+x^4)^{1/2}}$$
$$= 2\ln(x^3+2x^2-1) + \ln(\sin x) - \frac{1}{2}\ln(5+x^4)$$

And now we can apply the derivative operation, see where all the dy/dx's pop up, and collect them together to solve. The set up is:

$$\frac{d}{dx}(\ln y) = \frac{d}{dx}\left(2\ln(x^3 + 2x^2 - 1) + \ln(\sin x) - \frac{1}{2}\ln(5 + x^4)\right)$$

Applying the derivative operation to both sides,

$$\begin{aligned}
\frac{1}{y}\frac{dy}{dx} &= \frac{2}{x^3 + 2x^2 - 1}(3x^2 + 4x) + \frac{1}{\sin x}(\cos x) - \frac{1}{2(5 + x^4)}(4x^3) \\
\frac{dy}{dx} &= y\left(\frac{2(3x^2 + 4x)}{x^3 + 2x^2 - 1} + \frac{\cos x}{\sin x} - \frac{2x^3}{5 + x^4}\right) \\
&= \frac{(x^3 + 2x^2 - 1)^2 \sin(x)}{\sqrt{5 + x^4}} \cdot \left(\frac{2(3x^2 + 4x)}{x^3 + 2x^2 - 1} + \frac{\cos x}{\sin x} - \frac{2x^3}{5 + x^4}\right)
\end{aligned}$$

■

---

(7) Find the derivative of $z = 2^x$.

□ The only exponential function we have dealt with so far is e^x, but this has a different base. Let's try logarithmic differentiation on this one to change the exponent to a multiplication. Since $z = 2^x$, taking the natural log of both sides gives:

$$\ln(z) = \ln(2^x) = 2\ln(x)$$

And now we're ready for the derivative operator:

$$\begin{aligned}
\frac{d}{dx}\ln(z) &= \frac{d}{dx}(x \ln 2) \\
\frac{1}{z}\frac{dz}{dx} &= \ln 2 \\
\frac{dz}{dx} &= z \ln 2 = 2^x \ln 2
\end{aligned}$$

Note in the last step we converted z back into 2^x so that the derivative is completely in terms of x. FFT: Can you design a derivative formula for a similar function with any base, i.e. $z = a^x$? ■

---

(8) Find the derivative of $5/x$ with respect to z if (a) x does not depend on z, and (b) if x does depend on z.

□ If x does not depend on z, then $\frac{d}{dz}\frac{5}{x} = 0$.

If x does depend on z, then

$$\frac{d}{dz}\frac{5}{x} = \frac{d}{dz}(5x^{-1}) = -5x^{-2} \cdot \frac{d}{dz}(x) = -\frac{5}{x^2} \cdot \frac{dx}{dz} \quad \blacksquare$$

---

(9) Find the derivative of $\pi r^2 h$ with respect to t if (a) neither h nor r depends on t, and (b) if only h depends on t.

☐ If neither h nor r depend on t, then

$$\frac{d}{dt}(\pi r^2 h) = 0$$

If h depends on t, then we can treat r as a constant:

$$\frac{d}{dt}(\pi r^2 h) = \pi r^2 \frac{d}{dt}(h) = \pi r^2 \frac{dh}{dt} \quad \blacksquare$$

---

(10) Given a circle with radius r and circumference C, what is the relationship between the rates of change of r and C (with respect to time)?

☐ We relate radius to circumference of a circle by $C = 2\pi r$. Thus, to relate their rates of change with respect to time,

$$\frac{d}{dt}(C) = \frac{d}{dt}(2\pi r)$$

$$\frac{dC}{dt} = 2\pi \frac{d}{dt}(r) = 2\pi \frac{dr}{dt} \quad \blacksquare$$

---

(11) Given a sphere with radius r and surface area S, what is the relationship between the rates of change of r and S (with respect to time)?

☐ We relate radius to surface area of a sphere by $S = 4\pi r^2$. Thus, to relate their rates of change with respect to time,

$$\frac{d}{dt}(S) = \frac{d}{dt}(4\pi r^2)$$

$$\frac{dS}{dt} = 4\pi \frac{d}{dt}(r^2)$$

$$\frac{dS}{dt} = 4\pi \cdot 2r \frac{dr}{dt} = 8\pi r \frac{dr}{dt} \quad \blacksquare$$

4.6 Antiderivatives Using Substitution

Introduction

Please recall fondly that we examined two versions of the chain rule. Both ultimately say the same thing, but each is amenable to a different secondary interpretation. One version of the chain rule, Eq. (4.35), lays out the immediate procedure for finding the derivative of a composition:

$$\frac{d}{dx}f(g(x)) = f'(g(x)) \cdot g'(x)$$

and the other version, Eq. (4.37), helps us relate multiple derivatives to each other:

$$\frac{dy}{dx} = \frac{dy}{du} \cdot \frac{du}{dx}$$

You may remember that we generated the second by applying the substitution $u = g(x)$ to the first. When $y = f(g(x))$ and $u = g(x)$, we can write $f'(g(x))$ as $f'(u)$ or dy/du, and $g'(x)$ as du/dx. These exchanges converted the first version of the chain rule into the second.

Both versions of the chain rule help us navigate something like the derivative of $y = \sin(x^2 + 1)$. Using the first version, we identify $f(x) = \sin(x)$ and $g(x) = x^2 + 1$, then write:

$$\begin{aligned}\frac{d}{dx}\sin(x^2+1) &= \cos(x^2+1) \cdot \frac{d}{dx}(x^2+1) \\ &= \cos(x^2+1) \cdot (2x) = 2x\cos(x^2+1)\end{aligned}$$

To use the second version, we can make the substitution $u = x^2 + 1$; then $y = \sin(u)$ and

$$\frac{dy}{du} = \cos(u) \qquad \text{and} \qquad \frac{du}{dx} = 2x$$

so, the second version of the chain rule gives us:

$$\frac{dy}{dx} = \frac{dy}{du} \cdot \frac{du}{dx} = \cos(u) \cdot (2x) = 2x\cos(x^2+1)$$

Either way, we get the same result.

This same sort of substitution procedure used just above in the second version can also help us find *antiderivatives* — particularly, when we have an antiderivative of something produced as a derivative using the chain rule.

Substitution for Antiderivatives

Consider the following antiderivatives, which are all presented in the most common format — with the smaller bits first, and the "bulkiest" part of the integrand second:

$$\int 2x\cos(x^2+1)\,dx \quad ; \quad \int 2xe^{x^2+1}\,dx$$
$$\int 2x\sqrt{x^2+1}\,dx \quad ; \quad \int 2x\,\text{sech}^2(x^2+1)\,dx$$

Each represents the antiderivative of something that would have been produced using the chain rule. If we rearrange them a bit, that might be more clear:

$$\int \cos(x^2+1)\cdot 2x\,dx \quad ; \quad \int e^{x^2+1}\cdot 2x\,dx$$
$$\int \sqrt{x^2+1}\cdot 2x\,dx \quad ; \quad \int \text{sech}^2(x^2+1)\cdot 2x\,dx$$

Do you recognize that each is in the form $\int F(g(x))g'(x)\,dx$?

In each antiderivative, the inner function $g(x)$ is the same: $g(x)=x^2+1$, so that $g'(x)=2x$. The outer functions $F(x)$ are all different. The main strategy of our solution technique for these is to "hide" the inside function for a few steps so that we can concentrate on the antiderivative of the outside function. That new solution strategy involves the same type of substitutions that we thought about with the second version of the chain rule.

Let's plan to make the substitution $u=x^2+1$ in each antiderivative. This substitution $u=x^2+1$ comes automatically with $du/dx=2x$. To fit this information with the antiderivatives a bit better, we will rewrite:

$$\frac{du}{dx}=2x \longrightarrow du=2x\,dx$$

With $u=x^2+1$ and $du=2x\,dx$, we are ready to give a mathematical makeover to our antiderivatives. For example,

$$\int 2x\cos(x^2+1)\,dx=\int \cos(x^2+1)\cdot(2x\,dx)=\int \cos(u)\,du$$

Similarly,

$$\int 2xe^{x^2+1}\,dx = \int e^{x^2+1}\cdot 2x\,dx = \int e^u\,du$$

$$\int 2x\sqrt{x^2+1}\,dx = \int \sqrt{x^2+1}\cdot 2x\,dx = \int \sqrt{u}\,du$$

$$\int 2x\,\operatorname{sech}^2(x^2+1)\,dx = \int \operatorname{sech}^2(x^2+1)\cdot 2x\,dx = \int \operatorname{sech}^2(u)\,du$$

The new version of each antiderivative looks much simpler than the original because we've temporarily masked the details from the inner function $g(x)$ to concentrate on the outer function:

$$\int \cos(u)\,du = \sin(u) + C$$

$$\int e^u\,du = e^u + C$$

$$\int \sqrt{u}\,du = \frac{2}{3}u^{3/2} + C$$

$$\int \operatorname{sech}^2(u)\,du = \tanh(u) + C$$

Now that the substitution has done its job, we need to revert back to the original variable by replacing each instance of u in the final results with the original substitution $u = x^2 + 1$:

$$\int 2x\cos(x^2+1)\,dx = \sin(x^2+1) + C$$

$$\int 2xe^{x^2+1}\,dx = e^{x^2+1} + C$$

$$\int 2x\sqrt{x^2+1}\,dx = \frac{2}{3}(x^2+1)^{3/2} + C$$

$$\int 2x\,\operatorname{sech}^2(x^2+1)\,dx = \tanh(x^2+1) + C$$

When you were looking at those examples, you may have been thinking, "Gosh, we were really lucky that there was a $2x$ sitting there in each antiderivative to help make the substitution $u = x^2 + 1$ work out!" And you're right. Normally, that doesn't happen so smoothly. But with a chosen substitution $u = g(x)$, as long as what's available in the integrand only differs from the true $g'(x)$ by a multiplicative constant, we're good to go. To see this, let's consider this slightly different version of one of those four, and its rearranged version:

$$\int x\cos(x^2+1)\,dx \longrightarrow \int \cos(x^2+1)\cdot x\,dx$$

The integrand does not *quite* match the form $F(g(x))g'(x)$, but it's close. The true $g'(x)$ should be $2x$, but we only see x. We can fix that by using our rule, "if you need something, put it there!" There are two slightly different ways to accomplish that.

Method 1: We can recognize that if would be really great to have a $2x\,dx$ at the end of the integrand to facilitate the substitution $u = x^2 + 1$. So, let's put a 2 in there! We do this by also inserting $1/2$ to be sure we don't actually change the function:

$$\int x\cos(x^2+1)\,dx = \int \cos(x^2+1)\cdot x\,dx = \int \cos(x^2+1)\cdot\frac{1}{2}\cdot 2x\,dx$$
$$= \frac{1}{2}\int \cos(x^2+1)\cdot 2x\,dx$$

Now we can directly implement the substitution $u = x^2+1$ with $du = 2x\,dx$:

$$\int x\cos(x^2+1)\,dx = \frac{1}{2}\int \cos(x^2+1)\cdot 2x\,dx = \frac{1}{2}\int \cos(u)\,du = \frac{1}{2}\sin(u)+C$$

And then we revert u back to x^2+1 so that:

$$\int x\cos(x^2+1)\,dx = \frac{1}{2}\sin(x^2+1)+C \tag{4.38}$$

Method 2 does basically the same thing, but by messing with the du and dx terms rather than inserting constants. We start again with

$$\int x\cos(x^2+1)\,dx = \int \cos(x^2+1)\cdot x\,dx$$

and we recognize that with a substitution $u = x^2+1$, we would really like to see the integral end with $2x\,dx$. But it ends with $x\,dx$ instead. So, how does $x\,dx$ relate to du? Well, when we install $u = x^2+1$, we generate the relation $du = 2x\,dx$. Using that:

$$2x\,dx = du \longrightarrow x\,dx = \frac{1}{2}du$$

And now we can complete the substitution,

$$\int x\cos(x^2+1)\,dx = \int \cos(x^2+1)\cdot(x\,dx) \overset{u=x^2+1}{\longrightarrow} \int \cos(u)\cdot\frac{1}{2}du$$
$$= \frac{1}{2}\int \cos(u)\,du$$

As before, this new version of the antiderivative is easy:

$$\frac{1}{2}\int \cos(u)\,du = \frac{1}{2}\sin(u)+C$$

Reversing the substitution (i.e. replacing u back with x^2+1),

$$\int x\cos(x^2+1)\,dx = \frac{1}{2}\sin(x^2+1)+C \tag{4.39}$$

By comparing (4.38) and (4.39), we see the same result by both methods. The methods really aren't that different, but Method 2 tends to be a bit more streamlined once you get used to all this, and so it's the default procedure in most cases.

Antiderivatives Using Substitution — The Procedure

The substitution process is one of the most important antiderivative techniques we have. It is usually called "u-substitution"; historically, the letter u has been the typical name we give to the substituted expression. That's certainly not required, though, you can use any letter you'd like. Here is the general procedure:

Given an antiderivative that looks likely to be related to the chain rule,

$$\int \hat{g}'(x)F(g(x))\,dx \longrightarrow \int F(g(x))\cdot\hat{g}'(x)\,dx$$

we do the following:

- Play "spot the derivative". The inner function $g(x)$ should be easy to identify; that $g(x)$ provides $g'(x)$. The term $\hat{g}'(x)$ in the integrand should be *close* to the true $g'(x)$. By "close" we mean different only by a multiplicative constant.
- With that identification made, prepare the substitution $u = g(x)$.
- Determine the *true* $du/dx = g'(x)$. Write this as $du = g'(x)\,dx$. And remember our notation: $g'(u)$ is what we'd *like* to see in the integrand, but $\hat{g}'(x)$ is what's actually sitting there.
- Use a multiplicative constant A to fix the difference between what we have, $\hat{g}'(x)$, and what we would like to see, $g'(x)$:

$$\hat{g}'(x)\,dx = Ag'(x)\,dx = A\,du$$

- Convert all parts of the antiderivative:

$$\begin{aligned}\int \hat{g}'(x)F(g(x))\,dx &= \int F(g(x))\cdot\hat{g}'(x)\,dx\\ &= \int F(u)\cdot A\,du = A\int F(u)\,du\end{aligned}$$

- Solve the new, easier antiderivative $A \int F(u)\, du$.
- Take the result, and reverse the substitution: replace u with the original $g(x)$.

Here is an important rule of thumb:

- You MUST be able to convert ALL instances of x in the antiderivative to an equivalent in terms of u. There cannot be a mix of x and u in the new version of the antiderivative.
- If you cannot convert all instances of x with your chosen substitution, then your substitution is not going to work.
- You usually find out very quickly if your substitution is going to fail. If it is, try something else!

This rule of thumb calls back to the part of the Preface of this book, where I say that *failure is a good thing.* In many problems, you might select a couple of substitutions before you come across the one that actually works. This is expected, so don't be discouraged by not finding the right substitution on your first try. Working through multiple attempts of substitution will give you practice with the manipulations, and greatly improve your intuition for making the proper substitutions in later problems.

EX 1 Find $\int 3x \sin(x^2 - 1)\, dx$.

Let's regroup the antiderivative:

$$\int 3x \sin(x^2 - 1)\, dx = \int \sin(x^2 - 1) \cdot 3x\, dx$$

We identify the inner function as $g(x) = x^2 - 1$. If things were going to work perfectly, we should also then see $2x\, dx$ in the integrand. But we don't, we see $3x\, dx$. So, how does $3x\, dx$ relate to du? Well, if $du = 2x\, dx$, then:

$$du = 2x\, dx \longrightarrow x\, dx = \frac{1}{2}\, du \longrightarrow 3x\, dx = \frac{3}{2}\, du$$

And now we can replace everything in the original antiderivative:

$$\int 3x \sin(x^2 - 1)\, dx = \int \sin(x^2 - 1) \cdot 3x\, dx = \int \sin(u) \cdot \frac{3}{2}\, du = \frac{3}{2} \int \sin(u)\, du$$

Solving the new, simpler antiderivative:

$$\frac{3}{2} \int \sin(u)\, du = \frac{3}{2}(-\cos(u)) + C = -\frac{3}{2} \cos(u) + C$$

Finally, we reverse the substitution $u = x^2 - 1$ to present the result:

$$\frac{3}{2}\int \sin(u)\, du == -\frac{3}{2}\cos(x^2-1) + C \quad \blacksquare$$

EX 2 Find $\displaystyle\int (3x-6)^5\, dx$.

The substitution $u = 3x - 6$ will change $(3x-6)^5$ to u^5. That's nice. But the choice $u = 3x - 6$ comes with $du = 3\, dx$. But we don't see $3\, dx$. we only see dx. This can be fixed:

$$u = 3x - 6 \longrightarrow du = 3\, dx \longrightarrow dx = \frac{1}{3}\, du$$

And now we're ready to roll: with $u = 3x - 6$ and $dx = \frac{1}{3}\, du$,

$$\begin{aligned}\int (3x-6)^5 dx &= \int u^5 \cdot \frac{1}{3}\, du = \frac{1}{3}\int u^5\, du \\ &= \frac{1}{3}\cdot\frac{1}{6}u^6 + C = \frac{1}{18}(3x-6)^6 + C \quad \blacksquare\end{aligned}$$

You Try It

(1) Find $\displaystyle\int (3x-2)^{20}\, dx$.

(2) Find $\displaystyle\int \frac{dx}{5-3x}$.

EX 3 Find $\displaystyle\int \frac{e^{-\sqrt{x}}}{\sqrt{x}}\, dx$.

Looking at the numerator, let's try the substitution $u = -\sqrt{x}$; here's a more streamlined handling of the rest of the integrand via the differentials du and dx:

$$u = -\sqrt{x} \longrightarrow du = -\frac{1}{2\sqrt{x}}\, dx$$

Now, in the integrand, we don't see the full $-\dfrac{1}{2\sqrt{x}}\, dx$, we only see $\dfrac{1}{\sqrt{x}}\, dx$. So, to patch that up,

$$du = -\frac{1}{2\sqrt{x}}\, dx \longrightarrow \frac{1}{\sqrt{x}}\, dx = -2\, du$$

Now we can put it all together:

$$\int \frac{e^{-\sqrt{x}}}{\sqrt{x}}\,dx = \int e^{-\sqrt{x}} \cdot \frac{1}{\sqrt{x}}\,dx = \int e^u\,(-2\,du)$$
$$= -2\int e^u\,du = -2e^u + C = -2e^{-\sqrt{x}} + C \quad \blacksquare$$

You Try It

(3) Find $\int \sqrt{4-t}\,dt$.

(4) Find $\int xe^{-x^2}\,dx$.

(5) Find $\int \cos\theta \sin^6\theta\,d\theta$.

(6) Find $\int \frac{\sin\theta}{\cos^2\theta}\,d\theta$.

Here's an example of an antiderivative that doesn't look suitable for substitution at first.

EX 4 Find $\int \tan(x)\,dx$.

This doesn't look like an integral where substitution would help at all. What are we going to do, say $u = x$? All that does is change the letters. If we try $u = \tan x$, then we would need to have something close to $du = \sec^2 x\,dx$ also sitting around in the integrand to complete the substitution. However, when dealing with trig functions, one useful strategy is: "When in doubt, turn everything into sines and cosines." Here,

$$\int \tan(x)\,dx = \int \frac{\sin(x)}{\cos(x)}\,dx$$

Then a choice $u = \cos(x)$ leads to

$$du = -\sin(x)\,dx \longrightarrow \sin(x)\,dx = -du$$

and now we rewrite the antiderivative:

$$\int \tan(x)\,dx = \int \frac{1}{\cos(x)} \cdot \sin(x)\,dx = \int \frac{1}{u}\,(-\,du) = -\int \frac{1}{u}\,du$$
$$= -\ln|u| + C = -\ln|\cos(x)| + C = \ln|\sec(x)| + C$$

(Make sure you understand the last step.) $\blacksquare$

We might as well box this one to go with all our other derivatives and antiderivatives involving trig functions:

$$\boxed{\int \tan(x)\,dx = \ln|\sec(x)| + C} \tag{4.40}$$

You Try It

(7) Find $\int \cot\theta\,d\theta$.

It's just as important to know what you *cannot* solve as it is to know what you can. Here is a "non-example" to illustrate a case where substitution will NOT work.

EX 5 Find $\int 2x^2\cos(x^2)\,dx$.

The natural choice of substitution would be $u = x^2$, in order to convert $\cos(x^2)$ to $\cos(u)$. With $u = x^2$, we have $du = 2x\,dx$. Is there a $2x\,dx$ avaiable in the integrand? Sure! We just have to split up the leading x^2:

$$\int 2x^2\cos(x^2)\,dx = \int x\cdot\cos(x^2)\cdot(2x\,dx)$$

So now, with $u = x^2$,

- $\cos(x^2)$ becomes $\cos(u)$. That's good.
- $2x\,dx$ becomes du. That's good.
- But, we still have that one lingering x. That's not good.

Hey, maybe we can use $u = x^2$ to write this leftover x as $x = \sqrt{u}$. That makes the resulting antiderivative look like this:

$$\int x\cdot\cos(x^2)\cdot(2x\,dx) = \int \sqrt{u}\cos(u)\,du$$

Wow. That's worse than the original. Maybe we can start over and use $u = x^2$ to change $x^2\cos(x^2)$ into $u\cos(u)$. But if we do that, we've used up all our x's, and there is nothing left to help with $du = 2x\,dx$. No matter what we do, the substitution fails. This antiderivative cannot be solved by means known to us. ■

Substitution Unrelated to the Chain Rule

Sometimes substitution can help out even for antiderivatives that have nothing to do with the chain rule. Here is an example where substitution will reconstruct an integrand in a helpful way.

EX 6 Find $\displaystyle\int \frac{x+1}{x-1}\,dx$.

The integrand doesn't really follow the form $\hat{g}'(x)F(g(x))$, but we'll try a substitution anyway. Looking at the denominator (I think I would always rather simplify a denominator than a numerator), setting $u = x - 1$ not only links $du = dx$, but it also affects the numerator:

$$x + 1 = (x - 1) + 2 = u + 2$$

And now the antiderivative is solvable; with $u = x - 1$,

$$\begin{aligned}\int \frac{x+1}{x-1}\,dx &= \int \frac{u+2}{u}\,du = \int 1 + \frac{2}{u}\,du \\ &= u + 2\ln|u| + C = (x-1) + 2\ln|x-1| + C \quad \blacksquare\end{aligned}$$

You Try It

(8) Find $\displaystyle\int t\sqrt{t+3}\,dt$.

In any problem involving substitution, the worst thing you can do is say, "I can't think of a good substitution, so I can't start the problem." That's the wrong way to look at it. If you can't think of the *best* substitution at first, TRY SOMETHING. You may fail. You'll discover pretty quickly if your substitution won't work out. But each attempt gives you some experience that, even though it may not help with the immediate problem, will build your intuition in future problems. A quarterback still learns something from every incomplete pass. A basketball player still learns something from every missed free throw. A student studying calculus will still learn something from every failed substitution. TRY SOMETHING.

Have You Learned...

- How to use substitution to help rewrite the antiderivative of an expression produced using the chain rule?
- How to adjust multiplicative constants to make your substitution work?

- How to recognize when your substitution will *not* work?
- How to use substitution to reorganize antiderivatives that may not have anything to do with the chain rule?

Antiderivatives Using Substitution — Problem List

Antiderivatives Using Substitution — You Try It

These appeared above; solutions begin on the next page.

Find the following antiderivatives:

(1) $\int (3x-2)^{20}\,dx.$

(2) $\int \frac{dx}{5-3x}.$

(3) $\int \sqrt{4-t}\,dt.$

(4) $\int xe^{-x^2}\,dx.$

(5) $\int \cos\theta \sin^6\theta\,d\theta.$

(6) $\int \frac{\sin\theta}{\cos^2\theta}\,d\theta.$

(7) $\int \cot\theta\,d\theta.$

(8) $\int t\sqrt{t+3}\,dt.$

Antiderivatives Using Substitution — Practice Problems

Try these as you get the hang of the You Try It problems. Solutions to these problems are available in Sec. A.4.6.

Find the following antiderivatives:

(1) $\int (2-x)^6\,dx.$

(2) $\int \frac{x}{x^2+1}\,dx.$

(3) $\int y^3\sqrt{2y^4-1}\,dy.$

(4) $\int \sqrt{x}e^{x^{3/2}}\,dx.$

(5) $\int (1+\tanh(\theta))^5\,\mathrm{sech}^2(\theta)\,d\theta.$

(6) $\int \frac{\sin x}{1+\cos x}\,dx$ and $\int \frac{\sinh x}{1+\cosh x}\,dx.$

(7) $\int \frac{x}{(x^2+1)^2}\,dx.$

(8) $\int \cos x \sin(\sin x)\,dx.$

Antiderivatives Using Substitution — Challenge Problems

Try these problems to test your skills with the ideas in this section. Solutions to these problems are available in Sec. B.4.6.

(1) Find $\int e^{2\cosh(t)}\sinh(t)\,dt.$

(2) Find $\int \frac{4t^3}{t^2-3}\,dt.$ (Hint: Write $4t^3$ as $2t^2 \cdot 2t$.)

(3) Find $\int \frac{x^5}{1+x^3}\,dx.$

Antiderivs Using Substitution — You Try It — Solved

(1) Find $\int (3x-2)^{20}\,dx$.

☐ With the substitution $u = 3x - 2$, we get

$$du = 3\,dx \quad \longrightarrow \quad dx = \frac{1}{3}du$$

So that

$$\int (3x-2)^{20}dx = \int u^{20}\cdot\frac{1}{3}du = \frac{1}{3}\int u^{20}\,du = \frac{1}{3}\cdot\frac{1}{21}u^{21}+C$$
$$= \frac{1}{63}(3x-2)^{21}+C \quad \blacksquare$$

(2) Find $\int \frac{dx}{5-3x}$.

☐ With $u = 5 - 3x$, we get

$$du = -3\,dx \quad \longrightarrow \quad dx = -\frac{1}{3}du$$

So that

$$\int \frac{dx}{5-3x} = -\frac{1}{3}\int\frac{du}{u} = -\frac{1}{3}\ln|u|+C = -\frac{1}{3}\ln|5-3x|+C \quad \blacksquare$$

(3) Find $\int \sqrt{4-t}\,dt$.

☐ We use the substitution $4 - t$ to get $dt = -du$, and so

$$\int\sqrt{4-t}\,dt = -\int\sqrt{u}\,du = -\frac{2}{3}u^{3/2}+C = -\frac{2}{3}(4-t)^{3/2}+C \quad \blacksquare$$

(4) Find $\int xe^{-x^2}\,dx$.

☐ We use the substitution $u = -x^2$ to get

$$du = -2x dx \quad \longrightarrow x\,dx = -\frac{1}{2}du$$

And so

$$\int xe^{-x^2}\,dx = \int e^{-x^2}\cdot x\,dx = -\frac{1}{2}\int e^u\,du = -\frac{1}{2}e^u+C = -\frac{1}{2}e^{-x^2}+C \quad \blacksquare$$

(5) Find $\int \cos\theta \sin^6\theta \, d\theta$.

☐ We use the substitution $u = \sin\theta$ to get $du = \cos\theta d\theta$, which is already a match to what's available:

$$\int \cos\theta \sin^6\theta \, d\theta = \int \sin^6\theta \cdot \cos\theta \, d\theta$$
$$= \int u^6 du = \frac{1}{7}u^7 + C = \frac{1}{7}\sin^7\theta + C \quad \blacksquare$$

(6) Find $\int \frac{\sin\theta}{\cos^2\theta} \, d\theta$.

☐ We use the substitution $u = \cos\theta$ to give

$$du = -\sin\theta \, d\theta \quad \longrightarrow \quad \sin\theta \, d\theta = -du$$

and

$$\int \frac{\sin\theta}{\cos^2\theta} d\theta = -\int \frac{du}{u^2} = -\frac{1}{u} + C = -\frac{1}{\cos\theta} + C = -\sec\theta + C \quad \blacksquare$$

(7) Find $\int \cot\theta \, d\theta$. (Hint: Remember, when in doubt, change all trig functions to ...)

☐ We can write

$$\int \cot\theta d\theta = \int \frac{\cos\theta}{\sin\theta} d\theta$$

and so with the choice $u = \sin\theta$, we have directly $du = \cos\theta d\theta$ and

$$\int \frac{\cos\theta}{\sin\theta} d\theta = \int \frac{du}{u} = \ln|u| + C = \ln|\sin\theta| + C \quad \blacksquare$$

(8) Find $\int t\sqrt{t+3} \, dt$.

☐ Let's select $u = t+3$, which (1) gives $du = dt$ and (2) makes $t = u-3$. Altogether,

$$\int t\sqrt{t+3} \, dt = \int (u-3)\sqrt{u} \, du = \int (u-3)u^{1/2} \, du$$
$$= \int u^{3/2} - 3u^{1/2} \, du = \frac{2}{5}u^{5/2} - 3 \cdot \frac{2}{3}u^{3/2} + C$$
$$= \frac{2}{5}u^{5/2} - 2u^{3/2} + C = \frac{2}{5}(t+3)^{5/2} - 2(t+3)^{3/2} + C \quad \blacksquare$$

Chapter 5

Calculus Has Its Ups and Downs

5.1 The Calc 1 Boss Fight: Related Rates

Introduction

The Chain Rule (and its evil clone "implicit differentiation") showed us how we can relate rates of change of variables which do not at first appear together in a relationship. Recall, for example, that if the variables x and y both depend on t, but there is no equation directly relating x to t or y to t, we can still write:

$$\frac{dy}{dt} = \frac{dy}{dx} \cdot \frac{dx}{dt}$$

Similarly, even if t itself does not appear in an initial relationship, we can introduce a rate of change with respect to t; for example, consider the area of a circle, πr^2:

$$\frac{d}{dt}(\pi r^2) = 2\pi r \frac{dr}{dt}$$

In physical problems, we often are faced with the need to relate rates of change with respect to *time* t of physical parameters such as length, area, volume, and so on. The Chain Rule and implicit differentiation are the keys to making progress in these situations.

Rates That Are Related to Other Rates

The most common "hidden quantity" that does not initially appear in equations describing physical relationships yet is needed within rates of change is time t. Imagine a spherical balloon which is being inflated. We can say it has a radius r, and the radius is *changing* at a given rate; if r is in centimeters, then the rate of increase could be described in units of centimeters per second, or cm/s. This rate of change is described mathematically as dr/dt. Clearly the volume of this spherical balloon is increasing as well, and we might like to find the rate of change in volume, dV/dt, that is due to the changing radius. There is no geometric formula that relates r, V, and t all together.

Our best starting point is the standard equation for a sphere, $V = 4\pi r^3/3$. Note that this comes automatically with the rate of change $dV/dr = 4\pi r^2$. But even in that expression, there is no t to be found. We can introduce rates of change with respect to time with either the Chain Rule or implicit differentiation. I'll show both techniques, then name the one we'll stick with for the majority of future examples.

The Chain Rule says that we can relate the rates of change of radius and volume with respect to time like this:

$$\frac{dV}{dt} = \frac{dV}{dr} \cdot \frac{dr}{dt}$$

We know dV/dt just from the regular equation for the volume of a sphere; for the other two rates, dr/dt and dV/dt, we must be given one so we can calculate the other using this Chain Rule equation. If we are told the rate of increase of the radius, then we know dr/dt, and can proceed to compute dV/dt using

$$\frac{dV}{dt} = 4\pi r^2 \frac{dr}{dt} \tag{5.1}$$

Implicit differentiation allows us to start with the static geometric formula for the volume of a sphere, and then "power it up" by applying the derivative with respect to t:

$$\begin{aligned} V &= \frac{4}{3}\pi r^3 \\ \frac{d}{dt}V &= \frac{d}{dt}\left(\frac{4}{3}\pi r^3\right) \\ \frac{dV}{dt} &= \frac{4}{3}\pi\frac{d}{dt}\left(r^3\right) = \frac{4}{3}\pi\left(3r^2\frac{dr}{dt}\right) \end{aligned}$$

If we tidy this up a bit, we find:

$$\frac{dV}{dt} = 4\pi r^2 \frac{dr}{dt} \tag{5.2}$$

A quick look at both (5.1) and (5.2) shows that we get exactly the same relation between dr/dt and dV/dt no matter which method we choose. So:

pick which method you prefer, and use it. In most cases, we will follow implicit differentiation.

Because we are handling physical dimensions and rates of change, units must be handled carefully. One nice thing about these computations is that — in most cases — units are developed naturally, according to the given related rates of change. In this spherical balloon example, we wanted to know dV/dt. If the radius is in centimeters, then we would expect the rate of change of radius to be in cm/sec, and we would expect the corresponding rate of change of volume to be in cm^3/sec. Equations (5.1) and (5.2) will produce units of cm^3/sec for dV/dt automatically:

$$\frac{dV}{dt} = 4\pi r^2 [cm]^2 \frac{dr}{dt}\left[\frac{cm}{sec}\right] = 4\pi r^2 \frac{dr}{dt}\left[\frac{cm^3}{sec}\right]$$

This becomes a good diagnostic tool for a proposed relation that ties together rates of change: do the units you predict agree with the units that are produced? If not, your relationship may not be the right one ... yet. Units are usually not something that you just wait until the end of the problem to slap on.

Related Rates Scenarios

In problems where we must juggle rates of change of given variables with respect to time, the framing story can often sound complicated. Just remember that no matter how complicated the story sounds, the problem will reduce to geometry: you will be dealing with a circle, triangle, sphere or other familiar geometric shape. All you have to do is identify the given underlying simple geometry, and come up with a static equation (usually a familiar one!) that relates the dimensions in the problem, like length, radius, area, volume, etc. NEVER try to build time t directly into any formula. Always introduce time t through the derivatives. The overall strategy follows like this:

(1) Find a known formula tying together the static geometry of the problem (is it founded on a triangle, rectangle, circle, etc.?).
(2) Power up the static formula by applying d/dt to both sides; this will "turn on" the geometry so that the given dimensions can start changing as needed; rates of change are introduced through implicit differentiation.
(3) Take a snapshot of the action at the right instant and provide all the known values; most problems will be structured as: "Here

is a bunch of geometric information, find the rate of change of ⟨ some variable ⟩ at the instant when ⟨ some variable hits a selected value ⟩". That "instant when" is when you take your snapshot.

Here are some purely geometric examples to get started.

EX 1 The radius of a sphere is increasing at $2.5\,cm/sec$. How fast is the volume of the sphere changing when the radius itself is $1\,cm$? $10\,cm$?

- "The radius of the sphere is increasing at $2.5\,cm/sec$" translates to $dr/dt = 2.5\,in/sec$.
- "How fast is the volume of the sphere increasing?" translates to "What is dV/dt?"

So, we need to tie dV/dt to dr/dt. The static relation between r and V for a sphere is $V = 4\pi r^3/3$. We can "power up" this relation by applying the operator d/dt to both sides, which is what we did to generate Eq. (5.2):

$$\frac{dV}{dt} = 4\pi r^2 \frac{dr}{dt}$$

Then, we take two snapshots. At the instant when $r = 1\,in$, we have

$$\frac{dV}{dt} = 4\pi(1\,in)^2\left(2.5\,\frac{in}{sec}\right) = 10\pi\,\frac{in^3}{sec}$$

When $r = 100\,in$, we have

$$\frac{dV}{dt} = \pi(10\,in)^2\left(2.5\,\frac{in}{sec}\right) = 1000\pi\,\frac{in^3}{sec}$$

Note that when the sphere is larger, the change in volume due to a given change in radius is larger than what the same change in radius generates for a smaller sphere. And just for emphasis, notice that the units work out automatically. We know to expect units of L^3/T for dV/dt, and the equation we used — with the given data — produced cm^3/sec. ■

You Try It

(1) The Great Clock of Westminster, which houses the giant bell named Big Ben, is really large! The minute hand of this clock is 14 feet long. How fast (in ft/min) is the tip of the minute hand tracing out a circle at any time? (Hint: do you remember the equation $s = r\theta$?)

(B) Snapshot:

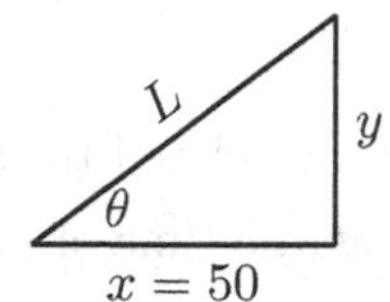

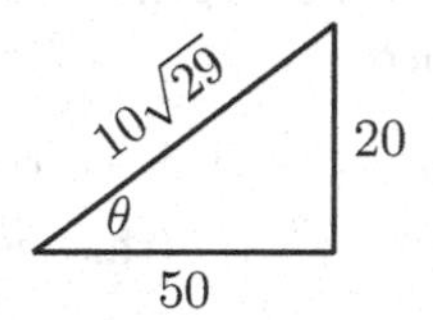

Fig. 5.1 A right triangle with changing dimensions.

EX 2 Suppose the height of a right triangle is increasing at $3\,cm/sec$, while the base remains constant at $50\,cm$. How fast is the corresponding opposite angle changing when the height of the triangle is $20\,cm$?

A picture is often a good idea, so let's draw a triangle and give some names / variables to dimensions. Figure 5.1(A) shows the base labeled as x, the height as y, the hypotenuse as L, and the angle of interest as θ. We also note that the base remains at constant $50\,cm$, whereas y, L, and θ may all change. With the given labeling and data in the problem, we have this interpretation: we are told that $dy/dt = 3\,cm/sec$, and we are asked for the value of $d\theta/dt$ at the instant when $y = 20\,cm$. We need a formula that relates the active variables / dimensions x (constant), y, and θ; the tangent function does this. Because x is constant at $50\,cm$ for the entire problem, it is safe to plug in that value immediately:

$$\tan\theta = \frac{y}{x} = \frac{y}{50}$$

Now we can power this up by a derivative with respect to time.

$$\begin{aligned}\frac{d}{dt}\tan\theta &= \frac{d}{dt}\left(\frac{y}{50}\right)\\ \sec^2\theta\frac{d\theta}{dt} &= \frac{1}{50}\frac{dy}{dt}\\ \frac{d\theta}{dt} &= \frac{1}{50}\frac{dy}{dt}\cdot\cos^2\theta\end{aligned}$$

There are three "mystery terms" here: dy/dt, $d\theta/dt$, and $\cos^2\theta$. The rate of change dy/dt is given, we are asking for the final value of $d\theta/dt$, and so we have to figure out what to do with $\cos^2\theta$. However, we only need to know the value of $\cos^2\theta$ at the time of our "snapshot".

Figure 5.1(B) shows our triangle at the time of the snapshot: $x = 50\,cm$ just like always, $y = 20\,cm$ at the snapshot. To determine $\cos\theta$ at the snapshot, we need the length of the hypotenus, but that's just a job for the Pythagorean Theorem:

$$\begin{aligned} L^2 &= (50)^2 + y^2 \quad \leftarrow \text{ (this is true at any time related to the problem)} \\ L^2 &= (50)^2 + (20)^2 \quad \leftarrow \text{ (this is true only at the time of the snapshot)} \\ L &= \sqrt{2900} = 10\sqrt{29} \end{aligned}$$

With the lengths of all three sides of the triangle at the snapshot, we can compute *any* trig function — specifically $\cos\theta$.

$$\cos\theta = \frac{\text{hyp}}{\text{adj}} = \frac{50}{10\sqrt{29}} = \frac{5}{\sqrt{29}}$$

(If you object to having a radical in the denominator, feel free to adjust that.) Now we have all the information we need:

$$\begin{aligned} \frac{d\theta}{dt} &= \frac{1}{50}\frac{dy}{dt} \cdot \cos^2\theta \\ \frac{d\theta}{dt} &= \frac{1}{50}(3)\left(\frac{5}{\sqrt{29}}\right)^2 \\ \frac{d\theta}{dt} &= \frac{3}{50} \cdot \frac{25}{29} = \frac{3}{58} \end{aligned}$$

Having made a big deal about how units often work out automatically in related rates problems, here's a case where they don't quite do that. Let's track them:

- The units of the value $1/50$ are $[-]/[cm]$ (where $[-]$ signifies a dimensionless quantity). Do you know why? Look back at the expression for $\tan\theta$ and see how the 50 was factored out of the denominator.
- The units of the term dy/dt are $[cm]/[sec]$.
- The term $\cos\theta$ is dimensionless, because of its definition as a ratio of two lengths, $[cm]/[cm]$.

If we embed these units into the calculation for $d\theta/dt$,

$$\frac{d\theta}{dt} = \frac{1}{50}\frac{[-]}{[cm]}\frac{dy}{dt}\frac{[cm]}{[sec]}\left(\frac{5}{\sqrt{29}}\right)^2\frac{[-]}{[-]}$$

which means the units of $d\theta/dt$ are $[-]/[sec]$ — or, verbally just "per second". Shouldn't there be a unit in the numerator there? The rate of change of our angle should be in either degrees per second or radians per second. The natural units for trigonometric functions are radians, not degrees (degrees are phony). A radian is understood to be a "dimension substitute", or a "unit to be named later". So we can put it in where needed — and we'd report, finally, that:

$$\frac{d\theta}{dt} = \frac{3}{58} \text{ radians per second} \quad \blacksquare$$

You Try It

(2) A baseball player runs from home plate to first base at $12\,ft/sec$. How fast is his straight-line distance to second base changing when he is half-way from home to first? (In baseball, base-paths are 90 ft.)

Now let's see some examples in which the story does not start out as a purely geometric statement, but still eventually reduces to a geometric problem.

EX 3 An employee at a pet store did not completely turn off a hose being used to fill a rectangular aquarium. The cross sectional area of the tank is $3\,ft^2$, and its total height is $1.5\,ft$. If the hose is leaking water into the aquarium at a rate of $0.1\,ft^3/sec$, how fast is the height of water in the tank changing when the tank is half full?

This is a strange story, but it reduces to a geometry problem involving a rectangular box. The box, though, represents the water itself — the length and width of the water match that of the aquarium, but the height of the water is changing as the tank fills. So let's interpret the data given in the problem.

- "The hose is leaking water into the aquarium at a rate of $0.1\,ft^3/sec$" is giving us a rate of change of the volume of water in the tank, say, dV/dt.
- "...how fast is the height of water in the tank changing ... " means we are asking for the rate of change in height of water, say, $\frac{dh}{dt}$.

We must come up with a static relationship between the height h and volume V of a "box" of water. This can just be the regular old $V = xyh$, where x and y are the horizontal length and width of the tank, and h is the height of the water — and this changes with time. Note that we're not told x and y individually, we're told that their product (area) is a constant $xy = 3\,ft^2$. Altogether, we can write our static equation as $V = 3h$. Then, we can power up this relationship,

$$V = 3h$$
$$\frac{d}{dt}(V) = \frac{d}{dt}(3h)$$
$$\frac{dV}{dt} = 3\frac{dh}{dt}$$

We are told dV/dt is a constant $0.1\,ft^3/sec$, so solving for $\frac{dh}{dt}$ gives

$$\frac{dh}{dt} = \frac{1}{30}\frac{ft}{min}$$

Note that we never had to take a snapshot when the tank was half-full. The relation between dh/dt and dV/dt did not depend on h itself. This makes sense: since the cross sectional area of the tank is constant, a constant increase in volume will cause a constant increase in height. Also, note that the units of ft/min are not just invented at the end, those units arise from the calculations — although they are not shown here from step to step. Make sure you can follow the units through the problem! ■

You Try It

(3) Jimbo is having a party to celebrate passing Calculus I, and he has a $56\,cm$-wide cylindrical dispenser full of iced tea. Sadly, the iced tea is leaking out, causing the height of the liquid in the tank to drop at a rate of $4\,cm/min$. How fast is the volume of the iced tea changing when the height is $14\,cm$?

At this point, I am sure you are saying, "Enough of these stupid stories, let's talk about graphs!" I agree.

EX 4 An object is moving in the positive direction along the line $y = 3x$. The x coordinate of the object is increasing at 2 units per second. How fast is the distance of the object from the point $(1, 1)$ increasing when the object is at the point $(2, 6)$?

We are told that $dx/dt = 2$. If D is distance from the point at any location (x, y) to the point (1,1), then we are asked to find dD/dt when $x = 2$.

Since the object is always on $y = 3x$, we can refine its coordinates to $(x, y) = (x, 3x)$. The distance D from the object at any point $(x, 3x)$ to the point $(1, 1)$ is then given by:

$$D^2 = (x-1)^2 + (3x-1)^2$$

This is our "static geometry". Let's power it up:

$$\begin{aligned}
\frac{d}{dt} D^2 &= \frac{d}{dt}\left((x-1)^2 + (3x-1)^2\right) \\
2D\frac{dD}{dt} &= 2(x-1)\frac{dx}{dt} + 6(3x-1)\cdot\frac{dx}{dt} \\
&= (2x - 2 + 18x - 6)\frac{dx}{dt} \\
\frac{dD}{dt} &= \frac{10x-4}{D}\frac{dx}{dt}
\end{aligned}$$

We take our snapshot when $x = 2$; we know dx/dt is always 2, but what's D when $x = 2$? When the point is at $(2, 6)$, the distance from the object to $(1, 1)$ is

$$D = \sqrt{(2-1)^2 + (6-1)^2} = \sqrt{1+25} = \sqrt{26}$$

so when the object is at $(2, 6)$,

$$\frac{dD}{dt} = \frac{10(2)-4}{\sqrt{26}} \cdot 2 = \frac{32}{\sqrt{26}}$$

The distance from the object to $(1, 1)$ at the instant in question is increasing at $32/\sqrt{26}$ units per second. As usual, if you're offended by radicals in a denominator, go ahead and fix it. ■

You Try It

(4) An object is moving along the parabola $y = x^2$. The x coordinate of the object is increasing at 1 unit per second. How fast is the object's distance from the origin increasing when the object is at the point $(2, 4)$?

Have You Learned...

- How to relate rates of change with respect to time using the Chain Rule or implicit differentiation?
- How to identify which derivatives are being given in a problem statement?

- How to identify which derivatives are being asked for in a problem statement?
- How to "power up" a geometric formula by taking the derivative of both sides with respect to time?
- How to take a "snapshot" of a dynamic process to embed information about a particular instant, and solve for an unknown rate?

Related Rates — Problem List

Related Rates — You Try It

These appeared above; solutions begin on the next page.

(1) The Great Clock of Westminster, which houses the giant bell named Big Ben, is really large! The minute hand of this clock is 14 feet long. How fast (in ft/min) is the tip of the minute hand tracing out a circle at any time? (Hint: do you remember the equation $s = r\theta$?)
(2) A baseball player runs from home plate to first base at $12\,ft/sec$. How fast is his straight-line distance to second base changing when he is half-way from home to first? (In baseball, base-paths are 90 ft.)
(3) Jimbo is having a party to celebrate passing Calculus I, and he has a $56\,cm$-wide cylindrical dispenser full of iced tea. Sadly, the iced tea is leaking out, causing the height of the liquid in the tank to drop at a rate of $4\,cm/min$. How fast is the volume of the iced tea changing when the height is $14\,cm$?
(4) An object is moving along the parabola $y = x^2$. The x coordinate of the object is increasing at 1 unit per second. How fast is the object's distance from the origin increasing when the object is at the point $(2, 4)$?

Related Rates — Practice Problems

Try these as you get the hang of the You Try It problems. Solutions to these problems are available in Sec. A.5.1.

(1) The volume of a cone increases at $5\,m^3/min$. Assuming the height of the cone stays a constant $10\,m$, how fast is the radius of the cone increasing when the volume of the cone is $500\,m^3$?
(2) A conspiracy theorist trying to get to the bottom of chemtrails is filming a rocket launch with a cell phone. He sits $4000\,ft$ from the launch pad. At launch, the rocket rises vertically at $600\,ft/sec$. How fast is the distance from this spectator to the rocket changing at the instant the rocket is $3000\,ft$ up?
(3) In 2018, golfer Kelly Kraft's tee shot on the 14th hole of a PGA tournament course hit a bird; the deflected shot caused him to miss out on advancing to the next round. Let's say that Kelly's tee is at the origin, and the trajectory of his ball follows the curve $y = -x^2/5 + 2x$. This clearly does not reflect real world distance units. In these new mystery

units, the ground speed of the tee shot is $dx/dt = 2$, and the bird was hovering at the point $(7.5, 3.75)$. What was the rate of change of the straight line distance D from the ball to the bird when the ball was at the peak of its trajectory?

Related Rates — Challenge Problems

Try these problems to test your skills with the ideas in this section. Solutions to these problems are available in Sec. B.5.1.

(1) If the length of one of the sides a pentagon is decreasing at $10\,cm/sec$, how fast is the area of the pentagon changing when the five sides of the pentagon are each $50\,cm$ long? Hint: The area of a pentagon is given by

$$A = \frac{5}{4}\cot\left(\frac{\pi}{5}\right)x^2$$

where x is the length of any one of the five equal sides.

(2) The surface area of a melting snowball decreases at a rate of $1\,cm^2/min$. Find the rate at which the diameter of the snowball is decreasing at the instant the diameter is $10\,cm$.

(3) An object is moving counterclockwise along the unit circle $x^2 + y^2 = 1$. The x coordinate of the object is changing at 1 unit per second. How fast is the object's distance from the point $(1, 0)$ changing when the object is at the point $(0, 1)$?

Related Rates — You Try It — Solved

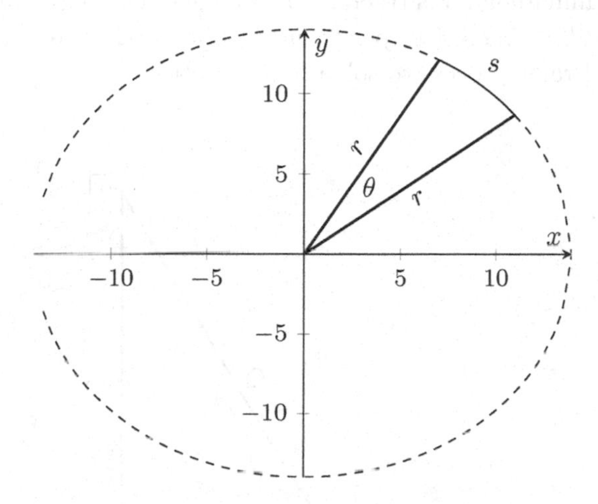

Fig. 5.2 Arc length, angle, and radius.

---

(1) The Great Clock of Westminster, which houses the giant bell named Big Ben, is really large! The minute hand of this clock is 14 feet long. How fast (in ft/min) is the tip of the minute hand tracing out a circle at any time? (Hint: do you remember the equation $s = r\theta$?)

☐ The static geometry relates to the piece of circumference swept out by the tip of the minute hand in a given amount of time. You may recall the circular geometry formula $s = r\theta$: for a given radius r and angular window θ, s represents the arc length of the piece of circular circumference inside the angular window — see Fig. 5.2. When we power this up, we get (as long as r is constant, as it is for a minute hand of a clock),

$$\frac{ds}{dt} = r\,\frac{d\theta}{dt}$$

There are 60 minutes in an hour, and the minute hand sweeps out a complete circle in that hour. Therefore, the minute hand sweeps out

$$\frac{d\theta}{dt} = \frac{2\pi}{60} = \frac{\pi}{30}$$

radians per minute. We don't really need a snapshot because we'd get the same results at any instant:

$$\frac{ds}{dt} = (14)\left(\frac{\pi}{30}\right) = \frac{7\pi}{15}$$

The units of r are feet and the units of $d\theta/dt$ are rad/min, and so the tip of the minute hand is tracing out a large circle's circumference at a rate of $ds/dt = 7\pi/15$ feet per minute. FFT: Did we really need the related rates process to solve this? ■

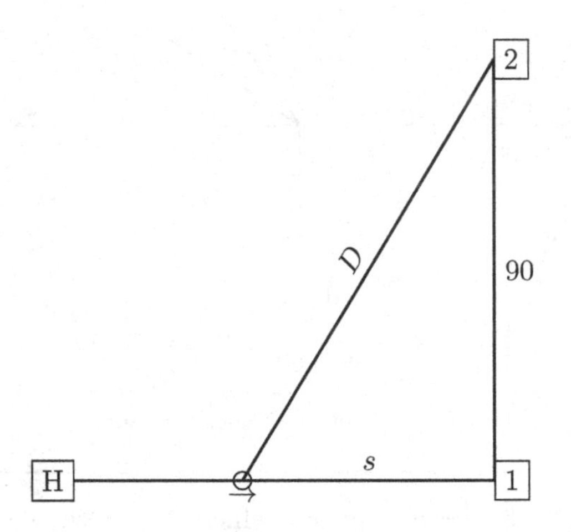

Fig. 5.3 Home plate, first base, and second base.

---

(2) A baseball player runs from home plate to first base at $12\,ft/sec$. How fast is his straight-line distance to second base changing when he is half-way from home to first? (In baseball, base-paths are 90 ft.)

□ We'll make a triangle out of the runner, first base, and second base — see Fig. 5.3. The parameters we have are:

- Distance of the runner from first base, variable s
- Distance from first to second base, constant 90 feet
- Distance of the runner to second base, variable D

Note that there is a right angle at first base. We are given the rate of change $ds/dt = -12\,ft/sec$ (do you know why it is negative?); we are hoping to find dD/dt.

We need a static relationship between D and s. For this simple right triangle, the Pythagorean Theorem will do: $D^2 = s^2 + 90^2$. Let's power it up!

$$\begin{aligned} \frac{d}{dt}(D^2) &= \frac{d}{dt}(s^2 + 90^2) \\ 2D\frac{dD}{dt} &= 2s\frac{ds}{dt} + 0 \\ \frac{dD}{dt} &= \frac{s}{D}\frac{ds}{dt} \end{aligned}$$

We take our snapshot at the instant the runner is half-way to first base, i.e. when $s = 45$ ft. At that instant, we can use the (static) Pythagorean Theorem to find that $D^2 = (45)^2 + (90)^2$, or $D = 45\sqrt{5}$. So now we have

$$\frac{dD}{dt} = \frac{s}{D}\frac{ds}{dt} = \frac{45}{45\sqrt{5}}(-12) = -\frac{12}{\sqrt{5}}$$

The distance to second is decreasing at a rate of $dD/dt = -12/\sqrt{5}$ feet per second. ■

---

(3) Jimbo is having a party to celebrate passing Calculus I, and he has a 56 *cm*-wide cylindrical dispenser full of iced tea. Sadly, the iced tea is leaking out, causing the height of the liquid in the tank to drop at a rate of 4 *cm/min*. How fast is the volume of the iced tea changing when the height is 14 *cm*?

□ We name the height h, the radius r, and the volume V. We are told that $dh/dt = -4\, cm/min$ (it's negative since the height is *decreasing*), and we want to compute dV/dt at the instant when $h = 14\, cm$. (We expect that dV/dt will have units of cm^3/min.) Our static geometry is established with usual equation for the volume of a cylinder, $V = \pi r^2 h$. The value of the radius is constant throughout, and so can be plugged in right away to give $V = \pi(28)^2 h$. Next, we power this up using a derivative with respect to time:

$$\begin{aligned} \frac{d}{dt}V &= \frac{d}{dt}(\pi(28)^2 h) \\ \frac{dV}{dt} &= 784\pi\,\frac{dh}{dt} \end{aligned}$$

With the known value of $dh/dt = -4$, then,

$$\frac{dV}{dt} = 784\pi(-4) = -3136\pi$$

The volume is decreasing at $-3136\pi\, cm^3/min$. While units are not shown throughout the calculations so as to not clutter things up, we do not just plop those units on at the end. Rather, we can confirm they are properly generated with our related rates relationship (units are embedded in brackets):

$$\frac{d}{dt}V = \frac{d}{dt}\left[\frac{-}{sec}\right](\pi(28[cm])^2 h[cm]) = 784\pi\,\frac{dh}{dt}\left[\frac{cm^3}{sec}\right] \blacksquare$$

---

(4) An object is moving along the parabola $y = x^2$. The x coordinate of the object is increasing at 1 unit per second. How fast is the object's distance from the origin increasing when the object is at the point $(2, 4)$?

□ The object will always be at points described by $(x, y) = (x, x^2)$. The distance from any point in the universe to the origin is

$$D^2 = (x-0)^2 + (y-0)^2$$

The distance from any point specifically on the curve $y = x^2$ is: Thus, the distance D to the origin from any point on our parabola $y = x^2$ is

$$D^2 = (x-0)^2 + (x^2-0)^2 = x^2 + x^4$$

This is our static equation. In the problem statement, we are given $dx/dt = 1$ and asked to find the corresponding dD/dt at the instant $x = 2$. We can power up our static relationship to introduce these rates of change. (If you'd like to, you can solve first for D — but I don't recommend it. Do you see why?)

$$\begin{aligned}\frac{d}{dt}D^2 &= \frac{d}{dt}\left(x^2 + x^4\right)\\ 2D\frac{dD}{dt} &= 2x\,\frac{dx}{dt} + 4x^3\frac{dx}{dt}\\ \frac{dD}{dt} &= \frac{2x+4x^3}{2D}\,\frac{dx}{dt}\\ \frac{dD}{dt} &= \frac{x+2x^3}{D}\,\frac{dx}{dt}\end{aligned}$$

Our snapshot is at the instant when the point is at $(2, 4)$. While we are given $dx/dt = 1$, we need to find the value of D at this instant. When the point is at $(2, 4)$, the distance from the object to $(0, 0)$ is

$$D = \sqrt{(2-0)^2 + (4-0)^2} = \sqrt{20} = 2\sqrt{5}$$

So when the object is at $(2, 4)$,

$$\frac{dD}{dt} = \frac{2+2(2)^3}{2\sqrt{5}} \cdot 1 = \frac{9}{\sqrt{5}}$$

The distance from the object to the origin at the instant in question is increasing at $9/\sqrt{5}$ units per second. ■

5.2 Derivatives and Graphs

Introduction

Here is where we start to put together some of the patterns you may have noticed while using derivatives to consider slopes of tangent lines.

Critical Points and Inflection Points

One of the most important things we can do with derivatives is discover where a function takes on an *extreme*, i.e. a maximum or minimum. Based on our experience so far, we can expect that if a function $f(x)$ attains a maximum or minimum at some point $x = c$, then the derivative of $f(x)$ should be zero at this point. So, here's a definition:

Definition 5.1. *A* ***critical point*** *of a function $f(x)$ is a point $x = c$ where $f'(c) = 0$ or is undefined.*

Note how this definition is interpreted. A maximum point or minimum point of a function $f(x)$ is a critical point, but not all critical points are maximums or minimums. For example, on the graph of $y = x^3$, we find a critical point at the origin, but the origin is not a max or min of the function. We will call this kind of critical point a *plateau*. When we find all of the critical points of a function, we have found a comprehensive list of all the points which *could be* maximums or minimums of the function. Similarly, we have

Definition 5.2. *An* ***inflection point*** *of a function $f(x)$ is a point $x = c$ where $f''(c) = 0$ or is undefined.*

EX 1 Find the critical points and inflection points of $f(x) = 4x^3 + 15x^2 - 18x$.

We find critical points where $f'(x) = 0$. Since

$$f'(x) = 12x^2 + 30x - 18 = 6(x+3)(2x-1)$$

we have $f'(x) = 0$ at $x = -3$ and $x = 1/2$. These are the critical points. We find inflection points where $f''(x) = 0$. Since $f''(x) = 24x + 30$, we have $f''(x) = 0$ at $x = -6/5$. This is the inflection point. (Note that neither $f'(x)$ nor $f''(x)$ is undefined at any point.) ■

EX 2 Find the critical points and inflection points of $f(x) = \sqrt{x^2 - 1}$.

We find critical points where $f'(x) = 0$. Since $f'(x) = x/\sqrt{x^2-1}$ (details omitted, since by now doing derivatives should be old news) we have $f'(x) = 0$ at $x = 0$ and $f'(x)$ undefined at $x = -1, 1$. So these two points are the critical points. We find inflection points where $f''(x) = 0$. Since $f''(x) = -(x^2-1)^{-3/2}$ (details omitted) we have that $f''(x)$ is never zero, but $f''(x)$ undefined at $x = -1, 1$. So both of the critical points are also inflection points. ■

You Try It

(1) Find the critical points and inflection points of $f(x) = x^3 + 3x^2 - 24x$.
(2) Find the critical points and inflection points of $f(t) = 5t^{2/3} + t^{5/3}$.

Graphical Analysis

We can extract a lot information about the (graphical) behavior of a function $f(x)$ from its first and second derivatives. For example, knowing that the first derivative represents slopes of tangent lines along $f(x)$, we have:

Useful Fact 5.1.

- *If $f(x)$ is increasing, $f'(x)$ is positive*
- *If $f(x)$ is decreasing, $f'(x)$ is negative*

It fits into this scheme that critical points of $f(x)$ are where functions could change from increasing to decreasing, or vice versa (note: *could* change, not *must* change).

We can make this more general by recognizing that $f'(x)$ is a perfectly good function itself, and so in a zone where the first derivative of $f(x)$ is increasing, then we would expect the *derivative of the derivative*, i.e. the second derivative, to be positive; similarly, where the first derivative is decreasing, the second derivative must be negative. These graphical "zones" have special meaning in terms of the shape of the graph of $f(x)$:

Definition 5.3. *As long as the first and second derivatives of $f(x)$ exist, then on an interval where $f'(x)$ is increasing, we will have $f''(x) > 0$; the graph of $f(x)$ displays a shape called* **concave up**; *on an interval where*

$f'(x)$ is decreasing, we will have $f''(x) < 0$; the graph of $f(x)$ displys a shape called ***concave down****.*

This fits nicely with Def. 5.2, as the locations where the sign of $f''(x) = 0$ are called inflection points. These are often where the graph's shape changes from concave up to concave down or vice versa — although that isn't completely necessary — we could see $f''(x) = 0$ at a location where the sign of $f''(x)$ is the same on both sides.

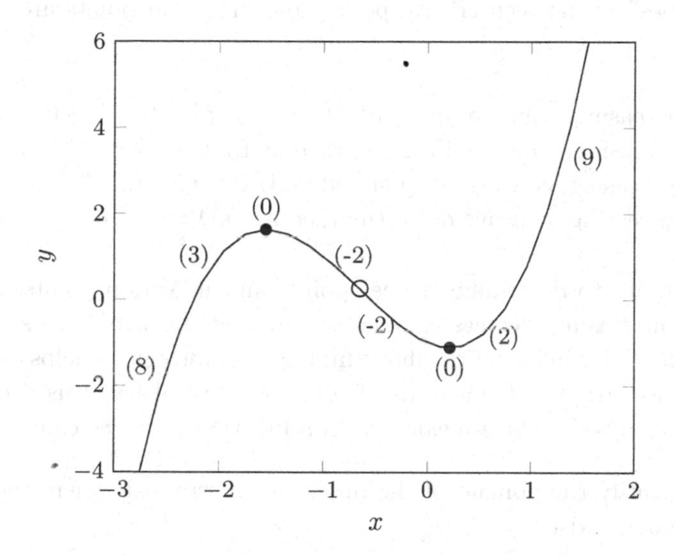

Fig. 5.4 The effect of magnitude and sign of f'.

There are four possible combinations of sign of $f'(x)$ and $f''(x)$: positive / positive, positive / negative, negative / positive, and negative / negative. Figure 5.4 shows a function which displays behaviors that arise from these combinations. In the figure, numbers in parentheses at various locations display the slope of a tangent line near that location. On the left half of the curve, you can see f' decreasing through 8, 3, 0, -2, and the curve has a concave down shape in this zone. On the right half of the curve, you can see f' increasing through -2, 0, 2, 9, and the curve has a concave up shape in this zone. Do not confuse values of f' being positive or negative with zones of increase or decrease; zones of increase and decrease can include both signs. The key behaviors are:

- Where $f'(x)$ is increasing, and so $f''(x)$ is positive, we expect the graph to be concave up.
- Where $f'(x)$ is decreasing, and so $f''(x)$ is negative, we expect the graph to be concave down.

An inflection point, where $f'' = 0$, is marked with an open circle; this is where the graph changes from concave down on the left to concave up on the right. Critical points, where $f' = 0$, are maked with solid circles; in this case, these coincide with maximums and minimums of the function. The "shapes" in between critical points and inflection points are labelled as follows:

- increasing, concave up (marked as I,U): $f' > 0$, $f'' > 0$
- increasing, concave down (marked as I,D): $f' > 0$, $f'' < 0$
- decreasing, concave up (marked as D,U): $f' < 0$, $f'' > 0$
- decreasing, concave down (marked as D,D): $f' < 0$, $f'' < 0$

The ability to determine critical points and inflection points, and intervals of increasing, decreasing, concave up, and concave down shapes — along with earlier information about finding asymptotes — helps us create very detailed graphs of functions of all types. We'll call this "complete graphical analysis", which means, given a function $f(x)$, we can:

(1) Identify the domain of the function, to establish where the graph should exist.
(2) Identify any horizontal and vertical asymptotes, to form a framework for the graph.
(3) Find all critical points and inflection points, and their coordinates.
(4) Break the domain into zones using the critical points and inflection points; in each zone, determine the signs of $f'(x)$ and $f''(x)$.
(5) Use the signs of $f'(x)$ and $f''(x)$ to determine where $f(x)$ is increasing, decreasing, concave up, and concave down.
(6) Sketch the graph by plotting the critical and inflection points, and joining them with pieces of $f(x)$ in the proper shapes.

EX 3 Perform a complete graphical analysis of $f(x) = x^4 + 4x^3$ and sketch its graph.

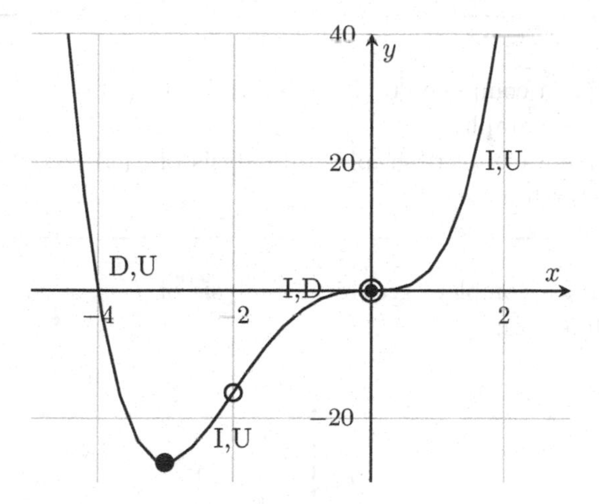

Fig. 5.5 Graphical analysis of $f(x) = x^4 + 4x^3$.

$f(x)$ has no asymptotes and its domain is all real numbers. To find critical and inflection points, we have

$$f'(x) = 4x^3 + 12x^2 = 4x^2(x+3) \; ; \; f'(x) = 0 \text{ at } x = -3, 0$$
$$f''(x) = 12x^2 + 24x = 12x(x+2) \; ; \; f''(x) = 0 \text{ at } x = -2, 0$$

The function's values at these critical points and inflection points are:

x	-3	-2	0
$f(x)$	-27	-16	0
type	CP	IP	CP/IP

Intervals set up by these critical and inflection points, the signs of the derivatives in each, and the resulting shape are:

int	$(-\infty, -3)$	$(-3, -2)$	$(-2, 0)$	$(0, \infty)$
$f'(x)$	$-$	$+$	$+$	$+$
$f''(x)$	$+$	$+$	$-$	$+$
shape	dec,ccu	inc,ccu	inc,ccd	inc,ccu

Note that the critical point $x = 3$ is a minimum, and the critical point at $x = 0$ is a plateau. The graph which reflects all this information is shown in Fig. 5.5. As is true for all graphs in this section, critical points are marked with solid circles, inflection points are marked with open circles, and any points which happen to be both will be marked with both. ■

You Try It

(3) Provide a complete graphical analysis of $f(x) = x^4 - 2x^2 + 3$ and sketch its graph.

(4) Provide a complete graphical analysis of $f(x) = 2x^3 - 3x^2 - 12x$ and sketch its graph.

EX 4 Perform a complete graphical analysis of $f(x) = x/(x^2+9)$ and sketch its graph.

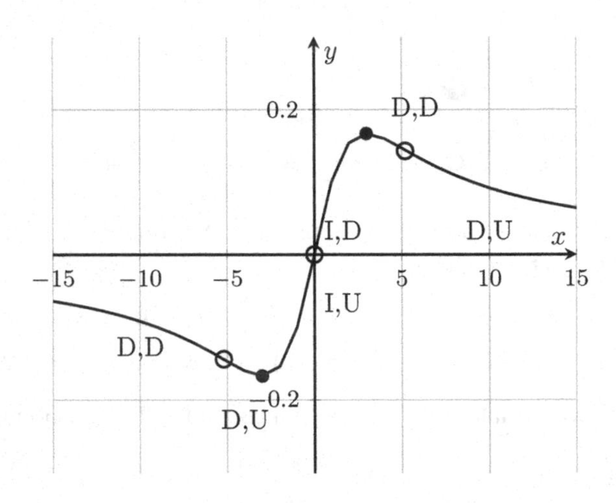

Fig. 5.6 Graphical analysis of $f(x) = x/(x^2+9)$.

There are no vertical asymptotes; the x-axis $y = 0$ is a horizontal asymptote. The domain is all real numbers. To find critical points and inflection points we have (details of derivatives omitted)

$$f'(x) = \frac{9 - x^2}{(x^2+9)^2}\ ;\ f'(x) = 0 \text{ at } x = -3, 3$$

$$f''(x) = \frac{2x(x^2 - 27)}{(x^2+9)^3}\ ;\ f''(x) = 0 \text{ at } x = 0, \pm 3\sqrt{3}$$

The function's values at these critical points and inflection points are:

x	$-3\sqrt{3}$	-3	0	3	$3\sqrt{3}$
$f(x)$	$-\sqrt{3}/12$	$-1/6$	0	$1/6$	$\sqrt{3}12$
type	IP	CP	IP	CP	IP

Intervals set up by these critical and inflection points, the signs of the derivatives in each, and the resulting shape are:

int	$(-\infty, -3\sqrt{3})$	$(-3\sqrt{3}, -3)$	$(-3, 0)$	$(0, 3)$	$(3, 3\sqrt{3})$	$(3\sqrt{3}, \infty)$
$f'(x)$	$-$	$-$	$+$	$+$	$-$	$-$
$f''(x)$	$-$	$+$	$+$	$-$	$-$	$+$
shape	dec,ccd	dec,ccu	inc,ccu	inc,ccd	dec,ccd	dec,ccu

The critical points at $x = -3$ and $x = 3$ are a minimum and maximum, respectively. Figure 5.6 shows the graph of this function with critical points and inflection points marked as usual. The shapes in between these marked points will match the information generated for this example. ■

You Try It

(5) Provide a complete graphical analysis of $f(x) = x(x+2)^3$ and sketch its graph.
(6) Provide a complete graphical analysis of $f(x) = x^2/(x^2+9)$ and sketch its graph.

EX 5 Perform a complete graphical analysis of $f(x) = x \ln x$ and sketch its graph.

There are no asymptotes and the domain is $x > 0$. For critical points and inflection points, we have

$$f'(x) = \ln x + 1 \; ; \; f'(x) = 0 \text{ at } x = 1/e$$
$$f''(x) = 1/x \; ; \; f''(x) \neq 0$$

The function's value at the critical point is $f(1/e) = -1/e$, and the coordinates of the critical point are $(1/e, -1/e)$. There are no inflection points.

Intervals set up by this critical point, the signs of the derivatives in each, and the resulting shape are:

int	$(0, 1/e)$	$(1/e, \infty)$
$f'(x)$	$-$	$+$
$f''(x)$	$+$	$+$
shape	dec,ccu	inc,ccu

Figure 5.7 shows the graph of this function. ■

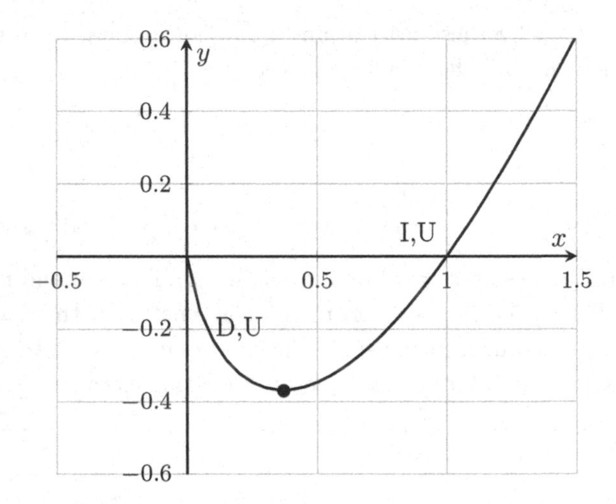

Fig. 5.7 Graphical analysis of $f(x) = x\ln(x)$.

You Try It

(7) Provide a complete graphical analysis of $f(x) = x\sqrt{x+3}$ and sketch its graph.

(8) Provide a complete graphical analysis of $f(x) = e^x/x$ and sketch its graph.

Have You Learned...

- How to identify critical points and inflection points of a function?
- How to identify where a function is increasing or decreasing?
- How to identify where a function is concave up and concave down?
- How combinations of increasing, decreasing, concave up, and concave down determine the shape of a function on any interval?
- How to determine which critical points are maximums or minimums of a function?

Derivatives and Graphs — Problem List

Derivatives and Graphs — You Try It

These appeared above; solutions begin on the next page.

(1) Find the critical points and inflection points of $f(x) = x^3 + 3x^2 - 24x$.
(2) Find the critical points and inflection points of $f(t) = 5t^{2/3} + t^{5/3}$
(3) Provide a complete graphical analysis of $f(x) = x^4 - 2x^2 + 3$ and sketch its graph.
(4) Provide a complete graphical analysis of $f(x) = 2x^3 - 3x^2 - 12x$ and sketch its graph.
(5) Provide a complete graphical analysis of $f(x) = x(x+2)^3$ and sketch its graph.
(6) Provide a complete graphical analysis of $f(x) = x^2/(x^2+9)$ and sketch its graph.
(7) Provide a complete graphical analysis of $f(x) = x\sqrt{x+3}$ and sketch its graph.
(8) Provide a complete graphical analysis of $f(x) = e^x/x$ and sketch its graph.

Derivatives and Graphs — Practice Problems

Try these as you get the hang of the You Try It problems. Solutions to these problems are available in Sec. A.5.2.

(1) Find the critical points and inflection points of $f(x) = x^3 + x^2 + x$.
(2) Find the critical points and inflection points of $f(t) = \sqrt{t}(1-t)$.
(3) Find the critical points and inflection points of $f(x) = \tanh(x)$ and confirm that these results are consistent with the graph you saw in Practice Problem 1 of Sec. 1.6.
(4) Find the critical points of $f(x) = xe^{2x}$.
(5) Provide a complete graphical analysis of $f(x) = 2 + 3x - x^3$ and sketch the graph.
(6) Provide a complete graphical analysis of $f(x) = x - 2\sin x$ on $(0, 3\pi)$ and sketch the graph of $f(x)$ on that interval.
(7) Provide a complete graphical analysis of $f(x) = x^2e^x$ and sketch its graph.
(8) Provide a complete graphical analysis of $B(x) = 3x^{2/3} - x$ and sketch its graph. (Hint: if you use tech to view the function, are you sure

you're seeing the whole thing? Be sure you've looked at the graph for You Try It 2 in this section.)

(9) Provide a complete graphical analysis of $f(x) = x/(x^2 - 9)$ and sketch its graph.

(10) Provide a complete graphical analysis of $f(x) = e^{2x} - e^x$ and sketch its graph.

Derivatives and Graphs — Challenge Problems

Try these problems to test your skills with the ideas in this section. Solutions to these problems are available in Sec. B.5.2.

(1) Provide a complete graphical analysis of $f(x) = \cos^2 x - 2\sin x$ on $(0, 2\pi)$ and sketch the graph of $f(x)$ on that interval.

(2) Provide a complete graphical analysis of $f(x) = (x^2 - 1)^3$ and sketch its graph.

(3) Provide a complete graphical analysis of $f(x) = x^2/(x-2)^2$ and sketch its graph.

Derivatives and Graphs — You Try It — Solved

(1) Find the critical points and inflection points of $f(x) = x^3 + 3x^2 - 24x$.

☐ We find critical points where $f'(x) = 0$, i.e. where

$$\begin{aligned} 3x^2 + 6x - 24 &= 0 \\ x^2 + 2x - 8 &= 0 \\ (x+4)(x-2) &= 0 \\ x = -4 \;,\; x &= 2 \end{aligned}$$

We find inflection points where $f''(x) = 0$, i.e. where $6x + 6 = 0$, or $x = -1$.

The critical points are $x = -4$ and $x = 2$; the inflection point is $x = -1$. Figure 5.8 shows a graph of this function with critical points (solid circles) and inflection points (open circles) marked. ■

(2) Find the critical points and inflection points of $f(t) = 5t^{2/3} + t^{5/3}$.

☐ We find critical points where $f'(t) = 0$, i.e. where

$$\begin{aligned} \frac{10}{3}t^{-1/3} + \frac{5}{3}t^{2/3} &= 0 \\ \frac{10}{3t^{1/3}} + \frac{5}{3}t^{2/3} &= 0 \\ \frac{2}{t^{1/3}} + t^{2/3} &= 0 \\ 2 + t &= 0 \end{aligned}$$

So $t = -2$ is a critical point. Note also, though, that $f'(t)$ is not defined for $t = 0$, so this is also a critical point. We find critical points where $f''(t) = 0$, i.e. where

$$\begin{aligned} -\frac{10}{9}t^{-4/3} + \frac{10}{9}t^{-1/3} &= 0 \\ -\frac{10}{9t^{4/3}} + \frac{10}{9t^{1/3}} &= 0 \\ -\frac{10t^{4/3}}{9t^{4/3}} + \frac{10t^{4/3}}{9t^{1/3}} &= 0 \\ -\frac{10}{9} + \frac{10t}{9} &= 0 \end{aligned}$$

From this, we can see that $f''(t) = 0$ at $t = 1$, and $f''(t)$ is undefined at $t = 0$. So these are the inflection points; $t = 0$ is both a critical point and an inflection point. Figure 5.9 shows a graph of this function with any critical points and inflection points marked. ■

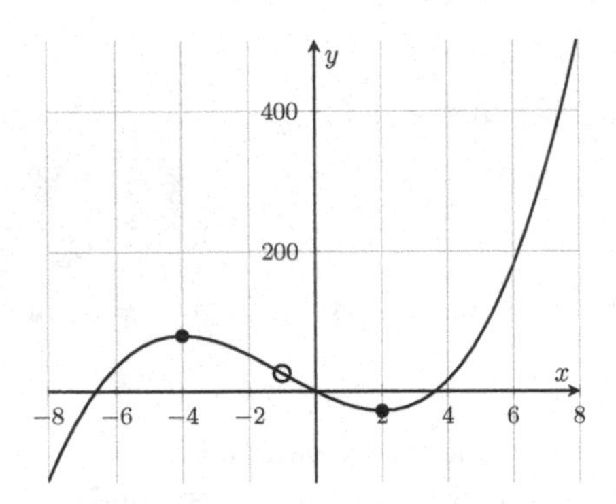

Fig. 5.8 Graphical analysis of $f(x) = x^3 + 3x^2 - 24x$.

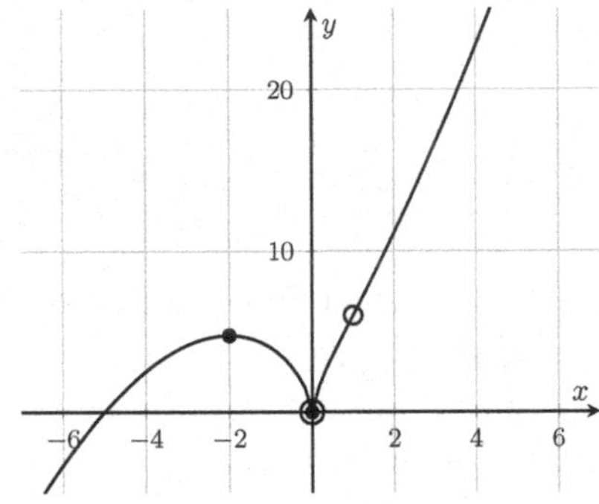

Fig. 5.9 Graphical analysis of $f(t) = 5t^{2/3} + t^{5/3}$.

---

(3) Provide a complete graphical analysis of $f(x) = x^4 - 2x^2 + 3$ and sketch its graph.

□ There are no asymptotes and the domain is all real numbers. To find critical points and inflection points, we have

$$f'(x) = 4x^3 - 4x = 4x(x^2 - 1)\ ;\ f'(x) = 0 \text{ at } x = 0, x = \pm 1$$

$$f''(x) = 12x^2 - 4 = 4(3x^2 - 1)\ ;\ f''(x) = 0 \text{ at } x = \pm 1/\sqrt{3}$$

The function's values at these critical points and inflection points, and resulting categorizations, are:

x	-1	$-1/\sqrt{3}$	0	$1/\sqrt{3}$	1
$f(x)$	2	22/9	3	22/9	2
type	CP	IP	CP	IP	CP
ext	min		max		min

Intervals set up by these critical points and inflection points are, in number-line ordering,

$$I_1 : (-\infty, -1) \quad I_2 : (-1, -1/\sqrt{3}) \quad I_3 : (-1/\sqrt{3}, 0)$$
$$I_4 : (0, 1/\sqrt{3}) \quad I_5 : (1/\sqrt{3}, 1) \quad I_6 : (1, \infty)$$

The signs of the derivatives in each, and the resulting shape are:

int	I_1	I_2	I_3	I_4	I_5	I_6
$f'(x)$	−	+	+	−	−	+
$f''(x)$	+	+	−	−	+	+
shape	dec,ccu	inc,ccu	inc,ccd	dec,ccd	dec,ccu	inc,ccu

Figure 5.10 shows a graph of this function with any critical points and inflection points marked. ■

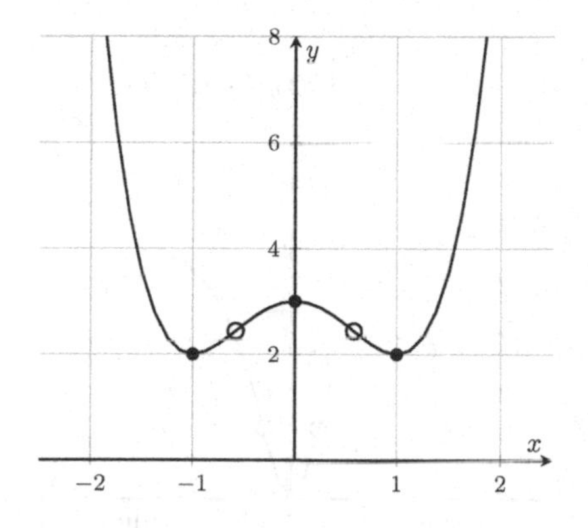

Fig. 5.10 Graphical analysis of $f(x) = x^4 - 2x^2 + 3$.

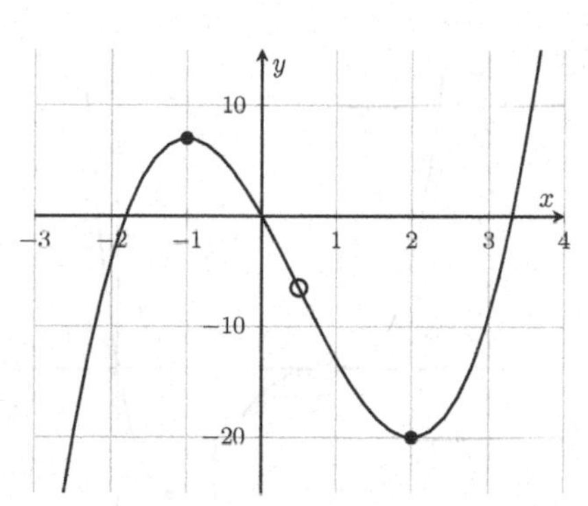

Fig. 5.11 Graphical analysis of $f(x) = 2x^3 - 3x^2 - 12x$.

(4) Provide a complete graphical analysis of $f(x) = 2x^3 - 3x^2 - 12x$ and sketch its graph.

□ There are no asymptotes and the domain is all real numbers. To find critical points and inflection points, we have:

$$f'(x) = 6x^2 - 6x - 12 = 6(x-2)(x+1) \; ; \; f'(x) = 0 \text{ at } x = -1, 2$$
$$f''(x) = 12x - 6 = 6(2x-1) \; ; \; f''(x) = 0 \text{ at } x = 1/2$$

The function's values at these critical points and inflection points, and resulting categorizations, are:

x	-1	$1/2$	2
$f(x)$	7	$-13/2$	-20
type	CP	IP	CP
ext	max		min

Intervals set up by these critical points and inflection points, the signs of the derivatives in each, and the resulting shape are:

int	$(-\infty,-1)$	$(-1,1/2)$	$(1/2,2)$	$(2,\infty)$
$f'(x)$	$+$	$-$	$-$	$+$
$f''(x)$	$-$	$-$	$+$	$+$
shape	inc,ccd	dec,ccd	dec,ccu	inc,ccu

Figure 5.11 shows a graph of this function with any critical points and inflection points marked. ■

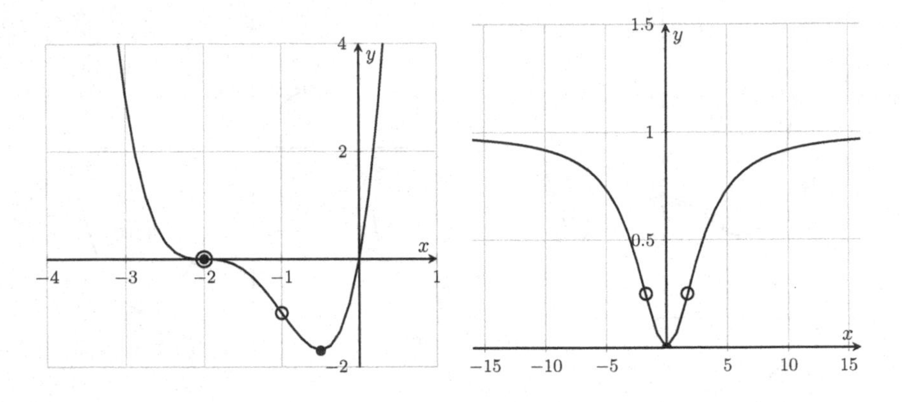

Fig. 5.12 Graphical analysis of $f(x) = x(x+2)^3$.

Fig. 5.13 Graphical analysis of $f(x) = x^2/(x^2+9)$.

(5) Provide a complete graphical analysis of $f(x) = x(x+2)^3$ and sketch its graph.

□ There are no asymptotes and the domain is all real numbers. We have (simplified)

$$f'(x) = 2(x+2)^2(2x+1)\ ;\ f'(x) = 0 \text{ at } x = -2, -1/2$$

$$f''(x) = 12(x+1)(x+2)\ ;\ f''(x) = 0 \text{ at } x = -2, -1$$

The function's values at these critical points and inflection points are:

x	-2	-1	$-1/2$
$f(x)$	0	-1	$-27/16$
type	CP/IP	IP	CP
ext	neither		min

Intervals set up by these critical and inflection points, the signs of the derivatives in each, and the resulting shape are:

int	$(-\infty,-2)$	$(-2,-1)$	$(-1,-1/2)$	$(-1/2,\infty)$
$f'(x)$	$-$	$-$	$-$	$+$
$f''(x)$	$+$	$-$	$+$	$+$
shape	dec,ccu	dec,ccd	dec,ccu	inc,ccu

Figure 5.12 shows a graph of this function with any critical points and inflection points marked. ■

(6) Provide a complete graphical analysis of $f(x) = x^2/(x^2+9)$ and sketch its graph.

□ There are no vertical asymptotes; there is a horizontal asymptote at $y = 1$. The domain is all real numbers. To find critical points and inflection points, we have (simplified)

$$f'(x) = \frac{18x}{(x^2+9)^2} \ ; \ f'(x) = 0 \text{ at } x = 0$$

$$f''(x) = \frac{54(3-x^2)}{(x^2+9)^3} \ ; \ f''(x) = 0 \text{ at } x = \pm\sqrt{3}$$

The function's values at these critical points and inflection points, and the resulting categorizations, are:

x	$-\sqrt{3}$	0	$\sqrt{3}$
$f(x)$	$1/4$	0	$1/4$
type	*IP*	*CP*	*IP*
ext		min	

Intervals set up by these critical and inflection points, the signs of the derivatives in each, and the resulting shape are:

int	$(-\infty,-\sqrt{3})$	$(-\sqrt{3},0)$	$(0,\sqrt{3})$	$(\sqrt{3},\infty)$
$f'(x)$	$-$	$-$	$+$	$+$
$f''(x)$	$-$	$+$	$+$	$-$
shape	dec,ccd	dec,ccu	inc,ccu	inc,ccd

Figure 5.13 shows a graph of this function with any critical points and inflection points marked. ■

(7) Provide a complete graphical analysis of $f(x) = x\sqrt{x+3}$ and sketch its graph.

□ There are no asymptotes and the domain is $x \geq -3$. To find critical and inflection points, we have

$$f'(x) = \frac{3(x+2)}{2\sqrt{x+3}} \; ; \; f'(x) = 0 \text{ at } x = -2$$

$$f''(x) = \frac{3(x+4)}{4(x+3)^{3/2}} \; ; \; f''(x) \neq 0 \text{ inside its domain}$$

The function's values at the critical points (there are no inflection points) is $(-2, -2)$.

Intervals set up by this critical point, the signs of the derivatives in each, and the resulting shape are:

int	$(-3, -2)$	$(-2, \infty)$
$f'(x)$	$-$	$+$
$f''(x)$	$+$	$+$
shape	dec,ccu	inc,ccu

Figure 5.14 shows a graph of this function with any critical points and inflection points marked. ■

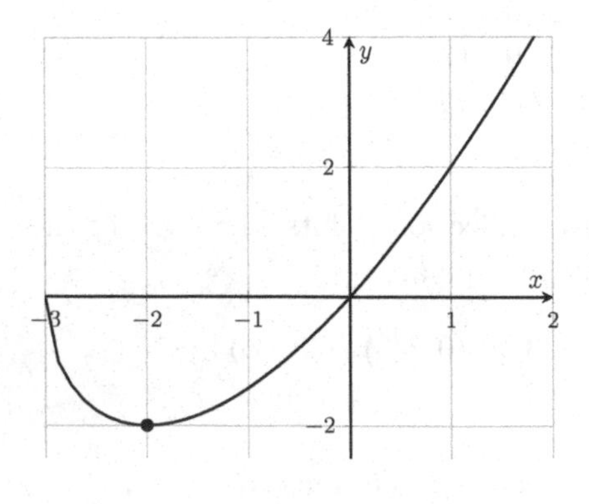

Fig. 5.14 Graphical analysis of $f(x) = x\sqrt{x+3}$.

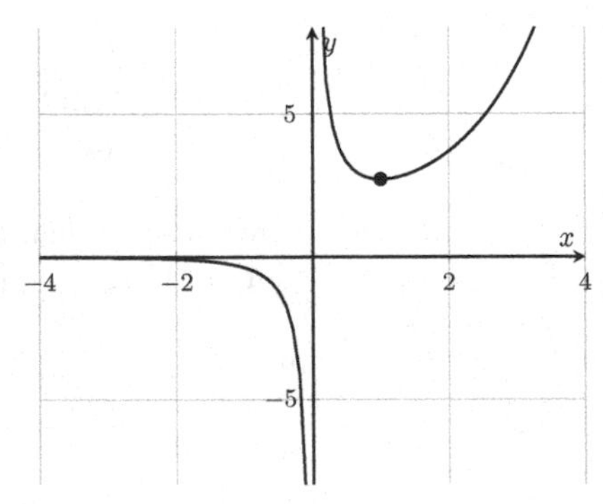

Fig. 5.15 Graphical analysis of $f(x) = e^x/x$.

(8) Provide a complete graphical analysis of $f(x) = e^x/x$ and sketch its graph.

□ There is a vertical asymptote at $x = 0$. The domain is $x \neq 0$. To find critical and inflection points, we have (simplified)

$$f'(x) = \frac{e^x(x-1)}{x^2}\ ;\ f'(x) = 0 \text{ at } x = 1$$

$$f''(x) = \frac{e^x(x^2-2x+2)}{x^3}\ ;\ f''(x) \neq 0$$

The function's value at the critical point is $(1, e)$.

Intervals set up by the asymptote and the critical point, the signs of the derivatives in each, and the resulting shape are:

int	$(-\infty, 0)$	$(0, 1)$	$(1, \infty)$
$f'(x)$	$-$	$-$	$+$
$f''(x)$	$-$	$+$	$+$
shape	dec,ccd	dec,ccu	inc,ccu

Figure 5.15 shows a graph of this function with any critical points and inflection points marked. ■

5.3 Optimization

Introduction

Optimization is the process of seeking an optimal scenario. Is that vague? You bet! Depending on the situation, it may be necessary to find where some quantity is the largest of feasible options. Or smallest. Or some other -est. If you run a business and have a profit function, you likely want to maximize it. If you have a cost function, you likely want to minimize it. Maybe you own a shoe company, and you need to make a box of a certain volume, but want to use the least amount of cardboard in doing so.

An optimization problem usually has two primary components: the "objective function" is a recipe for the quantity we're trying to maximize or minimize. "Constraints" are rules that apply to the ingredients of the objective function; these often come in the form of inequalities. Let's say we just adopted a new dog, who needs an outdoor space in which she can run around. Someone has given us a certain length of fencing we can use to enclose an open area. Some geometric formula for the total available area would be our objective function, which we want to maximize; our constraint is that the amount of fence we can use must be less than the amount we were given. There is no one magic strategy to setting up and solving optimization problems, but if you guessed these involve derivatives ... you're right!

Local vs Global Extremes

Suppose the function shown in Fig. 5.16 represents a hypothetical profit function for a business that can produce anywhere from 0 to 105 gizmos per day. This curve shows the net profit P as a function of the number x of gizmos produced. Let's say you work for this business, and your boss has asked for your advice. Rather than give you this plot, though, she gives you the actual profit function $P(x)$. Having taken Calculus, you're all excited to use your knowledge, so you do a critical point analysis of this function, and you (correctly) find a critical point at $x = 25$ which is a maximum. You then return to your boss and tell her that the company should produce 25 gizmos per day to maximize profit.

Well, guess what: you're fired! In reality, the business should make the maximum of 105 gizmos, since the biggest profit is attained there.

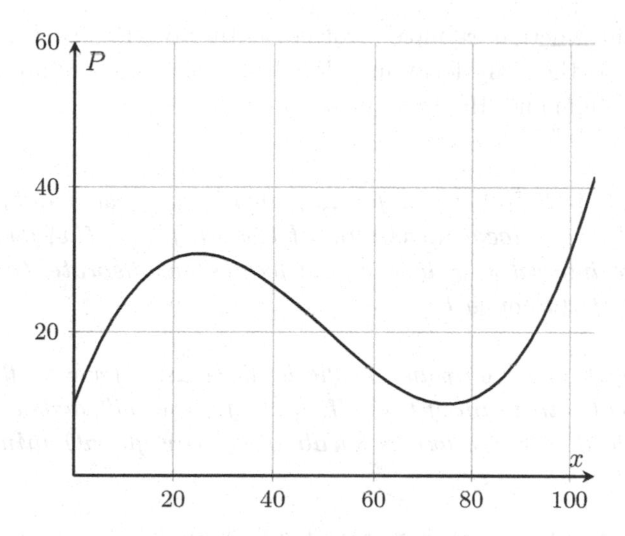

Fig. 5.16 A hypothetical profit function.

However, we would not discover this value of $x = 125$ as the true optimal value through an analysis of critical points since it isn't a critical point.

This is the difference between a local and a global extreme. The maximum at $x = 25$ is indeed a maximum relative to other nearby points, so it is a *local* maximum. But it is not the largest value the function attains on the entire interval we're considering. This latter value is called a *global* maximum. A function can have many local maximums on a given interval, but only one global maximum.[1]

Let's suppose you received extra information that, for the next month, the company would only be able to produce 90 gizmos per day. In that case, your original suggestion to make 25 gizmos per day is now the correct one. The profit function has remained the same, but the *constraint* has changed. More formally, we'd note that for this $P(x)$, there are different answers to the following prompts:

(1) Find the global maximum of $P(x)$ for $0 \leq x \leq 105$.
(2) Find the global maximum of $P(x)$ for $0 \leq x \leq 90$.

[1]Well, there can be a tie between multiple points that all hit the global maximum value.

The profit function example involved, naturally, the search for a maximum. But as there are local and global maximums, there are also local and global minimums. Here's a quick summary:

Definition 5.4. *If $f(c) \geq f(x)$ for all points in the local vicinity of $x = c$, then $x = c$ locates a* ***local maximum*** *of $f(x)$. If $f(c) \geq f(x)$ for all points in an entire interval $[a, b]$ then $x = c$ locates an* ***absolute (or global) maximum*** *of $f(x)$ on $[a, b]$.*

If $f(c) \leq f(x)$ for all points in the local vicinity of $x = c$, then $x = c$ locates a ***local minimum*** *of $f(x)$. If $f(c) \leq f(x)$ for all points in an entire interval $[a, b]$ then $x = c$ locates an* ***absolute (or global) minimum*** *of $f(x)$ on $[a, b]$.*

There has to be a better strategy for finding global extremes, though, than saying "plot your function and use your eyeballs to locate the maximum or minimum". Here is that better strategy, in the case where you're asked to determine the global extreme of an objective function over a given interval of the variable which provides the constraint:

(1) Find the critical points of the function within the given interval; compute the value of the function at each critical point.
(2) Compute the value of the function at each endpoint of the given interval.
(3) Search through the critical points and endpoints for the true global maximum and/or minimum.

EX 1 Find the global extremes of $f(x) = x^3/3 + x^2/2 - 6x + 8$ on $[-2, 4]$.

Since $f'(x) = x^2+x-6 = (x+3)(x-2)$, we have $f'(x) = 0$ at $x = -3, 2$. But $x = -3$ is not in the given interval $[-2, 4]$ so we toss it out. Candidates for the global extremes thus include the critical point $x = 2$ and the endpoints $x = -2$ and $x = 4$. Computing $f(x)$ at each candidate point, we get

$$f(-2) = \frac{58}{3} \quad , \quad f(2) = \frac{2}{3} \quad , \quad f(4) = \frac{40}{3}$$

On $[-2, 4]$, then, the global minimum is at the critical point $(2, 2/3)$ and the global maximum is at the endpoint $(-2, 58/3)$. ■

You Try It

(1) Find the global extremes of $f(x) = 3x^2 - 12x + 5$ on $[0, 3]$.
(2) Find the global extremes of $f(x) = x/(x^2 + 1)$ on $[0, 2]$.

Optimization

Moving beyond generic functional examples, it's much more interesting to do optimization in real contextual problems. In most of these examples, we are given a description of what we want to optimize, but will need to form our own function to use. Here is a general strategy:

- Identify all quantities involved and assign them variables
- Identify any known (given) values of these quantities
- Identifty the objective quantity: what needs to be optimized (maximized or minimized)
- Identify any constraints (usually some restriction of a variable to a certain interval)
- Design the objective function that links the objective quantity to the constraining variable(s)
- If your objective function has multiple independent variables, use ancillary information (such as a constraint) to eliminate all but one — this is usually the tricky part!
- Find the critical points of your resulting function; compare their values to those of the endpoints to locate the true global extreme(s)

Let's start with a somewhat dry example.

EX 2 Find two numbers whose difference is 100 and whose product is a minimum.

First of all, let's name our two numbers x and y. We want to minimize the product, which can be written $P = xy$. This is our objective function. Recognize that it has two independent variables (x and y), and so unless you've taken multivariable calculus already, the "derivative" of this is a mystery. However, we have ancillary information: the difference between x and y is 100, i.e. $x - y = 100$. This is our constraint. With the constraint $x - y = 100$, we can write $x = 100 + y$ and eliminate one of the two variables:

$$P = xy$$
$$P = (100 + y)y = 100y + y^2$$

The updated version of the objective function has only one independent variable. Critical points come from $dP/dy = 0$, i.e. where $100 + 2y = 0$, or at $y = -50$. There are no endpoints to check since y (and so also x) can be any real number. So the only point of interest is $y = -50$, and when $y = -50$, we have $x = 100 + y = 100 - 50 = 50$. The two numbers whose difference is 100 and whose product is a minimum are -50 and 50. Go on, test it out! You will not be able to find any other pair of numbers whose difference is 100 and whose product is smaller. (Note that "minimum" and "smaller" don't have to stop at zero, the product can certainly be negative.) FFT: How do we know we didn't find a *maximum* product? ■

You Try It

(3) Find two positive numbers whose product is 100 and whose sum is a minimum.

Here is a much more geometric example:

EX 3 A security company needs to close off a rectangular space of $1000\,m^2$. What dimensions of the area will minimize the amount of fence needed?

Let's name the dimensions of the rectangular area L (length) and W (width). Our *objective* is to minimize the total amount of fence needed to enclose this area — but the total amount needed is just the perimeter P of the rectangle, $P = 2L + 2W$. So, this is our objective function, and it currently has two independent variables. The constraint can be used to eliminate one of these two variables: since the total area enclosed must be exactly $1000\,m^2$, we can write $LW = 1000$, or $W = 1000/L$. Passing that to the objective function:

$$P = 2L + 2W = 2L + 2\left(\frac{1000}{L}\right) = 2L + \frac{2000}{L}$$

so then

$$v\frac{dP}{dL} = 2 - \frac{2000}{L^2}$$

Minimizing the perimeter P means finding where $dP/dL = 0$:

$$\begin{aligned} 0 &= 2 - \frac{2000}{L^2} \\ L^2 &= 1000 \\ L &= \sqrt{1000} = 10\sqrt{10} \end{aligned}$$

So to minimize the perimeter we need a length of $L = 10\sqrt{10}$; the width that goes with this comes from the original area equation:

$$
\begin{aligned}
LW &= 1000 \\
(10\sqrt{10})W &= 1000 \\
W &= \frac{1000}{10\sqrt{10}} = 10\sqrt{10}
\end{aligned}
$$

It turns out that the necessary length and width are equal! Thus, to enclose a given area with minimum perimeter, we need a square; in this case, $L = W = 10\sqrt{10}\,m$. FFT: How do we know we didn't find a maximum perimeter? ■

You Try It

(4) If 1200 cm^2 of material is available to make a box with a square base and open top, find the largest possible volume of the box.

EX 4 A race consists of two legs, one on land and one in water. The object is to get from Start to End in the least time; a contestant can jump from land into water at any location along the shoreline. Where is the best location?

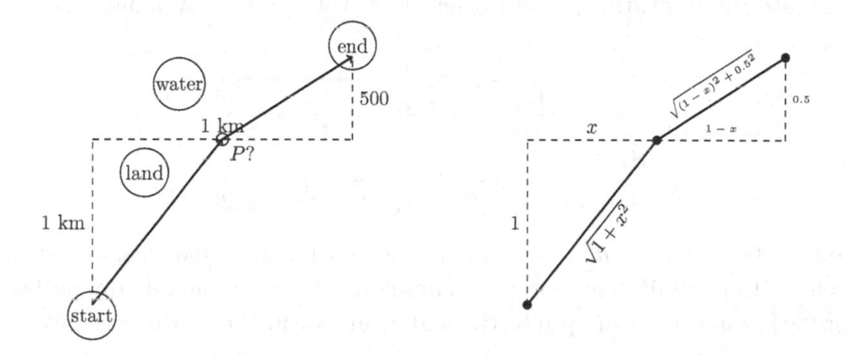

Fig. 5.17 A race over land and water.

Fig. 5.18 Labeling with variables.

Figure 5.17 shows the general set up. The Start point is on land, 1 km from shore, the End point is $1/2\,km$ into the water. A contestant in the race can run on land at 6 km/hr and can swim at 1.5 km/hr. At what point

P along the shoreline should she jump from land to water to minimize her overall time?

First, let's make the units consistent by doing everything in kilometers. Next, let's name some variables. We'll name as x the distance along the shore from the Start to where she jumps into the water; the remaining horizontal distance is then $1 - x$. The actual distances covered over land and through water are, using the Pythagorean theorem,

$$\sqrt{1+x^2} \qquad \text{on land and} \qquad \sqrt{(1-x)^2+(0.5)^2} \qquad \text{on water}$$

The variables and these resulting geometric measurements are shown in Fig. 5.18. Our job is to come up with an expression for total time taken, and minimize it. (This will be the objective function.) But given a distance d, rate r, and time t, we all know the familiar formula $d = rt$, which means $t = d/r$. We have a distance and rate for each leg of the race, and so we can write a formula for the total time taken:

$$T = T_{land} + T_{water} = \frac{\sqrt{1+x^2}}{6} + \frac{\sqrt{(1-x)^2+0.25}}{1.5}$$

To minimize this time, we'll look for where $dT/dx = 0$. I'll omit the details of the derivative, as these steps should be old news. (OK, I'll admit I used tech — and so can you! Is it worth taking the time to track down a computer, turn it on, open a program or browser, type in the function, and compute the derivative — all rather than just doing it by hand? You be the judge.)

$$T = \frac{\sqrt{1+x^2}}{6} + \frac{2\sqrt{(1-x)^2+0.25}}{3}$$

$$\frac{dT}{dx} = \frac{x}{6\sqrt{1+x^2}} + \frac{2x-2}{3\sqrt{(1-x)^2+0.25}}$$

Solving the equation $dT/dx = 0$ is not a task for mere mortals, and using tech for that result gives $x \approx 0.9$. Therefore, the racer should stay on land for $0.9\,km$ and then jump into the water and swim the rest of the way.

FFT: (1) What was the constraint in this problem? (2) How do we know we found a minimum for the total time and not a maximum? (3) If the racer's running speed and swimming speed were the same, where would be find P? ■

You Try It

(5) A piece of wire 10 m long is cut into two pieces. One piece is bent into a square and the other is bent into an equilateral triangle. How should the wire be cut so that the total area enclosed by both shapes is a minimum?

Have You Learned...

- The difference between a local or global (absolute) extreme of a function?
- How to identify the objective function in an optimization problem?
- How to determine a constraint in an optimization problem?
- How to use a constraint to simplify an objective function?
- How to use derivatives to determine the (global?) maximum or minimum of the equation written for the quantity in question?

Optimization — Problem List

Optimization — You Try It

These appeared above; solutions begin on the next page.

(1) Find the absolute extremes of $f(x) = 3x^2 - 12x + 5$ on $[0, 3]$.
(2) Find the absolute extremes of $f(x) = x/(x^2 + 1)$ on $[0, 2]$.
(3) Find two positive numbers whose product is 100 and whose sum is a minimum.
(4) If 1200 cm^2 of material is available to make a box with a square base and open top, find the largest possible volume of the box.
(5) A piece of wire 10 m long is cut into two pieces. One piece is bent into a square and the other is bent into an equilateral triangle. How should the wire be cut so that the total area enclosed by both shapes is a minimum?

Optimization — Practice Problems

Try these as you get the hang of the You Try It problems. Solutions to these problems are available in Sec. A.5.3.

(1) Find the absolute extremes of $f(x) = x^3 - 3x + 1$ on $[0, 3]$.
(2) Find the absolute extremes of $f(x) = (x^2 - 4)/(x^2 + 4)$ on $[-4, 4]$.
(3) Find the absolute extremes of $f(x) = (x^2 - 1)^3$ on $[-1, 2]$.
(4) Find a positive number such that the sum of the number and its reciprocal is as small as possible.
(5) A box with a square base and open top must have a volume of $32000\, cm^3$. Find the dimensions of the box that minimize the amount of material used.
(6) Find the point on the line $6x+y = 9$ that is closest to the point $(-3, 1)$.
(7) A man in a rowboat is 5 miles from the nearest point on shore. He needs to reach, in the shortest time possible, a second point on shore that is 6 miles east of the current closest landing spot. Where should he land if he can row $2\, mi/hr$ and walk $4\, mi/hr$?

Optimization — Challenge Problems

Try these problems to test your skills with the ideas in this section. Solutions to these problems are available in Sec. B.5.3.

(1) A box with a rectangular base and open top must have a volume of $10\,m^3$. The length of the base is twice the width. Material for the base costs \$10 / m^2 and the sides cost \$6 / m^2. Find the cost of materials for the cheapest such container.
(2) Estimate the point on $y = \tan x$ that is closest to the point $(1, 1)$. (Plan on using Maple or a calculator to estimate the final result.)
(3) A piece of wire 10 m long is cut into two pieces. One piece is bent into a square and the other is bent into a circle. How should the wire be cut so that the total area enclosed by both shapes is a minimum?

Optimization — You Try It — Solved

(1) Find the global extremes of $f(x) = 3x^2 - 12x + 5$ on $[0, 3]$.

□ Since $f'(x) = 6x - 12$, we have $f'(x) = 0$ at $x = 2$. Candidates for the global extremes thus include the critical point $x = 2$ and the endpoints $x = 0$ and $x = 3$. Now $f(2) = -7$, $f(0) = 5$, and $f(3) = -4$. The points to choose from are thus $(2, -7)$, $(0, 5)$ and $(3, -4)$. On $[0, 3]$, then, the global minimum is located at $(2, -7)$ and the global maximum is located at at $(0, 5)$. ■

(2) Find the global extremes of $f(x) = x/(x^2 + 1)$ on $[0, 2]$.

□ By the quotient rule (details omitted) we have

$$f'(x) = \frac{1 - x^2}{(x^2 + 1)^2}$$

So $f'(x) = 0$ at $x = \pm 1$. But $x = -1$ is not within the interval $[0, 2]$ so we throw it out. Candidates for the global extremes thus include the critical point $x = 1$ and the endpoints $x = 0$ and $x = 2$. The full points at these locations are: $(1, 1/2), (0, 0)$, and $(2, 2/5)$. On the interval $[0, 2]$, then, the global minimum and maximum are at $(0, 0)$ and $(1, 1/2)$, respectively. ■

(3) Find two positive numbers whose product is 100 and whose sum is a minimum.

□ We want to minimize $f(x, y) = x + y$. We know $xy = 100$ so that $y = 100/x$. We must reduce the function to one variable, so we write:

$$f(x, y) = x + y$$
$$f(x) = x + \frac{100}{x}$$
$$f'(x) = 1 - \frac{100}{x^2}$$

Then $f'(x) = 0$ when $x = 10$ and if $x = 10$, then also $y = 10$. TOI
FFT: How do we know $x = 10$ does not provide a maximum? TOI ■

(4) If 1200 cm^2 of material is available to make a box with a square base and open top, find the largest possible volume of the box.

□ Let x be the length of the sides of the square base, and h the height of the box. We want to maximize $V = x^2 h$. We must reduce this function to one variable x or h. But we know that the total amount (area) of material required to make the box is 1200 cm^2. With one square base of area x^2 and four sides of area xh, we know that $1200 = x^2 + 4xh$. Then we can solve for h and write:

$$\begin{aligned} V(x,h) &= x^2 h \\ V(x) &= x^2\left(\frac{1200 - x^2}{4x}\right) \\ &= \frac{1}{4}(1200x - x^3) \\ V'(x) &= \frac{1}{4}(1200 - 3x^2) \end{aligned}$$

Then $V'(x) = 0$ when $x^2 = 400$ or when $x = 20$. If $x = 20$ then $1200 = (20)^2 + 4(20)h$ or $h = 10$. The values $x = 20$, $h = 10$ maximize the volume, and that maximum volume is $V = (20)^2(10) = 4000\,cm^3$.

■

---

(5) A piece of wire 10 m long is cut into two pieces. One piece is bent into a square and the other is bent into an equilateral triangle. How should the wire be cut so that the total area enclosed by both shapes is a minimum?

□ Let's cut the wire in two pieces, of length x and $10 - x$. We'll use the wire of length x to make (random choice) the triangle. This means each side of the equilateral triangle (and specifically the base b) will have length $x/3$. The height h of the triangle is then $x/(2\sqrt{3})$ (do you know why?), and we can write the area of the triangle as

$$A_T = \frac{1}{2}\,bh = \frac{1}{2}\left(\frac{x}{3}\right) + \left(\frac{x}{2\sqrt{3}}\right) = \frac{x^2}{12\sqrt{3}}$$

Since the length x got used for the triangle, we have the length $10 - x$ to form the square. So, each side of the square has length $(10 - x)/4$, and

$$A_s = \left(\frac{10 - x}{4}\right)^2$$

Altogether, then, the the TOTAL enclosed area is the sum of the areas of the triangle and square:

$$A(x) = A_T + A_S = \frac{x^2}{12\sqrt{3}} + \left(\frac{10 - x}{4}\right)^2$$

so then

$$A'(x) = \frac{x}{6\sqrt{3}} - \frac{2}{4}\left(\frac{10-x}{4}\right)$$

Looking for where $A'(x) = 0$:

$$\begin{aligned} 0 &= \frac{x}{6\sqrt{3}} - \left(\frac{10-x}{8}\right) \\ x &= \frac{270(120-\sqrt{3})}{11} \\ &\approx 5.65 \end{aligned}$$

You can test x values to the left and right of this point to see that $A'(x) < 0$ to the left and and $A'(x) > 0$ to the right, so we know this x value provides a minimum total area. So we obtain the minimum total area by using about $5.65m$ for the triangle and $4.35m$ for the square.

■

5.4 Local Linear Approximation and L-Hopital's Rule

Introduction

When you read this section, you may think that you've lost your place and traveled back to Chapter 2. That is not the case. Rather, we are now in a position to use derivatives to evaluate some limits in special forms we could not deal with before. This particular topic doesn't need to be in any one place in our sequence of applications of derivatives, so it was held off until we got the really good stuff taken care of.

Local Linear Approximation

There are many situations where a functions tangent line at a point can act as a surrogate for the function in neighborhoods very close to the point of tangency. Let's put some language to that.

Definition 5.5. *If $f'(x_0)$ exists, then the* **local linear approximation** *of $f(x)$ at $x = x_0$ is:*

$$L(x) = f(x_0) + f'(x_0)(x - x_0)$$

If you are thinking to yourself, "Gosh, that looks an awful lot like the equation of the line tangent to $f(x)$ at x_0!" you are right: it is precisely that thing. The idea is that as long as $f'(x_0)$ exists, if you zoom in enough on x_0, the function will become indistinguishable from its tangent line there. We have an approximation to $f(x)$ that is a linear thing (the tangent line), in close proximity to the point of tangency (local). Thus: "local linear approximation". It may seem like overkill to make up another name for what's essentially the tangent line, but our ability to form a local linear approximation implies we may be able to build a local quadratic approximation, or a local cubic approximation, and so on. This idea is explored in detail in several chapters. But for now ...

EX 1 Find the local linear approximation of $f(x) = \sqrt{x}$ at $x_0 = 1$.

We have the following ingredients:

- $f(x_0) = f(1) = \sqrt{1} = 1$
- $f'(x) = \dfrac{1}{2\sqrt{x}}$, so $f'(x_0) = f'(1) = \dfrac{1}{2}$

Therefore by Def. 5.5,

$$L(x) = f(x_0) + f'(x_0)(x - x_0) = 1 + \frac{1}{2}(x - 1) = \frac{1}{2}x + \frac{1}{2}$$

Figure 5.19 shows the function $f(x)$ and its local linear approximation $L(1)$ on two scales; on the closer scale (b), it is hard to distinguish the two. ■

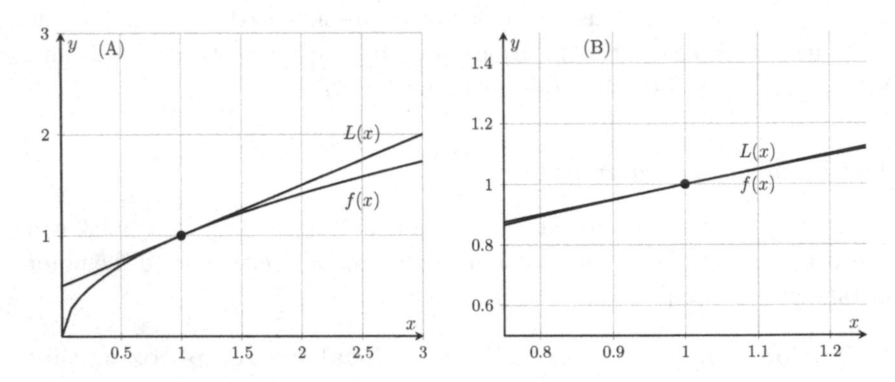

Fig. 5.19 The function $f(x) = \sqrt{x}$ and $L(1)$ on (A) wide, and (B) zoomed scales.

You Try It

(1) Find the local linear approximation of $f(x) = \sin(3x)$ at $x_0 = 0$.

A situation where swapping out a function for its local linear approximation might be useful is in determination of limits. Consider the limit

$$\lim_{x \to 0} \frac{\sin 3x}{\sin x}$$

If we think through the process, we see a quotient in which both the numerator and denominator are approaching zero as x approaches zero. Since the quotient $0/0$ is not defined, we cannot resolve this limit yet. The ultimate value of the limit will determined by the outcome of the "race" to zero between the numerator and denominator. Is the numerator approaching zero faster than the denominator, or is the denominator approaching zero faster? Either way, how does the final value of the limit come as a result of this "race" to zero? This limit presents what is called an *indeterminate form*, because it ... can't be determined. There are many types of indeterminate forms, as we'll see below.

For now, though, here's a conjecture: if $\sin 3x$ and $\sin x$ both have a local linear approximation near $x = 0$, and these local linear approximations are indistinguishable from the functions themselves when we're really close to $x = 0$, wouldn't the limit of the quotient of these approximations be the same as the limit of the quotient of the actual functions? Let's play that game.

For $f(x) = \sin 3x$, we have $f(0) = 0$. Also, since $f'(x) = 3\cos 3x$, we have $f'(0) = 3$. So then the local linear approximation of $f(x)$ is $L_f = 0 + 3(x - 0) = 3x$.

For $g(x) = \sin x$, we have $g(0) = 0$. Also, since $g'(x) = \cos x$, we have $g'(0) = 1$. So then the local linear approximation of $g(x)$ is $L_g = 0 + 1(x - 0) = x$.

Let's retry the limit when we exchange $f(x)$ and $g(x)$ with $L_f(0)$ and $L_g(0)$:

$$\lim_{x \to 0} \frac{\sin 3x}{\sin x} = \lim_{x \to 0} \frac{L_f(0)}{L_g(0)} = \lim_{x \to 0} \frac{3x}{x} = 3$$

Now, I happen to know that the true value of the limit is indeed 3, and so this technique worked out in this case. After a bit of practice, we'll take a look at this technique made a bit more formal.

EX 2 Give the value of

$$\lim_{x \to -1} \frac{x + 1}{x^2 + 4x + 3}$$

using established limit rules. Then determine this limit by replacing the numerator and denominator with their local linear approximations at the limit point. Compare the results.

Using some factoring then regular limit rules,

$$\lim_{x \to -1} \frac{x - 1}{x^2 + 4x + 3} = \lim_{x \to -1} \frac{x + 1}{(x + 1)(x + 3)} = \lim_{x \to -1} \frac{1}{x + 3} = \frac{1}{2}$$

Let's name $f(x) = x + 1$; since this is already a linear function, it is its own local linear approximation at $x = -1$. (If you don't believe that, just go through the motions of creating L_f.) Let's also name $g(x) = x^2 + 4x + 3$. Then $g(-1) = 0$. Since $g'(x) = 2x + 4$, then $g'(-1) = 2$. The local linear approximation of $g(x)$ at the limit point $x_0 = -1$ is, then, $L_g =$

$0+2(x-(-1)) = 2x+2$. So when we trade f and g for their local linear approximations at the limit point $x_0 = -1$,

$$\lim_{x\to-1}\frac{x-1}{x^2+4x+3} = \lim_{x\to-1}\frac{x+1}{2x+2} = \lim_{x\to-1}\frac{x+1}{2(x+1)} - \lim_{x\to-1}\frac{1}{2} = \frac{1}{2}$$

We have found the same limit value with both methods. ■

You Try It

(2) Give the value of $\lim\limits_{x\to1}\dfrac{x^2}{e^x}$ using established limit rules. Then determine this limit by replacing the numerator and denominator with their local linear approximations at the limit point. Compare the results.

A side note on local linear approximations: although we've seen a direct use of local linear approximations, they fall into a category of mathematical constructions that are often more useful by the fact that we know we are allowed to form them, than is the actual forming of them in itself. That may not make much sense right now, but later on, when you read, "Hey, remember back in Chapter 5 when I pointed out that sometimes the potential of forming a construct is more important than the construct itself?" That's this!

Indeterminate Forms

You should remember that finding a limit of a function involves considering the behavior of the function as its variable approaches some specific target value — such as, "the limit of $\sin x$ as x approaches 0." In many cases, that limit is detectable, or can be made detectable, after some algebraic rearrangement of the function. But there are some functions which are built with competing influences that cannot be resolved. For example, consider the different behaviors of

$$f(x) = \frac{x}{1} \quad , \quad g(x) = \frac{1}{e^x} \quad , \quad h(x) = \frac{1}{e^x} \quad , \quad Q(x) = \frac{x}{e^x}$$

as x goes to $+\infty$.

- The function $f(x)$ is a quotient of x itself and a constant; so, as x goes to infinity, the numerator "wins"; we know that $\lim\limits_{x\to\infty} f(x) = \infty$.
- The function $g(x)$ is a quotient of a constant and a function which grows literally exponentially; so, as x goes to infinity, the denumerator "wins"; we know that $\lim\limits_{x\to\infty} g(x) = 0$.

- The function $h(x)$ is a quotient of two functions which both grow towards ∞, but after algebraic simplification, we can resolve that $\lim_{x\to\infty} h(x) = 0$.
- The function $Q(x)$ is a quotient of two functions which both grow towards ∞. We cannot do any algebraic simplification. We know that the denominators grows "faster" than the numerator, but how does that affect $\lim_{x\to\infty} Q(x)$? We don't know for sure; we can *guess*, or we can start cranking out large tables of the values of the function for larger and larger values of x — but neither method is satisfying.

A function built of expressions which provide competing influences when coupled with a limit process is said to be in an indeterminate form. In the examples above, f and g do not lead to indeterminate forms. The function h leads to a pretty lame indeterminate form, since it is resolved after one step of algebraic simplification. But Q leads to a true indeterminate form. It cannot be resolved through algebraic means. By contrast, the function $R(x) = xe^x$ is not an indeterminate form because the tendencies of x and e^x as x goes towards ∞ compound each other rather than compete against each other.

Useful Fact 5.2. *Genreally speaking, you have found an* ***indeterminate form*** *if a function tends towards one of these structures when attacked by a limit:*

$$\frac{0}{0}\,,\ \frac{\infty}{\infty}\,,\ 0\times\infty\,,\ 0^{\infty}\,,\ 0^{0}\,,\ 1^{\infty}$$

In all of the cases shown in Useful Fact 5.2, there are two competing influences on the limit, and the final value of the limit (if any) cannot be determined by techniques we've seen so far. The problem with indeterminate forms is that we cannot resolve (yet) a final value. Consider the following two limits — both of which show the indeterminate form ∞/∞, but which have very different outcomes:

$$\lim_{x\to\infty}\frac{x^2}{x^3} \qquad \text{vs} \qquad \lim_{x\to\infty}\frac{x^3}{x^2}$$

Much like $h(x)$ above, these are lame indeterminate forms — they can be resolved immediately with algebraic simplification. But what about these?

$$\lim_{x\to 0}\frac{\sin x}{x} \quad ; \quad \lim_{x\to 0}\frac{\sin 2x}{x} \quad ; \quad \lim_{x\to 0}\frac{\sin 5x}{x} \tag{5.3}$$

You may recognize the first limit; all three of these are indeterminate forms $0/0$, and all have different final values. We will see those final values later.

As we start handling indeterminate forms, never forget these fundamental facts:

- A limit of form ∞/∞ does not necessarily have a value of 1.
- A limit of form $0/0$ does not necessarily have a value of 1.
- A limit of form 0^∞ does not necessarily have a value of 0 or ∞.
- A limit of form ∞^0 does not necessarily have a value of 0 or ∞.

These four facts are what make indeterminate forms so much fun! Limits which resolve to these forms must be investigated further to determine their final values.

EX 3 Two of these are in indeterminate form, and one is not. Which is which?

$$(a) \lim_{x\to\infty} \frac{x^2+e^x}{x^3+1} \quad ; \quad (b) \lim_{x\to\infty} \frac{e^{-x}}{x^3+1} \quad ; \quad (c), \lim_{x\to 0^+} x^2 \ln x$$

- Limit (a) is in indeterminate form. (Alright, before going further, let's agree that we won't worry about the difference between the phrase "this expression is IN indeterminate form" and the phrase, "this expression is AN indeterminate form." We can use both interchangeably. We're all friends here.) In (a), both the numerator and denominator go to ∞ as x goes to ∞. There is no algebraic simplification that can resolve this. Limit (a) is in indeterminate form ∞/∞.
- Limit (b) can be rearranged:

$$\lim_{x\to\infty} \frac{e^{-x}}{x^3+1} = \lim_{x\to\infty} \frac{1}{e^x(x^3+1)} = 0$$

The limit can be resolved directly, it is not in indeterminate form.

- In limit (c), we have a product between one term (x^2) which goes to 0 as x approaches 0, and another term ($\ln x$) which goes to $-\infty$ as x approaches 0. We cannot resolve these competing influences yet. It is in indeterminate form $0 \cdot \infty$. ■

You Try It

(3) One of these is an indeterminate form, the other two are not. Which is the indeterminate form and why?

$$(a)\ \lim_{x\to 0}\frac{x-\frac{1}{2}}{\cos(\pi x)}\ ;\quad (b)\ \lim_{x\to 1}\frac{x-\frac{1}{2}}{\cos(\pi x)}\ ;\quad (c)\ \lim_{x\to 1/2}\frac{x-\frac{1}{2}}{\cos(\pi x)}$$

Please look back at some Examples and You Try Its from above.

- The purpose of YTI 1 is to illustrate that the distinction between a "regular" limit and one in indeterminate form can have just as much to do with the target value of x as it does with the function itself.
- YTI 2 shows a regular limit which can be resolved normally, but we also solve it by replacement of the numerator and denominator by their local linear approximations at the limit point.
- EX 2 shows a limit which presents an indeterminate form that is resolved by replacement of the numerator and denominator by their local linear approximations at the limit point.

The scheme of evaluating limits via replacement of component functions by local linear approximations will not be taken further, now that we've seen that we can do it. If a limit does not present an indeterminate form and can be resolved by established rules and techniques, we don't need to do this extra step. For limits which present indeterminate forms, we will streamline the replacement process a bit further. This new strategy is called L-Hopital's Rule.

L-Hopital's Rule

When you have a nasty quotient
But you just don't have the time
Then trade the pieces f and g
With f prime and g prime

(I just made that up, and sure it's no "Lo d-hi over hi-d-lo, all over the square of the bottom we go!",[2] but maybe it will help.)

Putting this in more useful terms:

[2]The quotient rule!

Useful Fact 5.3. *When a limit expression resolves to* $0/0$ *or* ∞/∞ *form, try trading the numerator and denominator with their derivatives:*

$$\lim_{x \to a} \frac{f(x)}{g(x)} = \lim_{x \to a} \frac{f'(x)}{g'(x)}$$

This works only when the limit on the left is indeterminate, but the limit on the right exists and can be resolved.

This is called **L-Hopital's Rule**. L-Hopital's Rule streamlines the strategy of replacement of the numerator and denominator by their local linear approximations with an even more direct replacement by only their derivatives. A brief overview of why this rule works is at the end of this section. For now, let's put it to work.

EX 4 Let's evaluate the limits we saw in Eq. (5.3),

$$\lim_{x \to 0} \frac{\sin x}{x} \quad ; \quad \lim_{x \to 0} \frac{\sin 2x}{x} \quad ; \quad \lim_{x \to 0} \frac{\sin 5x}{x}$$

as well as the general case

$$\lim_{x \to 0} \frac{\sin nx}{x}$$

Each of these is in indeterminate form $0/0$. We can swap the functions with their derivatives, per L-Hopital's Rule. It is important to remember that we don't ask for the derivative of the whole quotient! (This is a common error.) Rather, we exchange the numerators and denominators individually.

$$\lim_{x \to 0} \frac{\sin x}{x} = \lim_{x \to 0} \frac{\cos x}{1} = \lim_{x \to 0} \cos x = 1$$

$$\lim_{x \to 0} \frac{\sin 2x}{x} = \lim_{x \to 0} \frac{2\cos 2x}{1} = \lim_{x \to 0} 2\cos 2x = 2$$

$$\lim_{x \to 0} \frac{\sin 5x}{x} = \lim_{x \to 0} \frac{5\cos 5x}{1} = \lim_{x \to 0} 5\cos 5x = 5$$

$$\lim_{x \to 0} \frac{\sin nx}{x} = \lim_{x \to 0} \frac{n\cos nx}{1} = \lim_{x \to 0} n\cos nx = n \quad \blacksquare$$

EX 5 Evaluate $\lim_{x \to 0} \frac{\sin(2x)}{\cos(4x)}$.

Now that we're getting on a roll, it's tempting to try this:

$$\lim_{x \to 0} \frac{\sin(2x)}{\cos(4x)} = \lim_{x \to 0} \frac{2\cos(2x)}{-4\sin(4x)}$$

DON'T DO THAT. Do you know the old cliche, "When you have a hammer, every problem looks like a nail"? That trap is seen here. For some reason, once students learn L-Hopital's Rule, they like to use it to hammer out the limit of every quotient encountered from that point forward. If you use L-Hopital's Rule here, you will get the wrong answer. In order for L-Hopital's Rule to work, the original limit must be presenting an indeterminate form. That isn't the case here. Our limit can be resolved immediately. As x approaches 0, the numerator $\sin(2x)$ approaches 0, but the denominator approaches 1. Therefore, the limit approaches 0/1 ... which is not indeterminate, it's very much *determinate*:

$$\lim_{x\to 0} \frac{\sin(2x)}{\cos(4x)} = \frac{0}{1} = 0$$

L-Hopital's Rule would lead us to (incorrectly) conclude that the limit is $-\infty$. ■

You Try It

(4) One of these two limits can be resolved with L-Hopital's Rule. Identify it and solve it.

$$\lim_{x\to 1} \frac{x^9 - 1}{x^5 - 1} \qquad \text{or} \qquad \lim_{x\to 1} \frac{x^9 - 1}{x^5 - 1}$$

Repeated Use of L-Hopital's Rule

Direct applications of L-Hopital's rule are fairly straightforward; it's almost harder to know when you can and cannot apply the technique. But there are cases in which things get a little more dicey.

For example, what if L-Hopital's Rule doesn't help on the first try? If using L-Hopital's Rule on one indeterminate form hands you back a second indeterminate form, just use it again!

EX 6 Evaluate $L = \lim_{x\to\infty} \frac{e^x}{x^3 + 2x^2 + x + 5}$.

This presents an indeterminate form ∞/∞, so we can use L-Hopital's Rule:

$$\lim_{x\to\infty} \frac{e^x}{x^3 + 2x^2 + x + 5} = \lim_{x\to\infty} \frac{e^x}{3x^2 + 4x + 1}$$

This has returned a second indeterminate form. Well, we can just treat this new limit as a brand new problem. It presents an indeterminate form

∞/∞, so we can use L-Hopital's Rule:

$$\lim_{x\to\infty} \frac{e^x}{3x^2+4x+1} = \lim_{x\to\infty} \frac{e^x}{6x+4}$$

This has returned a thurd indeterminate form. Well, we can just treat this new limit as a brand new problem. It presents an indeterminate form ∞/∞, so we can use L-Hopital's Rule (this is getting to be like L-Hopital's Rule Inception):

$$\lim_{x\to\infty} \lim_{x\to\infty} \frac{e^x}{6x+4} = \lim_{x\to\infty} \frac{e^x}{6} = \frac{1}{6} \lim_{x\to\infty} e^x$$

Finally, we've reached the end of the line; this newest pass through L-Hopital's Rule lets us determine $L = \infty$. ∎

You Try It

(5) Evaluate $\lim_{t\to 0} \frac{e^t - 1}{t^3}$.

(6) Evaluate $\lim_{x\to 0} \frac{1-\cos x}{x^2}$.

Indeterminate Forms That Aren't Quotients

Another secondary benefit of L-Hopital's Rule is that we can use it to solve some limits which may not be in the form of a quotient at first. Now, certainly, L-Hopital's Rule is designed to work on quotients, so — if you are presented with a different indeterminate form such as $0 \times \infty$, which is not (yet) a quotient, can you rearrange it to present an equivalent quotient that's still an indeterminate form?

EX 7 Evaluate $L = \lim_{x\to 0^+} \ln x \sin x$.

The current limit presents an indeterminate form $-\infty \cdot 0$, which is not set up for L-Hopital's Rule. Here, we can invoke the rule of "make something look worse before we can make it look better." We are in charge of this limit, and if we want it to look like a quotient, then dang it, we can make it look like a quotient — as long as we do it properly. Note that we can rewrite

$$\ln x \sin x = \frac{\sin x}{1/\ln x}$$

Since $\ln x$ heads for $-\infty$ as x approaches 0 from the right, then $1/\ln x$ heads for 0; the latter form, now a quotient, presents the indeterminate form $0/0$ — and now L-Hopital's Rule can be applied.

$$\lim_{x\to 0^+} \ln x \sin x = \lim_{x\to 0^+} \frac{\sin x}{1/\ln x} = \lim_{x\to 0^+} \frac{\cos x}{1/x} = \lim_{x\to 0^+} x\cos x = 0 \quad \blacksquare$$

You Try It

(7) Evaluate $\lim_{x\to 0^+} \sqrt{x}\ln(x)$.

L-Hopital's Rule With Logarithmic Differentiation

Wow, this sounds awful — but it's just another method for taking one indeterminate form and manipulating it into another one that can be solved using L-Hopital's Rule. Here's the overall decision tree for L-Hopital's Rule in general:

(A) If your indeterminate form is already a quotient, try L-Hopital's Rule immediately.

(B) If one use of L-Hopital's Rule leads to a new (quotient) indeterminate form, just use (A) again.

(C) If your indeterminate form is a product, try converting it to a quotient and then see (B).

(D) If your indeterminate form is in exponential form, using logarithmic differentiation will produce a product; then, see (A) or (C) as appropriate.

Given all this, be sure to not overuse L-Hopital's Rule; a common mistake is to go on autopilot and keep using L-Hopital's Rule every time a quotient shows up, but you have to double or triple check that your quotient presents a genuine indeterminate form; if you arrive at a quotient that can be solved "normally", you have to eject from L-Hopital's Rule. Again, don't make L-Hopital's Rule a hammer that makes you see every limit problem as a nail.

EX 8 Evaluate $L = \lim_{x\to\infty} x^{\frac{1}{x}}$.

The current limit presents an indeterminate form ∞^0 (which is not necessarily equal to either ∞ or 1). It would be great to drop that $1/x$ out of the exponent, and that's something logarithms can do. Are we allowed to just take the logarithm of both sides of this limit? Sure! Remember, we're in charge. Let $E = x^{\frac{1}{x}}$; then

$$\ln(E) = \ln x^{\frac{1}{x}} = \frac{1}{x}\ln x = \frac{\ln x}{x}$$

This new quotient presents the indeterminate form ∞/∞ as x approaches ∞, and we can now use L-Hopital's Rule directly:

$$\lim_{x\to\infty}\frac{\ln x}{x} = \lim_{x\to\infty}\frac{1/x}{1} = \lim_{x\to\infty}\frac{1}{x} = 0$$

Now we have to be careful; this limit we just found is not the original limit, it is the limit of the natural logarithm of our original function. When we defined $E = x^{\frac{1}{x}}$, we named the original problem as $\lim_{x\to\infty} E$. But what we just computed was $\lim_{x\to\infty} \ln E$. In general, though, getting from the latter to the former is easy using the inverse of the logarithm:

$$\text{If } L_1 = \lim_{x\to\infty} \ln E \text{ then } \lim_{x\to\infty} E = e^{L_1}$$

So, with $\lim_{x\to\infty} \ln E = 0$

$$\lim_{x\to\infty} x^{\frac{1}{x}} = e^0 = 1 \quad \blacksquare$$

You Try It

(8) Evaluate $\lim_{x\to 0}(\sin x)^x$.

Into the Pit!!

The Origin of L-Hopital's Rule

Most functions we deal with here are continuous and differentiable over their domains, except perhaps at some isolated points; for the most part, we expect our functions to have local linear approximations at almost any point. Remember that the "local linear approximation" of a function $f(x)$ at a given point x_0 is just a dressed-up version of the tangent line at that point:

$$L(x_0) = f(x_0) + f'(x_0)(x - x_0)$$

Let's say we have a limit that presents an indeterminate form $0/0$:

$$\lim_{x\to a}\frac{f(x)}{g(x)} = \frac{0}{0}$$

When our functions $f(x)$ and $g(x)$ meet the ideal conditions of continuity and differentiability at and near $x = a$, then this limit produces the given indeterminate form because $f(a) = 0$ and $f(b) = 0$ (such as in EX 2). When we replace the numerator $f(x)$ and denominator $g(x)$ by their local linear approximations, we get

$$\lim_{x \to a} \frac{f(x)}{g(x)} = \lim_{x \to a} \frac{f(a) + f'(a)(x-a)}{g(a) + g'(a)(x-a)}$$

But since we're getting the indeterminate form $0/0$, we know $f(a) = g(a) = 0$, and that kicks off a bit of simplification:

$$\begin{aligned}\lim_{x \to a} \frac{f(x)}{g(x)} = \lim_{x \to a} \frac{f(a) + f'(a)(x-a)}{g(a) + g'(a)(x-a)} &= \lim_{x \to a} \frac{0 + f'(a)(x-a)}{0 + g'(a)(x-a)} \\ &= \lim_{x \to a} \frac{f'(a)(x-a)}{g'(a)(x-a)} = \lim_{x \to a} \frac{f'(a)}{g'(a)} = \frac{f'(a)}{g'(a)}\end{aligned}$$

If we carry our assertion that f and g are "nice functions", meaning they have well behaved derivatives at and near $x = a$, then $f'(a)$ is the limit of $f'(x)$ as x approaches a, and $g'(a)$ is the limit of $g'(x)$ as x approaches a. This means that the final step of our limit calculation, we can back off of the constants $f'(a)$ and $g'(a)$ within the limit expression like so:

$$\frac{f'(a)}{g'(a)} = \lim_{x \to a} \frac{f'(x)}{g'(x)}$$

Putting it all together,

$$\lim_{x \to a} \frac{f(x)}{g(x)} = \lim_{x \to a} \frac{f'(x)}{g'(x)}$$

This is L-Hopital's Rule. In the process laid out here, this evolved for a limit of "nice" functions presenting an indeterminate form $0/0$ for a limit point $x = a$. It turns out that this rule is more widely general; it applies for limits as $x \to \pm\infty$, and for the indeterminate form ∞/∞. However, to demonstrate all of this in a mathematical court of law is not something we can do there. The demonstration at hand is here just to show how the idea of local linear approximations can lead to good stuff. If you want to see a deeper and broader proof of L-Hopital's Rule, then continue taking more math classes until you're eligible for the proof-based coursse often called Advanced Calculus or Real Analysis.

Have You Learned...

- How to reimagine tangent lines as local linear approximations to $f(x)$ at and near a point?
- How to identify a limit in indeterminate form?
- How to apply L-Hopital's Rule to help with an indeterminate limit in the form of a quotient?
- How to rearrange an indeterminate limit that is not a quotient, into the form of a quotient?
- How to use L-Hopital's rule more than once on a single limit, but know when to stop?
- How to apply logarithmic differentiation to rearrange a function in exponential form so that L-Hopital's rule can help with a limit?

LLA and LHR — Problem List

LLA and LHR — You Try It

These appeared above; solutions begin on the next page.

(1) Find the local linear approximation of $f(x) = \sin(3x)$ at $x_0 = 0$.

(2) Give the value of $\lim_{x\to 1} \frac{x^2}{e^x}$ using established limit rules. Then determine this limit by replacing the numerator and denominator with their local linear approximations at the limit point. Compare the results.

(3) One of these is an indeterminate form, the other two are not. Which is the indeterminate form and why?

$$(a) \lim_{x\to 0} \frac{x - \frac{1}{2}}{\cos(\pi x)} \quad ; \quad (b) \lim_{x\to 1} \frac{x - \frac{1}{2}}{\cos(\pi x)} \quad ; \quad (c) \lim_{x\to 1/2} \frac{2x - 1}{\cos(\pi x)}$$

(4) Evaluate $\lim_{x\to 1} \frac{x^9 - 1}{x^5 - 1}$.

(5) Evaluate $\lim_{t\to 0} \frac{e^t - 1}{t^3}$.

(6) Evaluate $\lim_{x\to 0} \frac{1 - \cos x}{x^2}$.

(7) Evaluate $\lim_{x\to 0^+} \sqrt{x}\ln(x)$.

(8) Evaluate $\lim_{x\to 0} (\sin x)^x$.

LLA and LHR — Practice Problems

Try these as you get the hang of the You Try It problems. Solutions to these problems are available in Sec. A.5.4.

(1) Find the local linear approximation of $f(x) = 2/x$ at $x_0 = 1$.

(2) Find the local linear approximation of $f(x) = e^x$ at $x_0 = 0$.

(3) Evaluate $\lim_{x\to 1} \frac{x^a - 1}{x^b - 1}$ $(a, b \neq 0)$.

(4) Evaluate $\lim_{t\to 0} \frac{e^{3t} - 1}{t}$.

(5) Evaluate $\lim_{x\to 0} \frac{\sin x - x}{x^3}$.

(6) Evaluate $\lim_{x\to -\infty} x^2 e^x$.

LLA and LHR — Challenge Problems

Try these problems to test your skills with the ideas in this section. Solutions to these problems are available in Sec. B.5.4.

Evaluate the following limits:

$$(1) \lim_{x\to 0} \frac{x + \sin x}{x + \cos x} \qquad (2) \lim_{x\to 0} \frac{\tan x - x}{x^3} \qquad (3) \lim_{x\to 0^+} \sin x \ln x$$

Local Lin Approx & L-Hopital's Rule — You Try It — Solved

(1) Find the local linear approximation of $f(x) = \sin(3x)$ at $x_0 = 0$.

☐ We need these ingredients:

- $f(x_0) = f(0) = \sin(3 \cdot 0) = 0$
- $f'(x_0) = f'(0) = 3\cos(3 \cdot 0) = 3$

so that

$$L(0) = f(0) + f'(0)(x - 0) = 0 + 3(x - 0) = 3x$$

The local linear approximation of $f(x) = \sin(3x)$ at $x = 0$ is $L(0) = 3x$.

■

(2) Give the value of $\lim\limits_{x \to 1} \frac{x^2}{e^x}$ using established limit rules. Then determine this limit by replacing the numerator and denominator with their local linear approximations at the limit point. Compare the results.

☐ Using established limit laws, since x^2 and e^x are continuous everywhere,

$$\lim_{x \to 1} \frac{x^2}{e^x} = \frac{1^2}{e^1} = \frac{1}{e}$$

Now we can find the local linear approximations of $f(x) = x^2$ and $g(x) = e^x$ at $x_0 = 1$:

- $f(x_0) = f(1) = 1^2 = 1$
- $f'(x_0) = f'(1) = 2(1) = 2$

so $L_f(1) = f(1) + f'(1)(x - 1) = 1 + 2(x - 1) = 2x - 1$. Next,

- $g(x_0) = g(1) = e^1 = e$
- $g'(x_0) = g'(1) = e^1 = e$

so $L_g(1) = g(1) + g'(1)(x - 1) = e + e(x - 1) = ex$. Now we reconstruct the limit using these local linear approximations:

$$\lim_{x \to 1} \frac{x^2}{e^x} = \lim_{x \to 1} \frac{2x - 1}{ex} = \frac{2(1) - 1}{e} = \frac{1}{e}$$

By both methods, we find $\lim\limits_{x \to 1} \frac{x^2}{e^x} = \frac{1}{e}$. ■

(3) One of these is an indeterminate form, the other two are not. Which is the indeterminate form and why?

$$(a)\ \lim_{x\to 0}\frac{2x-1}{\cos(\pi x)}\quad ;\quad (b)\ \lim_{x\to 1}\frac{x-\frac{1}{2}}{\cos(\pi x)}\quad ;\quad (c)\ \lim_{x\to 1/2}\frac{2x-1}{\cos(\pi x)}$$

□ Limit (a) is not indeterminate, and it resolves immediately to

$$\lim_{x\to 0}\frac{2x-1}{\cos(\pi x)}=\frac{-1}{\cos(0)}=-1$$

Limit (b) is not indeterminate, and it resolves immediately to

$$\lim_{x\to 1}\frac{2x-1}{\cos(\pi x)}=\frac{1}{\cos(\pi)}=\frac{1}{(-1)}=-1$$

Limit (c) presents the indeterminate form 0/0, because

- $2x-1\to 0$ as $x\to 1/2$
- $\cos(\pi x)\to\cos(\pi/2)=0$ as $x\to 1/2$ *blacksquare*

(4) Evaluate $\lim_{x\to 1}\frac{x^9-1}{x^5-1}$.

□ The given limit presents the indeterminate form 0/0. Using L-Hopital's Rule,

$$\lim_{x\to 1}\frac{x^9-1}{x^5-1}=\lim_{x\to 1}\frac{9x^8}{5x^4}=\frac{9}{5}\quad ■$$

(5) Evaluate $\lim_{t\to 0}\frac{e^t-1}{t^3}$.

□ The given limit presents the indeterminate form 0/0. Using L-Hopital's Rule,

$$\lim_{t\to 0}\frac{e^t-1}{t^3}=\lim_{t\to 0}\frac{e^t}{3t^2}=\infty\quad ■$$

(6) Evaluate $\lim_{x\to 0}\frac{1-\cos x}{x^2}$.

□ The given limit presents the indeterminate form 0/0. Using L-Hopital's Rule,

$$\lim_{x\to 0}\frac{1-\cos x}{x^2}=\lim_{x\to 0}\frac{\sin x}{2x}=\lim_{x\to 0}\frac{\cos x}{2}=\frac{1}{2}$$

(Note we had to apply L-Hopital's Rule twice since the first application resulted in another indeterminate form.) ■

(7) Evaluate $\lim_{x\to 0+} \sqrt{x}\ln(x)$.

☐ The given limit presents the indeterminate form $-0 \cdot \infty$. It is not yet in a form which L-Hopital's Rule can handle, so we must rearrange things:

$$\lim_{x\to 0+} \sqrt{x}\ln(x) = \lim_{x\to 0+} \frac{\ln x}{1/\sqrt{x}} = \lim_{x\to 0+} \frac{\ln x}{x^{-1/2}} \cdots$$

This is where we apply L-Hopital's Rule:

$$\cdots = \lim_{x\to 0+} \frac{1/x}{(-1/2)x^{-3/2}}$$

and now it's just simplification

$$\cdots = -2 \lim_{x\to 0+} x^{-1}x^{3/2} = -2 \lim_{x\to 0+} x^{1/2} = 0 \quad \blacksquare$$

(8) Evaluate $\lim_{x\to 0}(\sin x)^x$.

☐ The given limit presents the indeterminate form 0^0. It is not yet in a form which L-Hopital's Rule can handle, but if we set $E = (\sin x)^x$, then $\ln(E) = x \ln \sin(x)$. This is *still* not in a form which L-Hopital's Rule can handle, so we apply one more rearrangement:

$$\ln(E) = x \ln \sin(x) = \frac{\ln \sin x}{1/x}$$

If we seek the limit of $\ln(E)$, we can use L-Hopital's Rule on that limit,

$$\begin{aligned}\lim_{x\to 0} \frac{\ln \sin x}{1/x} &= \lim_{x\to 0} \frac{\cos x/\sin x}{-1/x^2} \\ &= -\lim_{x\to 0} \frac{x^2 \cos x}{\sin x} \quad \text{which itself needs L-Hopital's Rule} \\ &= -\lim_{x\to 0} \frac{2x\cos x - x^2 \sin x}{\cos x}\end{aligned}$$

This is no longer an indeterminate form and resolves directly to 0. Don't forget that this is not the final answer! This limit of 0 is the limit of $\ln(E)$, not the original limit (which is the limit of E itself). Since $\lim_{x\to 0} \ln(E) = 0$, then

$$\lim_{x\to 0} E = \lim_{x\to 0}(\sin x)^x = e^0 = 1 \quad \blacksquare$$

5.5 Calc Offers Good Value When It Gets on a Rolle

Introduction

As you know, we don't lean heavily into the more theoretical side of things in this text, we're more focused on the "toolkit" side of things. But, there are some results that bear mentioning because they are valuable as the foundations of some useful tools. You've seen one of them already, in Sec. 2.4 — the Intermediate Value Theorem. There is another which sounds an awful lot like the IVT, but is based on derivatives instead — the Mean Value Theorem. And, the MVT has a little brother called Rolle's Theorem that encapsulates a very useful and intuitive concept. In this section, we'll use Rolle's Theorem as a lead-in to the Mean Value Theorem, and also show a computational technique based on the IVT.

Rolle's Theorem

In a proper theoretical treatment of Calculus, we would "invent" the Mean Value Theorem first, and generate Rolle's Theorem as a consequence of the MVT. Since this is not a proper theoretical treatment of Calculus, we'll do it the other way. Let's see Rolle's Theorem first, then see the bigger picture in the MVT.

First, as a reminder, here is the Intermediate Value Theorem, seen in Sec. 2.4:

Theorem 5.1. *Suppose that $f(x)$ is continuous on the closed interval $[a, b]$ and W is any number between $f(a)$ and $f(b)$. Then there is a number c in $[a, b]$ such that $f(c) = W$.*

Theorem 5.2. *Rolle's Theorem.* *Suppose that $f(x)$ is continuous on the closed interval $[a, b]$ and differentiable on the open interval (a, b). If $f(a) = f(b)$, then there is a number c in (a, b) such that $f'(c) = 0$.*

The Intermediate Value Theorem and Rolle's Theorem are very similar, but with one fundamental difference: the payoff to Rolle's Theorem involves the *derivative* of $f(x)$ rather than $f(x)$ itself. The idea of Rolle's Theorem is pretty simple. If a function's graph has the same y value at two different locations, and it's not allowed to have any discontinuities, then there must be a critical point of the function somewhere in between the two locations. There are only a few options:

- The function is a straight horizontal line connecting $(a, f(a))$ to $(b, f(b))$ (remember, $f(a)$ and $f(b)$ are the same number) — in this case, every point $x = c$ between $x = a$ and $x - b$ provides a critical point, where $f'(c) = 0$. Or,
- The function is decreasing as it passes through $(a, f(a))$; well, then it must turn and start going uphill to get back to $(b, f(b))$; the location of the "turn" is the critical point.
- The function is increasing as it passes through $(a, f(a))$; well, then it must turn and start going downhill to get back to $(b, f(b))$; the location of the "turn" is the critical point.

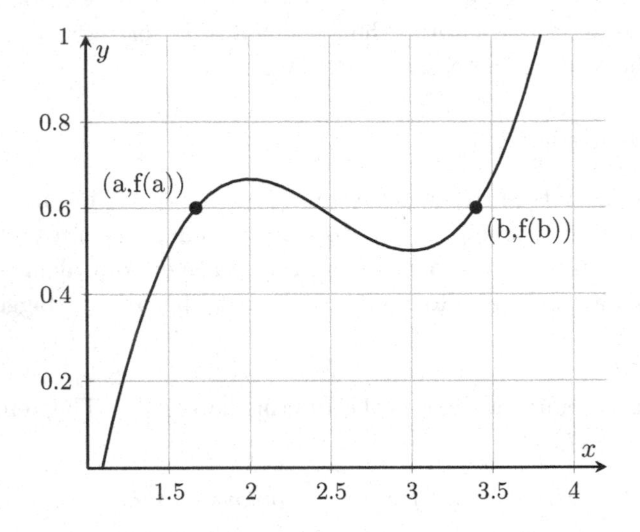

Fig. 5.20 Rolle's Theorem in action.

Figure 5.20 shows a scenario where we have two points $(a, f(a))$ and $(b, f(b))$ such that $f(a) = f(b)$. We can clearly see *at least* one point in between $x = a$ and $x = b$ (two, actually!) where $f'(x) = 0$. Rolle's Theorem insists there must be "a number" where this happens; it does not rule out there being multiple places where $f'(x) = 0$. Regardless of the shape of the graph of $f(x)$, if $f(a) = f(b)$, the graph simply cannot get from $(a, f(a))$ to $(b, f(b))$ without bending to change direction, unless the function has one or more discontinuities.

EX 1 Show that $f(x) = x^4 + 6x^3 + 8x^2 - 6x - 9$ meets the conditions of Rolle's Theorem on $[-1, 1]$. Then find the value $x = c$ guaranteed by Rolle's Theorem to exist on that interval. Numerical estimates are OK.

Rolle's Theorem poses some requirements, so let's check to be sure they're met:

- $f(x)$ is a polynomial, so it is continuous and differentiable everywhere, particularly on $[-1, 1]$. ✓
- for $a = -1$ and $b = 1$, we have $f(a) = 0$ and $f(b) = 0$, so $f(a) = f(b)$ on the interval $[a, b]$. ✓

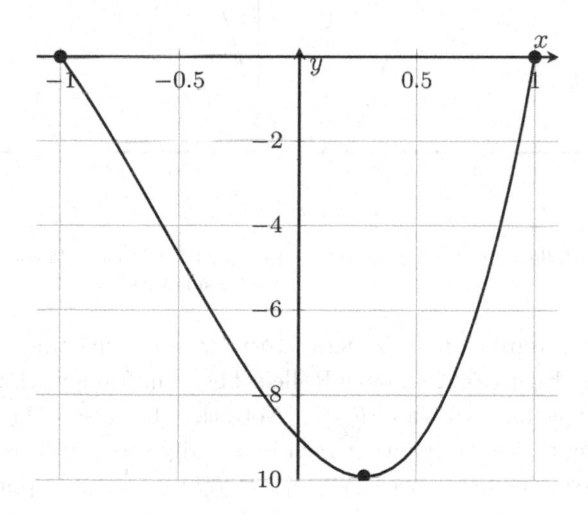

Fig. 5.21 Rolle's Theorem framework, with EX 1.

So Rolle's Theorem is good to go, and it insists we should be able to find a location $x = c$ somewhere on the interval $[-1, 1]$ where $f'(c) = 0$. Figure 5.21 shows $f(x)$ on the interval $[-1, 1]$ and we can *see* the location where $f'(c) = 0$. Since $f'(x) = 4x^3 + 18x^2 + 16x - 6 = 2(x + 3)(2x^2 + 3x - 1)$, the location in the interval $[-1, 1]$ where $f'(c) = 0$ must come from the term $2x^2 + 3x - 1$. This quadratic has roots at $-\dfrac{3}{4} \pm \dfrac{\sqrt{17}}{4}$; the root that is within in our interval $[-1, 1]$ is $-\dfrac{3}{4} + \dfrac{\sqrt{17}}{4} \approx 0.28$. ■

You Try It

(1) Show that $f(x) = -5x^3 + 21x^2 - 28x + 30$ meets the conditions of Rolle's Theorem on $[1, 2]$. There are *two* values $x = c$ guaranteed by Rolle's Theorem to exist on that interval; find them both. Numerical estimates are OK.

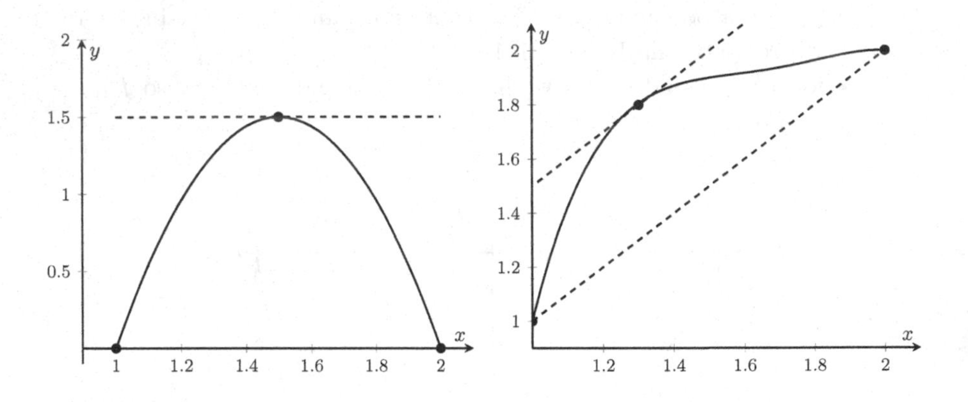

Fig. 5.22 A Rolle's Theorem scenario on $[1, 2]$.

Fig. 5.23 A Rolle's Theorem scenario tipped sideways.

Let's work towards a more general story by re-interpreting what Rolle's Theorem says. Figure 5.22 shows a Rolle's Theorem framework in which we have $f(a) = f(b)$ for $a = 1$ and $b = 2$. Note that because $f(1) = f(2)$, the line connecting these two points (which is actually the x-axis) is horizontal. Now by Rolle's Theorem, because $f(a) = f(b)$, there is a point $(c, f(c))$ where $f'(c) = 0$ — this happens to be $(1.5, 1.5)$. At this point, we see the (dashed) line tangent to $f(x)$ is also horizontal. Let's use this to rephrase Rolle's Theorem as follows,

Useful Fact 5.4. *Suppose that $f(x)$ is continuous on the closed interval $[a, b]$ and differentiable on the open interval (a, b). If $f(a) = f(b)$, then there is a value c in (a, b) such that the line tangent to $f(x)$ at $(c, f(c))$ is horizontal.*

Next, remember that a line connecting two points on a graph of a function $f(x)$ is called a *secant* line between those two points. When $f(a) = f(b)$, the secant line connecting $(a, f(a))$ and $(b, f(b))$ is horizontal. And if we then find a point $(c, f(c))$ where the line tangent to $f(x)$ is horizontal, we have a location where the line tangent to $f(x)$ at $(c, f(c))$ is *parallel* to the secant line connecting $(a, f(a))$ to $(b, f(b))$. This story is still told by Fig. 5.22. Rolle's Theorem can be rephrased one last time:

Useful Fact 5.5. *Suppose that $f(x)$ is continuous on the closed interval $[a, b]$ and differentiable on the open interval (a, b). If $f(a) = f(b)$, then there is a value c in (a, b) such that the line tangent to $f(x)$ at $(c, f(c))$ is parallel to the secant line from $(a, f(a))$ to $(b, f(b))$.*

Figure 5.23 shows a scenario just like the one in Fig. 5.22, except that is has been "tipped" on its side. Here, we see points $(a, f(a))$ and $(b, f(b))$ for which $f(a)$ is not equal to $f(b)$. There is still, though, a (dashed) secant line connecting $(a, f(a))$ and $(b, f(b))$. There is still a location $(c, f(c))$ where the (dashed) line tangent to $f(x)$ is seen to be parallel to the secant line between the endpoints. In this case, it just so happens that these lines are not horizontal.

So was the requirement that $f(a) = f(b)$ given in Rolle's Theorem really necessary? Well, for Rolle's Theorem itself, which poses the existence of a point $x = c$ where $f'(c) = 0$, then yes, it was necessary. But if we're talking about the rewording of Rolle's Theorem in Useful Fact 5.5, then no, it is not necessary. If we drop the requirement that $f(a) = f(b)$, everything else in the final interpretation still holds up. Useful Fact 5.5 can be updated without harm by the removal of the requirement $f(a) = f(b)$.

Useful Fact 5.6. *Suppose that $f(x)$ is continuous on the closed interval $[a, b]$ and differentiable on the open interval (a, b). Then whatever the values of $f(a)$ and $f(b)$ are, there is guaranteed to be a value c in (a, b) such that the line tangent to $f(x)$ at $(c, f(c))$ is parallel to the secant line from $(a, f(a))$ to $(b, f(b))$.*

The Mean Value Theorem

Useful Fact 5.6 is an informal statement, and we should tighten it up. Let's put our language about lines being parallel into measurements of their slopes. The slope of the tangent line at $(c, f(c))$ is $f'(c)$, and the slope of the secant line between $(a, f(a))$ and $(b, f(b))$ is $\dfrac{f(b) - f(a)}{b - a}$. This is old news. With these quantities, we can rephrase Useful Fact 5.6 into the Mean Value Theorem:

Theorem 5.3. *Suppose that $f(x)$ is continuous on the closed interval $[a,b]$ and differentiable on the open interval (a,b). Then there is a number c in (a,b) such that*

$$f'(c) = \frac{f(b) - f(a)}{b - a}$$

If the Mean Value Theorem is just tossed into your lap in this form, it can be hard to really see its importance. Okay great, there's a formula involving a value of the derivative. So what? Well, first, when you understand the geometry behind it, then it becomes a bigger deal. But even so, it turns out that the Mean Value Theorem itself is not used by itself as a tool very often. It's real power lies in the fact that it often becomes a building block in the development of even stronger ideas. It's like a powerhouse relief pitcher of Calculus — maybe you don't see it often, but when you do, you're really glad you have it.

Let's put it to work and get familiar with it.

EX 2 Does the function $f(x) = x^3/3 + 1$ on $[0,2]$ satisfy the conditions of the Mean Value Theorem? If so, find a value c which is guaranteed by the MVT.

Since $f(x)$ is a polynomial, it is certainly continuous on $[0,2]$ and differentiable on $(0,2)$, since polynomials are both continuous and differentiable everywhere. On the interval $[a,b] = [0,2]$, we have $f(0) = 1$ and $f(2) = 11/3$, so that

$$\frac{f(b) - f(a)}{b - a} = \frac{4}{3}$$

The MVT guarantees existence of a point $x = c$ where $f'(c) = 4/3$, i.e. $x^2 = 4/3$. Algebraically, this equation provides two values, $x = \pm 2/\sqrt{3}$, but only one of those values lies within the interval of interest. Therefore, we've found $c = 2/\sqrt{3}$. Figure 5.24 shows $f(x)$, the secant line between endpoints on the interval $[0,2]$, and the line tangent to $f(x)$ at $x = c$. ■

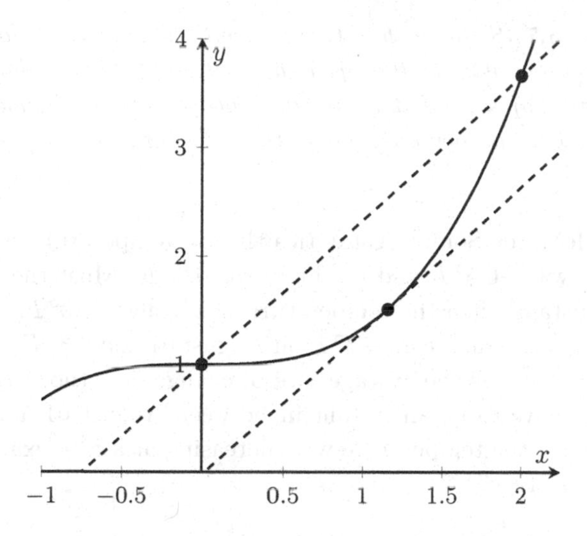

Fig. 5.24 A Mean Value Theorem scenario, with EX 2.

> **You Try It**
>
> (2) Does the function $f(x) = |x|$ on $[-2, 2]$ satisfy the conditions of the Mean Value Theorem? If so, find a value c which is guaranteed by the MVT.
>
> (3) Does the function $f(x) = \sin x$ on $[0, \pi/2]$ satisfy the conditions of the Mean Value Theorem? If so, find a value c which is guaranteed by the MVT.

Another interpretation of the Mean Value Theorem can be generated if we recall the tie between slopes of tangent and secant lines and rates of change. Specifically, you should remember that:

- The slope of a secant line between two points $(a, f(a))$ and $(b, f(b))$ on the graph of $f(x)$ measures the average rate of change of $f(x)$ between $x = a$ and $x = b$. That is, the average rate of change is $\dfrac{f(b) - f(a)}{b - a}$.
- The slope of a tangent line at a point $(c, f(c))$ on the graph of $f(x)$ measures the instantaneous rate of change of $f(x)$ there.

If we use these two ideas, we can present:

Useful Fact 5.7. *Suppose that $f(x)$ is continuous on the closed interval $[a, b]$ and differentiable on the open interval (a, b). Then there will be a location c in (a, b) such that the instantaneous rate of change of $f(x)$ at $x = c$ is equal to the overall average rate of change on the interval from $x = a$ to $x = b$.*

EX 3 In McMurdo Sound, Antarctica the low temperature on August 01, 2020 was $-6.5° F$ and the high was $0.4° F$. Must there have been an instant when the temperature was exactly $-3° F$ on that day? If the temperature hit $-3° F$ on August 01 and $6.3° F$ on November 01,[3] what was the average rate of change over those 92 days? Did there have to be an instant in between August 01 and November 01 when the temperature was increasing at a rate exactly equal to that average?

The first part of this question is a flashback to the Intermediate Value Theorem (Sec. 2.4). Since a function representing temperature is necessarily continuous, if the temperature function had endpoint values of -6.5 and 0.4, then there must have been an instant when the temperature hit every value in between — specifically, $-3° F$.

Between Aug 01 and Nov 01, the temperature increased from $-3° F$ to $6.3° F$ over those 92 days, and so the average rate of change (increase) was $\Delta T = \dfrac{6.3 - (-3)}{92} \approx 0.101$ degrees per day. Again, since a temperature function must be continuous and differentiable, then by the Mean Value Theorem, yes, there must be an instant between Aug 01 and Nov 01 when the instantaneous rate of change of temperature was also equal to 0.101 degrees per day. ■

You Try It

(4) According to the *NY Times*, the number of *new* cases of Covid-19 reported in the US was 33,970 on May 01, 2020 and 184,286 on December 01, 2020. Calculate the average rate of change of this number over those 214 days. Must there have been a time between May 01 and December 01 on which the instantaneous rate of change of this number was equal to this average?

[3]Temperature data from weatherbase.com

Root Finding: Bisection Method and Newton's Method

As long as we're thinking about the Intermediate Value Theorem again, and we also now know about derivatives and tangent lines, it's a good time to see in action some numerical techniques that help find roots which the IVT can guarantee are there. You should recall from Sec. 2.4 this simple story: if $f(x)$ is continous on an interval $[a, b]$, and if $f(a)$ and $f(b)$ have different signs, then there must be a root of $f(x)$ somewhere between $x = a$ and $x = b$.

The *bisection method* exploits this idea to zoom in on a root. If we've tested $f(x)$ on $[a, b]$ and we find that $f(a)$ and $f(b)$ have opposite signs, then if we drop in the midpoint of the interval, $c = (a + b)/2$ and look at the sign of $f(c)$, then we will have narrowed down the location of the root. We can repeat this process as many times as we'd like, to find the root with as much accuracy as we'd like. It's a simple, but inefficient process.

EX 4 The function $f(x) = e^x - 2$ has a root somewhere in the interval $[0, 1]$. Perform 5 steps of the bisection method to zoom in on the location of the root.

First, we get the signs of the original endpoints: $f(0) < 0$ and $f(1) > 0$. That's why the IVT guarantees a root on the interval $[0, 1]$. Now we can start bisecting (follow along in Fig. 5.25):

- Step 1: The midpoint of $[0, 1]$ is 0.5, and $f(0.5) = e^{0.5} - 2 \approx -0.35$. As $f(0.5) < 0$, we now know the root is in the interval $[0.5, 1]$.
- Step 2: The midpoint of $[0.5, 1]$ is 0.75, and $f(0.75) \approx 0.12$. As $f(0.75) > 0$, we now know the root is in the interval $[0.5, 0.75]$.
- Step 3: The midpoint of $[0.5, 0.75]$ is 0.625, and $f(0.625) \approx -0.13$. As $f(0.625) < 0$, we now know the root is in the interval $[0.625, 0.75]$.
- Step 4: The midpoint of $[0.625, 0.75]$ is 0.6875, and $f(0.6875) \approx -0.011$. As $f(0.6875) < 0$, we now know the root is in the interval $[0.6875, 0.75]$.
- Step 5: The midpoint of $[0.6875, 0.75]$ is 0.71875, and $f(0.71875) \approx 0.0752$. As $f(0.7175) > 0$, we now know the root is in the interval $[0.6875, 0.71875]$.

So by now, we have found the root to be in between $x = 0.6875$ and $x = 0.71875$. In Fig. 5.25, each step is shown graphically (from the top, down); the half-interval in which the root is trapped by the Intermediate Value Theorem (shown by a thick line segment) becomes the full interval in the next step; in this way, we repeatedly cut the size of the interval containing the root in half. Of course, we know the true root of $f(x) = e^x - 2$ is $x = \ln 2 \approx 0.69$, which is indeed in the interval $[0.6875, 0.71875]$. Our search technique is zooming in on this very slowly. ■

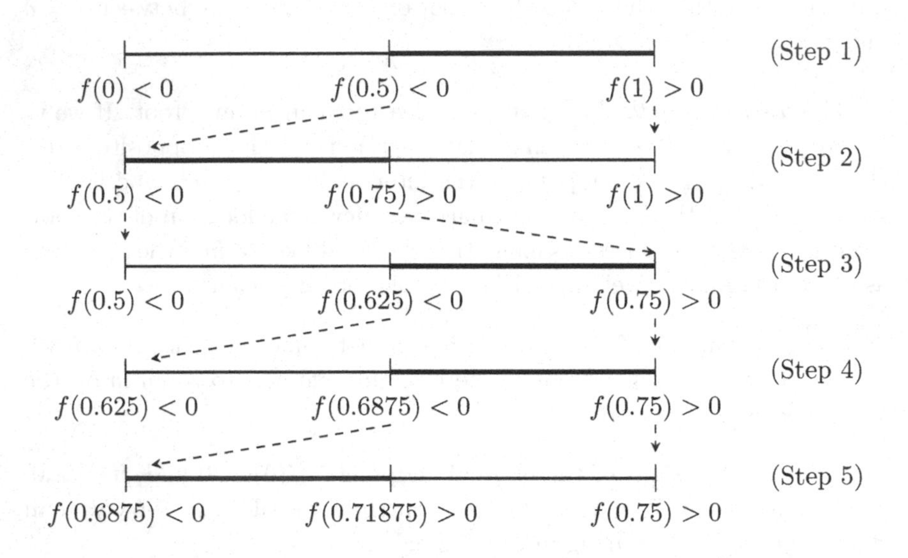

Fig. 5.25 The Bisection Method, with EX 4.

You Try It

(5) The function $f(x) = \cos^2(x) - \sqrt{x}$ has a root somewhere in the interval $[1/2, 1]$. Perform 3 steps of the bisection method to zoom in on the location of the root.

The bisection method is a glacially slow method of tracking down a root of a function, although you can't really go wrong with it. A different technique, called **Newton's Method**, does the job more quickly; it relies on the graph of $f(x)$ and tangent lines on that graph to perform "numerical target practice" for finding a root. There are two levels at which we can watch the process operate: first, we follow an excruciating step by step process to understand what is happening. But then, from this excruciating process, we develop a quick and easy formula that leaps us from one guess

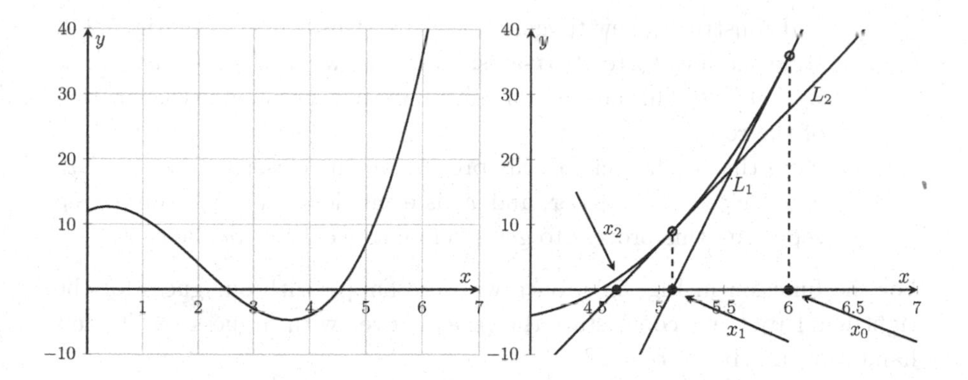

Fig. 5.26 $f(x) = x^3 - 6x^2 + 4x + 12$, with two visible roots.

Fig. 5.27 Two steps of Newton's Method.

as to the location of the root to a next, more accurate guess. Then we can repeat the process many times, very quickly. In fact, using a spreadsheet to automate the process is highly recommended.

Here is an example of the slow, step by step, process. Figure 5.26 shows a graph of the function $f(x) = x^3 - 6x^2 + 4x + 12$. Clearly, there are two roots: one has a value a bit larger than 2, and one has a value a bit larger than 4. To get our process started, we need a guess as to the value of a root. A bad guess as to the value of the larger root is $x = 6$. This is a bad guess, but that's fine. Here's how this works (follow along on Fig. 5.27, which is zoomed in a bit on the right hand root from Fig. 5.26):

- Start on the x-axis at the initial guess of the root, named $x_0 = 6$. This is not the root. Go to the graph of $f(x)$ at the point $(x_0, f(x_0)) = (6, 36)$ (follow the dashed line from (6,0) to $f(x)$ on Fig. 5.27).
- Construct the line tangent to the graph at this point $(x_0, f(x_0)) = (6, 36)$, and follow this line down to the x-axis (see the line L_1 on Fig. 5.27). You know how to do this, so I can just tell you that the equation of this tangent line is $L_1(x) = 40(x - 6) + 36$, and it intersects the x-axis at $x = 5.1$. The location where this tangent line hits the x-axis is our next "guess" as to the location of the root; and we'll name it $x_1 = 5.1$.
- Now start at x_1 and repeat the above two steps. Since $f(x_1) = f(5.1) = 8.991$, we go to the point $(5.1, 8.991)$ on the graph of $f(x)$

and construct a new tangent line there (this is line L_2 in Fig. 5.27). Ride the line L_2 to the x-axis; the point where L_2 hits the axis is $x_2 = 4.668$. This now becomes our new new guess as to the location of the root.

- Note that each cycle of this process got us closer to the root. x_0 is not close; x_1 is closer, and x_2 is even closer yet. We could keep repeating this process to get as close as we'd like to the root.

For the function in Fig. 5.26, once we were happy with our guess for the right hand root, we could start the process over with a guess for the left hand root, maybe at $x_0 = 2$?

Performing Newton's Method in this step by step manner is fun and all, but is very slow and cumbersome. However, it is very systematic: pick a point on the x-axis, go to the graph of $f(x)$, fire off a tangent line, ride that tangent line back to the x-axis, repeat. All of this can be wrapped up in one calculation, so that x_0 generates x_1, then x_1 generates x_2, and so on. This is called an *iterative* process, and in general, we can say that x_n produces x_{n+1}. Here's how we get there:

Suppose our current guess as to the location of the root of $f(x)$ is $x = x_n$. If we hop to the graph of $f(x)$ at the point $(x_n, f(x_n))$, the line $L(x)$ tangent to $f(x)$ has the equation

$$L(x) - f(x_n) = f'(x_n)(x - x_n)$$

(this is just $y - y_0 = m(x - x_0)$ tailored to our notation). This line intersects the x-axis when $L(x) = 0$, or when

$$\begin{aligned} 0 - f(x_n) &= f'(x_n)(x - x_n) \\ -\frac{f(x_n)}{f'(x_n)} &= x - x_n \\ x &= x_n - \frac{f(x_n)}{f'(x_n)} \end{aligned}$$

This new x coordinate is our next guess as to the location of the root. So, we can rename it x_{n+1},

$$\boxed{x_{n+1} = x_n - \frac{f(x_n)}{f'(x_n)}} \tag{5.4}$$

Equation (5.4) is Newton's Method rolled up into one calculation which gets repeated over and over. If we have $f(x)$ (and so also $f'(x)$), this cascade of values of x should (hopefully) zoom in on the root. As much as everyone would love to crank out such repetetive calculations by hand, this is the sort of thing that is ideal for a spreadsheet.

FFT: There are conditions which must be satisfied in order for Newton's Method to be applicable. They are not stated here. Can you think of one or two conditions that would be necessary?

EX 5 Use five iterations of Newton's Method with an initial value of $x_0 = 2$ to discover the root of $f(x) = e^x - 2$. Compare to EX 4.

We are searching for x_5, starting with $x_0 = 2$. The ingredients we need are $f(x) = e^x - 2$ and $f'(x) = e^x$. We can code Eq. (5.4) into a spreadsheet as follows:

- Plan one column to hold values of $f(x)$ for each x_n.
- Plan another column to hold values of $f'(x)$ for each x_n.
- Plan a third column to hold successive values of x.

Figures 5.28 to 5.30 show this implemented in Google Sheets. In Fig. 5.28, Column A is to contain values of $f(x)$, Column B is to contain values of $f'(x)$, and Column C contains will contain the sequence of successive estimates of the root, i.e. x_0, x_1, x_2, ..., x_5. We see the initial value $x_0 = 2$ in place in cell C2, and the value of $f(x_0) = e^{x_0} - 2$ being created in cell A3 with the calculation / command " `=EXP(C2)-2` ". While not shown, you can imagine that cell B3, which holds $f'(x_0)$, is filled using " `=EXP(C2)` ". Then, with these values in place, Fig. 5.29 shows how Newton's Method, Eq. (5.4), is implemented by reference to cells C2, A3, B3. This creates the updated guess as to the root, named x_1, held in cell C3. Then, by copying the rows down a few times, we can generate the remaining iterations — see Fig. 5.30.

In EX 4, we used the Bisection Method to determine this root, and after 5 steps, we were not very close: we narrowed the root down to the interval [0.6875, 0.71875]. In Newton's Method, after only 5 steps, we have already fixed the root to four places after the decimal. That's pretty good! ■

You Try It

(6) Use five iterations of Newton's Method with an initial value of $x_0 = 2$ to discover the smaller positive root of $f(x) = x^3 - 6x^2 + 4x + 12$ (see Fig. 5.26).

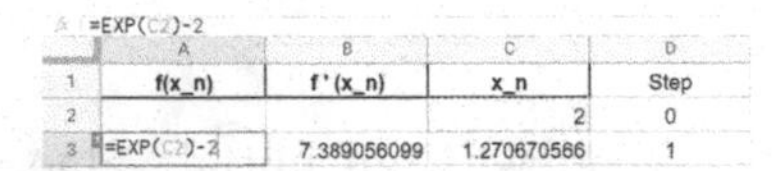

=EXP(C2)-2

	A	B	C	D
1	f(x_n)	f'(x_n)	x_n	Step
2			2	0
3	=EXP(C2)-2	7.389056099	1.270670566	1

Fig. 5.28 Spreadsheet cells for Newton's Method (a), w/ EX 5.

=C2-A3/B3

	A	B	C	D
1	f(x_n)	f'(x_n)	x_n	Step
2			2	0
3	5.389056099	7.389056099	=C2-A3/B3	1

Fig. 5.29 Spreadsheet cells for Newton's Method (b), w/ EX 5.

B	C	D	E
f(x_n)	f'(x_n)	x_n	Step
		2	0
5.389056099	7.389056099	1.270670566	1
1.563241151	3.563241151	0.8319573037	2
0.2978118574	2.297811857	0.702350584	3
0.01849177	2.01849177	0.6931894023	4
0.000084445163	2.000084445	0.6931471815	5

Fig. 5.30 Spreadsheet cells for Newton's Method (c), w/ EX 5.

You can imagine that the Newton's Method process is not perfect. You have to have somewhat of an idea where the root is to start with. When graphs have oddities in their shapes, you can actually end up moving away from the root instead of towards it, or oscillating back and forth around it. Newton's Method certainly does not work when there are discontinuities in the region you are searching. If tinkering with this sort of things sounds fun, then a class in *Numerical Analysis* should be in your future. On the other hand, you may find this uninteresting — after all, if you're going to find roots of a function, you're just going to use your calculator, or Wolfram Alpha, or some other tool, right? Well, these techniques provide the foundations for the algorithms used by your tech, and if you have not fought your way through some simple examples to start gaining intuition about how root finding works, then you will have no intuition to tell you when something has gone wrong. Use of tech should enhance your knowledge, not replace it.

FFT: We have considered two new theorems in this section: Rolle's Theorem, and the Mean Value Theorem. Here are a couple of things to puzzle over:

- We saw Rolle's Theorem first to ease into things, but really, Rolle's Theorem is an immediate consequence of the Mean Value Theorem. Why?
- Both Theorems start with the same phrase, "Suppose that $f(x)$ is continuous on the closed interval $[a, b]$ and differentiable on the open interval (a, b)." Why do we need to consider the closed interval $[a, b]$ and the open interval (a, b)? Why not just say, "Suppose that $f(x)$ is continuous and differentiable on the interval $[a, b]$"? Surely that would be simpler, right?

Have You Learned...

- The formal statement of Rolle's Theorem?
- Geometric interpretations of Rolle's Theorem?
- How Rolle's Theorem can generalize into the Mean Value Theorem?
- Geometric interpretations of the Mean Value Theorem?
- The formal statement of the Mean Value Theorem?
- An interpretation of the Mean Value Theorem relative to rates of change?
- How to (sometimes) find roots of a function with the Bisection Method?
- How to (sometimes) find roots of a function with Newton's Method?

Good Value — Problem List

Good Value — You Try It

These appeared above; solutions begin on the next page.

(1) Show that $f(x) = -5x^3 + 21x^2 - 28x + 30$ meets the conditions of Rolle's Theorem on $[1, 2]$. There are *two* values $x = c$ guaranteed by Rolle's Theorem to exist on that interval; find them both. Numerical estimates are OK.
(2) Does the function $f(x) = |x|$ on $[-2, 2]$ satisfy the conditions of the Mean Value Theorem? If so, find a value c which is guaranteed by the MVT.
(3) Does the function $f(x) = \sin x$ on $[0, \pi/2]$ satisfy the conditions of the Mean Value Theorem? If so, find a value c which is guaranteed by the MVT.
(4) The number of *new* cases of Covid-19 reported in the US was 33,970 on May 01, 2020 and 184,286 on December 01, 2020.[4] Calculate the average rate of change of this number over those 214 days. Must there have been a time between May 01 and December 01 on which the instantaneous rate of change of this number was equal to this average?
(5) The function $f(x) = \cos^2(x) - \sqrt{x}$ has a root somewhere in the interval $\left[\frac{1}{2}, 1\right]$. Perform 3 steps of the bisection method to zoom in on the location of the root.
(6) Use five iterations of Newton's Method with an initial value of $x_0 = 2$ to discover the smaller positive root of $f(x) = x^3 - 6x^2 + 4x + 12$ (see Fig. 5.26).

Good Value — Practice Problems

Try these as you get the hang of the You Try It problems. Solutions to these problems are available in Sec. A.5.5.

(1) Show that $f(x) = \sin(x)$ meets the conditions of Rolle's Theorem on $[\pi/4, 9\pi/4]$. How many appropriate values $x = c$ are guaranteed by Rolle's Theorem to exist on that interval? How many such values are there, really? Find them all (exactly).
(2) Does the function $f(x) = x^3 + x^2$ on $[-1, 1]$ satisfy the conditions of

[4]Data from The New York Times.

the Mean Value Theorem? If so, find a value c which is guaranteed by the MVT.

(3) Does the function $f(x) = |\cos(x)|$ on $[0, 2\pi]$ satisfy the conditions of the Mean Value Theorem? If so, find a value c which is guaranteed by the MVT.

(4) Does the function $f(x) = \sec(x)$ on $[0, \pi]$ satisfy the conditions of the Mean Value Theorem? If so, find a value c which is guaranteed by the MVT.

(5) Does the function $f(x) = \sinh(x)$ on $[0, \ln 2]$ satisfy the conditions of the Mean Value Theorem? If so, find a value c which is guaranteed by the MVT.

(6) On January 01, 2020, the price of one bitcoin was \$7194 USD. On June 02, 2020, the price was \$10,211. What is the average rate of change between those two days? Assuming the change in the conversion rate is a continuous function[5], did there have to be an instant in between January 01 and June 02 when the price was increasing at a rate exactly equal to the average rate of change?

(7) Use five iterations of Newton's Method with an initial value of $x_0 = 2$ to estimate a root of $f(x) = -x^4 + x^3 + 2$.

(8) Use Newton's Method with an initial value of $x_0 = -1$ to estimate a root of $f(x) = x^5 - x + 1$. How many iterations does it take so that your results are no longer changing for the first 5 digits after the decimal?

(9) Try to use Newton's Method to find the root of $f(x) = x^3$. Obviously the actual root is $x = 0$; pick a number close (but not equal to!) the actual root (like $x_0 = 1$ or $x_0 = -1$, etc.). How many iterations does it take to identify the root correct to three places after the decimal, i.e. $0.000nnn$? Why do you think this problem might be zooming in on the root so much slower than in, say, YTI 6 or PP 7?

Good Value — Challenge Problems

Try these problems to test your skills with the ideas in this section. Solutions to these problems are available in Sec. B.5.5.

(1) Does the function $f(x) = \ln x$ on $[0, e]$ satisfy the conditions of the Mean Value Theorem? If so, find a value c which is guaranteed by the MVT.

[5]Narrator: It's not...

(2) Suppose you are driving on a road that has a speed limit of 55mph. At one point, the police clock you driving at 50mph. Five minutes later, you pass another police car five miles away from the first one, and there you are clocked at 55mph. Why should you get a speeding ticket even though you were not speeding either time you passed a police car? (This question is related to the Mean Value Theorem — your answer must state what you know about this situation from the perspective of the MVT and what conclusion you can draw.)

(3) Try to use Newton's Method to find the a root of $f(x) = x^3 - 3x^2 + x - 1$ with an initial guess of $x_0 = 1$. What happens? Can you explain the behavior of the Newton's Method process? Try again with a different (better?) starting value. Are you able to detect a root? Find the value to a precision of $n.nnnn$.

Good Value — You Try It — Solved

(1) Show that $f(x) = -5x^3 + 21x^2 - 28x + 30$ meets the conditions of Rolle's Theorem on $[1, 2]$. There are *two* values $x = c$ guaranteed by Rolle's Theorem to exist on that interval; find them both. Numerical estimates are OK.

□ Since $f(x)$ is a polynomial, it is continuous on $[1, 2]$ and differentiable on $(1, 2)$. Also, the value of the function is the same at both endpoints: $f(1) = f(2) = 18$. So, the conditions of Rolle's Theorem are satisfied. Rolle's Theorem then guarantees at least one location $x = c$ in the interval (1.2) where $f'(c) = 0$. Since $f'(x) = -15x^2 + 42x - 28$, the quadratic equations gives roots of $f'(c)$ at

$$c = \frac{1}{15}\left(21 \pm \sqrt{21}\right)$$

These values are both in the given interval, $c \approx 1.09$ and $c \approx 1.71$. ■

(2) Does the function $f(x) = |x|$ on $[-2, 2]$ satisfy the conditions of the Mean Value Theorem? If so, find a value c which is guaranteed by the MVT.

□ This function is continuous on $[-2, 2]$, but it is *not* differentiable on $(-2, 2)$. The derivative of $|x|$ does not exist at $x = 0$. Do you remember why? So, $f(x)$ does not satisfy the conditions of the Mean Value Theorem on $[-2, 2]$. ■

(3) Does the function $f(x) = \sin x$ on $[0, \pi/2]$ satisfy the conditions of the Mean Value Theorem? If so, find a value c which is guaranteed by the MVT.

Since $f(x)$ is continuous and differentiable everywhere, it certainly satisfies the conditions of the Mean Value Theorem on $[0, \pi/2]$. Since we have

$$\frac{f(b) - f(a)}{b - a} = \frac{\sin\frac{\pi}{2} - \sin 0}{\frac{\pi}{2} - 0} = \frac{1 - 0}{\frac{\pi}{2}} = \frac{2}{\pi}$$

the Mean Value Theorem guarantees there is a point $x = c$ in the interval where $f'(c) = 2/\pi$. This value is the solution to $\cos c = 2/\pi$, or

$$c = \cos^{-1}\left(\frac{2}{\pi}\right) \approx 0.88$$

(Be sure to recognize that the MVT does not *provide* the value of c, we still have to use regular old algebra to do that. Rather, the MVT simply guarantees we are not wasting our time looking for that value within the given interval.) ■

---

(4) The number of *new* cases of Covid-19 reported in the US was 33,970 on May 01, 2020 and 184,286 on December 01, 2020.[6] Calculate the average rate of change of this number over those 214 days. Must there have been a time between May 01 and December 01 on which the instantaneous rate of change of this number was equal to this average?

□ The average rate of change over the 214 days was $(184286 - 33970)/214 \approx 702.5$ new cases per day.

Since these numbers are counting *people*, a function that describes these numbers should only have integer outputs; that is, the function is not continuous and so would not satisfy the conditions of the Mean Value Theorem, and we cannot make any conclusions about an instantaneous rate of change as compared to an overall average rate of change. Often, graphs of data such as this are visualized with a continuous graph, but technically, the underlying function cannot be continuous. ■

---

(5) The function $f(x) = \cos^2(x) - \sqrt{x}$ has a root somewhere in the interval $[1/2, 1]$. Perform 3 steps of the bisection method to zoom in on the location of the root.

□ The first bisection of the interval gives endpoints 1/2, 3/4, and 1, and (with help from tech), we have

$$f\left(\frac{1}{2}\right) \approx 0.062 \quad , \quad f\left(\frac{3}{4}\right) \approx -0.330 \quad , \quad f(1) \approx -0.708$$

and so the location of the root is now narrowed to the interval $[1/2, 3/4]$. The midpoint of this interval is 5/8, and

$$f\left(\frac{1}{2}\right) \approx 0.062 \quad , \quad f\left(\frac{5}{8}\right) \approx -0.132 \quad , \quad f\left(\frac{3}{4}\right) \approx -0.330$$

and so the location of the root is now narrowed to the interval $[1/2, 5/8]$. The midpoint of this interval is 9/16, and

$$f\left(\frac{1}{2}\right) \approx 0.062 \quad , \quad f\left(\frac{9}{16}\right) \approx -0.034 \quad , \quad f\left(\frac{5}{8}\right) \approx -0.132$$

[6]Data from The New York Times.

and so the location of the root is now narrowed to the interval $[1/2, 9/16]$, or in approximate form, the interval $[0.500, 0.563]$. ■

(6) Use five iterations of Newton's Method with an initial value of $x_0 = 2$ to discover the smaller positive root of $f(x) = x^3 - 6x^2 + 4x + 12$.

fx =D4-B5/C5

	A	B	C	D
1	Step	f(x)	f'(x)	x_n
2	0			2
3	1	4	-8	2.5
4	2	0.125000	-7.25000	2.51724
5	3	0.000451	-7.19738	=D4-B5/C5
6	4	0.000000	-7.19719	2.51730
7	5	0.000000	-7.19719	2.51730

Fig. 5.31 Spreadsheet cells for Newton's Method, with YTI 6.

□ We can code use a spreadsheet to iterate the Newton's Method formula, $x_{n+1} = x_n - f(x_n)/f'(x_n)$, with $f(x) = x^3 - 6x^2 + 4x + 12$ and $f'(x) = 3x^2 - 12x + 4$. Figure 5.31 shows the result. Column B contains the function entered as, for example, " `=D2^3-6*D2^2+4*D2+12` ". Column C has the derivative function entered similarly. Column D shows the successive iterated values of x_n computed according to Newton's Method. Note that after only three steps, we have identified the root to five places after the decimal. The root is $x \approx 2.51730$. ■

Chapter 6

The Best Mathematics Symbol There Is ... So Far

6.1 Sigma Notation

Introduction

In Sec. 6.2, we will see how to fire up the process of estimating the area under a curve. This will involve partitioning the area under the curve into rectangle-ish shapes, estimating the area of each "rectangle", and then adding all these little sub-areas to get an estimate of the total area. We will also note that the more rectangles we generate, the better the estimate should be. This process can be articulated much more efficiently if we utilize a notation that is built to express sums of many terms that are based on a pattern. In this very brief section, we'll see how summation notation (also called "sigma notation") works. This notation will also be the cornerstone of Chapter 10 (in Volume 2). However, this notation is *not* the notation referred to in the title of this chapter ... that's coming later!

Summation Notation

Sigma notation is a way to express a sum involving many terms in a brief manner. The generic format looks like this:

$$\sum_{k=1}^{n}(a_k)$$

In this expression, k is a counter which

- starts at the value given below the $\sum$
- ends at the value given above the $\sum$
- is always incremented by 1

The spot held by a_k contains the recipe that tells you what to do with each k. The $\sum$ directs us to add up all the results. For example, consider the sum

$$\sum_{k=1}^{5} k^2$$

The Greek letter $\sum$ indicates we're about to add some things up, there are 5 contributions to the sum associated with the counter (or *index*) $k = 1, 2, \ldots, 5$, and the recipe a_k says to square k for each contribution to the sum. Therefore,

$$\sum_{k=1}^{5} k^2 = (1)^2 + (2)^2 + (3)^2 + (4)^2 + (5)^2 = 1 + 4 + 9 + 16 + 25 = 55$$

A common component of a recipe for a sum is a multiple of -1 which causes the terms to alternate in sign. This little piece can come in many forms, like $(-1)^k$, $(-1)^{k+1}$, or $(-1)^{k-1}$. Whether this multiple causes the first term to be positive or negative depends on the starting value of k. But once the summation is underway, the ordering of signs of terms is either $+, -, +, -, \ldots$ or $-, +, -, +, \ldots$.

EX 1 Evaluate $\displaystyle\sum_{k=1}^{10} (-1)^k (k-1)$

We're going to have ten terms altogether, corresponding to $k = 1, 2, \ldots, 10$. The recipe for each term is $(-1)^k (k-1)$. The terms in our summation will alternate in sign because of the $(-1)^k$. Since the initial value for k is -1, the leading term in the summation should be negative (although the value of the lead term will be zero). We usually won't break the terms out into a table beforehand, but just for illustration purposes, the 10 terms in this summation are as follows; each k corresponds to a term $(-1)^k (k-1)$:

k	1	2	3	4	5	6	7	8	9	10
a_k	0	1	-2	3	-4	5	-6	7	-8	9

Therefore,

$$\sum_{k=1}^{10} (-1)^k (k-1) = 0 + 1 + (-2) + 3 + (-4) + 5 + (-6) + 7 + (-8) + 9 = 5 \quad \blacksquare$$

You Try It

(1) Evaluate $\displaystyle\sum_{k=0}^{4} \frac{k-1}{k+1}$.

We not only have to be able to evaluate a given sum, but to encode an extended sum into summation notation.

EX 2 Write the following sum in sigma notation:

$$\frac{1}{2}+\frac{2}{3}+\frac{3}{4}+\frac{4}{5}+\ldots+\frac{49}{50}$$

There are 49 terms in this sum. There is no one magic way to set up the index for a summation, but when there's no reason to do otherwise, it's best to start at 1. So our counter will start at $k=1$ and go to $k=49$. Each term is a fraction in which the numerator is k and the denominator is $k+1$. So we can write:

$$\frac{1}{2}+\frac{2}{3}+\frac{3}{4}+\frac{4}{5}+\ldots+\frac{49}{50}=\sum_{k=1}^{49}\frac{k}{k+1}$$

There is no one unique way to do this. We could also write the summation in any of these other forms, as well as numerous others:

$$\sum_{k=2}^{50}\frac{k-1}{k} \quad ; \quad \sum_{k=0}^{48}\frac{k+1}{k+2} \quad ; \quad \sum_{k=22}^{70}\frac{k-21}{k-20}$$

The right-most version is particularly silly, but it's still correct. ■

You Try It

(2) Write the following sum in sigma notation:

$$\frac{1}{2}-\frac{2}{3}+\frac{3}{4}-\frac{4}{5}+\ldots+\frac{49}{50}$$

A huge benefit of sigma notation is that we can sometimes develop a *closed form* value for the sum; this means that instead of having to calculate and add every single term, we have a small formula that will give us the final value simply by plugging in n, the final number of terms.

EX 3 Use trial and error to determine which of the following could be the correct closed form representation of $\sum_{i=1}^{n} i$.

$$\frac{n(n+1)}{2} \quad , \quad \frac{(n+1)(n+2)}{2} \quad , \quad \frac{(n-1)(n)}{2}$$

First, note that the index does not have to be k; we've changed it to i. The summation simply asks us to add a line of consecutive integers; for example,

$$\sum_{i=1}^{5} i = 1 + 2 + 3 + 4 + 5 = 15$$

$$\sum_{i=1}^{8} i = 1 + 2 + 3 + 4 + 5 + 6 + 7 + 8 = 36$$

$$\sum_{i=1}^{30} i = 1 + 2 + 3 + 4 + 5 + \cdots + 28 + 29 + 30 = 465$$

I computed the final 30-term summation using a spreadsheet. You'll get your own chance at that very soon. In these three samples, we had n values of 5, 8, and 30. Let's try each of those values with the candidates for the closed form of the sums. For $n = 5$, we get

$$\frac{5(5+1)}{2} = \frac{30}{2} = 15\,, \quad \frac{(5+1)(5+2)}{2} = \frac{42}{2} = 21\,, \quad \frac{(5-1)(5)}{2} = \frac{20}{2} = 10$$

For $n = 8$, we get

$$\frac{8(8+1)}{2} = \frac{72}{2} = 36\,, \quad \frac{(8+1)(8+2)}{2} = \frac{90}{2} = 45\,, \quad \frac{(8-1)(8)}{2} = \frac{56}{2} = 28$$

For $n = 30$, we get

$$\frac{30(30+1)}{2} = \frac{930}{2} = 465\,, \quad \frac{(30+1)(30+2)}{2} = \frac{992}{2} = 496\,,$$
$$\frac{(30-1)(30)}{2} = \frac{870}{2} = 435$$

It looks like the first candidate closed form got it right all three times, so we'd have to guess that it's the correct one. This is not a proof for a mathematical court of law, but it's the best we have right now! We can claim:

$$\sum_{i=1}^{n} i = \frac{n(n+1)}{2}$$

Note that the accuracy of the closed form depends just as much on the *starting* value of the counter; the starting value is not used in the closed form, but the closed form certainly counts on the summation starting with the indicated value of i. ■

You Try It

(3) Use trial and error to determine which of the following could be the correct closed form representation of $\sum_{i=1}^{n} i^2$.:

$$\frac{(n-1)(n+1)}{4}, \quad \frac{n(n+1)(2n+1)}{6}, \quad \frac{n^3-n}{5}$$

There are occasions when we want to change maintain the same sum, but change the indexing — that is, change the starting or ending value of the counter; but, we can't do one without the other. If there are supposed to be ten terms in a summation that currently starts with a counter value of $i = 1$, then the final counter value must be $i = 10$. But if, for some reason, we want to change the starting counter to $i = 0$, then the final counter value must also be changed, to $i = 9$. This is called **reindexing** a summation.

EX 4 Give equivalent (reindexed) versions of the summation $\sum_{k=1}^{5}(k-2)(k-1)(k+1)$ which start at (a) $k = 0$ and (b) $k = 2$.

There are five terms in this sum, which is currently set to start with $k = 1$ and go to $k = 5$:

k	1	2	3	4	5
$(k-2)(k-1)(k+1)$	$(-1)(0)(2)$	$(0)(1)(3)$	$(1)(2)(4)$	$(2)(3)(5)$	$(3)(4)(6)$

The second row in that table must remain the same regardless of the indexing; the numbers we're adding up have to stay the numbers we're adding up. So if we maintain that row as a target and provide a different set of 5 consecutive index values which generate them, the question becomes: what's the new recipe? Here are those five terms generated from $k = 0, 1, 2, 3, 4$:

0	1	2	3	4	k
$(-1)(0)(2)$	$(0)(1)(3)$	$(1)(2)(4)$	$(2)(3)(5)$	$(3)(4)(6)$	$(k-1)(k)(k+2)$

and if we use $k = 2, 3, 4, 5, 6$:

2	3	4	5	6	k
$(-1)(0)(2)$	$(0)(1)(3)$	$(1)(2)(4)$	$(2)(3)(5)$	$(3)(4)(6)$	$(k-3)(k-2)(k)$

Therefore, here are three equivalent (reindexed) representations of the same sum:

$$\sum_{k=1}^{5}(k-2)(k-1)(k+1)\,;\ \sum_{k=0}^{4}(k-1)(k)(k+2)\,;\ \sum_{k=2}^{6}(k-3)(k-2)(k) \quad \blacksquare$$

You Try It

(4) Give equivalent (reindexed) versions of the summation $\sum_{i=0}^{6} \frac{k^2}{k^2+1}$ which start at (a) $k = 1$ and (b) $k = 5$.

Summation Notation With Limits

This part of the section is either practice with, or motivation for learning how to use, spreadsheets. Microsoft Excel and Google Sheets are two common spreadsheet platforms. Spreadsheets are quite handy for evaluating sums, especially when the sums contain lots of terms. Later in Chapter 10 (in Volume 2) we're going to investigate series with lots of terms. And very shortly, in Sec. 6.3, we are going to combine summation notation with limits. I'm not telling you that now to make you toss this book aside and never come back to it (please don't!), but to make sure you have options for investigating such nasty things.

First, you have to learn not to be intimidated by an expression like this:

$$\lim_{n\to\infty} \sum_{k=1}^{n} \left(\frac{1}{2}\right)^k$$

If we remove the limit part of it, then the summation alone is not bad, we're just adding up powers of 1/2. One of the mind-blowing facts that comes in Calculus is that it's possible to add up an infinite number of terms and still get a finite result. So the inclusion of the limit sign here urges us to consider: what is the sum when $n = 10$? $n = 100$? $n = 100,000$? Do the results we get for the summation for larger and larger values of n look like they're closing in on some specific target?

A summation is said to **converge** if it gets closer to a finite number as n gets larger and larger, i.e. as we have $n \to \infty$. A summation **diverges** if it does not converge.

A spreadsheet is very handy to quickly generate a list of summations with more and more terms, like these:

$$\sum_{k=1}^{10} \left(\frac{1}{2}\right)^k \quad \sum_{k=1}^{20} \left(\frac{1}{2}\right)^k \quad \sum_{k=1}^{40} \left(\frac{1}{2}\right)^k \quad \sum_{k=1}^{80} \left(\frac{1}{2}\right)^k \quad \cdots \tag{6.1}$$

Figure 6.1 shows a screen shot of the ten-term summation shown in Eq. (6.1). The first column has the values of k. The second column has the corresponding term in the summation, generated with a cell formula; for example, the value in cell B7 is created using " `=(1/2)^(A7)`". The final sum in cell B12 is created with " `=SUM(B2:B11)`" FFT: If we extended the columns out to 20, 40, then 80 terms, do you think the final value of the sum might be getting closer and closer to a certain value? (Hint: It is!)

fx =(1/2)^(A7)

	A	B
1	k	(1/2)^k
2	1	0.5
3	2	0.25
4	3	0.125
5	4	0.0625
6	5	0.03125
7	6	0.015625
8	7	0.0078125
9	8	0.00390625
10	9	0.001953125
11	10	0.0009765625
12		0.9990234375

Fig. 6.1 Performing a summation in a spreadsheet.

When you plan to have a series involve a large upper value for the counter, intermediate values of the summation (for example, out to 10, then 20, then 30 terms) are called *partial sums*, and can be denoted s_{10}, s_{20}, and so on. This table shows the partial sums of the summation in (6.1) for increasing values of n:

n	10	20	40	80
s_n	0.999	0.999999	0.999999999999	1

(The actual value of s_{80} isn't quite 1, but at the normal numerical resolution, a spreadsheet will no longer be able to distinguish the value from 1, and so reports it as 1.) It looks like

$$\lim_{n\to\infty} \sum_{k=1}^{n} \left(\frac{1}{2}\right)^k = 1 \quad \blacksquare$$

Another useful technical tool is the web site Wolfram Alpha. You can ask this platform for partial sums by going there (just search, you'll find it!) and entering, for example, `sum (1/2)^k, k=1 to 80`. This platform can tell the difference between that result and the actual integer 1, and you will see the following result reported:

$$\sum_{k=1}^{80} \left(\frac{1}{2}\right)^k \approx 0.999\ldots999 \text{ 29 nines}$$

EX 5 Do some numerical experiments in a spreadsheet or Wolfram Alpha to determine if the following limit converges or diverges:

$$\lim_{n\to\infty} \sum_{k=1}^{n} \frac{1}{k(k+1)}$$

This table shows several estimated partial sums generated using Wolfram Alpha:

n	10	25	50	100
s_n	0.909	0.96	0.98	0.99

The partial sums are increase more slowly as n gets bigger, and it looks like there may be a cap of 1. I'm going to claim this limit / sum converges. ■

You Try It

(5) Do some numerical experiments in a spreadsheet or Wolfram Alpha to determine if the following limit converges or diverges:

$$\lim_{n\to\infty} \sum_{k=1}^{n} \frac{1}{k}$$

Have You Learned...

- How to write a summation in sigma notation, and read out a sum that is in sigma notation?
- How to adjust the starting value of the index in a summation that's in sigma notation?
- How to generate the final values of certain summations without doing the term by term addition?
- How to apply limits to some summations?
- How to impement a summation in a spreadsheet?

Sigma Notation — Problem List

Sigma Notation — You Try It

These appeared above; solutions begin on the next page.

(1) Evaluate $\sum_{k=0}^{4} \frac{k-1}{k+1}$.

(2) Write the following sum in sigma notation:

$$\frac{1}{2} - \frac{2}{3} + \frac{3}{4} - \frac{4}{5} + \ldots + \frac{49}{50}$$

(3) Use trial and error to determine which of the following could be the correct closed form representation of $\sum_{i=1}^{n} i^2$:

$$\frac{(n-1)(n+1)}{4} \quad , \quad \frac{n(n+1)(2n+1)}{6} \quad , \quad \frac{n^3 - n}{5}$$

(4) Give equivalent (reindexed) versions of the summation $\sum_{k=0}^{6} \frac{k^2}{k^2+1}$ which start at (a) $k = 1$ and (b) $k = 5$.

(5) Do some numerical experiments in a spreadsheet or Wolfram Alpha to determine if the following limit converges or diverges:

$$\lim_{n \to \infty} \sum_{k=1}^{n} \frac{1}{k}$$

Sigma Notation — Practice Problems

Try these as you get the hang of the You Try It problems. Solutions to these problems are available in Sec. A.6.1.

(1) Evaluate $\sum_{k=3}^{6} (k^2 - k)$.

(2) Write the following two sum in sigma notation:

$$2 + 4 + 6 + 8 + \ldots + 200 \qquad \text{and} \qquad 1 + 3 + 5 + 7 + \ldots + 201$$

(3) Use trial and error to determine which of the following could be the correct closed form representation of $\sum_{i=1}^{n} i^3$:

$$\left(\frac{n(n+1)}{2}\right)^2 \quad , \quad \left(\frac{(n+1)(n+2)}{2}\right)^2$$

(4) Give equivalent (reindexed) versions of the summation $\sum_{i=1}^{10} \sin\left(\frac{i\pi}{10}\right)^2$ which start at (a) $i = 0$ and (b) $i = -1$.

(5) Do some numerical experiments in a spreadsheet or Wolfram Alpha to determine if the following limit converges or diverges:

$$\lim_{n\to\infty} \sum_{k=1}^{n} \left(\frac{1}{2^k} - \frac{1}{2^{k+1}}\right)$$

Sigma Notation — Challenge Problems

Try these problems to test your skills with the ideas in this section. Solutions to these problems are available in Sec. B.6.1.

(1) Use the closed form values that we're confident with after EX 3, YTI 3, and PP 3 in this section to compute the final value of the sum $\sum_{k=1}^{100}(2k^3 - k^2 + k - 5)$.

(2) In PP 5 of this section, we saw that the following limit / sum likely converges to 1/2.

$$\lim_{n\to\infty} \sum_{k=1}^{n} \left(\frac{1}{2^k} - \frac{1}{2^{k+1}}\right)$$

Are we allowed to distribute the limit / sum operation? Is the following statement true?

$$\lim_{n\to\infty} \sum_{k=1}^{n} \left(\frac{1}{2^k} - \frac{1}{2^{k+1}}\right) = \lim_{n\to\infty} \sum_{k=1}^{n} \frac{1}{2^k} - \lim_{n\to infty} \sum_{k=1}^{n} \frac{1}{2^{k+1}}$$

(3) Repeat what you just did in PP5 and CP 2 for the following limit / sum:

$$\lim_{n\to\infty} \sum_{k=1}^{n} \left(\frac{1}{k} - \frac{1}{k+1}\right)$$

That is, investigate whether the following is true:

$$\lim_{n\to\infty} \sum_{k=1}^{n} \left(\frac{1}{k} - \frac{1}{k+1}\right) = \lim_{n\to\infty} \sum_{k=1}^{n} \frac{1}{k} - \lim_{n\to infty} \sum_{k=1}^{n} \frac{1}{k+1}$$

Sigma Notation — You Try It — Solved

---

(1) Evaluate $\sum_{k=0}^{4} \frac{k-1}{k+1}$.

☐ The value of $a_k = \frac{k-1}{k+1}$ for each $k = 0, 1, 2, 3, 4$ is given in this table:

k	0	1	2	3	4
$\frac{k-1}{k+1}$	$\frac{-1}{1}$	$\frac{0}{2}$	$\frac{1}{3}$	$\frac{2}{4}$	$\frac{3}{5}$

And so with those values simplified,

$$\sum_{k=0}^{4} \frac{k-1}{k+1} = -1 + 0 + \frac{1}{3} + \frac{1}{2} + \frac{3}{5} = \frac{-30 + 0 + 10 + 15 + 18}{30} = \frac{13}{30} \blacksquare$$

---

(2) Write the following sum in sigma notation:

$$\frac{1}{2} - \frac{2}{3} + \frac{3}{4} - \frac{4}{5} + \ldots + \frac{49}{50}$$

☐ There are 49 terms in this summation. If we start the index at $i = 1$, then the final counter value is 49. Each term has i divided by $i + 1$. Also, the terms alternate in sign, with the first term (for $i = 1$) positive, so each term needs a multiple of $(-1)^{i+1}$. Therefore,

$$\frac{1}{2} - \frac{2}{3} + \frac{3}{4} - \frac{4}{5} + \ldots + \frac{49}{50} \sum_{k=1}^{49} (-1)^{i+1} \frac{i}{i+1} \quad \blacksquare$$

---

(3) Use trial and error to determine which of the following could be the correct closed form representation of $\sum_{i=1}^{n} i^2$:

$$\frac{(n-1)(n+1)}{4} \quad , \quad \frac{n(n+1)(2n+1)}{6} \quad , \quad \frac{n^3 - n}{5}$$

☐ Some sample summations done the long way are:

$$\sum_{i=1}^{5} i^2 = 1 + 4 + 9 + 16 + 25 = 55$$

$$\sum_{i=1}^{10} i^2 = 1 + 4 + 9 + \cdots + 81 + 100 = 385$$

$$\sum_{i=1}^{20} i^2 = 1 + 4 + 9 + \cdots + 361 + 400 = 2870$$

Let's try the first one with the candidates for the closed form of the sums. For $n = 5$, we get

$$\frac{(n-1)(n+1)}{4} = \frac{(4)(6)}{4} = 6\,, \quad \frac{n(n+1)(2n+1)}{6} = \frac{5(6)(11)}{6} = 55\,,$$

$$\frac{5^3-5}{5} = 24$$

So we already know the only successful candidate is the second. Let's confirm it with the other two cases. For $n = 10$,

$$\frac{n(n+1)(2n+1)}{6} = \frac{10(11)(21)}{6} = (55)(7) = 385$$

For $n = 20$,

$$\frac{n(n+1)(2n+1)}{6} = \frac{20(21)(41)}{6} = (70)(41) = 2870$$

It sure looks like

$$\sum_{i=1}^{n} i^2 = \frac{n(n+1)(2n+1)}{6}$$

(I used Wolfram Alpha to generate the two largest sums.) ■

---

(4) Give equivalent (reindexed) versions of the summation $\displaystyle\sum_{k=0}^{6} \frac{k^2}{k^2+1}$ which start at (a) $k = 1$ and (b) $k = 5$.

□ There must be seven terms in each sum, starting with the term $\dfrac{0}{0+1}$ and ending with $\dfrac{6^2}{6^2+1}$.

If we start with $k = 1$, then we have to end with $k = 7$, and the terms are consistent with $\dfrac{(k-1)^2}{(k-1)^2+1}$.

If we start with $k = 5$, then we have to end with $k = 11$, and the terms are consistent with $\dfrac{(k-5)^2}{(k-5)^2+1}$.

Therefore, the following three sums are equivalent:

$$\sum_{i=0}^{6} \frac{k^2}{k^2+1}\,, \qquad \sum_{i=1}^{7} \frac{(k-1)^2}{(k-1)^2+1}\,, \qquad \sum_{i=5}^{11} \frac{(k-5)^2}{(k-5)^2+1}$$ ■

---

(5) Do some numerical experiments in a spreadsheet or Wolfram Alpha to determine if the following limit converges or diverges:

$$\lim_{n\to\infty} \sum_{k=1}^{n} \frac{1}{k}$$

□ This table shows several estimated partial sums generated using Wolfram Alpha:

n	10	50	100	1000
s_n	2.93	4.5	5.19	7.5

The terms are continuing to grow (very slowly) and there is no hint of a cap, so I'm going to claim this limit / sum diverges.

The fun part of working with Wolfram Alpha is that the platform will display the exact fractional representation of each partial sum; the result for $n = 1000$ is a quotient of two integers that are each over two hundred digits long! Go try it! ■

6.2 The Area Under a Curve

Introduction

Figure 6.2 shows the rate of discharge (total volumetric flow) of the Wabash River in Indiana, recorded at a data station of the United States Geological Survey.[1] That is, the graph shows the total amount of water in cubic feet per second flowing past this recording station for several months leading into early 2020. A measure of the grand total amount of water which flowed past this point for the entire duration shown here (about a year) would come from the area of the region under the curve. But how in the world can we find the area of such a strange region?

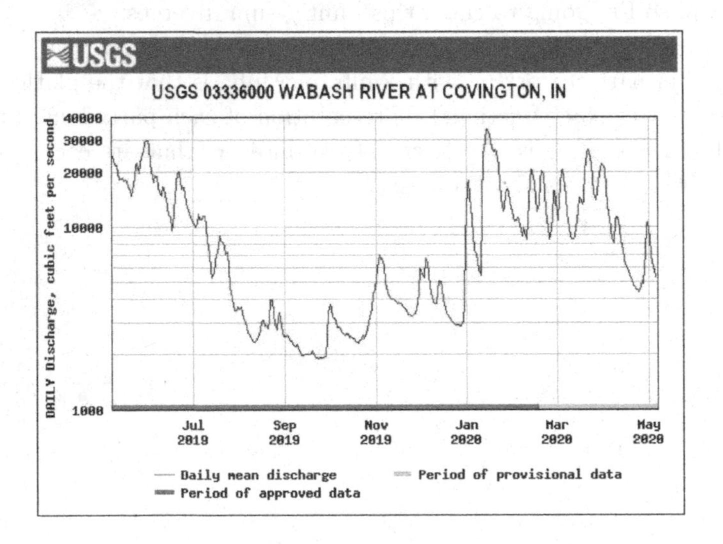

Fig. 6.2 Streamflow on the Wabash River (Indiana)

The Area Under a Curve

First, let's tackle the easy problems. If the graph of $f(x)$ is such that the area beneath it is a recognizable geometric shape, then finding the area under $f(x)$ over an interval $[a, b]$ is just a matter of using a known area formula. For example, the area under $f(x) = x$ on the interval $[0, 2]$ is shown in Fig. 6.3. The area in question is just a triangle, and its area is

[1] https://nwis.waterdata.usgs.gov/in/nwis/

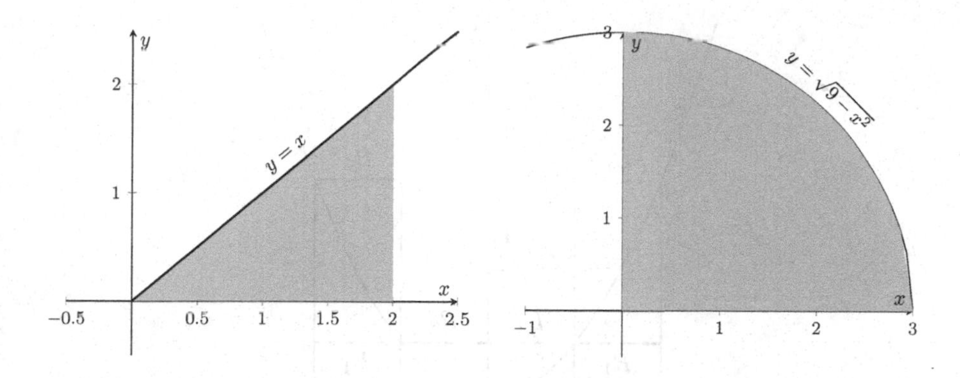

Fig. 6.3 Triangular area under $y = x$ on $[0, 2]$.

Fig. 6.4 Circular area under $y = \sqrt{9 - x^2}$ on $[0, 3]$.

$A = bh/2 = (2{\cdot}2)/2 = 2$. Easy! Or, consider the area under $f(x) = \sqrt{9 - x^2}$ on the interval $[0, 3]$, as shown in Fig. 6.4. This region a quarter of a circle. The area of the whole circle is $\pi(3)^2$, and so the (shaded) area we want, which is a quarter of that, is $A = 9\pi/4$. Again, easy!

With the easy (and therefore dull) cases out of the way, we can move on to the more interesting areas. Here is our general procedure. Given a function $f(x)$ and interval $[a, b]$, where we want the area under $f(x)$ over that interval, we

- Partition $[a, b]$ (which is on the x-axis) into n subintervals. (We get to pick n.)
- Treat each subinterval as the base of a standing rectangle.
- The height of each rectangle is determined by the function itself.
- Compute the area of each rectangle formed by the partition.
- Add up the areas of the rectangles to get an estimate of the total area.

During this description of our procedure, you may have noticed a certain vagueness in one item, which is illustrated in Fig. 6.5. The figure shows a partitioning[2] into four rectangles of the area under $f(x) = x^2 + 1$ on the interval $[-2, 2]$. Clearly, the width of each rectangle is 2. But what about the heights? We'll get a different measure of area of each rectangle depending on how we choose the heights. There are multiple options for

[2]This just means we chopped it up.

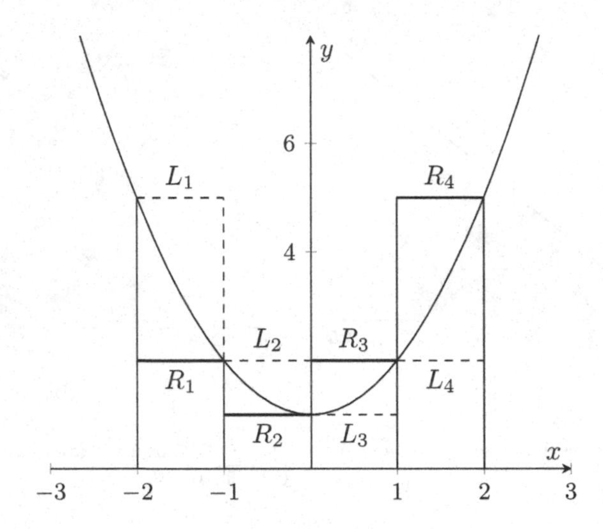

Fig. 6.5 Partitioning of the area under $y = x^2 + 1$ on $[-2, 2]$.

how we assign the height of each rectangle: use the left edges? The right edges? The middle? There is no correct or incorrect choice ... it's just a choice. As long as we do it consistently for all rectangles in a given region, and as long as we clearly *state* how we're picking them, we're good. Fortunately, calculus comes with some pre-defined descriptors which are fairly self-explanatory:

- The *left hand rule* uses the left edge of each rectangle as an estimate of its height. We use L_n to represent a left hand rule estimate of an area using n rectangles.
- The *right hand rule* uses the right edge of each rectangle as an estimate of its height. We use R_n to represent a right hand estimate of an area using n rectangles.

In Fig. 6.5, the dashed lines show rectangles formed according to the left hand rule; the thicker solid lines show the tops of rectangles formed according to the right hand rule. Let's finish estimating the area under $f(x) = x^2 + 1$ over $[-2, 2]$ in this example:

EX 1 The following table shows values of $f(x) = x^2 + 1$ tabulated over the interval $[-2, 2]$. Use these values to form left and right hand estimates of the area under $f(x)$ over $[-2,2]$.

x	-2	-1	0	1	2
$f(x)$	5	2	1	2	5

The area of each rectangle in an estimate is its base times its height. The base of every rectangle in this partitioning has width 1, for both the left and right hand rules. The left hand rule uses the left edge of each rectangle to assign the height. In the list of 5 endpoints in the partition, four of them correspond to left edges of rectangles; the right-most endpoint ($f(2) = 5$) is not used in the left hand estimate.

$$L_5 = 1(5) + 1(2) + 1(1) + 1(2) = 10$$

Since the right hand rule uses the right edge of each rectangle for its height, we will not use the left-most endpoint ($f(-2) = 5$) in the computation of the right hand estimate:

$$R_5 = 1(2) + 1(1) + 1(2) + 1(5) = 10$$

FFT: In this example, the left and right estimates came out the same, although that is not expected as a rule. Do you see why they came out the same? ■

You Try It

(1) Create left-hand and right-hand estimates for the area under the function $f(x)$ tabulated below. Identify which is a lower estimate and which is an upper estimate. (Hint: you'll have to infer from the table the interval $[a, b]$ and number n of rectangles.)

x	0	2	4	6	8	10
$f(x)$	1	3	4.25	5.35	6.25	7

Getting a tabulated function is nice because you don't have to compute the function's values yourself. And in many "real world" applications, data is collected by observation (i.e. tabulated) rather than extracted from a formula. The downside to tabulated data is that you're stuck with the resolution you have; that is, the table itself provides the interval $[a, b]$ and number n of rectangles you'll use. When we are not constrained by a table, we can vary n. Larger values of n should lead to better estimates of area.

EX 2 Create left-hand and right-hand estimates for the area under $f(x) = e^{-x}$ on the interval $[0, 2]$ using $n = 4$ rectangles. Identify which is a lower estimate and which is an upper estimate.

When we split $[0, 2]$ into 4 rectangles, we end up with 5 endpoints: $0, 1/2, 1, 3/2, 2$. We can make our own table of values, which are made into approximations:

x	0	1/2	1	3/2	2
$f(x)$	e^{-0}	$e^{-1/2}$	e^{-1}	$e^{-3/2}$	e^{-2}
$\approx$	1	0.6065	0.3678	0.2231	0.1353

The width of each rectangle is 1/2. The left hand estimate will use all but the last endpoint to find a height, and we have

$$L_4 = \frac{1}{2}(1) + \frac{1}{2}(0.6065) + \frac{1}{2}(0.3678) + \frac{1}{2}(0.2231) \approx 1.099$$

Since $f(x)$ is decreasing (going "downhill") over the interval $[0, 2]$, the left hand rule rectangles will include extra area from outside the function (do you see why?) and will give an overestimate of the true area.

The right hand estimate will use all but the first endpoint to find a height, and we have

$$R_4 = \frac{1}{2}(0.6065) + \frac{1}{2}(0.3678) + \frac{1}{2}(0.2231) + \frac{1}{2}(0.1353) + \approx 0.6664$$

Note that since L_4 overestimates the area under the curve and R_4 underestimates it, the true area under the curve is between R_4 and L_4. How could we improve our estimates? ■

You Try It

(2) Create left-hand and right-hand estimates for the area under $f(x) = 25 - x^2$ on the interval $[0, 5]$ using $n = 5$ rectangles. Identify which is a lower estimate and is an upper estimate.

Formalizing the Process

Left and right hand estimates to the area under $f(x)$ on $[a, b]$ for n rectangles are poor estimates for small values of n, and should improve as n increases. But using larger values of n will become cumbersome, so we need a better way to describe and generalize the process we're using.

Given a function $f(x)$, interval $[a, b]$, and number of rectangles n, we can immediately write many important components of the problem:

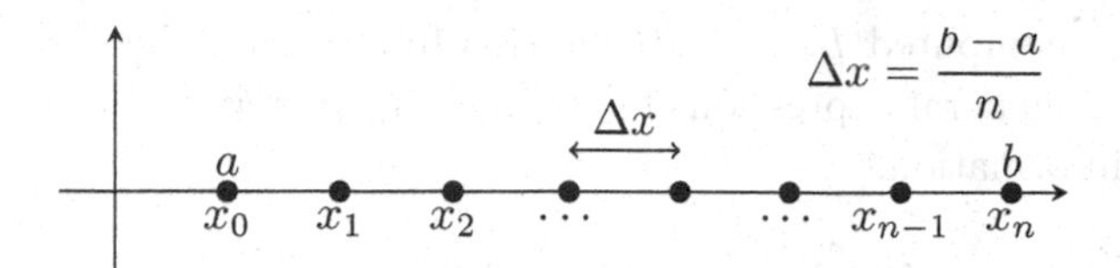

Fig. 6.6 Design of endoints x_i in a partitioning of n rectangles.

- Let's name the width of each of the n rectangles as Δx; then, $\Delta x = \dfrac{b-a}{n}$.
- Let's name the list of endpoints of the rectangles generated by the partitioning of $[a, b]$ into n rectangles as $x_0, x_1, \ldots, x_{n-1}, x_n$, where $x_0 = a$ and $x_n = b$. (These endpoints are shown in Fig. 6.6.) Since each endpoint is separated by Δx from the one before it, then $x_1 = a + \Delta x$, $x_2 = a + 2\Delta x$, $x_3 = a + 3\Delta x$, etc. Therefore, for a generic endpoint x_i (where $0 \leq i \leq n$),

$$x_i = a + i\Delta x = a + i\left(\frac{b-a}{n}\right)$$

- The heights at each endpoint are

$$h_i = f(x_i) = f(a + i\Delta x)$$

- The area of the ith rectangle is

$$A_i = h_i w_i = f(x_i)\Delta x$$

Now, the left and right hand rules are formed by the sums of *some* of the areas A_i. The left hand rule does not use endpoint x_n, and therefore

$$L_n = \sum_{i=0}^{n-1} f(x_i)\Delta x$$

The right hand rule does not use endpoint x_0, and therefore

$$R_n = \sum_{i=1}^{n} f(x_i)\Delta x$$

Writing these formulas for the left and right hand rules is much easier than writing out the summations explicitly.

EX 3 Write the left and right hand estimates L_n and R_n in summation form for the area under $f(x) = e^{-x}$ on the interval $[0, 2]$.

In EX 2, we computed L_4 and R_4 for this function on the given interval. But the more general expressions for L_n and R_n (for *any* n) are contructed using this information:

- The width of each rectangle will be $\Delta x = \dfrac{b-a}{n} = \dfrac{2}{n}$.
- The endpoints are given by $x_i = a + i\Delta x = 0 + i\left(\dfrac{2}{n}\right) = \dfrac{2i}{n}$.
- The heights are given by $f(x_i) = e^{-x_i} = e^{-2i/n}$.

Therefore,

$$L_n = \sum_{i=0}^{n-1} f(x_i)\Delta x = \sum_{i=0}^{n-1} e^{-2i/n} \cdot \frac{2}{n}$$

and

$$R_n = \sum_{i=1}^{n} f(x_i)\Delta x = \sum_{i=1}^{n} e^{-2i/n} \cdot \frac{2}{n}$$

We could now use these formulas to compute L_n and/or R_n for $n = 4$ or $n = 16$ or $n = 32$ and so on. ■

You Try It

(3) Write the left and right hand estimates L_n and R_n in summation form for the area under $f(x) = 25 - x^2$ on the interval $[0, 5]$.

Riemann Sums

OK, here's the *real* reason that we want to write the left and right hand rules in tidy summation notation: in order to improve our approximations to the area under $f(x)$ over $[a, b]$, we need to use larger values of n. The more rectangles we have, the closer our estimate will get to the true area under $f(x)$. If 100 rectangles gave a good estimate, then 1000 will give a better estimate. The estimate using 1000 rectangles will be improved by using 10,000 rectangles. And so on. In other words, we're asking: what's the trend in the value of the overall estimate as n gets larger and larger. Does that process sound familiar? That loud "Oh no!" I just heard from you suggests it does. We're letting n increase towards infinity. This is a *limit* process in which the value of the left and right hand estimates are closing in on (converging on) the true area as n gets bigger. If we let A

represent the *actual* area under $f(x)$ over $[a, b]$, we have

$$A = \lim_{n\to\infty} \sum_{i=1}^{n} f(x_i)\Delta x$$

This looks like the right hand estimate since i starts at 1 and not 0. But as n gets larger, Δx gets smaller, and in the limit, there won't be a difference anymore between L_n and R_n; so, we just use the set-up for R_n.

This combination of a limit with a sum to produce the area under a curve is called a *Riemann Sum.*

EX 4 Write the Riemann sum that describes the area under $f(x) = xe^x$ on $[1, 2]$.

All we have to do is form the generic form of R_n and slap a limit on it! To get R_n, we note that

- The width of each rectangle will be $\Delta x = \dfrac{b-a}{n} = \dfrac{2-1}{n} = \dfrac{1}{n}$.
- The endpoints are given by $x_i = a + i\Delta x = 1 + i\left(\dfrac{1}{n}\right) = 1 + \dfrac{i}{n}$.
- The heights are given by $f(x_i) = x_i e^{x_i} = \left(1 + \dfrac{i}{n}\right) e^{1+i/n}$.

Therefore,

$$R_n = \sum_{i=1}^{n} f(x_i)\Delta x = \sum_{i=1}^{n} \left(1 + \frac{i}{n}\right) e^{1+i/n} \cdot \frac{i}{n}$$

and the Riemann Sum is formed by the limit of R_n as $n \to \infty$,

$$A = \lim_{n\to\infty} \sum_{i=1}^{n} \left(1 + \frac{i}{n}\right) e^{1+i/n} \cdot \frac{i}{n} \quad \blacksquare$$

You Try It

(4) Write the Riemann sum that describes the area under $f(x) = \sqrt[4]{x}$ on $[1, 16]$.

You can test your comfort with Riemann sums by trying a problem the other way — given a Riemann sum, figure out the function and interval it is based on:

EX 5 The following Riemann sum represents the area under some $f(x)$ on an interval $[a, b]$. What are $f(x)$, a, and b?

$$\lim_{n\to\infty}\sum_{i=1}^{n}\frac{1}{\sqrt{2+2i/n}}\cdot\left(\frac{2}{n}\right)$$

We have to match the Riemann sum we're given to the general form

$$\lim_{n\to\infty}\sum_{i=1}^{n} f(x_i)\Delta x$$

Let's untangle this starting at the right. We see that $\Delta x = 2/n$. But we also know that in general $\Delta x = (b-a)/n$. So $b - a = 2$. This doesn't immediately tell us either a or b alone, though. So let's check out the rest; from the Riemann sum, we can identify

$$f(x_i) = \frac{1}{\sqrt{2+2i/n}}$$

which we can decode as

$$f(x) = \frac{1}{\sqrt{x}} \quad \text{for} \quad x_i = 2 + \frac{2i}{n} = 2 + i\cdot\frac{2}{n}$$

Since the general recipe for x_i is

$$x_i = a + i\cdot\Delta x$$

we can match up $a = 2$. And since we already determined $b - a = 2$, we now know $b = 4$. And that's it! The given Riemann sum represents the area under $f(x) = 1/\sqrt{x}$ on the interval $[2, 4]$.

FFT: What if instead of identifying

$$f(x) = \frac{1}{\sqrt{x}} \quad \text{for} \quad x_i = 2 + \frac{2i}{n}$$

we identified

$$f(x) = \frac{1}{\sqrt{2+x}} \quad \text{for} \quad x_i = \frac{2i}{n}$$

Why would the interval $[a, b]$ become $[0, 2]$ instead? The Riemann sum represents the same area no matter how we "decode" it, as long we we decode it properly. So how do we know that the area under $dspf(x) = \frac{1}{\sqrt{x}}$ on $[2, 4]$ is the same as the area under $f(x) = 1/\sqrt{2+x}$ on $[0, 2]$. ■

You Try It

(5) The following Riemann sum represents the area under some $f(x)$ on an interval $[a, b]$. What are $f(x)$, a, and b?

$$\lim_{n\to\infty} \sum_{i=1}^{n} \tan\left(\frac{i\pi}{4n}\right) \cdot \left(\frac{\pi}{4n}\right)$$

Have You Learned...

- How to approximate the area under a graph of some $f(x)$ using a whole bunch of rectangles?
- How to find the height of each rectangle using the function $f(x)$?
- How to determine the width of each rectangle using the two end-points of interest, and the number of rectangles to be created?
- How to estimate the area under a graph of some $f(x)$ by adding the areas of these rectangles?
- That there are choices to be made in the estimation process, which lead to slightly different estimates?
- How to describe the summation process using sigma notation, i.e. how to write Riemann sums?
- How to identify $f(x)$, a, b, and n for a given Riemann sum?
- That the word "under" in the phrase "area under a curve" has to be taken a bit loosely?

The Area Under A Curve — Problem List

The Area Under A Curve — You Try It

These appeared above; solutions begin on the next page.

(1) Create left-hand and right-hand estimates for the area under the function $f(x)$ tabulated below. Identify which is a lower estimate and which is an upper estimate.

x	0	2	4	6	8	10
$f(x)$	1	3	4.25	5.35	6.25	7

(2) Create left-hand and right-hand estimates for the area under $f(x) = 25 - x^2$ on the interval $[0, 5]$ using $n = 5$ rectangles. Identify which is a lower estimate and which is an upper estimate.

(3) Write the left and right hand estimates L_n and R_n in summation form for the area under $f(x) = 25 - x^2$ on the interval $[0, 5]$.

(4) Write the Riemann sum that describes the area under $f(x) = \sqrt[4]{x}$ on $[1, 16]$.

(5) The following Riemann sum represents the area under some $f(x)$ on an interval $[a, b]$. What are $f(x)$, a, and b?

$$\lim_{n\to\infty} \sum_{i=1}^{n} \tan\left(\frac{i\pi}{4n}\right) \cdot \left(\frac{\pi}{4n}\right)$$

The Area Under A Curve — Practice Problems

Try these as you get the hang of the You Try It problems. Solutions to these problems are available in Sec. A.6.2.

(1) Create left-hand and right-hand estimates for the area under the function $f(x)$ tabulated below. Identify which is a lower estimate and which is an upper estimate.

x	0	2	4	6	8	10	12
$f(x)$	9	8.8	8.2	7.25	6	4.1	1

(2) Create left-hand and right-hand estimates for the area under $f(x) = \sin(x)$ on the interval $[0, \pi]$ using $n = 4$ rectangles.

(3) Write the left and right hand estimates L_n and R_n in summation form for the area under $f(x) = \sin(x)$ on the interval $[0, \pi]$.

(4) Write the Riemann sum that describes the area under $f(x) = x \cos x$ on $[0, \pi/2]$.

(5) Write the Riemann sum that describes the area under $f(x) = \ln x/x$ on $[3, 10]$.

(6) The following Riemann sum represents the area under some $f(x)$ on an interval $[a, b]$. What are $f(x)$, a, and b?

$$\lim_{n\to\infty} \sum_{i=1}^{n} \left(5 + \frac{2i}{n}\right)^{10} \cdot \left(\frac{2}{n}\right)$$

The Area Under A Curve — Challenge Problems

Try these problems to test your skills with the ideas in this section. Solutions to these problems are available in Sec. B.6.2.

(1) Write the Riemann sum that describes the area under $f(x) = x/(1-x)$ on $[2, 5]$.

(2) Write the Riemann sum that describes the area under $f(x) = \cos(x^2)$ on $[0, \pi/2]$.

(3) The following Riemann sum represents the area under some $f(x)$ on an interval $[a, b]$. What are $f(x)$, a, and b?

$$\lim_{n\to\infty} \sum_{i=1}^{n} \ln\left(\left(\frac{i\pi}{n}\right)^2 + 1\right) \cdot \left(\frac{\pi}{n}\right)$$

The Area Under A Curve — You Try It — Solved

(1) Create left-hand and right-hand estimates for the area under the function $f(x)$ tabulated below. Identify which is a lower estimate and which is an upper estimate.

x	0	2	4	6	8	10
$f(x)$	1	3	4.25	5.35	6.25	7

☐ The interval presented by the x-values is $[a, b] = [0, 10]$. There are six endpoints, and therefore five rectangles. The width of each is 2. The left hand estimate will use the left edges of the rectangles as their heights; here is the sum of widths times heights for the five rectangles using their left edges as heights:

$$L_5 = 2(1) + 2(3) + 2(4.25) + 2(5.35) + 2(6.25) = 39.7$$

This will be an underestimate, since the function is increasing, and therefore the left hand rule will miss some area.

The right hand estimate will use the right edges of the rectangles as their heights:

$$R_5 \approx 2(3) + 2(4.25) + 2(5.35) + 2(6.25) + 2(7) = 51.7$$

This will be an overestimate, since the function is increasing, and therefore the right hand rule will include extra area. ■

(2) Create left-hand and right-hand estimates for the area under $f(x) = 25 - x^2$ on the interval $[0, 5]$ using $n = 5$ rectangles. Identify which is a lower estimate and which is an upper estimate.

☐ If $[0, 5]$ is partitioned into 5 rectangles, the endpoints are $x = 0, 1, 2, 3, 4, 5$. Then the width of each rectangle is 1. The function values at each of these endpoints is:

x	0	1	2	3	4	5
$f(x)$	25	24	21	16	9	0

The left hand estimate uses the left edges of the rectangles as their heights:

$$L_5 \approx 1(25) + 1(24) + 1(21) + 1(16) + 1(9) = 95$$

This will be an overestimate of the total area since the graph is concave down and the rectangles drawn from their left edges go above the curve itself and include small bits of area they shouldn't.

The right hand estimate uses the right edges of the rectangles as their heights:

$$R_5 \approx 1(24) + 1(21) + 1(16) + 1(9) + 1(0) = 70$$

This will be an underestimate of the total area since the graph is concave down and the rectangles drawn from their right edges go under the curve and miss small bits of area. ■

---

(3) Write the left and right hand estimates L_n and R_n in summation form for the area under $f(x) = 25 - x^2$ on the interval $[0, 5]$.

□ The more general expressions for L_n and R_n (for *any* n) are contructed using this information:

- The width of each rectangle will be $\Delta x = \dfrac{b-a}{n} = \dfrac{5}{n}$.
- The endpoints are given by $x_i = a + i\Delta x = 0 + i\left(\dfrac{5}{n}\right) = \dfrac{5i}{n}n$
- The heights are given by $f(x_i) = 25 - x_i^2 = 25 - \left(\dfrac{5i}{n}\right)^2$

Therefore,

$$L_n = \sum_{i=0}^{n-1} f(x_i)\Delta x = \sum_{i=0}^{n-1}\left(25 - \left(\frac{5i}{n}\right)^2\right) \cdot \frac{5i}{n}$$

and

$$L_n = \sum_{i=1}^{n} f(x_i)\Delta x = \sum_{i=1}^{n}\left(25 - \left(\frac{5i}{n}\right)^2\right) \cdot \frac{5i}{n} \blacksquare$$

---

(4) Write the Riemann sum that describes the area under $f(x) = \sqrt[4]{x}$ on $[1, 16]$.

□ We must construct

$$A = \lim_{n \to \infty} \sum_{i=1}^{n} f(x_i)\Delta x$$

For this problem, we have:

- $\Delta x = \dfrac{b-a}{n} = \dfrac{16-1}{n} = \dfrac{15}{n}$
- The individual endpoints are $x_i = a + i \cdot \Delta x = 1 + \left(\dfrac{15i}{n}\right)$
- The heights of the rectangles are $f(x_i) = \left(1 + \dfrac{15i}{n}\right)^{1/4}$

So we have

$$A = \lim_{n\to\infty} \sum_{i=1}^{n} f(x_i)\Delta x = \lim_{n\to\infty} \sum_{i=1}^{n} \left(1 + \frac{15i}{n}\right)^{1/4} \cdot \left(\frac{15}{n}\right)$$

We are not required to evaluate the limit. Hooray! ■

---

(5) The following Riemann sum represents the area under some $f(x)$ on an interval $[a, b]$. What are $f(x)$, a, and b?

$$\lim_{n\to\infty} \sum_{i=1}^{n} \tan\left(\frac{i\pi}{4n}\right) \cdot \left(\frac{\pi}{4n}\right)$$

☐ By comparing this to

$$\lim_{n\to\infty} \sum_{i=1}^{n} f(x_i)\Delta x$$

we can identify the following:

- Since $\Delta x = \dfrac{b-a}{n} = \dfrac{\pi}{4n}$ then we can identify $b - a = \dfrac{\pi}{4}$.
- $f(x_i) = \tan(x_i)$ where...
- $x_i = a + i \cdot \Delta x = i \cdot \dfrac{\pi}{4n}$. Writing this as $x_i = 0 + i\left(\dfrac{\pi}{4n}\right)$ and matching to the general form $x_i = a + i\Delta x$, we can identify $a = 0$

Since $a = 0$ and $b - a = \dfrac{\pi}{4}$, then $b = \dfrac{\pi}{4}$. This Riemann sum computes the area under $f(x) = \tan x$ on $[0, \pi/4]$. ■

6.3 Definite Integrals

Introduction

We'll introduce a new (or not so new) notation to represent the area under $f(x)$ over $[a, b]$, and learn to manipulate the notation.

The Area Problem Revisited

From now on, we will use the following notation to represent the area under $f(x)$ over the interval $[a, b]$:

$$\int_a^b f(x)\,dx$$

You should find a certain piece of that notation familiar, with slight modifications; for now, just accept that it's there and don't ask questions. The mystery will be revealed later, and we don't want to spoil it! This is called a *definite integral*, since the means of computing the area under $f(x)$ involves totalling (or integrating) the areas of many subregions.

Note that this is just the *symbol* we're using to represent the area under $f(x)$ over $[a, b]$ so that we don't have to keep writing "the area under $f(x)$ over $[a, b]$" again and a, since that gets tiresome. The symbol is independent of the means to compute the area, it's just the symbol for it. In fact, right now, we only know of two ways to find the area under a curve:

(1) By simple geometry, if we're lucky enough that the area under $f(x)$ forms a known geometric shape like a triangle, trapezoid, or circle.
(2) By setting up and computing a Riemann sum.

And even for the second one, note that we have never actually computed a Riemann sum to its full extent; we've been limited to only *estimating* a Riemann sum by using only a finite number of rectangles (i.e. finding the left and right hand estimates L_n and R_n).

But even though it seems we're pretty restricted in the ways in which we can compute an actual value of some $\int_a^b f(x)dx$, we can still start off by practicing with the use of the notation. Here are some you should already be able to do:

EX 1 Evaluate this definite integral using geometry: $\int_0^2 \frac{x}{2}\,dx$

Remember, this is asking for the area under $f(x) = x/2$ over the interval $[0, 2]$. Now the function $y = x/2$ is a straight line, with slope $1/2$. The area under it, over the interval $[0, 2]$ is a triangle with base 1 and height 1 (draw a picture to see), and so the area of this triangle is $1/2$. Therefore, we write:

$$\int_0^2 \frac{x}{2}\,dx = \frac{1}{2} \quad \blacksquare$$

You Try It

(1) Evaluate this definite integral using geometry: $\int_{-2}^2 \sqrt{4 - x^2}\,dx$.

Negative Areas and Combined Areas

So far, the phrase "the area under $f(x)$" has not caused any problems because our functions have been positive. But functions have negative values, too, and their graphs occupy entire regions below the x axis. And so we need to revise our understanding of the area "under" $f(x)$ to really mean the area *between* $f(x)$ and the x axis. When $f(x)$ is positive, the area "under" it is literally under it; when $f(x)$ is negative, the area "under" it is actually above it.[3] When a region "under" some $f(x)$ lies below the x-axis, it's area is assigned a negative value. For example, the area "under" $f(x) = x$ on the interval $[-1, 0]$ is assigned as $-1/2$.

EX 2 Evaluate this definite integral using geometry: $\int_{-2}^0 \frac{x}{2}\,dx$

The area "under" $f(x) = x/2$ over the interval $[-2, 0]$ is below the x axis. As in EX 1, the area is a triangle with base 1 and height 1 (draw a picture to see), and so the area of this triangle is $1/2$. But since this area comes from below the x-axis, we write:

$$\int_0^2 \frac{x}{2}\,dx = -\frac{1}{2} \quad \blacksquare$$

This interplay between positive and negative area leads to interesting results:

[3]People in the southern hemisphere may have to reverse this.

EX 3 Evaluate this definite integral using geometry: $\int_{-2}^{2} \frac{x}{2}\, dx$

This example puts together the two areas from EX 1 and EX 2; let's call these areas A_1 and A_2, as shown in Fig. 6.7. We already know that $A_1 = 1/2$ and $A_2 = -1/2$. The given definite integral is asking for the area "under" $f(x) = x/2$ over the interval $[-2, 2]$, which means we need to total of A_1 and A_2. Well, then, we have

$$\int_{-2}^{2} \frac{x}{2}\, dx = 0 \quad \blacksquare$$

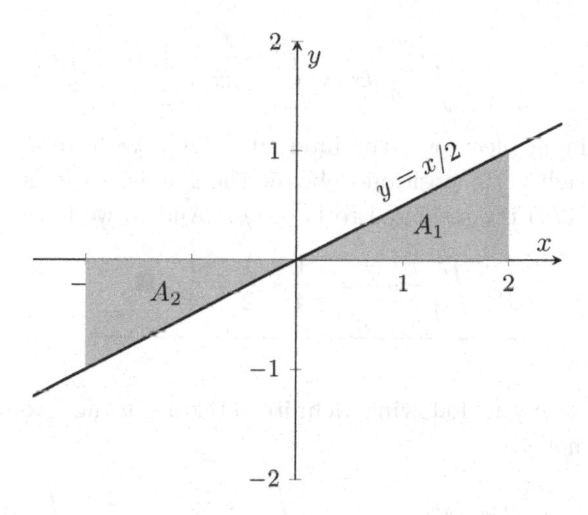

Fig. 6.7 The area "under" $y = x/2$ on $[-2, 2]$.

It may be strange to consider an area that we can see to be zero, but it happens a lot — and this fact leads to interesting results in applications. This is one example of a general idea: when the total area requested by a definite integral is composed of regions that have identical (symmetric) shapes but opposite signs, the total area will be zero. But even if the areas don't cancel out entirely, the net result is a combination of the positive and negative portions:

EX 4 Evaluate this definite integral using geometry: $\displaystyle\int_{-1}^{2} \frac{x}{2}\, dx$

Referring to the Fig. 6.7, perhaps it's clear that we can write this definite integral as:

$$\int_{-1}^{2} \frac{x}{2}\, dx = \int_{-1}^{0} \frac{x}{2}\, dx + \int_{0}^{2} \frac{x}{2}\, dx$$

The separation into two parts reflects the idea that the total area under $f(x) = \dfrac{x}{2}$ over $[-1, 2]$ is the net sum of the area over $[-1, 0]$ and the area over $[0, 2]$. We already know that the second integral is A_1 from EX 1, and so we have

$$\int_{-1}^{2} \frac{x}{2}\, dx = \int_{-1}^{0} \frac{x}{2}\, dx + \frac{1}{2}$$

For the remaining piece over the interval $[-1, 0]$, we have a triangle with base 1 and height $1/2$, positioned below the x axis, so its area (which is geometrically $1/4$) is considered to be $-1/4$. And so we have

$$\int_{-1}^{2} \frac{x}{2}\, dx = -\frac{1}{4} + \frac{1}{2} = \frac{1}{4} \quad \blacksquare$$

You Try It

(2) Evaluate the following definite integrals using geometry and/or symmetry:

$$\int_{0}^{4} 2 - x\, dx \quad ; \quad \int_{-1}^{1} x^3\, dx \quad ; \quad \int_{-\pi}^{\pi} \sin x\, dx$$

Linking Definite Integrals to Riemann Sums

It's nice the in rare cases (like the above examples) we can compute the area under a curve using simple geometry and symmetry, the harsh reality is that in general, the only way we know how to express the total area under $f(x)$ is (so far) using a Riemann sum. Therefore, the most general definition of a definite integral is

$$\int_{a}^{b} f(x)\, dx = \lim_{n\to\infty} \sum_{i=1}^{n} f(x_i)\Delta x \tag{6.2}$$

EX 5 Convert this definite integral into a Riemann sum: $\int_{-1}^{1}(1-2x^2)\,dx$

Using the procedure from Sec. 6.2, we have:

- $\Delta x = \dfrac{b-a}{n} = \dfrac{1-(-1)}{n} = \dfrac{2}{n}$
- The individual endpoints are $x_i = a + i \cdot \Delta x = -1 + \dfrac{2i}{n}$
- The heights of the rectangles are $f(x_i) = 1 - 2x_i^2 = 1 - 2\left(-1+\dfrac{2i}{n}\right)^2$

So

$$\int_{-1}^{1}(1-2x^2)dx = \lim_{n\to\infty}\sum_{i=1}^{n} f(x_i)\Delta x = \lim_{n\to\infty}\sum_{i=1}^{n}\left[1-2\left(-1+\frac{2i}{n}\right)^2\right]\cdot\frac{2}{n} \quad \blacksquare$$

You Try It

(3) Convert the definite integral into a Riemann sum: $\int_{-1}^{5}(1+3x)\,dx$.

(4) Convert the definite integral into a Riemann sum: $\int_{0}^{4}(x^2-3)\,dx$.

Since we also know how to take a given Riemann sum and deduce the $f(x)$ and $[a,b]$ involved, we should be able to handle a problem like this one:

EX 6 Convert the following Riemann sum into a definite integral:

$$\lim_{n\to\infty}\sum_{i=1}^{n}\left(1+\frac{3i}{n}\right)^2\sqrt{\pi+1+\frac{3i}{n}}\cdot\frac{3}{n}$$

Let's break it down like this:

- The very end of the expression tells us that $\Delta x = \dfrac{3}{n}$.
- The appearance in two places of $1 + \dfrac{3i}{n} = 1 + i\Delta x$ suggests that we have $x_i = 1 + \frac{3i}{n}$.
- The "fate" of each x_i seems to be placement into $f(x_i) = x_i^2\sqrt{\pi + x_i}$.
- Matching $x_i = 1 + i \cdot \dfrac{3}{n}$ to $x_i = a + i\Delta x$, we find $a = 1$.

- Matching $\Delta x = \frac{3}{n}$ to $\Delta x = \frac{b-a}{n}$, we see that $b-3$ is 3; since $a = 1$, this means $b = 4$.

Putting all this together,

$$\lim_{n\to\infty}\sum_{i=1}^{n}\left(1+\frac{3i}{n}\right)^2\sqrt{\pi+1+\frac{3i}{n}}\cdot\frac{3}{n} = \int_1^4 x^2\sqrt{\pi+x}\,dx$$

Note that this answer isn't the only one possible! By making some different choices (starting with x_i), we could also have determined

$$\lim_{n\to\infty}\sum_{i=1}^{n}\left(1+\frac{3i}{n}\right)^2\sqrt{\pi+1+\frac{3i}{n}}\cdot\frac{3}{n} = \int_0^3 (x+1)^2\sqrt{\pi+1+x}\,dx$$

Both options are correct. FFT: Why would the two possible resulting areas be the same? (Hint: recall what you know about graphs of $f(x)$ vs $f(x+c)$.) ■

You Try It

(5) Convert the following Riemann sum into a definite integral:

$$\lim_{n\to\infty}\sum_{i=1}^{n}\frac{i\pi}{n}\sin\frac{i\pi}{n}\cdot\frac{\pi}{n}$$

Computing Riemann Sums

Even though we've been writing down Riemann sums for two sections now, we haven't computed any! We have approximated some by using a finite number of rectangles, in the left and right hand estimates L_n and R_n, but we have never actually found a limit as $n \to \infty$. It turns out that it is very difficult to evaluate Riemann sums directly, except in the simple cases where the summations end up being over 1, i or i^2. The reason we can do such sums is because of these simple formulas that were first seen in Sec. 6.1:

$$\sum_{i=1}^{n}(1) = n \tag{6.3}$$

$$\sum_{i=1}^{n}(i) = \frac{n(n+1)}{2} \tag{6.4}$$

$$\sum_{i=1}^{n}(i^2) = \frac{n(n+1)(2n+1)}{6} \tag{6.5}$$

Wherever one of the sums on the left shows up in a Riemann sum, we can replace it by the closed form expression on the right. This will help us compute a small set of Riemann sums. Given a Riemann sum for a function consisting of, at worst, a combination of constant term, x, and x^2, we can build the Riemann sum, expand the terms, and use the expressions in (6.3).

EX 7 Compute this definite integral using a Riemann sum:
$$\int_{-1}^{1} (1 - 2x^2)\, dx$$

In EX 5, we found that the Riemann sum for this definite integral is:

$$\int_{-1}^{1} (1 - 2x^2) dx = \lim_{n \to \infty} \sum_{i=1}^{n} \left[1 - 2\left(-1 + \frac{2i}{n}\right)^2 \right] \cdot \frac{2}{n}$$

Do you remember our rule that, "sometimes you have to make something look worse before it can look better?" That applies here. We need to expand the terms in the summation so that we can group the terms with coefficients of 1, i, and i^2 together. This isn't pretty.

$$\begin{aligned}\left[1 - 2\left(-1 + \frac{2i}{n}\right)^2\right] \cdot \frac{2}{n} &= \left[1 - 2\left(1 - \frac{4i}{n} + \frac{4i^2}{n^2}\right)\right] \cdot \frac{2}{n} \\ &= \left(-1 + \frac{8i}{n} - \frac{8i^2}{n^2}\right) \cdot \frac{2}{n} \\ &= -\frac{2}{n} + \frac{16i}{n^2} - \frac{16i^2}{n^3}\end{aligned}$$

Now we can load this back into the summation; anything not involving the counter i can be factored out of the sums:

$$\begin{aligned}\sum_{i=1}^{n} \left[1 - 2\left(-1 + \frac{2i}{n}\right)^2\right] \cdot \frac{2}{n} &= \sum_{i=1}^{n} \left(-\frac{2}{n} + \frac{16i}{n^2} - \frac{16i^2}{n^3}\right) \\ &= -\sum_{i=1}^{n} \frac{2}{n} + \sum_{i=1}^{n} \frac{16i}{n^2} - \sum_{i=1}^{n} \frac{16i^2}{n^3} \\ &= -\frac{2}{n} \sum_{i=1}^{n} (1) + \frac{16}{n^2} \sum_{i=1}^{n} (i) - \frac{16}{n^3} \sum_{i=1}^{n} (i^2)\end{aligned}$$

Now we can replace each summation with the appropriate formula from (6.3). Once that's done, we need to make sure all the appearances of n are clumped together in each term:

$$-\frac{2}{n}\sum_{i=1}^{n}(1)+\frac{16}{n^2}\sum_{i=1}^{n}(i)-\frac{16}{n^3}\sum_{i=1}^{n}(i^2)$$
$$=-\frac{2}{n}(n)+\frac{16}{n^2}\cdot\frac{n(n+1)}{2}-\frac{16}{n^3}\cdot\frac{n(n+1)(2n+1)}{6}$$
$$=-2+\frac{16}{2}\cdot\frac{n(n+1)}{n^2}-\frac{16}{6}\cdot\frac{n(n+1)(2n+1)}{n^3}$$
$$=-2+8\cdot\frac{n(n+1)}{n^2}-\frac{8}{3}\cdot\frac{n(n+1)(2n+1)}{n^3}$$

Now that we've grouped appearances of n together inside each term we're ready to plug this mess back into the limit. We have definitely made these terms look horrible, it it will be quite satisfying to see how it all collapses back down in the limit:

$$\lim_{n\to\infty}\sum_{i=1}^{n}\left[1-2\left(-1+\frac{2i}{n}\right)^2\right]\cdot\frac{2}{n}$$
$$=\lim_{n\to\infty}\left(-2+8\cdot\frac{n(n+1)}{n^2}-\frac{8}{3}\cdot\frac{n(n+1)(2n+1)}{n^3}\right)$$
$$=-2+8(1)-\frac{8}{3}\cdot(2)=-2+8-\frac{16}{3}=\frac{2}{3}$$

and so altogether

$$\int_{-1}^{1}(1-2x^2)\,dx=\frac{2}{3}$$

Don't sell this calculation short, it's a big developmental step. For the last several years of your life, you've been able to compute areas enclosed by geometric shapes. You recently learned how to *estimate* the area of an irregular (meaning: not triangular, circular, or square) region under a curve. And now you have the ability to compute the area of many such irregular regions *exactly*. And at least in this case, it's pretty nifty that the exact value of the area turns out to be such a nice little number, 2/3. ■

Fortunately, these area calculations are going to get easier, but it's important to know what's going on behind the scenes before you learn how to bypass the nastier version of the calculations. If you don't know what's going on in the background and gain some intuition about results, you will never know when things are going wrong.

You Try It

(6) Compute this definite integral using a Riemann sum: $\int_{-1}^{5} (1 + 3x)\, dx$.

(7) Compute this definite integral using a Riemann sum: $\int_{0}^{4} (x^2 - 3)\, dx$.

Have You Learned...

- How to associate the area under a curve with a definite integral (or multiple definite integrals, if needed)?
- How to link definite integrals to Riemann sums?
- How some definite integrals can be evaluated using what you already know about geometry?
- How some very specific types of definite integrals can be evaluated using closed form final values for certain Riemann sums?

The Definite Integral — Problem List

The Definite Integral — You Try It

These appeared above; solutions begin on the next page.

(1) Evaluate this definite integral using geometry: $\int_{-2}^{2} \sqrt{4-x^2}\, dx$.

(2) Evaluate the following definite integrals using geometry and/or symmetry:

$$\int_0^4 2 - x\, dx \quad ; \quad \int_{-1}^{1} x^3\, dx \quad ; \quad \int_{-\pi}^{\pi} \sin x\, dx$$

(3) Convert this definite integral into a Riemann sum: $\int_{-1}^{5} (1+3x)\, dx$.

(4) Convert this definite integral into a Riemann sum: $\int_0^4 (x^2-3)\, dx$.

(5) Convert this following Riemann sum into a definite integral:

$$\lim_{n\to\infty} \sum_{i=1}^{n} \frac{i\pi}{n} \sin \frac{i\pi}{n} \cdot \frac{\pi}{n}$$

(6) Compute this definite integral using a Riemann sum: $\int_{-1}^{5} (1+3x)\, dx$.

(7) Compute this definite integral using a Riemann sum: $\int_0^4 (x^2-3)\, dx$.

The Definite Integral — Practice Problems

Try these as you get the hang of the You Try It problems. Solutions to these problems are available in Sec. A.6.3.

(1) Evaluate this definite integral using geometry: $\int_0^{10} |x-5|\, dx$.

(2) Evaluate these definite integrals using geometry and / or symmetry:

$$\int_0^3 2 - x\, dx \quad ; \quad \int_0^{\pi} \cos x\, dx$$

(3) Convert this integral into a Riemann sum: $\int_1^4 (x^2+2x-5)\, dx$.

(4) Convert the following Riemann sum into a definite integral.

$$\lim_{n\to\infty} \sum_{i=1}^{n} \frac{e^{1+4i/n}}{1+4i/n} \cdot \frac{4}{n}$$

(5) Compute this definite integral using a Riemann sum: $\int_1^4 (x^2 + 2x - 5)\,dx$.

(6) Find the value of $\int_0^1 (1 - 2x^2)\,dx$. (Hint: there's an easy way, and a hard way.)

The Definite Integral — Challenge Problems

Try these problems to test your skills with the ideas in this section. Solutions to these problems are available in Sec. B.6.3.

(1) Convert the following Riemann sum into a definite integral:

$$\lim_{n\to\infty} \sum_{i=1}^{n} \left[4 - 3\left(\frac{2i}{n}\right)^2 + 6\left(\frac{2i}{n}\right)^5\right] \cdot \frac{2}{n}$$

(2) Evaluate this definite integral using geometry $\int_{-3}^0 (1 + \sqrt{9 - x^2})\,dx$.

(3) Convert this definite integral into a Riemann sum, and compute the Riemann sum: $\int_0^2 (2 - x)^2\,dx$.

Definite Integrals — You Try It — Solved

(1) Evaluate this definite integral using geometry: $\int_{-2}^{2} \sqrt{4-x^2}\,dx$.

☐ We can recognize this as the area of the upper half of a circle of radius 2 centered at the origin. The entire circle would have area of 4π, so the upper half of the circle has area 2π, so

$$\int_{-2}^{2} \sqrt{4-x^2}\,dx = 2\pi \quad \blacksquare$$

(2) Evaluate the following definite integrals using geometry and / or symmetry:

$$\int_{0}^{4} 2 - x\,dx \quad ; \quad \int_{-1}^{1} x^3\,dx \quad ; \quad \int_{-\pi}^{\pi} \sin x\,dx$$

☐ Each definite integral has a combination of function and interval which has two equally sized, but oppositely sizes, regions contributing to the total area. So,

$$\int_{0}^{4} 2 - x\,dx = 0 \quad ; \quad \int_{-1}^{1} x^3\,dx = 0 \quad ; \quad \int_{-\pi}^{\pi} \sin x\,dx = 0 \quad \blacksquare$$

(3) Convert this definite integral into a Riemann sum: $\int_{-1}^{5} (1+3x)\,dx$.

☐ For this problem, we have:

- $\Delta x = \dfrac{b-a}{n} = \dfrac{5-(-1)}{n} = \dfrac{6}{n}$
- The individual endpoints are

$$x_i = a + i \cdot \Delta x = -1 + i \cdot \frac{6}{n} = -1 + \frac{6i}{n}$$

- The heights of the rectangles are

$$f(x_i) = 1 + 3x_i = 1 + 3\left(-1 + \frac{6i}{n}\right) = -2 + \frac{18i}{n}$$

So

$$\int_{-1}^{5} (1+3x)dx = \lim_{n\to\infty} \sum_{i=1}^{n} f(x_i)\Delta x = \lim_{n\to\infty} \sum_{i=1}^{n} \left(-2 + \frac{18i}{n}\right) \cdot \left(\frac{6}{n}\right) \quad \blacksquare$$

---

(4) Convert this definite integral into a Riemann sum: $\int_0^4 (x^2 - 3)\,dx$.

☐ For this problem, we have:

- $\Delta x = \dfrac{b-a}{n} = \dfrac{4-0}{n} = \dfrac{4}{n}$
- The individual endpoints are $x_i = a + i \cdot \Delta x = 0 + i \cdot \dfrac{4}{n} = \dfrac{4i}{n}$
- The heights of the rectangles are $f(x_i) = x_i^2 - 3 = \left(\dfrac{4i}{n}\right)^2 - 3$

So

$$\int_0^4 (x^2 - 3)\,dx = \lim_{n\to\infty} \sum_{i=1}^{n} f(x_i)\Delta x = \lim_{n\to\infty} \sum_{i=1}^{n} \left(\left(\frac{4i}{n}\right)^2 - 3\right) \cdot \frac{4}{n} \quad \blacksquare$$

---

(5) Convert the following Riemann sum into a definite integral.

$$\lim_{n\to\infty} \sum_{i=1}^{n} \frac{i\pi}{n} \sin \frac{i\pi}{n} \cdot \frac{\pi}{n}$$

☐ Let's break it down like this:

- The very end of the expression tells us that $\Delta x = \dfrac{\pi}{n}$.
- The appearance in two places of $\dfrac{i\pi}{n} = i\Delta x$ suggests that we have $x_i = \dfrac{i\pi}{n}$.
- The "fate" of each x_i seems to be placement into $f(x_i) = x_i \sin x_i$.
- Matching $x_i = 1 + i \cdot \dfrac{\pi}{n}$ to $x_i = a + i\Delta x$, we find $a = 0$.
- Matching $\Delta x = \dfrac{\pi}{n}$ to $\Delta x = \dfrac{b-a}{n}$, we see that $b - 3$ is π; since $a = 0$, this means $b = \pi$.

Putting all this together,

$$\lim_{n\to\infty} \sum_{i=1}^{n} \frac{i\pi}{n} \sin \frac{i\pi}{n} \cdot \frac{\pi}{n} = \int_0^{\pi} x \sin x\,dx \quad \blacksquare$$

(6) Compute this definite integral using a Riemann sum: $\int_{-1}^{5}(1+3x)\,dx$.

☐ We found the Riemann sum for this definite integral in YTI 3:

$$\int_{-1}^{5}(1+3x)dx = \lim_{n\to\infty}\sum_{i=1}^{n}\left(-2+\frac{18i}{n}\right)\cdot\left(\frac{6}{n}\right)\dots$$

If we multiply terms out and collect appearances of 1 and i, we can continue with:

$$\dots = \lim_{n\to\infty}\sum_{i=1}^{n}\left(-\frac{12}{n}+\frac{108i}{n^2}\right) = \lim_{n\to\infty}\left(-\frac{12}{n}\sum_{i=1}^{n}(1)+\frac{108}{n^2}\sum_{i=1}^{n}(i)\right)$$

We already know that $\sum_{i=1}^{n}(1) = n$. Using the closed form expression for the sum of i as seen in (6.3), we can then group n's together and apply the limit:

$$\begin{aligned}\dots &= \lim_{n\to\infty}\left(-\frac{12}{n}\cdot(n)+\frac{108}{n^2}\frac{n(n+1)}{2}\right)\\ &= \lim_{n\to\infty}\left(-12+\frac{54(n+1)}{n}\right) = -12+54 = 42\end{aligned}$$

Altogether, $\int_{-1}^{5}(1+3x)\,dx = 42$ ■

(7) Compute the definite integral using a Riemann sum: $\int_{0}^{4}(x^2-3)\,dx$.

☐ We found the Riemann sum for this definite integral in YTI 4. Using that result, we can expand the expression being summed, gather the terms 1 and i^2:

$$\begin{aligned}\int_{0}^{4}(x^2-3)dx &= \lim_{n\to\infty}\sum_{i=1}^{n}\left(\left(\frac{4i}{n}\right)^2-3\right)\cdot\frac{4}{n}\\ &= \lim_{n\to\infty}\sum_{i=1}^{n}\left(\frac{16i^2}{n^2}-3\right)\cdot\frac{4}{n} = \lim_{n\to\infty}\sum_{i=1}^{n}\left(\frac{64i^2}{n^3}-\frac{12}{n}\right)\\ &= \lim_{n\to\infty}\left(\sum_{i=1}^{n}\frac{64i^2}{n^3}-\sum_{i=1}^{n}\frac{12}{n}\right)\\ &= \lim_{n\to\infty}\left(\frac{64}{n^3}\sum_{i=1}^{n}(i^2)-\frac{12}{n}\sum_{i=1}^{n}(1)\right)\dots\end{aligned}$$

We already know that $\sum_{i=1}^{n}(1) = n$. Using the closed form expression for the sum of i^2 as seen in (6.3), we can then group n's together and apply the limit:

$$\begin{aligned}
\cdots &= \lim_{n\to\infty}\left(\frac{64}{n^3}\frac{n(n+1)(2n+1)}{6} - \frac{12}{n}(n)\right)\\
&= \lim_{n\to\infty}\left(\frac{64}{6}\frac{n(n+1)(2n+1)}{n^3} - 12\right)\\
&= \frac{64}{6}(2) - 12 = \frac{64}{3} - 12 = \frac{28}{3}
\end{aligned}$$

Altogether, $\int_0^4 (x^2 - 3)\,dx = \frac{28}{3}$. ■

6.4 The Fundamental Theorem of Calculus

Introduction

In this chapter, we seen definite integrals, their interpretation as areas, and the means by which we can write their values using Riemann sums and compute those values. But cleanly evaluating Riemann sums is feasible only in a few limited cases. The Fundamental Theorem of Calculus is going to allow us to make more headway in the evaluation of definite integrals. I'll let you judge from the name of this theorem whether it's a big deal or not.

The Fundamental Theorem of Calculus

The secret to the Fundamental Theorem of Calculus lies in the link between the area under a function $f(x)$ and the antiderivative of $f(x)$. It's not intuitive why this link should exist, so here's a brief story about that.

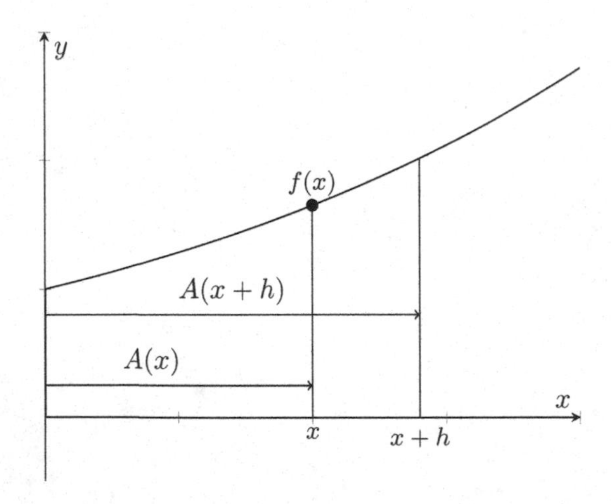

Fig. 6.8 A magic area function $A(x)$ gives area under $f(x)$ on $[0, x]$.

Let's suppose we had a magical area function $A(x)$ that told us the area under a function $f(x)$ from $x = 0$ out to any coordinate x, as illustrated in Fig. 6.8. Two locations are marked (by vertical lines) under the function in this figure; at a coordinate of x, and a coordinate of $x+h$. The region under the function between these lines is approximately a rectangle. The width of the rectangle is h and the height is (approximately) $f(x)$. So, measured one way, the area of that rectangle is about $f(x) \cdot h$. But using our magic area function, the area of this rectangle is also known as $A(x+h) - A(x)$. (Do you see why?) Therefore, we have two measures of the area of the same rectangle, and they need to be about the same:

$$A(x+h) - A(x) \approx f(x) \cdot h$$

Rearranging this, we get

$$\frac{A(x+h) - A(x)}{h} \approx f(x)$$

These two measures of the area under the curve get closer to each other as the width of the rectangle decreases. Ultimately, we have equality as $h \to 0$:

$$\lim_{h \to 0} \frac{A(x+h) - A(x)}{h} = f(x)$$

The left side is the definition of a derivative, and so

$$A'(x) = f(x)$$

What does this mean? Remember that $A(x)$ is the magic area function that gives the area under $f(x)$ on any interval $[0, x]$. And so this means that the derivative of the area function is the function $f(x)$ itself. Turning this around, the antiderivative of $f(x)$ is f's own area function.

As a quick example, then, consider that the area under $f(x) = x^2$ from $x = 0$ to $x = 2$ is now known to be $2^3/3$, or $8/3$.

What if we needed the area under $f(x)$, not from $x = 0$ to some other point, but between two arbitrary points $x = a$ and $x = b$? Well, our area function $A(x)$ (which is the antiderivative of $f(x)$) would tell us this: $A(b)$ is the area under $f(x)$ on $[0, b]$; similarly, $A(a)$ is the area under $f(x)$ on $[0, a]$. So the area under $f(x)$ only on $[a, b]$ would be $A(b) - A(a)$. And this is the hook we need. To find the area under $f(x)$ on the interval $[a, b]$, find the antiderivative of $f(x)$, plug in the two endpoints a and b, and subtract the results in the proper order. Here is a more legal way of saying that:

Theorem 6.1. *Let $f(x)$ be a function that's continuous on the interval $[a, b]$, and let $F(x)$ be any antiderivative of $f(x)$. Then*

$$\int_a^b f(x)\, dx = F(b) - F(a)$$

*This is called the **Fundamental Theorem of Calculus**.*

We have by no means even come close to a solid proof of this theorem. The story of the "magic area function" is only a *suggestion* that the Fundamental Theorem of Calculus may be valid. If you are interested in seeing something better, head for a deeper Calculus text, or — as you've seen suggested several times before this — hang out in the math curriculum until you get to Advanced Calculus or Real Analysis!

Here are a few things to note about the Fundamental Theorem of Calculus (often abbreviated FTOC to conserve letters). First, no mention has been made of the arbitrary constant that tags along with antiderivatives. Because the FTOC (see) says we can use any antiderivative of $f(x)$, and so we routinely pick the one which comes with the arbitrary constant $C = 0$. You could also argue your way through the omission of the arbitrary constant by saying, look, whichever antiderivative we pick will have a $+C$ that will substract out as soon as we form $F(b) - F(a)$. Second, the expression presented in the FTOC is not a *definition* of a definite integral; the actual definition of a definite integral is the one we saw in Eq. (6.2) that involves Riemann sums. Rather, the Fundamental Theorem of Calculus provides a means to *compute* the definite integral — as long as we're dealing with a function whose antiderivative can be found. Finally, since the difference of two values of a function (at $x = a$ and $x = b$) will become a common expression, we have a special notation for it:

$$F(x)\Big|_a^b \quad \text{means} \quad F(b) - F(a)$$

At this point, it's time for lots of examples! In each example, we will need to find an antiderivative. The required steps will not be shown, as finding antiderivatives is considered "old news". However, a list of some known and commonly appearing antiderivatives are given at the end of these notes.

EX 1 Evaluate $\displaystyle\int_0^2 \frac{x}{2}\,dx$.

We can choose the function $F(x) = x^2/4$ as an antiderivative of $f(x) = x/2$ (with $C = 0$), and

$$\int_0^2 \frac{x}{2}\,dx = \frac{x^2}{4}\Big|_0^2 = \frac{2^2}{4} - \frac{0^2}{4} = 1 \quad ■$$

EX 2 Evaluate $\int_{-1}^{1} x^3\,dx$.

We can choose the function $F(x) = x^4/4$ as an antiderivative of $f(x) = x^3$ (with $C = 0$), and

$$\int_{-1}^{1} x^3\,dx = \left.\frac{x^4}{4}\right|_{-1}^{1} = \frac{(-1)^4}{4} - \frac{1^1}{4} = 0$$

FFT: You could (should?) have predicted the final value of zero before doing any calculations. Why? ■

You Try It

(1) Evaluate $\int_{-1}^{3} x^5\,dx$.

(2) Evaluate $\int_{1}^{2} \frac{3}{t^4}\,dt$.

EX 3 Evaluate $\int_{-1}^{1} (1 - 2x^2)\,dx$.

Do you remember that we did this one with Riemann Sums in EX 7 of Sec. 6.3, and we found a value of 2/3. Will we get the same thing with this new technique? I hope so! The function $F(x) = x - 2x^3/3$ is an antiderivative (with $C = 0$) of $f(x) = 1 - 2x^2$, and

$$\int_{-1}^{1} (1 - 2x^2)\,dx = \left.\left(x - \frac{2x^3}{3}\right)\right|_{-1}^{1}$$
$$= \left(1 - \frac{2(1)^3}{3}\right) - \left((-1) - \frac{2(-1)^3}{3}\right) = 2 - \frac{4}{3} = \frac{2}{3} \quad ■$$

EX 4 Evaluate $\int_{-\pi}^{\pi} \sin x\,dx$.

I think it's time to stop posing the antiderivative we're going to use ahead of the integral, we can just introduce it in line. Meaning, the flow of the solution to this problem can simply be this:

$$\int_{-\pi}^{\pi} \sin x\,dx = \left.(-\cos(x))\right|_{-\pi}^{\pi} = -\cos\pi - (-\cos(-\pi)) = 1 + (-1) = 0$$

[IOI] FFT: Again, you could (should?) have predicted the final value of zero before doing any calculations. Why? [IOI] ■

You Try It

(3) Evaluate $\displaystyle\int_{\pi}^{2\pi} \cos\theta\, d\theta$.

(4) Evaluate $\displaystyle\int_{-1}^{0} (2x - e^x)\, dx$.

You Try It

(5) Evaluate $\displaystyle\int_{1}^{4} \sqrt{t}(1+t)\, dt$.

(6) Evaluate $\displaystyle\int_{0}^{\pi} (4\sin\theta - 3\cos\theta)\, d\theta$.

The Fundamental Theorem of Calculus should be used with caution; when confronted with a definite integral, don't plunge into antiderivatives and subtraction without checking that the Fundamental Theorem is really valid in that case.

EX 5 Why can't we use the Fundamental Theorem to evaluate $\displaystyle\int_0^2 \frac{1}{x-1}\, dx$?

The function $f(x) = 1/(x-1)$ has an antiderivative (it's $\ln|x-1|$), so what's the problem? Well, the problem is that the FTOC requires the function $f(x)$ to be continuous on the interval $[a, b]$; here, the function $f(x)$ is not continuous on the interval $[0, 2]$ — so the Fundamental Theorem does not apply. ■

You Try It

(7) Why can't we evaluate $\displaystyle\int_{\pi}^{2\pi} \csc^2\theta\, d\theta$?

The Alter Ego of the FTOC

A second version of the Fundamental Theorem comes from exploration of the definite integral as an area function. First, note that the variable written into a definite integral does not matter; for example, all of the following definite integrals would have the same value:

$$\int_0^1 \cos(x)\,dx \qquad \int_0^1 \cos(t)\,dt \qquad \int_0^1 \cos(s)\,ds \qquad \int_0^1 \cos(w)\,dw$$

In the process of evaluating a definite integral, the variable eventually gets replaced by the limits of integration and vanishes, so the variable inside the integral is irrelevant. On the other hand, note that all of the following definite integrals would have *different* values:

$$\int_a^1 \cos(t)\,dt \qquad \int_a^{1.1} \cos(t)\,dt \qquad \int_a^{1.2} \cos(t)\,dt \qquad \int_a^{1.3} \cos(t)\,dt$$

Keeping the left-most limit of integration a constant and varying the right-hand limit b will yield different areas, and thus different values of the integral. And what do you call something which yields different outputs upon being given different inputs? A function! The spot where the input goes gets marked by a variable, so we can write all four of the above definite integrals as instances of this function:

$$\int_a^x \cos(t)\,dt$$

What is the independent variable of the function? It's not t! Remember, t is the variable of integration, and does not matter. The above expression is, instead, a function of x, so we can write:

$$y(x) = \int_a^x \cos(t)\,dt$$

We have just created an entirely new beast, a function defined as a definite integral! And like any other function, it might have a derivative. So, what is dy/dx for

$$y = \int_a^x \cos(t)\,dt$$

Let's work it out: the antiderivative of $\cos(t)$ (with $C = 0$) is $\sin(t)$, and so

$$y = \int_a^x \cos(t)\,dt = \sin(t)\Big|_a^x = \sin(x) - \sin(a)$$

and then

$$\frac{dy}{dx} = \frac{d}{dx}\left(\sin(x) - \sin(a)\right) = \cos(x)$$

so that

$$\text{If } y = \int_a^x \cos(t)\,dt \quad \text{then} \quad \frac{dy}{dx} = \cos(x)$$

Hmmm. Looks like we just took the funtion inside the integral, and swapped t for x. And that's pretty much what happens. When you consider the question closely, you'll see that we're asking for the derivative of an integral, or rather, the derivative of an antiderivative. That just brings us back to where we started. The t's are gone, and x remains. Formally, we have

$$\text{If } y = \int_a^x f(t)\,dt \quad \text{then} \quad \frac{dy}{dx} = f(x)$$

This is a second version of the Fundamental Theorem of Calculus since it also deals with the relationship between a definite integral and antiderivatives. It works whether we can find the antiderivative of $f(t)$ or not ... in fact, we don't have to, since we know the final result. And, did you see that the specific value of a (the lower limit of integration) doesn't matter?

EX 6 Find dy/dx for each of the following functions $y(x)$:

$$y = \int_1^x te^{2t}\,dt \quad ; \quad y = \int_6^x te^{2t}\,dt \quad ; \quad y = \int_{245}^x te^{2t}\,dt$$

The answer to each is the same:

$$\text{If } y = \int_a^x te^{2t}\,dt \quad \text{then} \quad \frac{dy}{dx} = xe^{2x}$$

This will hold for $a = 1$, $a = 6$, and $a = 245$. ■

You Try It

(8) Find $g'(y)$ for $g(y) = \int_2^y t^2 \sin(t)\,dt$.

Just for fun, we can combine this other version of the Fundamental Theorem of Calculus with the Chain Rule. And yes, this *is* fun, don't argue with me.) What if the upper limit of integration is not just x but a little function itself, say, $g(x)$? Then

$$\text{If } y = \int_a^{g(x)} f(t)\,dt \quad \text{then} \quad \frac{dy}{dx} = f(g(x))g'(x)$$

EX 7 Find dy/dx for $y = \int_2^{x^3} \ln(t^2 + 1)\, dt$.

Recognizing $f(t) = \ln(t^2 + 1)$ and $g(x) = x^3$, we have $f(g(x)) = \ln((x^3)^2 + 1) = \ln(x^6 + 1)$ and $g'(x) = 3x^2$, so

$$\text{If } y = \int_1^{x^3} \ln(t^2 + 1)\, dt \quad \text{then} \quad \frac{dy}{dx} = \ln(x^6 + 1) \cdot 3x^2 \; ■$$

You Try It

(9) Find dy/dx for $y = \int_3^{\sqrt{x}} \frac{\cos t}{t}\, dt$.

From this point on in Calculus, you we will see a LOT of applications of the FTOC as originally posed, but very few of the second "alter ego" version. However, many of you will take more advanced classes in which the idea of a function defined as a definite integral can be a crucial one. Just wait until you see Laplace Transforms! So while this second facet of the FTOC may not have a big impact in the remaining chapters of this work, it can be very important down the road.

Have You Learned...

- How the area under $f(x)$ is linked to the antiderivative of $f(x)$?
- How to find the value of a definite integral using the Fundamental Theorem of Calculus?
- How to identify cases in which the Fundamental Theorem of Calculus is not appropriate?
- How a definite integral can be used to define a function, with the variable as one of the limits of integration?
- How to interpret such a function in terms of the area under some $f(x)$?

The Fundamental Theorem of Calculus — Problem List

The Fundamental Theorem of Calculus — You Try It

These appeared above; solutions begin on the next page.

Evaluate the following definite integrals:

(1) $\int_{-1}^{3} x^5\,dx.$

(2) $\int_{1}^{2} \frac{3}{t^4}\,dt.$

(3) $\int_{\pi}^{2\pi} \cos\theta\,d\theta.$

(4) $\int_{-1}^{0} (2x - e^x)\,dx.$

(5) $\int_{1}^{4} \sqrt{t}(1+t)\,dt.$

(6) $\int_{0}^{\pi} (4\sin\theta - 3\cos\theta)\,d\theta.$

(7) Why can't we use the Fundamental Theorem to evaluate $\int_{\pi}^{2\pi} \csc^2\theta\,d\theta.$

(8) Find $g'(y)$ for $g(y) = \int_{2}^{y} t^2\sin(t)dt.$

(9) Find dy/dx for $y = \int_{3}^{\sqrt{x}} \frac{\cos t}{t} dt.$

The Fundamental Theorem of Calculus — Practice Problems

Try these as you get the hang of the You Try It problems. Solutions to these problems are available in Sec. A.6.4.

Evaluate the following definite integrals:

(1) $\int_{-2}^{5} 6\,dx.$

(2) $\int_{-2}^{0} (u^5 - u^3 + u^2)\,du.$

(3) $\int_{1}^{9} \frac{1}{2x}\,dx.$

(4) $\int_{0}^{9} \sqrt{2t}\,dt.$

(5) $\int_{0}^{\pi/4} \frac{1+\cos^2\theta}{\cos^2\theta}\,d\theta.$

(6) Why can't we use the Fundamental Theorem to evaluate $\int_{-2}^{3} x^{-5}\,dx?$

(7) Why can't we evaluate $\int_{0}^{\pi/6} \csc\theta\cot\theta\,d\theta?$

(8) Find $g'(u)$ for $g(u) = \int_3^u \frac{1}{x + x^2}\,dx$.

(9) Find dy/dx for $y = \int_1^{\cos x} (t + \sin t)\,dt$.

The Fundamental Theorem of Calculus — Challenge Problems

Try these problems to test your skills with the ideas in this section. Solutions to these problems are available in Sec. B.6.4.

(1) Evaluate $\int_1^2 \frac{y + 5y^7}{y^3}\,dy$.

(2) Evaluate $\int_{\pi/4}^{\pi/3} \sec\theta \tan\theta\,d\theta$.

(3) Find dy/dx for $y = \int_1^{x^2} \sqrt{1 + r^3}\,dr$.

FTOC — You Try It — Solved

---

(1) Evaluate $\displaystyle\int_{-1}^{3} x^5\,dx$.

$$\square \int_{-1}^{3} x^5 dx = \frac{1}{6}x^6\bigg|_{-1}^{3} = \frac{1}{6}(3^6 - 1) = \frac{728}{6} = \frac{364}{4} \quad \blacksquare$$

---

(2) Evaluate $\displaystyle\int_{1}^{2} \frac{3}{t^4}\,dt$.

$$\square \int_{1}^{2} \frac{3}{t^4} dt = -\frac{1}{t^3}\bigg|_{1}^{2} = -\left(\frac{1}{2^3} - 1\right) = \frac{7}{8} \quad \blacksquare$$

---

(3) Evaluate $\displaystyle\int_{\pi}^{2\pi} \cos\theta\,d\theta$.

$$\square \int_{\pi}^{2\pi} \cos\theta d\theta = \sin\theta\bigg|_{\pi}^{2\pi} = 0 - 0 = 0 \quad \blacksquare$$

---

(4) Evaluate $\displaystyle\int_{-1}^{0} (2x - e^x)\,dx$.

$$\square \int_{-1}^{0} (2x - e^x) dx = (x^2 - e^x)\bigg|_{-1}^{0} = (0 - e^0) - [(-1)^2 - e^{-1}] = \frac{1}{e} - 2 \quad \blacksquare$$

---

(5) Evaluate $\displaystyle\int_{1}^{4} \sqrt{t}(1+t)\,dt$.

$$\begin{aligned}
\square \int_{1}^{4} \sqrt{t}(1+t) dt &= \int_{1}^{4} (t^{1/2} + t^{3/2}) = (\frac{2}{3}t^{3/2} + \frac{2}{5}t^{5/2})\bigg|_{1}^{4} \\
&= (\frac{2}{3}4^{3/2} + \frac{2}{5}4^{5/2}) - (\frac{2}{3} + \frac{2}{5}) \\
&= \frac{16}{3} + \frac{64}{5} - \frac{2}{3} - \frac{2}{5} \\
&= \frac{14}{3} + \frac{62}{5} = \frac{265}{15} \quad \blacksquare
\end{aligned}$$

(6) Evaluate $\int_0^{\pi} (4\sin\theta - 3\cos\theta)\, d\theta$.

$$\square \int_0^{\pi} (4\sin\theta - 3\cos\theta) d\theta = (-4\cos\theta - 3\sin\theta)\Big|_0^{\pi}$$
$$= (-4(-1) - 3(0)) - (-4(1) - 3(0))$$
$$= 8 \quad \blacksquare$$

(7) Why can't we use the Fundamental Theorem to evaluate $\int_{\pi}^{2\pi} \csc^2\theta\, d\theta$?

☐ Since $\csc^2\theta$ is undefined at $x = \pi$ and $x = 2\pi$, which are both within the interval of integration $[a, b]$, the given does not exist. ■

(8) Find $g'(y)$ for $g(y) = \int_2^y t^2 \sin(t)\, dt$.

☐ Adapting the Fundamental Theorem to these variables, we have

$$\frac{d}{dy} \int_a^y f(t)\, dt = f(y)$$

Setting $f(t) = t^2 \sin(t)$, we get

$$g'(y) = \frac{d}{dy} \int_2^y t^2 \sin(t) dt = y^2 \sin y \quad \blacksquare$$

(9) Find dy/dx for $y = \int_3^{\sqrt{x}} \frac{\cos t}{t}\, dt$.

☐ Adapting the Fundamental Theorem (with chain rule) to these variables, we have

$$\frac{d}{dx} \int_a^{g(x)} f(t)\, dt = f((g(x))g'(x)$$

Identifying $f(t) = \dfrac{\cos(t)}{t}$ and $g(x) = \sqrt{x}$, we have

$$\frac{dy}{dx} = \frac{\cos\sqrt{x}}{\sqrt{x}} \frac{d}{dx}\sqrt{x} = \frac{\cos\sqrt{x}}{2\sqrt{x}\cdot\sqrt{x}} = \frac{\cos\sqrt{x}}{2x} \quad \blacksquare$$

6.5 Definite Integrals With Substitution

Introduction

We now know from the Fundamental Theorem of Calculus that a "simple" way of finding the value of a definite integral is to use the antiderivative of the function being integrated. But, of course, this method is only as simple as finding the required antiderivative. The most common technique for finding antiderivatives is substitution, and here we will adapt it for use with definite integrals.

Review of Antiderivatives With Substitution

Recall that substitution is a technique which can "reverse" the chain rule. That is, given a derivative that was likely to have been produced by the chain rule, substitution can often help find where the derivative came from. Consider an antiderivative of the form

$$\int f(g(x)) \cdot g'(x)\, dx$$

The substitution $u = g(x)$ (for which $du = g'(x)\, dx$) turns this antiderivative into

$$\int f(u)\, du$$

and this new version might be easier to solve — much of the clutter is hidden. The tricky part about substitution is that it isn't always immediately obvious whether a given integrand really is in a form that can be morphed into $f(g(x))g'(x)$; For example, only two of the three following antiderivatives are immediately ready for substitution, do you know which two?

$$\int xe^{x^2}\, dx \quad ; \quad \int \frac{\sin(\sqrt{x})}{\sqrt{x}}\, dx \quad ; \quad \int \frac{x^3}{x^2+1}\, dx$$

For more review about how to do substitution, refer to Sec. 4.6.

Definite Integrals With Substitution

Using substitution to solve definite integrals proceeds in the same manner as for general antiderivatives, or *indefinite integrals*, with one extra step.

As mentioned in Sec. 4.6, the procedure of substitution is actually a change of variable. We take one integral written in terms of the variable, say x, and convert it into an integral written in terms of a new variable, say u. In a *definite* integral, the limits of integration a and b are x values, and so they must be converted to u values according to the substitution.

Here is an example showing how various constants can start flying around.

EX 1 Evaluate $\int_1^3 \frac{3}{7-2x}\,dx$.

First, we pick a substitution and make sure all the pieces of the integral have a match. With $u = 7 - 2x$ we have

$$du = -2\,dx \qquad \text{or} \qquad -\frac{1}{2}\,du = dx$$

Next, we take care of the limits of integration; we use the same substitution formula, $u = 7 - 2x$:

$$x = 1 \to u = 5$$
$$x = 3 \to u = 1$$

Now we convert the entire integral:

$$\int_1^3 \frac{3}{7-2x}\,dx = \int_5^1 \frac{3}{u}\left(-\frac{1}{2}\,du\right) = -\frac{3}{2}\int_5^1 \frac{du}{u} = -\frac{3}{2}\ln|u|\Big|_5^1$$
$$= -\frac{3}{2}(\ln 1 - \ln 5) = -\frac{3}{2}(0 - \ln 5) = \frac{3\ln 5}{2} \quad \blacksquare$$

Never be afraid to try different substitutions until you come across the right one:

EX 2 Evaluate $\int_{1/3}^{1/2} \frac{e^{1/x}}{x^2}\,dx$.

First, we pick a substitution and make sure all the pieces of the integral have a match. If we try $u = e^{1/x}$, then we have to go along with that $du = e^{1/x} \cdot (-1/x^2)\,dx$. Since the one instance of $e^{1/x}$ in the integral is already reserved for u, there is not another one to help with du, and so we can't get a complete match with that substitution. So let's try something a bit smaller. With $u = 1/x$ we have

$$du = -\frac{1}{x^2}\,dx \quad \Longrightarrow -du = \frac{1}{x^2}\,dx$$

That's better. Next, we take care of the limits of integration; we use the same substitution formula, $u = 1/x$ to change them:

$$x = \frac{1}{3} \rightarrow u = 3$$
$$x = \frac{1}{2} \rightarrow u = 2$$

Now we convert the entire integral:

$$\int_{1/3}^{1/2} \frac{e^{1/x}}{x^2}\,dx = -\int_{3}^{2} e^u\,du = -e^u\Big|_3^2 = -(e^2 - e^3) = e^3 - e^2 \quad \blacksquare$$

You Try It

(1) Evaluate $\int_2^3 (5x-1)^4\,dx$.

(2) Evaluate $\int_0^2 \frac{dx}{3x+1}\,dx$.

Here is an example with a different variable of integration, to illustrate that the variable being used has no bearing on the strategy.

EX 3 Evaluate $\int_0^{\pi/2} \cos^3\theta \sin\theta\,d\theta$.

With $u = \cos\theta$ we have $du = -\sin\theta\,d\theta$, or $\sin\theta\,d\theta = -du$. For the end-points of integration,

$$\theta = 0 \rightarrow u = 1$$
$$\theta = \frac{\pi}{2} \rightarrow u = 0$$

and

$$\int_0^{\pi/2} \cos^3\theta \sin\theta\,d\theta = -\int_1^0 u^3\,du = -\frac{1}{4}u^4\Big|_1^0 = -\frac{1}{4}(0^4 - 1^4) = \frac{1}{4} \quad \blacksquare$$

You Try It

(3) Evaluate $\int_{-2}^0 t\sqrt{4-t^2}\,dt$.

Solving definite integrals is like taking a stroll through a spooky house. There you are minding your own business, and bam, out jumps an inverse tangent function. Get used to this, as inverse tangents are a common occurence in Calc II!

EX 4 Evaluate $\int_0^{\pi/2} \frac{\sin x}{1+\cos^2 x}\, dx$.

With $u = \cos x$ we have $du = -\sin x\, dx$ or $\sin x\, dx = -du$. For the endpoints of integration,

$$x = 0 \to u = 1$$
$$x = \frac{\pi}{2} \to u = 0$$

So then,

$$\int_0^{\pi/2} \frac{\sin x}{1+\cos^2 x}\, dx = -\int_1^0 \frac{du}{1+u^2} = -\tan^{-1}(u)\Big|_1^0$$
$$= -(\tan^{-1}(0) - \tan^{-1}(1)) = -\left(0 - \frac{\pi}{4}\right) = \frac{\pi}{4} \quad \blacksquare$$

You Try It

(4) Evaluate $\int_0^1 \frac{e^x}{e^{2x}+1}\, dx$.

More practice is always a good thing, right?

You Try It

(5) Evaluate $\int_0^{\pi/2} \cos\theta e^{\sin\theta}\, d\theta$.

(6) Evaluate $\int_{\pi/4}^{\pi/2} \cot\theta\, d\theta$.

(7) Evaluate $\int_0^{\pi/3} \frac{\sin\theta}{\cos^2\theta}\, d\theta$.

If you've been through this section and still feel like you need more practice on integration by substitution (referring to both this section and Sec. 4.6), then don't worry — the next chapter starts off with a review of integration by substitution in both indefinite and definite integrals, together. If you are confident in your substitution skills, perhaps you can jump ahead to the section after that.

Have You Learned...

- How to identify u and du in an integrand which likely links to a composition?

- How to adjust constants to account for an imperfect match to u and / or du?
- What you are *not* allowed to do in your quest to fix the match to u and u?
- To adjust limits of integration according to a selection for u?

Definite Integrals With Substitution — Problem List

Definite Integrals With Substitution — You Try It

These appeared above; solutions begin on the next page.

Evaluate the following definite integrals:

(1) $\displaystyle\int_2^3 (5x-1)^4\,dx.$

(2) $\displaystyle\int_0^2 \frac{dx}{3x+1}\,dx.$

(3) $\displaystyle\int_{-2}^0 t\sqrt{4-t^2}\,dt.$

(4) $\displaystyle\int_0^1 \frac{e^x}{e^{2x}+1}\,dx.$

(5) $\displaystyle\int_0^{\pi/2} \cos\theta e^{\sin\theta}\,d\theta.$

(6) $\displaystyle\int_{\pi/4}^{\pi/2} \cot\theta\,d\theta.$

(7) $\displaystyle\int_0^{\pi/3} \frac{\sin\theta}{\cos^2\theta}\,d\theta.$

Definite Integrals With Substitution — Practice Problems

Try these as you get the hang of the You Try It problems. Solutions to these problems are available in Sec. A.6.5.

Evaluate the following definite integrals:

(1) $\displaystyle\int_{-3}^{-2} (4-x)^2\,dx.$

(2) $\displaystyle\int_{\sqrt{2}}^{\sqrt{3}} \frac{x}{x^2+1}\,dx.$

(3) $\displaystyle\int_0^1 y^3 e^{y^4}\,dy.$

(4) $\displaystyle\int_0^{\pi/4} (1+\tan\theta)^2\sec^2\theta\,d\theta.$

(5) $\displaystyle\int_0^{\pi/4} \frac{\cos x}{1+\sin^2 x}\,dx.$

(6) $\displaystyle\int_0^{\pi/2} \cos x\sin(\sin x)\,dx.$

(7) $\displaystyle\int_{-1}^1 \cosh x\,dx.$

(8) Which integral has the largest value? Give that value:

$$\int_{-1}^1 \sinh(x)\,dx \quad , \quad \int_{-1}^1 \cosh(x)\,dx \quad , \quad \text{or} \quad \int_{-1}^1 \tanh(x)\,dx$$

Definite Integrals With Substitution — Challenge Problems

Try these problems to test your skills with the ideas in this section. Solutions to these problems are available in Sec. B.6.5.

Evaluate the following definite integrals:

(1) $\displaystyle\int_4^9 \frac{1}{\sqrt{x}(\sqrt{x}+1)^2}\,dx$ (2) $\displaystyle\int_0^1 \frac{x}{1+x^4}\,dx$ (3) $\displaystyle\int_0^1 xe^{-x^2}\,dx$

Definite Integrals With Substitution — You Try It — Solved

---

(1) Evaluate $\int_2^3 (5x-1)^4\,dx$.

□ Using $u = 5x-1$, we get $du = 5\,dx$, or $dx = du/5$. For the endpoints,

$$x = 2 \to u = 9$$
$$x = 3 \to u = 14$$

Then altogether,

$$\int_2^3 (5x-1)^4\,dx = \int_9^{14} u^4 \cdot \frac{1}{5}\,du = \frac{1}{5}\int_9^{14} u^4\,du = \frac{1}{5}\cdot\frac{1}{5}u^5\Big|_9^{14}$$
$$= \frac{1}{25}(14^5 - 9^5) = \frac{1}{25}(14^5 - 9^5) = 19151 \quad \blacksquare$$

---

(2) Evaluate $\int_0^2 \frac{dx}{3x+1}$.

□ With $u = 3x+1$, we have $du = 3\,dx$ or $dx = \frac{1}{3}\,du$.

$$x = 0 \to u = 1$$
$$x = 2 \to u = 7$$

Then,

$$\int_0^2 \frac{dx}{3x+1} = \frac{1}{3}\int_1^7 \frac{du}{u} = \frac{1}{3}\ln|u|\Big|_1^7 = \frac{1}{3}(\ln 7 - \ln 1) = \frac{\ln 7}{3} \quad \blacksquare$$

---

(3) Evaluate $\int_{-2}^0 t\sqrt{4-t^2}\,dt$.

□ With $u = 4 - t^2$, we have $du = -2t\,dt$ or $t\,dt = -du/2$. For the endpoints,

$$t = -2 \to u = 0$$
$$t = 0 \to u = 4$$

Then altogether,

$$\int_{-2}^0 t\sqrt{4-t^2}\,dt = -\frac{1}{2}\int_0^4 \sqrt{u}\,du = -\frac{1}{2}\cdot\frac{2}{3}u^{3/2}\Big|_0^4$$
$$= -\frac{1}{3}(4^{3/2} - 0^{3/2}) = -\frac{8}{3} \quad \blacksquare$$

(4) Evaluate $\displaystyle\int_0^1 \frac{e^x}{e^{2x}+1}\,dx$.

☐ Remember that $e^{2x} = (e^x)^2$. With $u = e^x$, then $du = e^x\,dx$. For the endpoints,

$$x = 0 \to \qquad u = e^0 = 1$$
$$x = 1 \to \qquad u = e^1 = e$$

Then altogether,

$$\int_0^1 \frac{e^x}{e^{2x}+1}\,dx = \int_1^e \frac{du}{u^2+1} = \tan^{-1}(u)\Big|_1^3$$
$$= \tan^{-1}(e) - \tan^{-1}(1) = \tan^{-1}(e) - \frac{\pi}{4} \quad \blacksquare$$

(5) Evaluate $\displaystyle\int_0^{\pi/2} \cos\theta e^{\sin\theta}\,d\theta$.

☐ With $u = \sin\theta$ we have $du = \cos\theta d\theta$, and for endpoints,

$$\theta = 0 \to u = 0$$
$$\theta = \frac{\pi}{2} \to u = 1$$

And,

$$\int_0^{\pi/2} \cos\theta e^{\sin\theta}\,d\theta = \int_0^1 e^u\,du = e^u\Big|_0^1 = e^1 - e^0 = e - 1 \quad \blacksquare$$

(6) Evaluate $\displaystyle\int_{\pi/4}^{\pi/2} \cot\theta\,d\theta$.

☐ First, we write

$$\int_{\pi/4}^{\pi/2} \cot\theta\,d\theta = \int_{\pi/4}^{\pi/2} \frac{\cos\theta}{\sin\theta}\,d\theta$$

Then using $u = \sin\theta$ gives $du = \cos\theta d\theta$; and,

$$\theta = \frac{\pi}{4} \to u = \frac{1}{\sqrt{2}}$$
$$\theta = \frac{\pi}{2} \to u = 1$$

So that,

$$\int_{\pi/4}^{\pi/2} \frac{\cos\theta}{\sin\theta}\,d\theta = \int_{1/\sqrt{2}}^{1} \frac{du}{u}$$

$$= \ln|u|\Big|_{1/\sqrt{2}}^{1} = \ln 1 - \ln\frac{1}{\sqrt{2}}$$

$$= \ln\sqrt{2} = \frac{1}{2}\ln 2 \quad \blacksquare$$

---

(7) Evaluate $\displaystyle\int_0^{\pi/3} \frac{\sin\theta}{\cos^2\theta}\,d\theta$.

□ The substitution $u = \cos\theta$ gives $du = -\sin\theta\,d\theta$ or $\sin\theta\,d\theta = -du$. The endpoints are adapted as follows:

$$x = 0 \to u = 1$$

$$x = \pi/3 \to u = 1/2$$

Altogether,

$$\int_0^{\pi/3} \frac{\sin\theta}{\cos^2\theta}\,d\theta = -\int_1^{1/2} \frac{du}{u^2} = \int_{1/2}^{1} \frac{du}{u^2} = -\frac{1}{u}\Big|_{1/2}^{1}$$

$$= -\left(\frac{1}{1} - \frac{1}{1/2}\right) = -(1-2) = 1 \quad \blacksquare$$

Appendix A

Solutions to All Practice Problems

A.1 Chapter 1: Practice Problem Solutions

A.1.1 *Basics of Functions — Practice — Solved*

(1) Given the function $f(x) = x/\sqrt{x^2+c}$, identify the name of the function and the name of the independent variable, and then find and simplify $f(1)$, $f(c)$, and $f(c-1)$. Would $f(c) - f(1)$ be equal to $f(c-1)$?

☐ The name of the function is f. The independent variable is x. The parameter c must be a constant since it is not specified as the variable on the left side. The expressions are as follows:

$$f(1) = \frac{1}{\sqrt{1^2+c}} = \frac{1}{\sqrt{1+c}}$$

$$f(c) = \frac{c}{\sqrt{c^2+c}} = \frac{c}{\sqrt{c(c+1)}} = \frac{\sqrt{c}}{\sqrt{c+1}}$$

$$f(c-1) = \frac{(c-1)}{\sqrt{(c-1)^2+c}} = \frac{c-1}{\sqrt{c^2-2c+1+c}} = \frac{c-1}{\sqrt{c^2-c+1}}$$

$f(c) - f(1)$ would not be equal to $f(c-1)$. ■

(2) Given the function $t(p) = (p-2)^2/2$, identify the name of the function and the name of the independent variable, and then find and simplify $t(3)$, $t(2p)$, and $t(3+2p)$. Is $t(3+2p) = t(3) + t(2p)$?

☐ The name of the function is t and the independent variable is p.

Then,

$$t(3) = \frac{1}{2}(3-2)^2 = \frac{1}{2}$$

$$t(2p) = \frac{1}{2}(2p-2)^2 = \frac{1}{2}(4p^2 - 8p + 4) = 2p^2 - 4p + 2$$

$$t(3+2p) = \frac{1}{2}((3+2p)-2)^2 = \frac{1}{2}(1+2p)^2 = \frac{1}{2}(1+4p+4p^2)$$

Comparing results, it's evident that $t(3+2p) \neq t(3) + t(2p)$. ■

---

(3) If $f(x) = \sqrt{x}$, and a, b are constants, which of the following statements are true?

(a) $f(2x) = 2f(x)$
(b) $f(x-1) = \sqrt{x} - 1$
(c) $f(x) - 1 = \sqrt{x} - 1$
(d) $f(a+b) = f(a) + f(b)$
(e) $f(a \cdot b) = f(a) \cdot f(b)$

□

(a) $f(2x) = 2f(x)$: This is FALSE, since $f(2x) = \sqrt{2x} = \sqrt{2} * \sqrt{x}$.
(b) $f(x-1) = \sqrt{x} - 1$: This is FALSE, since $f(x-1) = \sqrt{x-1}$ and this is NOT $\sqrt{x} - 1$!!
(c) $f(x) - 1 = \sqrt{x} - 1$: This is TRUE, using the definition of $f(x)$
(d) $f(a+b) = f(a) + f(b)$: This is FALSE. $\sqrt{a+b}$ is NOT equal to $\sqrt{a} + \sqrt{b}$!!
(e) $f(a \cdot b) = f(a) \cdot f(b)$. This is TRUE: $f(ab) = \sqrt{ab} = \sqrt{a}\sqrt{b} = f(a)f(b)$. ■

---

(4) Find the domain and range of $h(p) = \frac{1}{p} - \sqrt{2p}$.

□ Since we can use any p value except 0, the domain of h is all real numbers except 0. We can get any number back from this function including 0, so the range is all real numbers. ■

---

(5) Four curves are shown in Fig. A.1 (next page). Only two of them are actually functions. Identify them (using labels A,B,C,D). Of those two, one is odd, and one is even. Identify them.

□ (B) and (D) fail the vertical line test, so (A) and (C) are the functions. (B) is symmetric across the y-axis (left/right symmetry) and so is the even function. (A) is symmetric across the origin (inverted left / right symmetry) and so is the odd function. ■

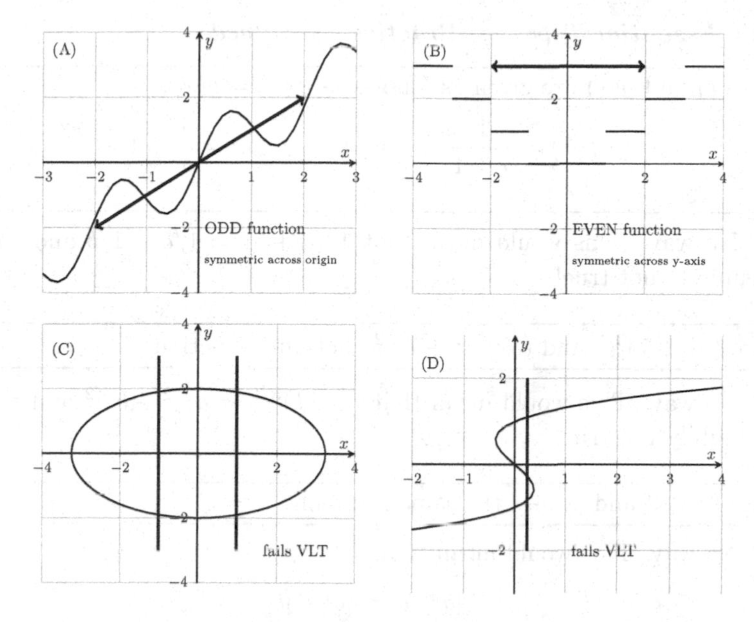

Fig. A.1 Find the functions, and in particular, identify the even and odd functions (w/ PP 6).

---

(6) Determine algebraically whether $f(x) = (x-1)(x^2-2)^{3/2}$ is an even function, an odd function, or neither.

□ To determine if f is even or odd, we have to see if $f(-x)$ is equal to either $f(x)$ or $-f(x)$. But,

$$f(-x) = ((-x)-1) * ((-x)^2-2)^{3/2} = (-x-1) * (x^2-2)^{3/2}$$
$$= -(x+1) * (x^2-2)^{3/2}$$

So $f(-x)$ is certainly not equal to $f(x)$ or $-f(x)$. The function is neither even nor odd. ■

A.1.2 *Essential Types — Practice — Solved*

(1) If $f(x)$ and $g(x)$ are given as follows, is $f(x) = g(x)$?

$$f(x) = \frac{1}{x+1} \qquad \text{and} \qquad g(x) = \frac{1}{x} + 1$$

☐ No way. This would mean that $1/(a+b) = 1/a + 1/b$ and that's definitely not true! ∎

(2) Are $(3x+7)^{3/2}$ and $(3x)^{3/2} + 7^{3/2}$ the same function?

☐ No way. This would mean that $(a+b)^{3/2} = a^{3/2} + b^{3/2}$ and that's definitely not true! ∎

(3) Are $|x+4|$ and $|x|+4$ the same function?

☐ No way. This would mean that

$$|a+b| = |a| + |b|$$

and that's definitely not true! ∎

(4) Find the equation of the line through $(-2,-3)$ and parallel to $y = 2x + 7$.

☐ A line parallel to $y = 2x + 7$ will also have a slope of 2, and so with the given point, we have

$$\begin{aligned} y - y_0 &= m(x - x_0) \\ y - (-3) &= (2)(x - (-2)) \\ y + 3 &= 2(x+2) \quad \blacksquare \end{aligned}$$

(5) Find the domain and range of $f(x) = (x+2)^{-1/2}$. How does the graph behave as x gets bigger and bigger?

☐ First, let's rewrite the function as $f(x) = \dfrac{1}{\sqrt{x+2}}$. The domain of $\sqrt{x+2}$ is $x \geq -2$ but we must exclude $x = -2$ so that the denominator does not become 0. So, the domain of f is $x > -2$. Given that domain, we can expect any positive number (not 0) as output, and so the range is all positive real numbers. As x gets bigger and bigger, $f(x)$ gets smaller and smaller (positive) and closer to zero. ∎

(6) Find the graphical oddities (asymptotes, holes) of $h(t) = \frac{t^3+t^2}{t^3+t^2-2t-2}$.

☐ The numerator factors as $t^3+t^2 = t^2(t+1)$, so hopefully $(t+1)$ will also be a factor of the denominator. With a bit of trial and error, you can determine that it indeed is a factor of the denominator, and so the entire function factors and simplifies as:

$$h(t) = \frac{t^3+t^2}{t^3+t^2-2t-2} = \frac{t^2(t+1)}{(t^2-2)(t+1)} = \frac{t^2}{t^2-2}$$

From the original factoring of the denominator, we see that the domain of this function is all reals except $t = \pm\sqrt{2}$ and $t = -1$. Since the term $t+1$ eventually cancels from the denominator, the function behaves just fine in the vicinity of $t = -1$, but we still can't use $t = -1$ itself. Therefore, there is a hole in the graph at $t = -1$. There are vertical asymptotes at the remaining roots of the denominator, $t = \pm\sqrt{2}$. ■

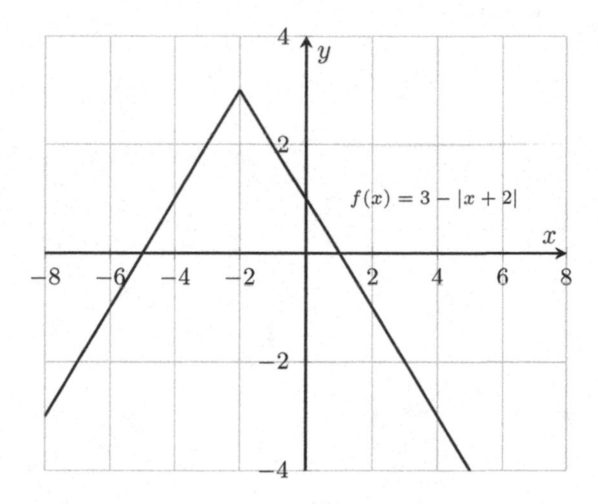

Fig. A.2 The graph of $f(x) = 3 - |x+2|$.

(7) Put the function $f(x) = 3 - |x+2|$ into piecewise form and sketch its graph.

☐ The absolute value bars do nothing when $x+2 \geq 0$ and "turn on" when $x+2 < 0$. So, the split point of this function is at $x = -2$. When

$x < -2$, i.e. when $x + 2$ is negative, the term $|x + 2|$ becomes $-(x + 2)$ as the absolute value bars do their thing. The piecewise form of this function is then,

$$f(x) = \begin{cases} 3 - (x + 2) & \text{for } x \geq -2 \\ 3 + (x + 2) & \text{for } x < -2 \end{cases}$$

or simplifying,

$$f(x) = \begin{cases} 1 - x & \text{for } x \geq -2 \\ 5 + x & \text{for } x < -2 \end{cases}$$

Therefore, the graph of this function is made of two lines, $y = -x + 1$ and $y = x + 5$ which switch from one to the other at $x = -2$ — see Fig. A.2. ■

A.1.3 *Exp and Log Functions — Practice — Solved*

(1) Sketch the graph of $y = 4^x - 3$; identify at least two points.

□ This is the graph of $y = 4^x$ shifted down by 3 units. Its y intercept is found where $x = 0$, and is therefore $(0, -2)$. A second point comes easily from $x = 1$, and is therefore $(1, 1)$. ■

(2) What are the domain and range of the function $f(x) = 3e^{-2x}$?

□ The graphs of plain ol' $y = e^x$ and $y = e^{-x}$ have domain of all reals and range of all positive reals. The inclusion of the extra constants 3 and 2 don't change that. ■

(3) Write $2\ln 4 - \ln 2$ as a single logarithm.

$$\square\, 2\ln 4 - \ln 2 = \ln 4^2 - \ln 2 = \ln 16 - \ln 2 = \ln\frac{16}{2} = \ln 8 \quad ■$$

(4) Find the value of $2^{\log_2 3 + \log_2 5}$.

□ There are a couple of ways to reduce this:

option 1: $2^{\log_2 3+\log_2 5} = 2^{\log_2 3\cdot 5} = 2^{\log_2 15} = 15$

option 2: $2^{\log_2 3+\log_2 5} = 2^{\log_2 3}\cdot 2^{\log_2 5} = 3\cdot 5 = 15$ ■

(5) Solve the equation $e^{2x+3} - 7 = 0$.

□ Follow the steps in order *(i)* to *(v)*:

i) $e^{2x+3} - 7 = 0$

ii) $e^{2x+3} = 7$ $\longrightarrow$ *iv)* $2x + 3 = \ln 7$

iii) $\ln e^{2x+3} = \ln 7$ *v)* $x = \frac{1}{2}(\ln 7 - 3)$ ■

(6) Solve the equation $\ln(4 - 2x) - \ln(2) = -3$.

□ We need to form a single logarithm before applying an exponential function. Follow the steps in order *(i)* to *(vi)*:

i) $\ln(4 - 2x) - \ln(2) = -3$ *iv)* $e^{\ln(2-x)} = e^{-3}$

ii) $\ln\frac{4-2x}{2} = -3$ $\longrightarrow$ *v)* $2 - x = e^{-3}$

iii) $\ln(2 - x) = -3$ *vi)* $x = 2 - e^{-3}$

And note that $x = 2 - e^{-3}$ can also be presented as $x = 2 - 1/e^3$. ■

(7) If as population $P(t)$ starts at 100 and doubles every 3 hours, write a general formula for the population as a function of time, $P(t)$. What will the population be after 15 hours?

□ If the population $P(t)$ starts at 100 and doubles every 3 hours, we have

$$P(0) = 100 \quad P(3) = 200 = 2 \cdot 100 \quad P(6) = 400 = 2^2 \cdot 100 \ldots$$
$$P(15) = 3200 = 2^5 \cdot 100$$

After any t hours, this pattern above shows that $P(t) = 2^{t/3} \cdot 100$. So, after 15 hours, we get $P(15) = 2^{15/3} \cdot 100 = 2^5(100) = 3200$. We can also find this by continuing the above pattern since 15 is a nice multiple of 3. ■

A.1.4 *Compositions & Inverses — Practice — Solved*

(1) If $f(x) = \sqrt{1+x}$ and $g(x) = \sqrt{1-x}$, then create and find the domains of $f+g$, $f-g$, fg, f/g.

□ The domain of $f(x)$ is $x \geq -1$; the domain of $g(x)$ is $x \leq 1$. Then:

- $f+g = \sqrt{1+x} + \sqrt{1-x}$; its domain is $-1 \leq x \leq 1$.
- $f-g = \sqrt{1+x} - \sqrt{1-x}$; its domain is $-1 \leq x \leq 1$.
- $fg = \sqrt{1+x} \cdot \sqrt{1-x}$; its domain is $-1 \leq x \leq 1$.
- $\dfrac{f}{g} = \dfrac{\sqrt{1+x}}{\sqrt{1-x}}$ Its domain is $-1 \leq x < 1$ (which is the same as the others in the problem except we also remove $x = 1$ as that point would make the denominator zero). ■

(2) If $f(x) = 1 - 3x$ and $g(x) = 5x^2 + 3x + 2$, then create and find the domains of $f(g(x))$, $g(f(x))$, $f(f(x))$, $g(g(x))$.

□ The domain of $f(x) = 1 - 3x$ is all reals; the domains of $g(x) = 5x^2 + 3x + 2$ is all reals. Then,

- $f(g(x)) = 1 - 3(5x^2 + 3x + 2)$; its domain is all reals.
- $g(f(x)) = 5(1-3x)^2 + 3(1-3x) + 2$; its domain is all reals
- $f(f(x)) = 1 - 3(1-3x) = -2 + 9x$; its domain is all reals.
- $g(g(x)) = 5(5x^2 + 3x + 2)^2 + 3(5x^2 + 3x + 2) + 2$; its domain is all reals. ■

(3) If $f(x) = 2/(x+1)$, $g(x) = x^3$, and $h(x) = \sqrt{x+3}$, what is $f(g(h(x)))$?

□ We have $g(h(x)) = (\sqrt{x+3})^3 = (x+3)^{3/2}$, and so

$$f(g(h(x))) = \frac{2}{(x+3)^{3/2}} = 2(x+3)^{-3/2} \quad ■$$

(4) What two functions $f(t)$ and $g(t)$ form $u(t) = e^t/(1+e^t)$ as the composition $f(g(t))$?

□ With $f(t) = t/(1+t)$ and $g(t) = e^t$, we get $u(t) = f(g(t)) = e^t/(1+e^t)$. ■

(5) What three functions $f(x)$, $g(x)$, and $h(x)$ form $H(x) = \log_2(\sqrt{x^4 - 1})$ when composed as $f(g(h))$?

☐ With $f(x) = \log_2(x)$, $g(x) = \sqrt{x}$, and $h(x) = x^4 - 1$, we get $g(h(x)) = \sqrt{x^4 - 1}$, and so $H(x) = f(g(h(x))) = \log_2(\sqrt{x^4 - 1})$. ■

---

(6) How can the graph of $y = x^2$ be adapted to create the graph of $y = 1 - x^2$?

☐ The graph of $y = 1 - x^2$ can be found by taking the graph of $y = x^2$, flipping it upside down ($-x^2$) and vertically raising the result up by 1 unit — as shown in Fig. A.3(A). ■

---

(7) How can the graph of $y = \sqrt[3]{x}$ be adapted to create the graph of $y = 1 + \sqrt[3]{x-1}$?

☐ The graph of $y = 1 + \sqrt[3]{x-1}$ can be found by taking the graph of $y = \sqrt[3]{x}$ and then performing two operations: (1) shifting to the right by one unit ($\sqrt[3]{x-1}$); (2) raising the result by one unit ($1 + \sqrt[3]{x-1}$) — as shown in Fig. A.3(B). ■

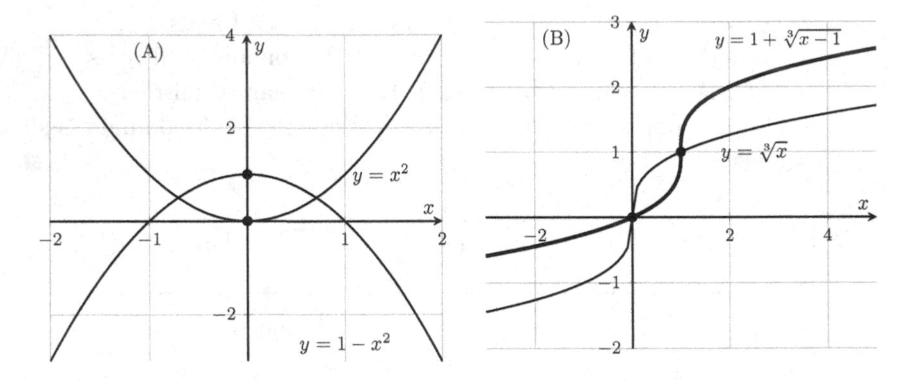

Fig. A.3 Graphs for Practice Problems 6 (A) and 7 (B).

---

(8) How can the graph of $y = e^x$ be adapted to create the graph of $y = 1 + 2e^x$?

☐ The graph of $y = 1 + 2e^x$ is formed by taking the graph of e^x, vertically stretching it by a factor of two and raising the result by 1 unit so that the y-intercept is $(0, 3)$ — as shown in Fig. A.5(A). ■

(9) How can the graph of $y = \ln(x)$ be adapted to create the graphs (a) $y = -\ln(x)$, (b) $y = \ln(-x)$, (c) $y = \ln|x|$?

☐ As seen in Fig. A.4, (a) $y = -\ln x$ is the graph of $y = \ln x$ flipped across the x-axis.
(b) $y = \ln(-x)$ is the graph of $y = \ln x$ reflected across the y-axis.
(c) $y = \ln|x|$ is the graph of $y = \ln x$ duplicated in mirror image across the y-axis; there are values of $x < 0$ now allowed. This graph is indeed visible in Fig. A.4, as it's the combination of the graphs of $\ln(x)$ and $\ln(-x)$. ■

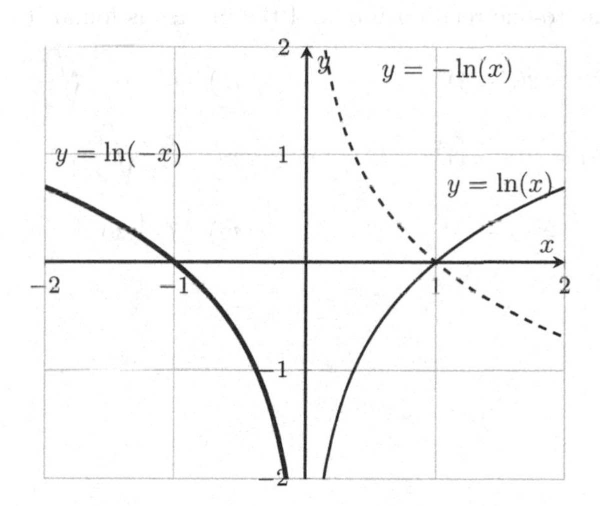

Fig. A.4 Variations on the graph of $ln(x)$.

(10) Find a domain over which $f(x) = 2(x-1)^4$ is one-to-one, and find its inverse function there.

☐ The entire graph of $f(x)$ is comprised of both the dashed and solid curves in Fig. A.5(B), and as such, fails the horizontal line test. However, if we restrict the domain to $x \geq 1$ (the solid curve only), then we

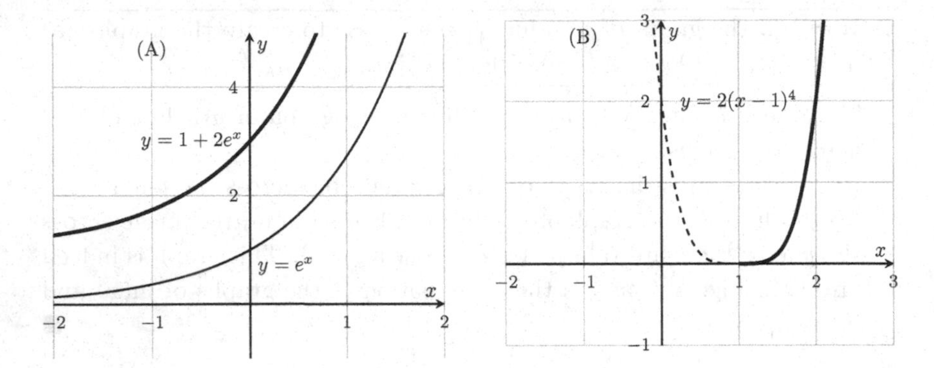

Fig. A.5 Graphs for Practice Problems 8 (A) and 10 (B).

have a one-to-one relationship, and the inverse is found by:

i) $y = 2(x-1)^4$		*iv)* $y - 1 = \sqrt[4]{\frac{x}{2}}$
ii) $x = 2(y-1)^4$	$\longrightarrow$	*v)* $y = \sqrt[4]{\frac{x}{2}} + 1$
iii) $\frac{x}{2} = (y-1)^4$		*vi)* $f^{-1}(x) = \sqrt[4]{\frac{x}{2}} + 1$ ■

A.1.5 *Trig Functions & Inverses — Practice — Solved*

(1) Curves D1 and D2 in Fig. A.6 are a trigonometric function (on its restricted domain) and its inverse. Identify them.

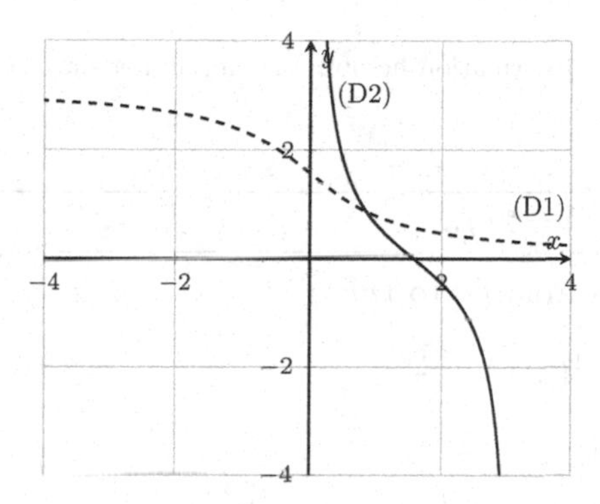

Fig. A.6 Name the trig function and its inverse.

□ D1 is $\cot^{-1}(x)$ and D2 is $\cot(x)$. ■

(2) Evaluate $\arctan(-1)$ and $\csc^{-1}(2)$.

□ The expression $\arctan(-1)$ represents the angle between $-\pi/2$ and $\pi/2$ whose tangent is -1. That angle is $-\pi/4$. The expression $\csc^{-1}(2)$ represents the angle on $[-\pi/2, 0) \bigcup (0, \pi/2]$ whose cosecant is 2. Note that angle is the same angle as $sin^{-1}(1/2)$. That angle is $\pi/6$. So:

$$\arctan(-1) = -1 \qquad \text{and} \qquad \csc^{-1}(2) = \frac{\pi}{6} \quad ■$$

(3) Simplify $\sin(\tan^{-1}(x))$ into its algebraic form.

□ The expression $\sin(\tan^{-1}(x))$ represents the sine of the angle whose tangent is x. If the tangent of this angle is x, then we have an opposite side of length x, and an adjacent side of length 1. This angle (A) and associated lengths are indicated in Fig. A.7. We can then complete the triangle by labeling the hypotenuse with length $\sqrt{1+x^2}$. So the sine of this angle is opp / hyp $= x/\sqrt{1+x^2}$ and

$$\sin(\tan^{-1}(x)) = \frac{x}{\sqrt{1+x^2}} \quad ■$$

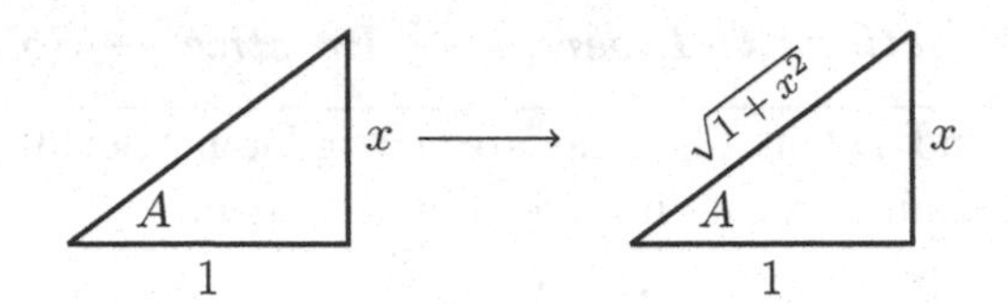

Fig. A.7 Information helpful for simplifying $\sin(\tan^{-1}(x))$.

---

(4) Solve $4\cos^{-1}(x^2) = 2\pi/3$.

☐ Follow along from *(i)* to *(v)*:

$$i)\quad 4\cos^{-1}(x^2) = \frac{2\pi}{3}$$

$$ii)\quad \cos^{-1}(x^2) = \frac{2\pi}{12} \qquad \longrightarrow \qquad iv)\quad x^2 = \frac{\sqrt{3}}{2}$$

$$iii)\quad \cos(\cos^{-1}(x^2)) = \cos\frac{\pi}{6} \qquad\qquad v)\quad x = \pm\sqrt{\frac{\sqrt{3}}{2}}$$

In some cases we elect to toss out one of the solutions in a "plus / minus" case like this. But here, the first thing that happens to a value of x in the equation being solved is that it gets squared before being handed to the inverse cosine. And upon squaring, either of these solution values of x does indeed produce a value that's within the "legal" restricted domain of the inverse cosine. So let's keep them both! ■

A.1.6 *Hyperbolic Funcs — Practice — Solved*

---

(1) State the domain and range of $\tanh(x)$ and give an approximate sketch its graph. (You should be able to identify some trends by its definition in terms of $\sinh(x)$ and $\cosh(x)$, and can plot a few points, right?)

☐ We know that $\tanh(x) = \dfrac{\sinh(x)}{\cosh(x)}$, and

- the range of $\sinh(x)$ is all real numbers
- the range of $\cosh(x)$ is all real numbers greater than or equal to one; more generally, $\cosh(x) > 0$
- since the denominator will never be zero and the domains of the numerator and denominator separately are all real numbers, the domain of $\tanh(x)$ is all real numbers
- since the numerator of $\tanh(x)$ can be any real number and the denominator is always positive, there are no evident restrictions on the range of $\tanh(x)$
- $tanh(x)$ will be negative where $\sinh(x) < 0$; also, $\tanh(x) = 0$ only where $\sinh(x) = 0$
- there will be no asymptotes on the graph of $\tanh(x)$
- gosh, the graph of $\tanh(x)$ might look a lot like the graph of $\sinh(x)$!

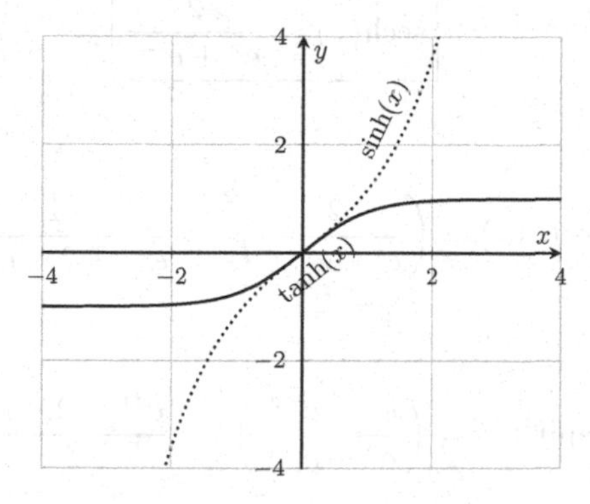

Fig. A.8 The graphs of $\tanh(x)$.

Figure A.8 shows the graph of $\tanh(x)$, along with $\sinh(x)$ (dotted) for comparison. The graph of $\tanh(x)$ does look like $\sinh(x)$ near the origin, but it levels off as we move away from the origin.

FFT: In fact, the graph of $\tanh(x)$ starts to creep in on either $y = 1$ or $y = -1$, depending on whether $x > 0$ or $x < 0$. Why is that? (This is a sneak preview of the next chapter!) Specifically, the range of $\tanh(x)$ has a floor and ceiling of of ± 1: $-1 \leq \tanh(x) \leq 1$. ■

(2) Give the following values as rational numbers: $\tanh(-\ln 3)$, $\operatorname{sech}(\ln 10)$, $\coth(\ln 3)$.

□

$$\tanh(-\ln 3) = \frac{e^{-\ln 3} - e^{-(-\ln 3)}}{e^{-\ln 3} + e^{-(-\ln 3)}} = \frac{\frac{1}{3} - 3}{\frac{1}{3} + 3} = \frac{1-9}{1+9} = -\frac{4}{5}$$

$$\operatorname{sech}(\ln 10) = \frac{2}{e^{\ln 10} + e^{-(\ln 10)}} = \frac{2}{10 + \frac{1}{10}} = \frac{20}{101}$$

$$\coth(\ln 3) = \frac{e^{\ln 3} + e^{-\ln 3}}{e^{\ln 3} - e^{-\ln 3}} = \frac{3 + \frac{1}{3}}{3 - \frac{1}{3}} = \frac{5}{4} \quad ■$$

(3) Give a definition of $\operatorname{sech}(x)$ using exponential functions, and see if there is a relationship between $\operatorname{sech}^2(x)$ and $\tanh^2(x)$.

□ Since $\sec(x) = 1/\cos(x)$, we can take a crazy guess that $\operatorname{sech}(x) = 1/\cosh(x)$, and so

$$\boxed{\operatorname{sech}(x) = \frac{2}{e^x + e^{-x}}}$$

Then,

$$\operatorname{sech}^2(x) = \left(\frac{2}{e^x + e^{-x}}\right)^2 = \frac{4}{e^{2x} + 2 + e^{-2x}}$$

Also,

$$\tanh^2(x) = \left(\frac{e^x - e^{-x}}{e^x + e^{-x}}\right)^2 = \frac{e^{2x} - 2 + e^{-2x}}{e^2 + 2 + e^{-2x}}$$

Then

$$\operatorname{sech}^2(x) + \tanh^2(x) = \frac{4 + e^{2x} - 2 + e^{-2x}}{e^2 + 2 + e^{-2x}} = \frac{e^{2x} + 2 + e^{-2x}}{e^2 + 2 + e^{-2x}} = 1$$

or $\operatorname{sech}^2(x) + \tanh^2(x) = 1$. ■

(4) Let $f(x) = \sinh(x)/(\cosh(x)+1)$. Does the graph of $f(x)$ have any vertical asymptotes? How do you know?

☐ The graph of $f(x)$ will only have vertical asymptotes if $\cosh(x)+1 = 0$, i.e. if $\cosh(x) = -1$. But we know the domain of this function is $\cosh(x) \geq 1$, so there are no points where $\cosh(x) = -1$, and there are no vertical asymptotes on the graph of $f(x)$. ■

(5) There is a trigonometric identity $\cos(A+B) = \cos A \cos B - \sin A \sin B$. Can you develop a similar identity for $\cosh(A+B)$?

☐ By Def. 1.3, $\cosh(A+B) = \frac{1}{2}(e^{A+B} + e^{-(A+B)})$. Let's build $\cosh A \cosh B$ and $\sinh A \sinh B$ and see if we can combine them somehow into $\cosh(A+B)$.

$$\begin{aligned}\cosh(A)\cosh(B) &= \frac{e^A + e^{-A}}{2} \cdot \frac{e^B + e^{-B}}{2} \\ &= \frac{1}{4}(e^{A+B} + e^{A-B} + e^{-A+B} + e^{-(A+B)})\end{aligned}$$

$$\begin{aligned}\sinh(A)\sinh(B) &= \frac{e^A - e^{-A}}{2} \cdot \frac{e^B - e^{-B}}{2} \\ &= \frac{1}{4}(e^{A+B} - e^{A-B} - e^{-A+B} + e^{-(A+B)})\end{aligned}$$

Might as well add them up and see what happens!

$$\begin{aligned}&\cosh(A)\cosh(B) + \sinh(A)\sinh(B) \\ &\quad = \frac{1}{4}(e^{A+B} + e^{A-B} + e^{-A+B} + e^{-(A+B)}) \\ &\qquad + \frac{1}{4}(e^{A+B} - e^{A-B} - e^{-A+B} + e^{-(A+B)}) \\ &\quad = \frac{1}{4}(2e^{A+B} + 2e^{-(A+B)}) = \frac{1}{2}(e^{A+B} + e^{-(A+B)})\end{aligned}$$

and the latter expression is $\cosh(A+B)$. Altogether, $\cosh(A+B) = \cosh A \cosh B + \sinh A \sinh B$ — which is not exactly the same as the trigonometric version, but is really close! ■

(6) Does the equation $\tanh^2 x + 4\tanh(x) + 4 = 0$ have any solutions? Why or why not? (Hint: Use the result of Practice Problem 1.)

☐ By factoring, we can rewrite $\tanh^2 x + 4\tanh(x) + 4 = 0$ to $(\tanh x + 2)^2 = 0$, which means we have solutions when $\tanh(x) = -2$, or $x = \tanh^{-1}(2)$. However, there are no such values; that is, $\tanh^{-1}(2)$ does not exist. By Practice Problem 1, we know that $-1 \leq \tanh(x) \leq 1$. ■

A.2 Chapter 2: Practice Problem Solutions

A.2.1 *Intro to Limits — Practice — Solved*

(1) Describe the following limits for the function $g(t)$ shown in Fig. A.9.

$$\text{(a)} \lim_{t\to 0^-} g(t) \quad \text{(b)} \lim_{t\to 0^+} g(t) \quad \text{(c)} \lim_{t\to 2^-} g(t) \quad \text{(d)} \lim_{t\to 2^+} g(t)$$

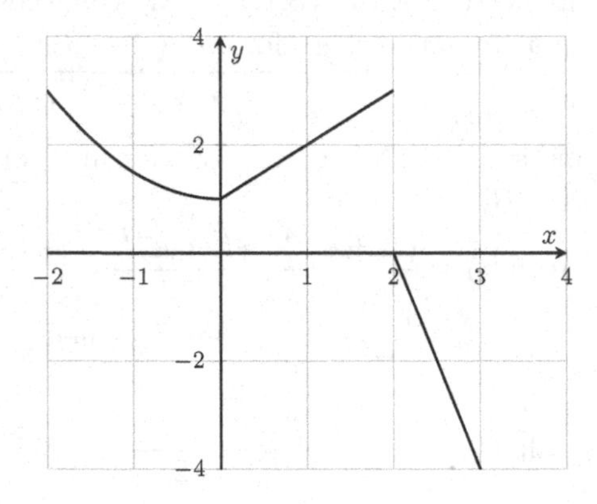

Fig. A.9 A piecewise function for PP 1.

☐ The limits are:

$$\text{(a)} \lim_{t\to 0^-} g(t) = 1 \quad \text{(b)} \lim_{t\to 0^+} g(t) = 1$$

$$\text{(c)} \lim_{t\to 2^-} g(t) = 3 \quad \text{(d)} \lim_{t\to 2^+} g(t) = 0 \quad \blacksquare$$

(2) Describe the following limits for the function $R(x)$ shown Fig. A.10.

$$\text{(a)} \lim_{x\to 2} R(x) = \quad \text{(b)} \lim_{x\to 5} R(x) \quad \text{(c)} \lim_{x\to -3^-} R(x) \quad \text{(d)} \lim_{x\to -3^+} R(x)$$

☐ The limits are:

$$\text{(a)} \lim_{x\to 2} R(x) = -\infty \quad \text{(b)} \lim_{x\to 5} R(x) = +\infty$$

$$\text{(c)} \lim_{x\to -3^-} R(x) = +\infty \quad \text{(d)} \lim_{x\to -3^+} R(x) = -\infty \quad ,\blacksquare$$

(3) Estimate the limit as $x \to -1$ for $f(x) = (x^2 - 2x)/(x^2 - x - 2)$ by examining several values of $f(x)$ in the vicinity of $x = -1$.

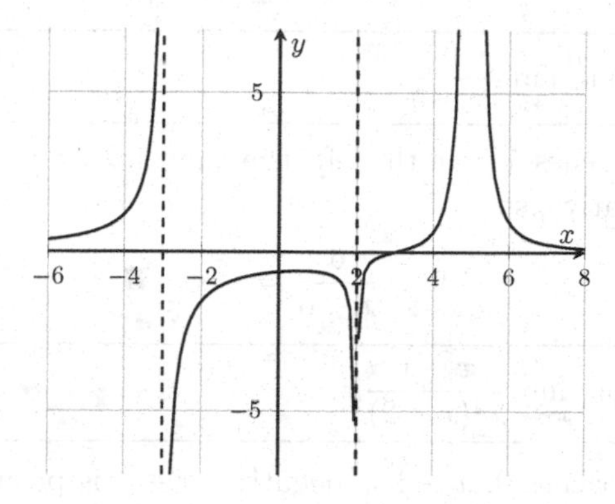

Fig. A.10 A piecewise function for PP 2.

☐ Here is a table of values of $f(x)$ at values in the vicinity of $x = -1$; the split occurs where $x = -1$ would sit:

x	-2	-1.5	-1.1	-1.01	-1.001
$f(x)$	2	3	11	101	1001

x	-0.999	-0.99	-0.95	-0.9	-0.5	0
$f(x)$	-999	-99	-19	-9	-1	0

So it looks like

$$\lim_{x\to -1^-} f(x) = +\infty \quad \text{and} \quad \lim_{x\to -1^+} f(x) = -\infty$$

so that $\lim_{x\to -1} f(x)$ does not exist. ■

(4) Estimate the limit as $x \to 0$ for $f(x) = \tan 3x/\tan 5x$ by examining several values of $f(x)$ in the vicinity of $x = 0$.

☐ Here is a table of values of $f(x)$ in the vicinity of $x = 0$:

x	-0.2	-0.1	-0.01	0.01	0.1	0.2
$f(x)$	0.4393	0.5662	0.5997	0.5997	0.5662	0.4393

and so conclude that

$$\lim_{x\to 0} \frac{\tan 3x}{\tan 5x} \approx 0.6 = \frac{3}{5} \quad ■$$

(5) Assess the limit $\lim_{x \to 5^-} \frac{6}{x-5}$.

□ As x approaches 5 from the left, the quantity $x - 5$ approaches zero, but stays negative, so

$$\lim_{x \to 5^-} \frac{6}{x-5} = -\infty \quad \blacksquare$$

(6) Assess the limit $\lim_{x \to 0} \frac{x-1}{x^2(x+2)}$.

□ As x approaches 0, $x - 1$ is negative, $x + 2$ is positive, and x^2 approaches zero but remains positive (regardless of whether x approaches 0 from the left or right). So

$$\lim_{x \to 0} \frac{x-1}{x^2(x+2)} = -\infty \quad \blacksquare$$

A.2.2 *Computation of Limits — Practice — Solved*

(1) If $\lim_{x\to a} f(x) = -3$, $\lim_{x\to a} g(x) = 0$, and $\lim_{x\to a} h(x) = 8$, then use the steps allowed by limit laws to determine the values of

(a) $\lim_{x\to a} \frac{f(x)}{h(x)}$ (b) $\lim_{x\to a} \frac{g(x)}{f(x)}$ (c) $\lim_{x\to a} \frac{f(x)}{g(x)}$ (d) $\lim_{x\to a} \frac{2f(x)}{h(x)-f(x)}$

The limits are

$$\text{(a) } \lim_{x\to a} \frac{f(x)}{h(x)} = \frac{\lim_{x\to a} f(x)}{\lim_{x\to a} h(x)} = \frac{-3}{8}$$

$$\text{(b) } \lim_{x\to a} \frac{g(x)}{f(x)} = \frac{\lim_{x\to a} g(x)}{\lim_{x\to a} f(x)} = \frac{0}{-3} = 0$$

$$\text{(c) } \lim_{x\to a} \frac{f(x)}{g(x)} = \frac{\lim_{x\to a} f(x)}{\lim_{x\to a} g(x)} \text{ which does not exist}$$

$$\text{(d) } \lim_{x\to a} \frac{2f(x)}{h(x)-f(x)} = \frac{2\lim_{x\to a} f(x)}{\lim_{x\to a} h(x) - \lim_{x\to a} f(x)} = \frac{2(-3)}{8-(-3)} = \frac{-6}{11} \quad \blacksquare$$

(2) Use the steps allowed by limit laws to determine $\lim_{t\to -1} (t^2+1)^3(t+3)^5$.

□
$$\begin{aligned}\lim_{t\to -1} (t^2+1)^3(t+3)^5 &= [\lim_{t\to -1} (t^2+1)]^3 \cdot [\lim_{t\to -1} (t+3)]^5 \\ &= [\lim_{t\to -1} t^2 + \lim_{t\to -1} 1]^3 \cdot [\lim_{t\to -1} t + \lim_{t\to -1} 3]^5 \\ &= (1+1)^3(-1+3)^5 \\ &= (8)(32) = 256 \quad \blacksquare\end{aligned}$$

(3) Determine $\lim_{x\to 3} \frac{x^2+2}{x-4}$.

□ This is a rational function, and $x = 3$ is in the domain ($x = 3$ doesn't make the denominator zero). So we're allowed to just plug it in:

$$\lim_{x\to 3} \frac{x^2+2}{x-4} = \frac{3^2+2}{3-4} = -11 \quad \blacksquare$$

(4) Determine $\lim_{x\to 4} \dfrac{x^2-4x}{x^2-3x-4}$.

$$\square \lim_{x\to 4} \frac{x^2-4x}{x^2-3x-4} = \lim_{x\to 4} \frac{x(x-4)}{(x-4)(x+1)} = \lim_{x\to 4} \frac{x}{x+1} = \frac{4}{5} \quad \blacksquare$$

(5) Determine $\lim_{x\to 1} \dfrac{x^3-1}{x^2-1}$.

$$\square \lim_{x\to 1} \frac{x^3-1}{x^2-1} = \lim_{x\to 1} \frac{(x-1)(x^2+x+1)}{(x-1)(x+1)} = \lim_{x\to 1} \frac{(x^2+x+1)}{x+1} = \frac{3}{2} \quad \blacksquare$$

(6) Determine $\lim_{h\to 0} \dfrac{\frac{1}{3+h}-\frac{1}{3}}{h}$.

$$\square \lim_{h\to 0} \frac{\frac{1}{3+h}-\frac{1}{3}}{h} = \lim_{h\to 0} \frac{\frac{3}{3(3+h)}-\frac{3+h}{3(3+h)}}{h} = \lim_{h\to 0} \frac{\frac{(3)-(3+h)}{3(3+h)}}{h}$$

$$= \lim_{h\to 0} \frac{\frac{-h}{3(3+h)}}{h} = \lim_{h\to 0} \frac{-1}{3(3+h)} = -\frac{1}{9} \quad \blacksquare$$

(7) Determine $\lim_{x\to -4^-} \dfrac{|x+4|}{x+4}$.

□ The numerator and denominator have the same magnitude at all points, but for points to the left of -4, the denominator is negative while the numerator is positive, so the quotient reduces to -1, and

$$\lim_{x\to -4^-} \frac{|x+4|}{x+4} = -1 \quad \blacksquare$$

(8) If $F(x) = \dfrac{x^2-1}{|x-1|}$, evaluate

(a) $\lim_{x\to 1^+} F(x)$ (b) $\lim_{x\to 1^-} F(x)$ (c) $\lim_{x\to 1} F(x)$

□ Since we can rewrite

$$F(x) = \frac{x^2-1}{|x-1|} = \frac{(x+1)(x-1)}{|x-1|}$$

note that the quotient $(x-1)/|x-1|$ is equal to 1 for $x > 1$ and is equal to -1 for $x < 1$. So,

$$\text{(a)} \lim_{x \to 1^+} F(x) = 2 \qquad \text{(b)} \lim_{x \to 1^-} F(x) = -2$$

so that (c) $\lim_{x \to 1} F(x)$ does not exist. ■

---

(9) Use the defining equations in Def. (1.3) to determine $\lim_{x \to 0} \sinh(x)$ and $\lim_{x \to 0} \cosh(x)$.

□

$$\lim_{x \to 0} \sinh(x) = \lim_{x \to 0} \frac{e^x - e^{-x}}{2} = \frac{e^0 - e^0}{2} = 0$$

$$\lim_{x \to 0} \cosh(x) = \lim_{x \to 0} \frac{e^x + e^{-x}}{2} = \frac{e^0 + e^0}{2} = 1 \quad ■$$

A.2.3 *Limits w/ Infinity — Practice — Solved*

(1) Determine all asymptotes of $f(x) = \dfrac{3x+5}{x-4}$.

□ There is a vertical asymptote at $x = 4$ due to the denominator. There is a horizontal asymptote, too:

$$\lim_{x\to\infty} \frac{3x+5}{x-4} = \lim_{x\to\infty} \frac{3+5/x}{1-4/x} = \frac{3+0}{1-0} = 3$$

(the limit as $x \to -\infty$ is the same). The horizontal asymptote is $y = 3$.

■

(2) Determine all asymptotes of $g(t) = \dfrac{t^2+2}{t^3+t^2-t-1}$.

□ Because the degree of the numerator is less than the degree of the denominator, the denominator "wins", and

$$\lim_{t\to-\infty} \frac{t^2+2}{t^3+t^2-t-1} = 0$$

Therefore the x-axis is a horizontal asmptote. To find any vertical asymptote, we have to factor the denominator. Now come on, stop complaining about that: all you have to do is head for Wolfram Alpha and type in "factor t^3+t^2-t-1", and you'll find that we can rewrite the function as

$$\frac{t^2+2}{t^3+t^2-t-1} = \frac{t^2+2}{(t^2-1)(t+1)} = \frac{t^2+2}{(t-1)(t+1)(t+1)}$$

Since none of the terms in the denominator will cancel with anything in the numerator, all terms contribute an asymptote, although one appears twice. The vertical asymptotes of this function are at $x = \pm 1$. ■

(3) Determine all asymptotes of $h(x) = \dfrac{x+2}{\sqrt{9x^2+1}}$.

□ We can compute the appropriate limit by factoring out the highest term from the numerator and denominator:

$$\lim_{x\to\infty} \frac{x+2}{\sqrt{9x^2+1}} = \lim_{x\to\infty} \frac{x(1+2/x)}{\sqrt{x^2(9+1/x^2)}} = \lim_{x\to\infty} \frac{1+2/x}{\sqrt{9+1/x^2}} = \frac{1}{3}$$

(The limit as $x \to -\infty$ is the same.) So there is a horizontal asymptote at $y = 1/3$. Since the denominator is never zero, there are no vertical asymptotes.

■

(4) Determine all asymptotes of $y = \dfrac{x^2 + 4}{x^2 - 1}$.

□ The graph has a vertical asymptote at $x = \pm 1$ since the left and right hand limits of the function are infinite at both locations. There is a horizontal asymptote at $y = 1$ because

$$\lim_{x \to \infty} \frac{x^2 + 4}{x^2 - 1} = \lim_{x \to \infty} \frac{1 + 4/x^2}{1 - 1/x^2} = 1 \quad \blacksquare$$

(5) Determine all asymptotes of $y = \dfrac{2x^2 - 1}{x - 2}$.

□ The graph has a vertical asymptote at $x = 2$ because the term $x - 2$ contributes a root that is not cancelled by anything in the numerator. Since the degree of the numerator is one more than the degree of the denominator, we expect to see a slant asymptote; further, because of this revision of $f(x)$:

$$\frac{2x^2 - 1}{x - 2} = \frac{x^2(2 - 1/x^2)}{x(1 - 2/x)} = x \cdot \frac{2 - 1/x^2}{1 - 2/x}$$

we see that the fractional term collapses to 2 as $x \to \pm\infty$, then the slant asymptote is $y = 2x$. ■

(6) Determine all asymptotes of $y = \tanh(x)$. (Hint: See Def. 1.4.)

□ Since

$$y = \tanh(x) = \frac{e^x - e^{-x}}{e^x + e^{-x}}$$

we see there will be no values of x for which the denominator will be zero. The function is continuous. So there are no vertical asymptotes. For horizontal asymptotes,

$$\lim_{x \to \infty} \tanh(x) = \lim_{x \to \infty} \frac{e^x - e^{-x}}{e^x + e^{-x}}$$

As x gets larger (while positive), the influence of e^{-x} diminishes, and so the limit starts to look like,

$$\lim_{x \to \infty} \tanh(x) = \lim_{x \to \infty} \frac{e^x}{e^x} = 1 \quad \blacksquare$$

A.2.4 *Limits & Continuity — Practice — Solved*

---

(1) Identify the locations and types of discontinuities in the graph shown in Fig. A.11, and also state the intervals on which the function is continuous.

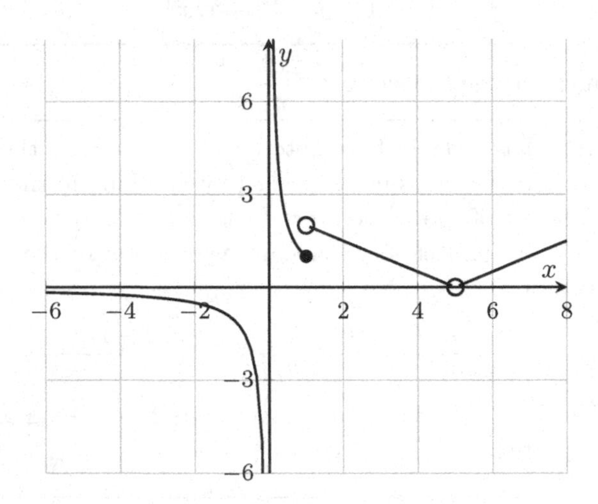

Fig. A.11 Discontinuous for Practice Problem 1.

☐ $f(x)$ is not continuous at $x = 0$, $x = 1$, and $x = 5$. The discontinuity at $x = 0$ is an infinite discontinuity; the discontinuity at $x = 1$ is a jump discontinuity; the discontinuity at $x = 5$ is a removable discontinuity. The intervals of continuity are $(-\infty, 0)$, $(0, 1]$, $(1, 5)$, and $(5, \infty)$. The endpoint $x = 1$ is included in the interval $(0, 1]$ because the left hand limit is equal to the value of the function at that point. ■

---

(2) If $f(x) = \sqrt[3]{x}$ and $g(x) = 1 + x^3$, where is the composition $f(g(x))$ continuous?

☐ First note that the functions $f(x) = \sqrt[3]{x}$ and $g(x) = 1 + x^3$ are both continuous everywhere. Therefore the composition $f(g(x)) = \sqrt[3]{1 + x^3}$ is continuous everywhere. ■

(3) Where is the function $H(x) = \cos(c^{\sqrt{x}})$ continuous?

□ The function $\sqrt{x}$ is continuous for all $x > 0$, and e^x and $\cos x$ are continuous everywhere. So the composition of these three functions, $H(x)$, is continuous for $x > 0$. ■

(4) Is this function continuous at $x = 1$?

$$f(x) = \begin{cases} \frac{x^2-x}{x^2-1} & \text{for } x \neq 1 \\ 1 & \text{for } x = 1 \end{cases}$$

□ We can simplify the top piece to

$$\frac{x^2 - x}{x^2 - 1} = \frac{x(x-1)}{(x+1)(x-1)} = \frac{x}{x+1}$$

As $x \to 1$, $f(x) \to 1/2$ but $f(1) = 1$. So the function is not continuous at $x = 1$. ■

(5) Describe everything you know about the continuity of

$$f(x) = \begin{cases} x+1 & \text{for } \quad x \leq 1 \\ 1/x & \text{for } 1 < x < 3 \\ \sqrt{x-3} & \text{for } \quad x \geq 3 \end{cases}$$

□ Since all three pieces of this function are continuous in their territory, the only places that $f(x)$ can be discontinuous are where the pieces connect (or not). So, since

$$\lim_{x \to 1^-} f(x) = 2 \quad \text{and} \quad \lim_{x \to 1^+} f(x) = 1$$

$$\lim_{x \to 3^-} f(x) = 1/3 \quad \text{and} \quad \lim_{x \to 3^+} f(x) = 0$$

$f(x)$ is discontinuous at both $x = 1$ and $x = 3$. At $x = 1$, $f(x)$ is continuous from the left. At $x = 3$, $f(x)$ is continuous from the right. ■

(6) Prove that the equation $\sqrt[3]{x} = 1 - x$ has a solution somewhere in the interval $(0, 1)$.

□ This is the same as showing the function $f(x) = 1 - x - \sqrt[3]{x}$ has a root on $(0, 1)$. (Do you see why?) The function $f(x) = 1 - x - \sqrt[3]{x}$ is continuous everywhere. Since $f(0) = 1$ and $f(1) = -1$, we see that $f(0) > 0$ and $f(1) < 0$. So by the Intermediate Value Theorem, there is a point somewhere between $x = 0$ and $x = 1$ where $f(x) = 0$, i.e. where $1 - x - \sqrt[3]{x} = 0$, i.e. where $\sqrt[3]{x} = 1 - x$. ■

(7) During the COVID-19 outbreak of 2020, the number of coronavirus cases in the United States on March 01 was reported (due to the spectacular testing program in the country) as 89. By March 12, the number of cases had increased to a reported 1,645. Can we use the IVT to prove that there must have been an instant between March 01 and March 12 when the number of cases was exactly 120?

□ If we had a function that gave us the number of people infected with respect to time, say $I(t)$, then this function would not be continuous, as its range would consist of only integers (number of people). Therefore, the IVT does not apply, and we can't draw any conclusions. ■

A.2.5 *Secant & Tangent Lines — Practice — Solved*

(1) Use a progression of secant lines and their slopes to estimate the slope of the line tangent to $y = 1/(1+x)$ at the point $(1, 1/2)$. Then find the equation of that tangent line.

□ We'll form secant lines by connecting $(1, 1/2)$ to several points to its left and right, and see the trend in slope as we get closer to $x = 1$. For any (x, y), the slope of the line connecting (x, y) to $(1, 0.5)$ is $m = (y - 0.5)/(x - 1)$. Here is a table (split at $x = 1$) of such slope values:

x :	0.5	0.9	0.99	0.999	1
y :	0.3333	0.4737	0.4975	0.4997	0.5
m :	0.3333	0.2631	0.2512	0.2501	?

x :	1	1.001	1.01	1.1	1.5
y :	0.5	0.5002	0.5025	0.5238	0.6000
m :	?	0.2499	0.2488	0.2381	0.2

So it looks like the slope of the line tangent to the curve at $(1, 1/2)$ should be 0.25. The equation of the tangent line is then:

$$y - \frac{1}{2} = \frac{1}{4}(x - 1) \quad \blacksquare$$

(2) Use limits to find the slope of the tangent line to $f(x) = 1 + 2x - x^2$ at (1,2), then find the equation of that line.

□ We're looking for the tangent line at $x = a = 1$, so the slope of the tangent line there is:

$$\begin{aligned} m_{tan} &= \lim_{h\to 0} \frac{f(a+h) - f(a)}{h} = \lim_{h\to 0} \frac{f(1+h) - f(1)}{h} \\ &= \lim_{h\to 0} \frac{[1 + 2(1+h) - (1+h)^2] - [1 + 2 - 1]}{h} \\ &= \lim_{h\to 0} \frac{1 + 2 + 2h - (1 + 2h + h^2) - 2}{h} \\ &= \lim_{h\to 0} \frac{-h^2}{h} = \lim_{h\to 0} -h = 0 \end{aligned}$$

So the equation of the tangent line is $y - 2 = 0(x - 1)$ or $y = 2$. ■

(3) Use limits to find the form of the slope of the tangent line to $f(x) = x^3 - 4x + 1$ at an unspecified point $x = a$. Then use that slope form to get the equations of the lines tangent to $f(x)$ at $x = 1$ and $x = 2$.

☐ The slope of the tangent line to $f(x) = x^3 - 4x + 1$ at $x = a$ comes from:

$$
\begin{aligned}
m_{tan} &= \lim_{h \to 0} \frac{f(a+h) - f(a)}{h} \\
&= \lim_{h \to 0} \frac{[(a+h)^3 - 4(a+h) + 1] - [a^3 - 4a + 1]}{h} \\
&= \lim_{h \to 0} \frac{(a^3 + 3a^2h + 3ah^2 + h^3 - 4a - 4h + 1) - (a^3 - 4a + 1)}{h} \\
&= \lim_{h \to 0} \frac{(3a^2h + 3ah^2 + h^3 - 4h}{h} \\
&= \lim_{h \to 0} 3a^2 + 3ah + h^2 - 4 = 3a^2 - 4
\end{aligned}
$$

At $x = 1$ we have $y = f(1) = -2$ and at $x = 2$ we have $y = f(2) = 1$. The slopes of the tangent lines at $(1, -2)$ and $(2, 1)$ are

$$3(1)^2 - 4 = -1 \qquad \text{and} \qquad 3(2)^2 - 4 = 8$$

so the tangent lines themselves are

$$y + 2 = -1(x - 1) \qquad \text{and} \qquad y - 1 = 8(x - 2) \quad \blacksquare$$

A.3 Chapter 3: Practice Problem Solutions

A.3.1 *The Rate of Change — Practice — Solved*

(1) The average number of hits per day each year on a popular web page about calculus is given in this table:

t	2012	2014	2016	2018
N	10,036	10,109	10,152	10,175

(a) Use the data in the table to estimate the average rate of change in the number of hits from (i) 2012 to 2016, (ii) 2014 to 2016, and (iii) 2016 to 2018. (b) Estimate the instantaneous rate of change in 2016.

□

(i) from 2012 to 2016: $\Delta N/\Delta t = (10,152 - 10,036)/4 = 29$ thousand hits per day

(ii) from 2014 to 2016: $\Delta N/\Delta t = (10,152 - 10,109)/2 = 21.5$ thousand hits per day

(iii) from 2016 to 2018: $\Delta N/\Delta t = (10,175 - 10,152)/2 = 11.5$ thousand hits per day

(b) We could estimate the instantaneous change in 2016 by averaging the average change from 2014 to 2016 and 2016 to 2018, which gives approximately $(21.5 + 11.5)/2 = 16.5$ thousand hits per day. ■

(2) If an arrow is shot upwards on the moon with a velocity of 58 m/s, its height in meters after t seconds is $y = 58t - 0.83t^2$. (a) Find the average velocity over time intervals starting at $t = 1$ and ending at (i) 2, (ii) 1.5, (iii) 1.1, (iv) 1.01, (v) 1.001. (b) Use those values to estimate the instantaneous velocity of the arrow at $t = 1$.

□ (a) The average velocity between any two points (t_1, y_1) and (t_2, y_2) is

$$v_{avg} = \frac{y_2 - y_1}{t_2 - t_1}$$

That is, v_{avg} is the slope of the secant line connecting the points. Note that in all time intervals, $t_1 = 1$. Since $y = 58t - 0.83t^2$, $t_1 = 1$ gives $y_1 = 57.17$, and to any other point, then, $v_{avg} = (y - 57.17)/(t - 1)$. So:

t	2	1.5	1.1	1.01	1.001	1
y	112.68	85.13	62.80	57.73	57.23	57.17
v_{avg}	55.51	55.93	56.26	56.33	56.34	?

(b) The instantaneous velocity at $t = 1$ is the slope of the tangent line to (t, y) at $t = 1$, which looks like it's about 56.34. ■

---

(3) Repeat part (b) of the previous Practice Problem, but use a limit to estimate the instantaneous velocity of the arrow at $t = 1$. Was your estimate of this instantaneous rate of change in the last problem close to this new, better, answer?

□ The function describing height of the arrow as a function of time is $y = 58t - 0.83t^2$. The instantaneous rate of change of y with respect to t at $t = 1$ will be:

$$\left(\frac{\Delta y}{\Delta t}\right)_{inst} = \lim_{h \to 0} \frac{y(1+h) - y(1)}{h}$$

Let's find the the pieces in the numerator of this expression:

$$\begin{aligned} y(1) &= 58(1) - 0.83(1)^2 = 57.17 \\ y(1+h) &= 58(1+h) - 0.83(1+h)^2 \\ &= 58 + 58h - 0.83(1 + 2h + h^2) \\ &= 57.17 + 56.34h - 0.83h^2 \end{aligned}$$

so

$$\begin{aligned} \left(\frac{\Delta y}{\Delta t}\right)_{inst} &= \lim_{h \to 0} \frac{y(1+h) - y(1)}{h} \\ &= \lim_{h \to 0} \frac{(57.17 + 56.34h - 0.83h^2) - (57.17)}{h} \\ &= \lim_{h \to 0} \frac{56.34h - 0.83h^2}{h} = \lim_{h \to 0} (56.34 - 0.83h) = 53.64 \end{aligned}$$

The instantaneous rate of change of height with respect to time at $t = 1$ is 53.64 m/s. The estimate from the previous practice problem was awesome! ■

A.3.2 *Derivative at a Point — Practice — Solved*

(1) If $f(x) = e^{-x}$, is there any point $x = a$ where $f'(a)$ is positive or zero?

☐ There is not. The graph of e^{-x} goes downhill everywhere; $f'(a)$ would be negative for every $x = a$ on the graph. ■

(2) If $g(x) = 1-x^3$, use a limit to determine $g'(0)$ and provide the equation of the line tangent to $g(x)$ at $x = 0$.

☐ The slope of the the tangent line to $g(x) = 1 - x^3$ at $(0, 1)$ is found via Def. 3.2. This will require building $f(0 + h) - f(0)$ and hoping we find an h to factor out of each term. We get

$$f(0+h) - f(0) = [1 - (0+h)^3] - [1 - 0^3] = -h^3$$

and so

$$g'(0) = \lim_{h\to 0} \frac{g(0+h) - g(0)}{h} = \lim_{h\to 0} \frac{-h^3}{h} = \lim_{h\to 0} -h^2 = 0$$

So at (0,1), the equation of the tangent line is $y - 1 = 0(x - 0)$, or, $y = 1$. ■

(3) If $f(t) = t^4 - 5t$, use a limit to determine $f'(a)$ where a is still unspecified.

☐ If $f(t) = t^4 - 5t$, then

$$\begin{aligned} f(a+h) - f(a) &= [(a+h)^4 - 5(a+h)] - [a^4 - 5a] \\ &= [a^4 + 4a^3h + 6a^2h^2 + 4ah^3 + h^4 - 5a - 5h] - [a^4 - 5a] \\ &= 4a^3h + 6a^2h^2 + 4ah^3 + h^4 - 5h \\ &= h(4a^3 + 6a^2h + 4ah^2 + h^3 - 5) \end{aligned}$$

and so,

$$\begin{aligned} f'(a) &= \lim_{h\to 0} \frac{f(a+h) - f(a)}{h} \\ &= \lim_{h\to 0} \frac{h(4a^3 + 6a^2h + 4ah^2 + h^3 - 5)}{h} \\ &= \lim_{h\to 0} (4a^3 + 6a^2h + 4ah^2 + h^3 - 5) = 4a^3 - 5 \end{aligned}$$ ■

(4) The following expression computes $f'(a)$ for some function $f(x)$ at some $x = a$. What are $f(x)$ and a?

$$\lim_{h\to 0} \frac{\sqrt[4]{16+h}-2}{h}$$

□ We must match this to the expression

$$\lim_{h\to 0} \frac{f(a+h)-f(a)}{h}$$

So evidently, $f(a+h) = \sqrt[4]{16+h}$ and $f(a) = 2$. This means the given expression is the derivative of $f(x) = \sqrt[4]{x}$ at $a = 16$. (As a check, we do confirm that $f(a) = \sqrt[4]{16} = 2$.) ■

(5) Use the identity $\sin(A+B) = \sin A\cos B + \cos A\sin B$ to find $f'(0)$ for $f(x) = \sin(x)$. Can you use this result to guess the value of $f'(a)$ at other locations, too?

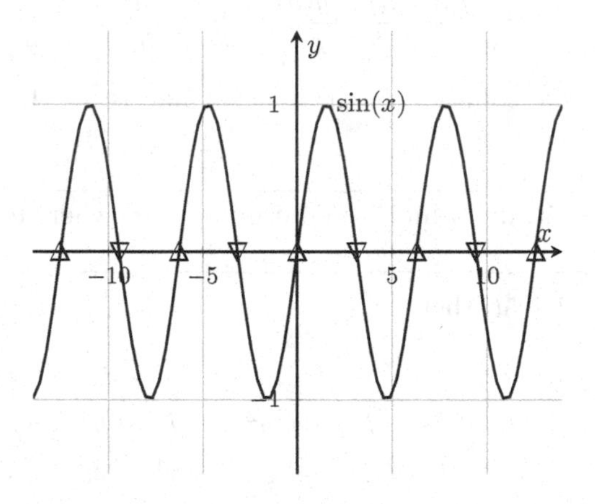

Fig. A.12 sin(x) and locations of max (Δ) and min (∇) derivative values.

□ We need to prepare for Def. 3.2 by building $\sin(0+h)-\sin(0)$. Using the identity given in the problem statement, this can be expanded as:

$$\begin{aligned}\sin(0+h)-\sin(0) &= [\sin(0)\cos(h)+\cos(0)\sin(h)]-\sin(0)\\ &= 0+(1)\sin(h)-0=\sin(h)\end{aligned}$$

and then,

$$f'(0) = \lim_{h\to 0}\frac{\sin(0+h)-\sin(0)}{h} = \lim_{h\to 0}\frac{\sin h}{h} = 1$$

And so we have $f'(0) = 1$ for $f(x) = \sin(x)$. Now, $x = 0$ is one of the locations where $\sin(x)$ is going uphill the most rapidly, and so we can expect that the max value of the derivative of $\sin(x)$ is 1. This happens not only at $x = 0$, but at all even multiples of π, such as $\ldots, -4\pi, -2\pi, 0, 2\pi, 4\pi, \ldots$ (marked with Δ in Fig. A.12). Plus, because of the symmetry of the graph, we can expect that the max slope or rate "downhill" on the graph $\sin(x)$ will be -1, and this will happen at odd multiples of π, such as $\ldots, -3\pi, -\pi, \pi, 3\pi, \ldots$ (marked with ∇ in Fig. A.12). ■

---

(6) Suppose the fuel consumption (in gallons per hour) c of a car is a function of velocity v (in mph), so that $c = f(v)$. (a) What is the meaning of $f'(20)$ in the context of the problem, and what are its units? (b) What would $f'(20) = -0.05$ mean?

□ (a) The derivative $f'(20)$ represents the rate of change of fuel consumption c induced by a small change in velocity while at velocity 20 mph; the units of this would be gallons per hour per miles per hour (phew!). And (b), the statement $f'(20) = -0.05$ means that if you are driving 20 mph and increase your velocity very slightly, your fuel consumption will decrease by -0.05 gallons per hour per mile per hour. ■

A.3.3 *The Derivative Function — Practice — Solved*

(1) Use the graph of $f(x) = x^3/3 - x$ to determine where $f'(x)$ is negative.

☐ The graph of $f(x)$ (shown in Fig. A.13) goes downhill between $x = -1$ and $x = 1$, and so $f'(x)$ is negative there. ■

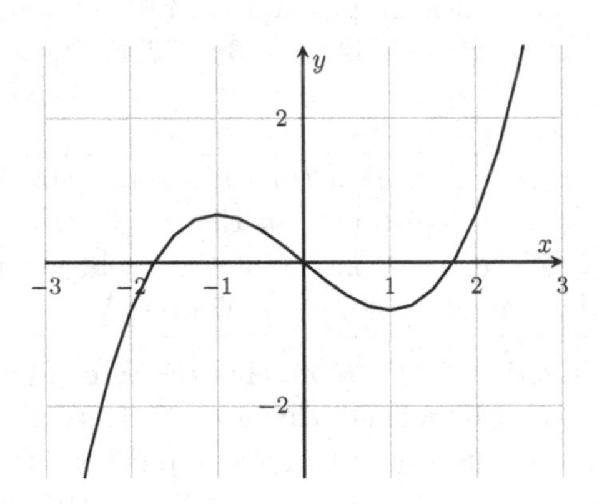

Fig. A.13 $f(x) = x^3/3 - x$ (with for PP 1).

(2) Use a limit to determine $f'(x)$ for $f(x) = 5x^2 + 3x - 2$.

☐ With this function, we have $f(x+h) = 5(x+h)^2 + 3(x+h) - 2 = 5(x^2 + 2xh + h^2) + 3x + 3h - 2$, so that

$$\begin{aligned} f(x+h) - f(x) &= [5(x^2 + 2xh + h^2) + 3x + 3h - 2] - [5x^2 + 3x - 2] \\ &= 5(2xh + h^2) + 3h = h(10x + 5h + 3) \end{aligned}$$

$$\begin{aligned} f'(x) &= \lim_{h \to 0} \frac{f(x+h) - f(x)}{h} \\ &= \lim_{h \to 0} \frac{h(10x + 5h + 3)}{h} = \lim_{h \to 0} (10x + 5h + 3) = 10x + 3 \end{aligned}$$ ■

(3) Use a limit to determine $f'(x)$ for $f(x) = 1/x^2$.

☐ With this function, we have

$$\begin{aligned} f(x+h) - f(x) &= \frac{1}{(x+h)^2} - \frac{1}{x^2} = \frac{x^2}{(x+h)^2 x^2} - \frac{(x+h)^2}{(x+h)^2 x^2} \\ &= \frac{x^2 - (x^2 + 2xh + h^2)}{(x+h)^2 x^2} = \frac{h(-2x - h)}{(x+h)^2 x^2} \end{aligned}$$

so that

$$f'(x) = \lim_{h\to 0} \frac{f(x+h) - f(x)}{h} = \lim_{h\to 0} \frac{h(-2x-h)}{(x+h)^2x^2} \cdot \frac{1}{h}$$

$$= \lim_{h\to 0} \frac{-2x-h}{(x+h)^2x^2} = \frac{-2x}{x^2x^2} = \frac{-2}{x^3} \quad \blacksquare$$

(4) At how many points would the function $f(x) = |\sin(x)|$ not be differentiable? Describe them.

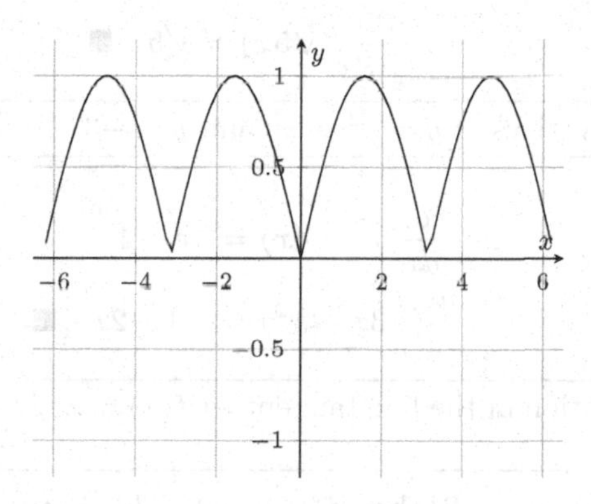

Fig. A.14 $f(x) = |\sin(x)|$ (with PP 4).

□ $f(x) = |\sin(x)|$ would not be differentiable at an infinite number of points. At each $x = n\pi$, where $\sin(x)$ itself crosses from positive to negative, the graph of $|\sin(x)|$ will have a cusp (some of them can be seen in Fig. A.14). ■

A.3.4 *Simple Derivs and Antiderivs — Practice — Solved*

(1) Find the derivatives of $y = x - 1$, $y = 12x - 5$, and $y = \sqrt{5}x$.

$$\square \quad \frac{d}{dx}(x-1) = 1$$
$$\frac{d}{dx}(12x-5) = 12$$
$$\frac{d}{dx}(\sqrt{5}x) = \sqrt{5} \quad \blacksquare$$

(2) Find the derivatives of $y = x^2 - 4x$ and $y = -3x + x^2$.

$$\square \quad \frac{d}{dx}(x^2 - 4x) = 2x - 4$$
$$\frac{d}{dx}(-3x + x^2) = -3 + 2x \quad \blacksquare$$

(3) Find the equation of the line tangent to $f(x) = x^2/2 - 2x - 1$ at $x = 2$.

□ Since $f'(x) = x - 2$, then $f'(2) = 0$. This is the slope of the tangent line. The point on the tangent line is $(2, f(2)) = (2, -3)$. So the equation of the tangent line is $y + 3 = 0(x - 2)$, or $y = -3$. ■

(4) Find the following antiderivatives:

$$\int (4)\, dx \quad ; \quad \int \frac{2}{3}x\, dx$$

$$\square \quad \int (4)\, dx = 4x + C$$
$$\int \frac{2}{3}x\, dx = \frac{1}{3}x^2 + C \quad \blacksquare$$

(5) Find the antiderivative $\int 1 - 6t\, dt$.

$$\square \quad \int 1 - 6t\, dt = t - 3t^3 + C \quad \blacksquare$$

A.3.5 *Power Rule — Practice — Solved*

(1) Find the derivative of $f(t) = \sqrt{t} - 1/\sqrt{t}$.

□ Note that

$$f(t) = \sqrt{t} - \frac{1}{\sqrt{t}} = t^{1/2} - t^{-1/2}$$

so by the power rule,

$$f'(t) = \frac{1}{2}t^{-1/2} - \left(-\frac{1}{2}t^{-3/2}\right) = \frac{1}{2\sqrt{t}} + \frac{1}{2t\sqrt{t}} \quad ■$$

(2) Find the derivative of $y = (x^2 - 2\sqrt{x})/x$.

□ Simplify first,

$$y = \frac{x^2 - 2\sqrt{x}}{x} = x - \frac{2}{\sqrt{x}} = x - 2x^{-1/2}$$

so by the power rule,

$$\frac{dy}{dx} = 1 - 2 \cdot \left(-\frac{1}{2}\right) \cdot x^{-3/2} = 1 + \frac{1}{x\sqrt{x}} \quad ■$$

(3) Find the equation of the line tangent to $y = (1 + 2x)^2$ at $x = 1$.

□ The slope of the tangent line to $y = (1 + 2x)^2$ at (1,9) is the value of the derivative at $x = 1$. To get the derivative, let's multiply the function out to see its terms, $y = 4x^2 + 4x + 1$, and get

$$\frac{dy}{dx} = 4\frac{d}{dx}x^2 + 4\frac{d}{dx}x + \frac{d}{dx}1 = 8x + 4$$

then $y'(1) = 8(1) + 4 = 12$. The equation of the tangent line is therefore $y - 9 = 12(x - 1)$. ■

(4) Find the locations where the tangent lines to $y = x^3 + 3x^2 + x + 3$ are horizontal.

□ The tangent line to $y = x^3 + 3x^2 + x + 3$ is horizontal where the derivative is zero. The derivative is $y' = 3x^2 + 6x + 1$. This doesn't factor well, so we need the quadratic formula to find where $y' = 0$:

$$x = \frac{-b \pm \sqrt{b^2 - 4ac}}{2a} = \frac{-6 \pm \sqrt{6^2 - 4(3)(1)}}{2(3)} = \frac{-6 \pm \sqrt{24}}{6}$$

$$= \frac{-6 \pm 2\sqrt{6}}{6} = -1 \pm \frac{\sqrt{6}}{3}$$

These are the values of x where the tangent line is horizontal. ■

(5) Find $\int x^{20} + 4x^{10} + 8\,dx$.

□

$$\int x^{20} + 4x^{10} + 8\,dx = \frac{1}{21}x^{21} + 4 \cdot \frac{1}{11}x^{11} + 8x + C$$
$$= \frac{1}{21}x^{21} + \frac{4}{11}x^{11} + 8x + C \quad ■$$

(6) Find $\int \frac{2x^4 - 3\sqrt{x}}{x^2}\,dx$.

□ Let's separate the terms of the integrand to get

$$\int \frac{2x^4 - 3\sqrt{x}}{x^2}\,dx = \int \frac{2x^4}{x^2} - \frac{3\sqrt{x}}{x^2}\,dx = \int 2x^2 - 3x^{-3/2}\,dx$$
$$= \frac{2}{3}x^3 - 3 \cdot (-2x^{-1/2}) + C = \frac{2}{3}x^3 + 6\frac{1}{\sqrt{x}} + C$$

(7) Find the antiderivative of $f(x) = 8x^3 + 12x + 3$ that goes through the point $(1, 6)$.

□ Let's name

$$F(x) = \int 8x^3 + 12x + 3\,dx = 2x^4 + 6x^2 + 3x + C$$

To make $F(1) = 6$, we need $6 = 2 + 6 + 3 + C$, so that $C = -5$, and $F(x) = 2x^4 + 6x^2 + 3x - 5$. ■

A.3.6 *Power Rule Pt 2 — You Try It — Solved*

(1) Find dy/dx for $y = (2x^2 - 5)^6$.

$$\square\ \frac{dy}{dx} = 6(2x^2-5)^5 \cdot \frac{d}{dx}(2x^2-5) = 24x(2x^2-5) \quad \blacksquare$$

(2) Find $h'(t)$ if $h(t) = \sqrt[4]{t^2+2}$.

□ Since $h(t) = (t^2+2)^{1/4}$, we have

$$h'(t) = \frac{1}{4}(t^2+2)^{-3/4} \cdot \frac{d}{dt}(t^2+2) = \frac{2t}{4}(t^2+2)^{-3/4} = \frac{t}{2(t^2+2)^{3/4}} \quad \blacksquare$$

(3) Find the instantaneous rate of change of $f(x) = \sqrt[5]{(2x^4+x+1)^3}$ at $x = 0$.

□ Writing $f(x) = (2x^4+x+1)^{3/5}$, we get $f'(x) = 3(2x^4+x+1)^{-2/5}/5$. So at $x = 0$, we have $f'(0) = 3(0+0+1)^{-2/5}/5 = 3/5$. The instantaneous rate of change of $f(x)$ at $x = 0$ is $3/5$. ■

(4) Find the equation of the line tangent to $y = \dfrac{3}{(x^2-4)^2+16}$ at $x = 2$.

□ The denominator can be rewritten as

$$(x^2-4)^2 + 16 = x^4 - 8x^2 + 16 + 16 = x^4 - 8x^2 + 32$$

and so

$$\begin{aligned}
\frac{dy}{dx} &= \frac{d}{dx}3(x^4-8x^2+32)^{-1} \\
&= -3(x^4-8x^2+32)^{-2}\frac{d}{dx}(x^4-8x^2+32) \\
&= -\frac{3}{(x^4-8x^2+32)^2}\cdot(4x^3-16x) = -\frac{12x(x^2-4)}{(x^2-4)^2+16}
\end{aligned}$$

(in which the denominator is returned to its original form). At $x = 2$, then,

$$\left.\frac{dy}{dx}\right|_{x=2} = -\frac{12(2)(2^2-4)}{(2^2-4)^2+16} = -\frac{0}{16} = 0$$

Hey, the tangent line is horizontal at $x = 2$! So the equation of this tangent line is determined by the y-coordinate,

$$y(2) = \frac{3}{(2^2-4)^2+16} = \frac{3}{0+16} = \frac{3}{16}$$

The equation of the tangent line at $x = 2$ is $y = \dfrac{3}{16}$. ■

(5) Analyze the heck out of the graph of $f(x) = 2/(x^3 + 1)$. ("Analyze the heck out of" is a very technical phrase meaning "find all horizontal and vertical asymptotes, roots, and locations of horizontal tangent lines".) Produce a sketch of the curve based on your work.

□ Before working on the derivative, we see that $f(x)$ will have a vertical asymptote at $x = -1$ since the denominator will be zero there. Also, the horizontal asymptote is $y = 0$, since that is the limit of $f(x)$ as $x \to \pm\infty$. There are no roots, since $f(x)$ itself is never zero. Now, onto the derivative for locations of horizontal tangent lines! Writing $f(x) = 2(x^3 + 1)^{-1}$, we get

$$f'(x) = -2(3x^2)(x^3 + 1)^{-2} = -\frac{6x^2}{(x^3 + 1)^2}$$

From this, we see that $f'(x) = 0$ when $x = 0$, so there is a horizontal tangent line at the point $(0, 2)$. Figure A.15 shows a graph of this function. ■

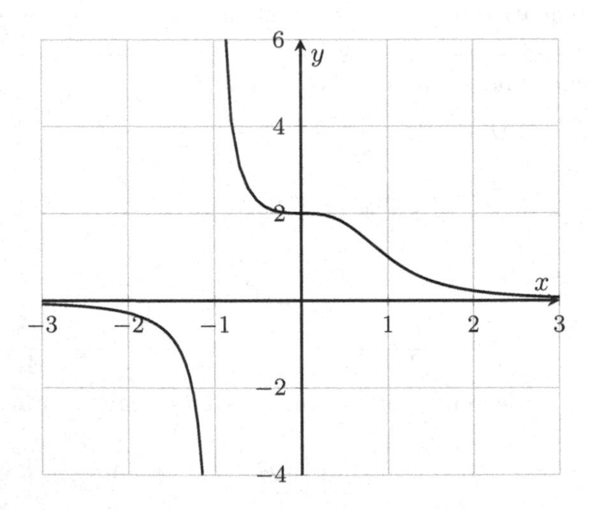

Fig. A.15 A graph of $f(x) = \dfrac{2}{x^3 + 1}$.

(6) Find the antiderivatives

$$(a) \int x^3(1-x^4)^3\,dx \quad ; \quad (b) \int 11x^3(1-x^4)^3\,dx$$

□ In matching the integrand of each to $g'(x)(g(x))^p$, we have $g(x) = 1 - x^4$, for which we need $g'(x) = -4x^3$. Neither integrand shows this exactly, but we can adjust:

$$\begin{aligned}(a) \int x^3(1-x^4)^3\,dx &= -\frac{1}{4}\int(-4x^3)(1-x^4)^3\,dx \\ &= -\frac{1}{4}\left(\frac{1}{4}(1-x^4)^4 + C\right) \\ &= -\frac{1}{16}(1-x^4)^4 + C\end{aligned}$$

$$\begin{aligned}(b) \int 11x^3(1-x^4)^3\,dx &= -\frac{11}{4}\int(-4x^3)(1-x^4)^3\,dx \\ &= -\frac{11}{4}\left(\frac{1}{4}(1-x^4)^4 + C\right) \\ &= -\frac{11}{16}(1-x^4)^4 + C \quad \blacksquare\end{aligned}$$

(7) Find the antiderivatives

$$(a) \int \frac{s^2}{\sqrt{s^3+1}}\,ds \quad ; \quad (b) \int \frac{s^2}{3\sqrt{s^3+1}}\,ds$$

□ Let's rewrite the function and also adjust constants to make perfect match between the integrands and the general form $g'(s)(g(s))^p$ (note the change in variable!). We have $g(s) = s^3 + 1$, and so need to make $g'(s) = 3s^2$ appear:

$$\begin{aligned}(a) \int \frac{s^2}{\sqrt{s^3+1}}\,ds &= \frac{1}{3}\int(3s^2)(s^3+1)^{-1/2}\,ds \\ &= \frac{1}{3}\cdot\frac{2}{1}(s^3+1)^{1/2} + C = \frac{2}{3}\sqrt{s^3+1} + C\end{aligned}$$

In (b), look out for the leading 1/3 due to the 3 in the denominator of the integrand:

$$\begin{aligned}\int \frac{s^2}{3\sqrt{s^3+1}}\,ds &= \frac{1}{3}\int \frac{s^2}{3\sqrt{s^3+1}}\,ds \\ &= \frac{1}{3}\cdot\frac{1}{3}\int (3s^2)(s^3+1)^{-1/2}\,ds \\ &= \frac{1}{9}\cdot\frac{2}{1}(s^3+1)^{1/2}+C = \frac{2}{9}\sqrt{s^3+1}+C \quad \blacksquare\end{aligned}$$

---

(8) Find $\int (2x^3+1)(x^4+2x+3)^3\,dx$.

☐ Trying to match the integrand to the usual form $g'(x)(g(x))^p$, we see $g(x) = x^4+2x+3$... so we'd really like to see $4x^3+2$ as well. By fiddling with constants, we can make that happen:

$$\begin{aligned}\int (2x^3+1)(x^4+2x+3)^3\,dx &= \int \frac{1}{2}\cdot 2(2x^3+1)(x^4+2x+3)^3\,dx \\ &= \frac{1}{2}\int (4x^3+2)(x^4+2x+3)^3\,dx\end{aligned}$$

Then,

$$\begin{aligned}\frac{1}{2}\int (4x^3+2)(x^4+2x+3)^3\,dx &= \frac{1}{2}\left(\frac{1}{4}(x^4+2x+3)^4+C\right) \\ &= \frac{1}{8}(x^4+2x+3)^4+C \quad \blacksquare\end{aligned}$$

A.4 Chapter 4: Practice Problem Solutions

A.4.1 *Exp & Log Function Derivs — Practice — Solved*

(1) Find the derivative and antiderivative of $y = 5e^x + \frac{3}{x}$.

$$\square \quad \frac{dy}{dx} = \frac{d}{dx}5e^x + \frac{3}{x} = 5e^x - \frac{3}{x^2}$$
$$\int 5e^x + \frac{3}{x}\,dx = 5e^x + 3\ln|x| + C \quad \blacksquare$$

(2) Find the derivative and antiderivative of $y = 7\sqrt{x} - e^x - \frac{2}{x}$.

$$\square \quad \frac{dy}{dx} = \frac{7}{2\sqrt{x}} - e^x + \frac{2}{x^2}$$
$$\int 7\sqrt{x} - e^x - \frac{2}{x}\,dx = \frac{7}{3/2}x^{3/2} - e^x - 2\ln|x| + C$$
$$= \frac{14}{3}x^{3/2} - e^x - 2\ln|x| + C \quad \blacksquare$$

(3) Find the equation of the line tangent to $y = \ln(x) + x$ at $x = 1$.

□ Since $dy/dx = 1/x + 1$, then at $x = 1$, $dy/dx = 2$. This is the slope of the tangent line. The point the line shares with the function is $(1, y(1)) = (1, \ln(1) + 1) = (1, 1)$. The equation of the tangent line is then $y - 1 = 2(x - 1)$. ■

(4) Find the derivative of $f(x) = e^{5x - \ln x^2}$.

$$\square \quad f'(x) = e^{5x-\ln x^2} \cdot \frac{d}{dx}(5x - \ln x^2)$$
$$= e^{5x-\ln x^2} \cdot \frac{d}{dx}(5x - 2\ln x) = e^{5x-\ln x^2}\left(5 - \frac{2}{x}\right)$$

Note that it may be tempting to simplify the function ahead of time:

$$e^{5x-\ln x^2} = e^{5x} \cdot e^{-\ln x^2} = e^{5x} \cdot e^{\ln x^{-2}} = e^{5x} \cdot x^{-2}$$

But, we don't know how to find the derivative of this new version of the function yet: it is written as the product of two functions of x, and its derivative is NOT just $5e^{5x} \cdot (-2x^{-3})$. This requires the *product rule*, which is coming up soon! ■

(5) Find $h'(t)$ if $h(t) = 2t^2 - e^{-t^3}$.

$$\square\, h'(t) = 4t - e^{-t^3} \cdot \frac{d}{dt}(-t^3) = 4t + 3t^2 e^{-t^3} \quad \blacksquare$$

(6) Find dy/dx if $y = \ln\left(\frac{1}{x^2} - 2\right)$.

□ The best way to tackle this is to rework the function, using properties of logs, before doing the derivative:

$$\ln\left(\frac{1}{x^2} - 2\right) = \ln\frac{1-2x^2}{x^2} = \ln(1-2x^2) - \ln(x^2) = \ln(1-2x^2) - 2\ln x$$

and so

$$\begin{aligned}\frac{d}{dx}\ln\left(\frac{1}{x^2} - 2\right) &= \frac{d}{dx}(\ln(1-2x^2) - 2\ln x) \\ &= \frac{1}{1-2x^2}\cdot\frac{d}{dx}(1-2x^2) - \frac{2}{x} \\ &= \frac{-4x}{1-2x^2} - \frac{2}{x} \quad \blacksquare\end{aligned}$$

(7) Find the derivative of $r(s) = \sqrt[4]{s} + \ln(5 - s^5)$.

$$\square \quad r'(s) = \frac{d}{ds}s^{1/4} + \frac{1}{5-s^5}\cdot\frac{d}{ds}(5-s^5) = \frac{1}{4}s^{-3/4} - \frac{5s^4}{5-s^5} \quad \blacksquare$$

(8) Find the antiderivative $\int x^{-3}e^{x^{-2}}\,dx$.

□ Recognize that when matching to Eq. (4.3) with an integrand $g'(x)e^{g(x)}$, we have $g(x) = x^{-2}$ and so we should look for $g'(x) = -2x^{-3}$. But we don't see $-2x^{-3}$, we only see x^{-3}. However, we can adjust:

$$\int x^{-3}e^{x^{-2}}\,dx = -\frac{1}{2}\int(-2x^{-3})e^{x^{-2}}\,dx = -\frac{1}{2}e^{x^{-2}} + C \quad \blacksquare$$

(9) Find the antiderivative $\int \frac{e^{3-\ln x}}{x}\,dx$.

□ Do we match the integrand to $g'(x)/g(x)$ or $g'(x)e^{g(x)}$? The second is better, because we will hit a dead end quickly if we try the first form. (Go ahead, try it!) Recognize that when matching to $g'(x)e^{g(x)}$, we have $g(x) = 3 - \ln x$ and so we should look for $g'(x) = -1/x$. We do see $1/x$ as part of the whole expression, but we are missing the negative $(-)$ sign, so let's adjust:

$$\int \frac{e^{3-\ln x}}{x}\,dx = -\int\left(-\frac{1}{x}\right)\cdot e^{3-\ln x}\,dx = -e^{3-\ln x} + C \quad \blacksquare$$

---

(10) Find the antiderivative $\displaystyle\int \frac{5}{2-3t}\,dt.$

☐ Recognize that when matching the integrand to $g'(t)/g(t)$, we have $g(t) = 2 - 3t$ and so we should look for $g'(x) = -3$. But we don't see -3, we see 5, so let's adjust:

$$\int \frac{5}{2-3t}\,dt = -\frac{5}{3}\int \frac{-3}{2-3t}\,dt = -\frac{5}{3}\ln|2-3t| + C \quad \blacksquare$$

---

(11) Find the antiderivative $\displaystyle\int \frac{2-x^2}{6x-x^3}\,dx.$

☐ Recognize that when matching the integrand to $g'(x)/g(x)$, we have $g(x) = 6x - x^3$ and so we should look for $g'(x) = 6 - 3x^2$. But we don't see $6 - 3x^2$, we see $2 - x^2$. However, $6 - 3x^2 = 3(2 - x^2)$, so we can adjust:

$$\int \frac{2-x^2}{6x-x^3}\,dx = \frac{1}{3}\frac{6-3^2}{6x-x^3}\,dx = \frac{1}{3}\ln|6x-x^3| + C \quad \blacksquare$$

---

(12) Find the derivative and antiderivative of $\sinh(6x)$.

☐ By (4.11) and (4.12),

$$\frac{d}{dx}\sinh(6x) = \cosh(6x)\frac{d}{dx}(6x) = 6\cosh(6x)$$

$$\int \sinh(6x)\,dx = \int \frac{1}{6}\cdot 6\sinh(6x)\,dx = \frac{1}{6}\cosh(6x) + C \quad \blacksquare$$

(13) Find the derivative and antiderivative of $\frac{1}{3}\sinh(3x) + x^2$.

□

$$\frac{d}{dx}\left(\frac{1}{3}\sinh(3x) + x^2\right) = \frac{1}{3}\frac{d}{dx}\sinh(3x) + \frac{d}{dx}(x^2)$$
$$= \frac{1}{3}\cdot 3\cosh(3x) + 2x = \cosh(3x) + 2x$$
$$\int\left(\frac{1}{3}\sinh(3x) + x^2\right)dx = \frac{1}{9}\int 3\sinh(3x)\,dx + \frac{1}{3}\int \cdot 3x^2\,dx$$
$$= \frac{1}{9}\cosh(3x) + \frac{1}{3}x^3 + C \quad \blacksquare$$

(14) If $g(x) = e^{-3\cosh(x)}$, what is $g'(x)$?

□ By (4.3),

$$g'(x) = e^{-3\cosh(x)} \cdot \frac{d}{dx}(-3\cosh(x)) = e^{-3\cosh(x)} \cdot (-3\sinh(x))$$
$$= -3\sinh(x)e^{-3\cosh(x)} \quad \blacksquare$$

(15) Evaluate $\int e^{2x}\cosh(e^{2x})\,dx$.

□ The derivative of e^{2x} is $2e^{2x}$, so doctoring up the integral to match (4.11) with $g(x) = e^{2x}$,

$$\int e^{2x}\cosh(e^{2x})\,dx = \frac{1}{2}\int 2e^{2x}\cosh(e^{2x})\,dx = \sinh(e^{3x}) + C \quad \blacksquare$$

A.4.2 *Trig Functions Derivs — Practice — Solved*

(1) Find the derivative and antiderivative of $y = 3\cos(x) - \sqrt{2}\sin(x)$.

$$\square \quad \frac{dy}{dx} = -3\sin(x) - \sqrt{2}\cos(x)$$

$$\int 3\cos(x) - \sqrt{2}\sin(x) + \sqrt{x}\,dx = 3\sin(x) - \sqrt{2}(-\cos x) + C$$
$$= 3\sin(x) + \sqrt{2}\cos x + C \quad \blacksquare$$

(2) At how many points will the graph of $y = \sin(x^2)$ have a horizontal tangent line? Find the three of those points that are closest to or at the origin.

□ The graph of the function will have a horizontal tangent line wherever $dy/dx = 0$. Since $dy/dx = 2x\cos(x^2)$, then this will happen at $x = 0$ and whenever x^2 is $\pm\pi/2, \pm 3\pi/2, \ldots$. The ones closest to or at the origin are:

$$x = 0 \quad ; \quad x = \sqrt{\frac{\pi}{2}} \quad ; \quad x = -\sqrt{\frac{\pi}{2}} \quad \blacksquare$$

(3) Find the equation of the line tangent to $y = \cos(\ln x)$ at $x = e$. All constants must be in exact form.

□ The point this tangent line shares with the function is $(e, y(e)) = (e, \cos(\ln e)) = (e, \cos 1)$. The slope of this tangent line is dy/dx at $x = e$:

$$\frac{dy}{dx} = -\sin(\ln x)\frac{d}{dx}(\ln x) = -\frac{\sin(\ln x)}{x}$$

so when $x = e$, we have

$$m_{tan} = -\frac{\sin(\ln e)}{e} = -\frac{\sin 1}{e}$$

So the equation of the tangent line is:

$$y - \cos 1 = -\frac{\sin 1}{e}(x - e) \quad \blacksquare$$

(4) Find the antiderivative of $y = 2\sin x\cos^2(x) + 3\cos x$ that goes through the point $(\pi, 0)$.

□ We have

$$\int 2\sin x\cos^2(x) + 3\cos x\,dx = -\frac{2}{3}\cos^3 x + 3\sin x + C$$

To make this go through the point $(\pi, 0)$ we need:

$$\begin{aligned} -\frac{2}{3}\cos^3\pi + 3\sin\pi + C &= 0 \\ -\frac{2}{3}(-1) + 3(0) + C &= 0 \\ \frac{2}{3} + C &= 0 \\ C &= -\frac{2}{3} \end{aligned}$$

Therefore the specific antiderivative we need is:

$$\int 2\sin x\cos^2(x) + 3\cos x\,dx = -\frac{2}{3}\cos^3 x + 3\sin x - \frac{2}{3} \quad \blacksquare$$

(5) Find the derivative of $g(t) = e^{\sec t} + \sec(e^t)$.

$$\begin{aligned} \square \quad g'(t) &= e^{\sec t}\frac{d}{dt}(\sec t) + \sec(e^t)\tan(e^t)\frac{d}{dt}(e^t) \\ &= e^{\sec t}\sec t\tan t + e^t\sec(e^t)\tan(e^t)\frac{d}{dt}(e^t) \quad \blacksquare \end{aligned}$$

(6) Find the derivative of $y(x) = \ln(\csc x) + \csc(\ln x)$.

$$\begin{aligned} \square \quad y'(x) &= \frac{1}{\csc x}\frac{d}{dx}(\csc x) - \csc(\ln x)\cot(\ln x)\frac{d}{dx}(\ln x) \\ &= \frac{1}{\csc x}(-\csc x\cot x) - \csc(\ln x)\cot(\ln x)\frac{1}{x} \\ &= -\cot x - \frac{1}{x}\csc(\ln x)\cot(\ln x) \quad \blacksquare \end{aligned}$$

(7) One of these antiderivatives cannot be solved using the formulas discussed in this section. Identify it, then solve the other two.

A) $\int \sin x\cos^2 x\,dx$ B) $\int \sin^2 x\cos x\,dx$ C) $\int \sin^2 x\cos^2 x\,dx$

□ The first two can be solved:

$$\text{(A)} \int \sin x \cos^2 x \, dx = -\frac{1}{2}\cos^3 x + C$$
$$\text{(B)} \int \sin^2 x \cos x \, dx = \frac{1}{3}\sin^3 x + C$$

In both cases, we match to the extended power rule with integrand $g'(x)(g(x))^p$ using $g(x) = \cos x$ or $g(x) = \sin x$. The third, (C) cannot be solved since it does not match the extended power rule in either case. There would be an extra $\sin x$ or $\cos x$ once $g(x)$ and $g'(x)$ were identified. For example if we tried to identify $g(x) = \sin x$ and so $g'(x) = \cos x$, we'd have:

$$\text{(C)} \int \sin^2 x \cos^2 x \, dx = \int \sin^2 x \cos x \cdot \cos x \, dx$$
$$= \int (g(x))^2 g'(x) \cdot \cos x \, dx$$

We cannot proceed with the extra $\cos x$ still in place. ■

---

(8) One of these antiderivatives cannot be solved using the formulas discussed in this section. Identify it, then solve the other two.

A) $\int \tan^2 x \, dx$ B) $\int \sec^2 x \tan^2 x \, dx$ C) $\int \sec^2 x \tan x \, dx$

$$\Box \text{(A)} \int \tan^2 x \, dx$$

cannot be solved. It is not a power rule problem since there is no $g'(x)$ for $g(x) = \tan x$. It does not match any other direct formula. (B) is solved using the extended power rule with $g(x) = \tan x$ and $g'(x) = \sec^2 x$:

$$\text{(B)} \int \sec^2 x \tan^2 x \, dx = \frac{1}{3}\tan^3 x + C$$

(C) is another power rule with $g(x) = \sec x$ and $p = 1$:

$$\text{(C)} \int \sec^2 x \tan x \, dx = \int \sec x(\sec x \tan x) \, dx = \frac{1}{2}\sec^2 x + C$$

(In the middle expression, we see $(g(x))^1 = \sec x$ and $g'(x) = \sec x \tan x$.) ■

(9) Find the antiderivative $dsp \int \sec^2 x(\tan^2 x + \tan x + 1)\,dx$.

☐ Each term is its own little power rule problem with $g(x) = \tan x$ and $g'(x) = \sec^2 x$:

$$\int \sec^2 x(\tan^2 x + \tan x + 1)\,dx = \frac{1}{3}\tan^3 x + \frac{1}{2}\tan^2 x + \tan x + C \quad ■$$

(10) Find the derivative of $y = \ln(\sin^2 x + \cos^2 x) + e^{\sec^2 x - \tan^2 x}$. (Hint: Don't just plunge in without looking closely at the function.)

☐ If we remember our trigonometric identities $\sin^2 x + \cos^2 x = 1$ and $\sec^2 x - \tan^2 x = 1$, then the function is $y = \ln(1) + e^1$, a constant, and so $dy/dx = 0$. If you do this the long way for practice, you should still get 0! ■

A.4.3 *Product & Quotient Rules — Practice — Solved*

(1) Find the derivative of $f(x) = \sqrt{x}e^x$.

☐ By the product rule,

$$f'(x) = \sqrt{x}\frac{d}{dx}e^x + \frac{d}{dx}\sqrt{x} \cdot e^x = \sqrt{x}e^x + \frac{1}{2\sqrt{x}} \cdot e^x$$
$$= e^x\left(\sqrt{x} + \frac{1}{2\sqrt{x}}\right) = e^x\left(\frac{2x+1}{2\sqrt{x}}\right) \quad \blacksquare$$

(2) Find the derivative of $y = e^t(\cos t + 5t)$.

☐ The independent variable is t, so the derivative we want is dy/dt. By the product rule, if $y = e^t(\cos t + 5t)$, then

$$\frac{dy}{dt} = e^t\frac{d}{dt}(\cos t + 5t) + \frac{d}{dt}e^t \cdot (\cos t + 5t)$$
$$= e^t(-\sin t + 5) + e^t(\cos t + 5t) = e^t(-\sin t + 5 + \cos t + 5t)$$
$$= e^t(\cos t - \sin t + 5t + 5) \quad \blacksquare$$

(3) Find the derivative of $f(t) = \dfrac{2t}{4+t^2}$.

☐ Using the quotient rule,

$$f'(t) = \frac{(4+t^2)\frac{d}{dx}(2t) - (2t)\frac{d}{dx}(4+t^2)}{(4+t^2)^2} = \frac{(4+t^2)(2) - (2t)(2t)}{(4+t^2)^2}$$
$$= \frac{8+2t^2-4t^2}{(4+t^2)^2} = \frac{8-2t^2}{(4+t^2)^2} \quad \blacksquare$$

(4) Find the derivative of $y = \dfrac{\tan x - 1}{\sec x}$.

☐ Using the quotient rule,

$$\frac{dy}{dx} = \frac{\sec x\frac{d}{dx}(\tan x - 1) - 10(\tan x - 1)\frac{d}{dx}\sec x}{(\sec x)^2}$$
$$= \frac{\sec x \sec^2 x - (\tan x - 1)\sec x \tan x}{\sec^2 x}$$
$$= \frac{\sec^2 x - (\tan x - 1)\tan x}{\sec x} = \frac{\sec^2 x - \tan^2 x + \tan x}{\sec x} = \frac{1 + \tan x}{\sec x}$$

Alternately, we can multiply the original equation up and down by $\cos x$ to convert it to $y = \sin x - \cos x$, with $dy/dx = \cos x + \sin x$. Be sure you know why the two versions of dy/dx are equivalent. ■

(5) Find the equation of the line tangent to $y = \dfrac{e^x}{x}$ at $(1, e)$.

□ This requires the slope of that tangent line, i.e. the value of dy/dx at $x = 1$. So since

$$\frac{dy}{dx} = \frac{x\frac{d}{dx}e^x - e^x\frac{d}{dx}x}{x^2} = \frac{xe^x - e^x}{x^2} = e^x\left(\frac{x-1}{x^2}\right)$$

we see that at $x = 1$, $\dfrac{dy}{dx} = 0$. So the equation of the tangent line is $y - e = 0(x - 1)$, i.e. the horizontal line $y = e$. ■

(6) Find the equation of the line tangent to $y = \tanh(x)$ where $x = 1$.

□ The point of tangency is $(1, \tanh(1))$. Since we found in (4.33) that the derivative of $\tanh(x)$ is $\operatorname{sech}^2(x)$, then at $x = 1$, the slope of the line tangent to $\tanh(x)$ will be $\operatorname{sech}^2(1)$. Altogether, the equation of this tangent line is

$$y - \tanh(1) = \operatorname{sech}^2(1)(x - 1)$$

Depending on your needs, you could write the hyperbolic values using their exponential form to get

$$y - \frac{e^1 - e^{-1}}{e^1 + e^{-1}} = \frac{2}{e^1 + e^{-1}}(x - 1)$$

or

$$y - \frac{e^2 - 1}{e^2 + 1} = \frac{2e}{e^2 + 1}(x - 1) \quad ■$$

(7) At what locations are the lines tangent to $y = xe^{-x^2}$ horizontal?

□ The derivative is

$$\frac{dy}{dx} = x\frac{d}{dx}e^{-x^2} + e^{-x^2}\frac{d}{dx}(x) = x(-2xe^{-x^2}) + e^{-x^2}(1) = e^{-x^2}(1 - 2x^2)$$

Since e^{-x^2} is never zero, the derivative is only zero when $1 - 2x^2 = 0$, or when $x = \pm\sqrt{1/2}$. These are the two locations where the graph of the function has horizontal tangent lines. ■

(8) Recall that velocity is the derivative of position. If an object's position as a function of time is given by $s(t) = \sin(t^2)/(t^2+1)$, estimate the first positive non-zero t value at which the object has a velocity of zero. (Note: This will require a numerical estimation.)

□ We have

$$v(t) = s'(t) = \frac{(t^2+1)\frac{d}{dt}(\sin(t^2)) - \sin(t^2)\frac{d}{dt}(t^2+1)}{(t^2+1)^2}$$

$$= \frac{(t^2+1)(2t\cos(t^2)) - \sin(t^2)(2t)}{(t^2+1)^2} = \frac{2t((t^2+1)\cos(t^2) - \sin(t^2))}{(t^2+1)^2}$$

From the numerator, it's clear that $v(0) = 0$, but we want the first non-zero positive t value at which the velocity is zero. This happens the first time $(t^2+1)\cos(t^2) - \sin(t^2) = 0$. Using Maple to estimate the solution, we get $t \approx 1.06$. ■

(9) Find the derivative of $f(x) = \tan\left(\frac{x}{x+1}\right)$.

□ Working from the outside in, we have

$$f'(x) = \sec^2\left(\frac{x}{x+1}\right) \cdot \frac{d}{dx}\left(\frac{x}{x+1}\right)$$

To evaluate the remaining derivative, we now need the quotient rule:

$$\frac{d}{dx}\left(\frac{x}{x+1}\right) = \frac{(x+1)\frac{d}{dx}(x) - (x)\frac{d}{dx}(x+1)}{(x+1)^2}$$

$$= \frac{(x+1)(1) - (x)(1)}{(x+1)^2} = \frac{1}{(x+1)^2}$$

All together, then,

$$f'(x) = \sec^2\left(\frac{x}{x+1}\right) \cdot \frac{1}{(x+1)^2} \quad ■$$

(10) Find the derivatives of $f(x) = \operatorname{csch}(x)$ and $g(x) = \coth(x)$.

□ Both are resolved by the quotient rule. Since $f(x) = \frac{1}{\sinh(x)}$,

$$f'(x) = \frac{\sinh(x) \cdot 0 - 1 \cdot \cosh(x)}{\sinh^2(x)} = \frac{\cosh(x)}{\sinh(x)} \cdot \frac{1}{\sinh(x)} = \coth(x)\operatorname{csch}(x)$$

Similarly, since $g(x) = 1/\tanh(x)$,

$$g'(x) = \frac{\tanh(x) \cdot 0 - 1 \cdot \operatorname{sech}^2(x)}{\tanh^2(x)} = -\operatorname{sech}^2(x)\coth^2(x)$$

$$= -\frac{1}{\cosh^2(x)} \cdot \frac{\cosh^2(x)}{\sinh^2(x)} = -\frac{1}{\sinh^2(x)} = -\operatorname{csch}^2(x) \quad \blacksquare$$

We used the newly derived Eq. (4.33) in this second part of the solution.

This might be a good time to prepare for yourself a nice table of all the derivatives of the six main hyperbolic functions, and any associated antiderivative formulas — since we've now found them all! ■

A.4.4 *Chain Rule — Practice — Solved*

(1) Rewrite the expression $2x/(3-5x)$ with the substitution $u = 3 - 5x$.

☐ If $u = 3 - 5x$, then $x = (3-u)/5$ and $2x = 2(3-u)/5$, so

$$\frac{2x}{3-5x} = \frac{2(3-u)/5}{u} = \frac{6-2u}{5u} \quad \text{with} \quad u = 3 - 5x \quad \blacksquare$$

(2) Rewrite the expression $e^{2x}\tan(2e^x)$ with the substitution $u = 2e^x$.

☐ Note that $e^{2x} = (e^x)^2$. And, if $u = 2e^x$ then $e^x = u/2$. So the substitution proceeds as:

$$e^{2x}\tan(2e^x) = (e^x)^2\tan(2e^x) = \left(\frac{u}{2}\right)^2\tan(u) = \frac{u^2}{4}\tan(u) \quad \blacksquare$$

(3) Rewrite the expression $x/(1-\sqrt{x})$ with the substitution $u = 1 - \sqrt{x}$.

☐ If $u = 1 - \sqrt{x}$, then $\sqrt{x} = 1 - u$ and $x = (1-u)^2$. So

$$\frac{x}{1-\sqrt{x}} = \frac{(1-u)^2}{u} \quad \text{with} \quad u = 1 - \sqrt{x} \quad \blacksquare$$

(4) Rewrite the expression $x^{3/2}$ with the substitution $u = \sqrt{x}$.

☐ Since $x^{3/2} = (\sqrt{x})^3$, we have

$$x^{3/2} = u^3 \quad \text{with} \quad u = \sqrt{x} \quad \blacksquare$$

(5) Find the derivatives of $y = (1+x^4)^{2/3}$ and $z = \sqrt[4]{1+2x+x^3}$.

☐ For the first,

$$\frac{dy}{dx} = \frac{2}{3}(1+x^4)^{-1/3}\frac{d}{dx}(1+x^4) = \frac{2}{3}(1+x^4)^{-1/3}(4x^3) = \frac{8x^3}{3\sqrt[3]{1+x^4}}$$

For the second, we can rewrite the function as $z = (1+2x+x^3)^{1/4}$, which is a composition $f(g(x))$ where $f(x) = x^{1/4}$ and $g(x) = 1 + 2x + x^3$. So,

$$\frac{dz}{dx} = f'(g(x)) \cdot g'(x) = \frac{1}{4}(1+2x+x^3)^{-3/4} \cdot \frac{d}{dx}(1+2x+x^3)$$

$$= \frac{1}{4}(1+2x+x^3)^{-3/4}(2+3x^2) = \frac{2+3x^2}{4(1+2x+x^3)^{3/4}} \quad \blacksquare$$

---

(6) Suppose y depends on x via $y = \sqrt{\tanh(x) + 1}$, and x depends on t via $x = t^2 - 2t + 3$, so that ultimately y depends on t. Use the Chain Rule to develop the derivative dy/dt (it's OK to leave a mix of x and t in the result).

☐ The overall Chain Rule for this will be

$$\frac{dy}{dt} = \frac{dy}{dx} \cdot \frac{dx}{dt}$$

From the individual relations, we have

$$\frac{dy}{dx} = \frac{1}{2}(\tanh(x) + 1)^{-1/2} \cdot \frac{d}{dx}(\tanh(x) + 1) = \frac{\text{sech}^2(x)}{2\sqrt{\tanh(x) + 1}}$$

and $dx/dt = 2t - 2$. Altogether, then,

$$\frac{dy}{dt} = \frac{\text{sech}^2(x)}{2\sqrt{\tanh(x) + 1}}(2t - 2)$$

For purity of the expression, we *should* replace x with $t^2 - 2t + 3$, but that would be really messy. ■

---

(7) Find the derivatives of $y = \tan(\sin^{-1}(x))$ and $y = \tan(\sinh(x))$.

☐ The first is a composition $f(g(x))$ where $f(x) = \tan x$ and $g(x) = \sin^{-1}(x)$. So,

$$\begin{aligned}\frac{dy}{dx} = f'(g(x)) \cdot g'(x) &= \sec^2(\sin^{-1}(x))\frac{d}{dx}\sin^{-1}(x) \\ &= \sec^2(\sin^{-1}(x)) \cdot \frac{1}{\sqrt{1 - x^2}}\end{aligned}$$

The second is similar, except with $\sinh(x)$ on the inside; using (4.11). For $y = \tan(\sinh(x))$,

$$\frac{dy}{dx} = \sec^2(\sinh(x)) \cdot \frac{d}{dx}(\sinh(x)) = \sec^2(\sinh(x))\cosh(x) \quad ■$$

---

(8) Find the derivative of $y = e^{-5x}\cos^{-1}(3x)$.

☐ This starts off as a product rule problem. In preparation for the product rule, we will need these individual derivatives, which each require the chain rule:

$$\frac{d}{dx}e^{-5x} = e^{-5x}\frac{d}{dx}(-5x) = -5e^{-5x}$$

$$\frac{d}{dx}\cos^{-1}(3x) = -\frac{1}{\sqrt{1 - (3x)^2}}\frac{d}{dx}(3x) = -\frac{3}{\sqrt{1 - 9x^2}}$$

so

$$\begin{aligned}\frac{dy}{dx} &= e^{-5x}\frac{d}{dx}\cos^{-1}(3x) + \cos^{-1}(3x)\frac{d}{dx}e^{-5x} \\ &= e^{-5x}\left(-\frac{3}{\sqrt{1-9x^2}}\right) + \cos^{-1}(3x)\left(-5e^{-5x}\right) \\ &= e^{-5x}\left(-\frac{3}{\sqrt{1-9x^2}} - 5\cos^{-1}(3x)\right) \\ &= -e^{-5x}\left(\frac{3}{\sqrt{1-9x^2}} + 5\cos^{-1}(3x)\right) \blacksquare\end{aligned}$$

---

(9) Find the derivative of $y = \sin^2 x/\cos x$.

☐ The sneaky way to find dy/dx is to rewrite the function as $y = \sin x \tan x$ and then use the product rule. But in the spirit of the chain rule, we'll tackle the problem as-is, which means we start with a quotient rule, and then use the chain rule in the process:

$$\begin{aligned}\frac{dy}{dx} &= \frac{\cos x \frac{d}{dx}\sin^2 x - \sin^2 x \frac{d}{dx}\cos x}{(\cos x)^2} \\ &= \frac{\cos x \cdot 2\sin x \frac{d}{dx}\sin x - \sin^2 x(-\sin x)}{\cos^2 x} \\ &= \frac{2\cos x \sin x \cos x + \sin^3 x}{\cos^2 x} = \frac{2\cos^2 x \sin x + \sin^3 x}{\cos^2 x} \\ &= 2\sin x + \sin x \frac{\sin^2 x}{\cos^2 x} = 2\sin x + \sin x \tan^2 x \quad \blacksquare\end{aligned}$$

---

(10) Find the derivative of $y = \sin(\sin(\sin x))$.

☐ This will require the chain rule twice in a row! Wheeee!

$$\begin{aligned}\frac{dy}{dx} &= \cos(\sin(\sin x)) \cdot \frac{d}{dx}\sin(\sin x) = \cos(\sin(\sin x)) \cdot \cos(\sin x)\frac{d}{dx}\sin x \\ &= \cos(\sin(\sin x)) \cdot \cos(\sin x) \cdot \cos x\end{aligned}$$

(This is a perfect example of why it's called the *chain* rule!) ■

---

(11) Find the equation of the line tangent to $y = \sin x + \sin^2 x$ at (0,0).

☐ We need the slope of that tangent line, i.e. the value of dy/dx at $x = 0$. Since

$$\frac{d}{dx}\sin^2(x) = \frac{d}{dx}(\sin x)^2 = 2\sin x \cos x$$

then

$$\frac{dy}{dx} = \cos x + 2\sin x \cos x$$

So at $x = 0$, $dy/dx = 1$ and the slope of the tangent line is $y - 0 = 1(x - 0)$, i.e. $y = x$. ■

---

(12) Find the velocity function for the position function $s(t) = A\cos(\omega t + \delta)$.

□ $$v(t) = s'(t) = -A\sin(\omega t + \delta)\frac{d}{dt}(\omega t + \delta) = -A\sin(\omega t + \delta)(\omega)$$
$$= -A\omega \sin(\omega t + \delta)$$ ■

A.4.5 *Implicit Differentiation — Practice — Solved*

(1) Find dy/dx for $x^2 - y^2 = 1$.

□ We recognize y is a function of x, so when we apply the derivative operator to both sides and invoke the chain rule, that's where dy/dx will appear:

$$\frac{d}{dx}\left(x^2 - y^2\right) = \frac{d}{dx}(1)$$
$$\frac{d}{dx}x^2 - \frac{d}{dx}(y)^2 = \frac{d}{dx}(1)$$
$$2x - 2y\frac{d}{dx}(y) = 0$$
$$2x - 2y\frac{dy}{dx} = 0$$
$$\frac{dy}{dx} = \frac{x}{y} \quad \blacksquare$$

(2) Find dy/dx for $1 + x = \sin(xy^2)$.

$$\square\ \frac{d}{dx}(1 + x) = \frac{d}{dx}\sin(xy^2)$$
$$0 + 1 = \cos(xy^2)\frac{d}{dx}(xy^2)$$
$$1 = \cos(xy^2)\left(x\frac{d}{dx}(y^2) + y^2\frac{d}{dx}x\right)$$
$$1 = \cos(xy^2)\left(x(2y)\frac{dy}{dx} + y^2\right)$$
$$1 - y^2\cos(xy^2) = 2xy\cos(xy^2)\frac{dy}{dx}$$
$$\frac{1 - y^2\cos(xy^2)}{2xy\cos(xy^2)} = \frac{dy}{dx} \quad \blacksquare$$

(3) Find dy/dx for $\sqrt{x + y} = 1 + x^2y^2$.

□ In this problem, there is an instance of y on each side of the function, so an instance of dy/dx will be generated on each side; we will have to gather them and solve, which might not be much fun:

$$\begin{aligned}
\frac{d}{dx}\sqrt{x+y} &= \frac{d}{dx}(1+x^2y^2) \\
\frac{1}{2\sqrt{x+y}}\frac{d}{dx}(x+y) &= 0 + \frac{d}{dx}(x^2y^2) \\
\frac{1}{2\sqrt{x+y}}\left(1+\frac{dy}{dx}\right) &= x^2\frac{d}{dx}(y^2) + y^2\frac{d}{dx}x^2 \\
\frac{1}{2\sqrt{x+y}}\left(1+\frac{dy}{dx}\right) &= x^2(2y)\frac{dy}{dx} + 2xy^2 \\
\frac{1}{2\sqrt{x+y}}\frac{dy}{dx} - 2x^2y\frac{dy}{dx} &= 2xy^2 - \frac{1}{2\sqrt{x+y}} \\
\frac{dy}{dx} - (2\sqrt{x+y})2x^2y\frac{dy}{dx} &= (2\sqrt{x+y})2xy^2 - 1 \\
\frac{dy}{dx}\left(1 - 4x^2y\sqrt{x+y}\right) &= 4xy^2\sqrt{x+y} - 1 \\
\frac{dy}{dx} &= \frac{4xy^2\sqrt{x+y} - 1}{1 - 4x^2y\sqrt{x+y}} \quad \blacksquare
\end{aligned}$$

(4) Find the equation of the line tangent to $x^{2/3} + y^{2/3} = 4$ at $(-3\sqrt{3}, 1)$.

☐ The equation of the tangent line requires the slope of that line, i.e. the value of dy/dx at $(-3\sqrt{3}, 1)$. We find dy/dx by taking the derivative of both sides.

$$\begin{aligned}
\frac{d}{dx}(x^{2/3} + y^{2/3}) &= \frac{d}{dx}(4) \\
\frac{2}{3}x^{-1/3} + \frac{2}{3}y^{-1/3}\frac{dy}{dx} &= 0 \\
y^{-1/3}\frac{dy}{dx} &= -x^{-1/3} \\
\frac{dy}{dx} &= -\frac{y^{1/3}}{x^{1/3}} \\
\frac{dy}{dx} &= -\sqrt[3]{\frac{y}{x}}
\end{aligned}$$

With $x = -3\sqrt{3}$ and $y = 1$, we get $dy/dx = \sqrt{3}/3$. So the equation of the tangent line is

$$y - 1 = \frac{\sqrt{3}}{3}(x + 3\sqrt{3}) \quad \blacksquare$$

(5) Find the derivative dy/dx for the inverse tangent function $y = \tan^{-1}(x)$.

☐ Rewrite the function as $\tan(y) = x$ then use implicit differentiation:

$$\begin{aligned}\frac{d}{dx}\tan(y) &= \frac{d}{dx}(x)\\ \sec^2(y)\frac{dy}{dx} &= 1\\ \frac{dy}{dx} &= \frac{1}{\sec^2 y} = \cos^2 y\end{aligned}$$

We can encode $\tan y = x$ into a right triangle by putting y in as an acute angle, with x and 1 being the lengths of the opposite and adjacent sides, respectively. Then the Pythagorean Theorem lets us write the length of the missing hypotenuse as $\sqrt{1+x^2}$. Then we have the lengths of all three sides, and can immediately find that $\cos y = 1/\sqrt{1+x^2}$. So,

$$\frac{dy}{dx} = \cos^2 y = \left(\frac{1}{\sqrt{1+x^2}}\right)^2 = \frac{1}{1+x^2} \quad ■$$

(6) Find dy/dx for $y = \dfrac{2(x^4+3x^2+x)(x^2+1)^5}{(x^3+1)^4}$.

☐ We can find the natural log of both sides to break the right side apart, and then proceed with differentiation:

$$\begin{aligned}\ln y &= \ln, \frac{2(x^4+3x^2+x)(x^2+1)^5}{(x^3+1)^4}\\ &= \ln 2 + \ln(x^4+3x^2+x) + 5\ln(x^2+1) - 4\ln(x^3+1)\\ \frac{d}{dx}(\ln y) &= \frac{d}{dx}\left(\ln 2 + \ln(x^4+3x^2+x) + 5\ln(x^2+1) - 4\ln(x^3+1)\right)\\ \frac{1}{y}\frac{dy}{dx} &= \frac{1}{x^4+3x^2+x}(4x^3+2x+1) + \frac{5}{x^2+1}(2x) - \frac{4}{x^3+1}(3x^2)\\ \frac{dy}{dx} &= y\left(\frac{4x^3+2x+1}{x^4+3x^2+x} + \frac{10x}{x^2+1} - \frac{12x^2}{x^3+1}\right)\\ &= \frac{2(x^4+3x^2+x)(x^2+1)^5}{(x^3+1)^4}\cdot\\ &\quad\left(\frac{4x^3+2x+1}{x^4+3x^2+x} + \frac{10x}{x^2+1} - \frac{12x^2}{x^3+1}\right) \quad ■\end{aligned}$$

(7) Find the slope of the line tangent to $f(x)$ at $x = 0$ where

$$f(x) = \frac{(x + \cos x)\sqrt[3]{(2-x)^2}}{5 + \sin^2 x}$$

☐ Rename $f(x)$ to y, and note that when $x = 0$,

$$y = \frac{(1)\sqrt[3]{4}}{5} = \frac{\sqrt[3]{4}}{5}$$

To generate dy/dx, we write the roots as exponents, and take the natural log of both sides to break the right side apart,

$$\begin{aligned}
\ln y &= \ln \frac{(x + \cos x)(2-x)^{2/3}}{5 + \sin^2 x} \\
&= \ln(x + \cos x) + \frac{2}{3}\ln(2 - x) - \ln(5 + \sin^2 x) \\
\frac{d}{dx}(\ln y) &= \frac{d}{dx}\left(\ln(x + \cos x) + \frac{2}{3}\ln(2 - x) - \ln(5 + \sin^2 x)\right) \\
\frac{1}{y}\frac{dy}{dx} &= \frac{1}{x + \cos x}(1 - \sin x) + \frac{2}{3}\cdot\frac{-1}{2-x} - \frac{1}{5 + \sin^2 x}(2\sin x\cos x) \\
\frac{dy}{dx} &= y\left(\frac{1 - \sin x}{x + \cos x} - \frac{2}{3(2-x)} - \frac{2\sin x\cos x}{5 + \sin^2 x}\right)
\end{aligned}$$

Now when $x = 0$ we get

$$\frac{dy}{dx} = \frac{\sqrt[3]{(2)^2}}{5}\left(\frac{1}{1} - \frac{1}{3} - 0\right) = \frac{\sqrt[3]{4}}{5}\left(\frac{2}{3}\right) = \frac{2\sqrt[3]{4}}{15} \quad ■$$

(8) Find the derivative of $\sqrt{1-x^2}$ with respect to r if (a) x does not depend on r, and (b) if x does depend on r.

☐ If x does not depend on r, then

$$\frac{d}{dr}\sqrt{1-x^2} = 0$$

If x does depend on r, then

$$\begin{aligned}
\frac{d}{dr}\sqrt{1-x^2} &= \frac{d}{dr}(1-x^2)^{1/2} \\
&= \frac{1}{2}(1-x^2)^{-1/2}\cdot\frac{d}{dr}(1-x^2) \\
&= \frac{1}{2}(1-x^2)^{-1/2}\cdot(-2x)\frac{dx}{dr} \\
&= -\frac{x}{\sqrt{1-x^2}}\frac{dx}{dr} \quad ■
\end{aligned}$$

(9) Find the derivative of $\sin^{-1} y$ with respect to t if (a) y does not depend on t, and (b) if y does depend on t.

□ If y does not depend on t, then

$$\frac{d}{dt}\sin^{-1} 2y = 0$$

If y does depend on t, then

$$\begin{aligned}\frac{d}{dt}\sin^{-1} 2y &= \frac{1}{\sqrt{1-(2y)^2}}\frac{d}{dt}(2y)\\ &= \frac{1}{\sqrt{1-4y^2}}\left(2\frac{dy}{dt}\right) = \frac{2}{\sqrt{1-4y^2}}\frac{dy}{dt} \quad \blacksquare\end{aligned}$$

(10) Given a cylinder with radius r, height h, and surface area S, what is the relationship between the rates of change of r and S (with respect to time)? (Assume h is constant.)

□ We relate radius and height to surface area of a cylinder by $S = 2\pi rh$. Thus, to relate the rates of change of r and S with respect to time (keeping h constant),

$$\begin{aligned}\frac{d}{dt}(S) &= \frac{d}{dt}(2\pi rh)\\ \frac{dS}{dt} &= 2\pi h\frac{d}{dt}(r) = 2\pi h\frac{dr}{dt} \quad \blacksquare\end{aligned}$$

(11) Given a sphere with radius r and volume V, what is the relationship between the rates of change of r and V (with respect to time)?

□ We relate radius and volume of a sphere by $V = \frac{4}{3}\pi r^3$. Thus, to relate the rates of change of r and V with respect to time,

$$\begin{aligned}\frac{d}{dt}(V) &= \frac{d}{dt}\left(\frac{4}{3}\pi r^3\right)\\ \frac{dV}{dt} &= \frac{4}{3}\pi\left(\frac{d}{dt}r^3\right)\\ \frac{dV}{dt} &= \frac{4}{3}\pi\left(3r^2\frac{dr}{dt}\right) = 4\pi r^2\frac{dr}{dt} \quad \blacksquare\end{aligned}$$

(12) Find dy/dx for $y\sin(x^2) = x\sin(y^2)$.

□ We treat y as a function of x. Since there is an instance of y on each side, then when we're done applying the chain rule and product rule, there will be an instance of dy/dx on both sides — these must be collected together and solved for:

$$\begin{aligned}
\frac{d}{dx}(y\sin(x^2)) &= \frac{d}{dx}(x\sin(y^2)) \\
y\frac{d}{dx}\sin(x^2) + \sin(x^2)\frac{d}{dx}y &= x\frac{d}{dx}\sin(y^2) + \sin(y^2)\frac{d}{dx}(x) \\
y(2x\cos(x^2)) + \frac{dy}{dx}\cdot\sin(x^2) &= x\cdot 2y\cos(y^2)\frac{dy}{dx} + \sin(y^2) \\
\frac{dy}{dx}(\sin(x^2) - 2xy\cos(y^2)) &= \sin(y^2) - 2xy\cos(x^2) \\
\frac{dy}{dx} &= \frac{\sin(y^2) - 2xy\cos(x^2)}{\sin(x^2) - 2xy\cos(y^2)} \quad \blacksquare
\end{aligned}$$

(13) Find the equation of the line tangent to $2(x^2 + y^2)^2 = 25(x^2 - y^2)$ at $(3, 1)$.

□ The equation of the tangent line requires the slope of that line, i.e. the value of dy/dx at $(3, 1)$. We find dy/dx by taking the derivative of both sides. Here I will be sneaky and not completely solve for dy/dx before inserting the data $x = 3$, $y = 1$. See if you think that makes the final manipulations for a numerical value of dy/dx easier. Let's start generally,

$$\begin{aligned}
\frac{d}{dx}2(x^2 + y^2)^2 &= \frac{d}{dx}25(x^2 - y^2) \\
4(x^2 + y^2)\frac{d}{dx}(x^2 + y^2) &= 25\left(2x - 2y\frac{dy}{dx}\right) \\
4(x^2 + y^2)\left(2x + 2y\frac{dy}{dx}\right) &= 50\left(x - y\frac{dy}{dx}\right)
\end{aligned}$$

Now that we've evolved a relation for dy/dx, we can move in to the point $(x, y) = (3, 1)$:

$$
\begin{aligned}
4(3^2 + 1^2)\left(2(3) + 2(1)\frac{dy}{dx}\right) &= 50\left(3 - 1\frac{dy}{dx}\right) \\
240 + 80\frac{dy}{dx} &= 150 - 50\frac{dy}{dx} \\
130\frac{dy}{dx} &= -90 \\
\frac{dy}{dx} &= -\frac{9}{13}
\end{aligned}
$$

So at $(3, 1)$ we get $dy/dx = -3$ and the equation of the tangent line is

$$
y - 1 = -\frac{9}{13}(x - 3) \quad \blacksquare
$$

A.4.6 *Antiderivs Using Substitution — Practice — Solved*

---

(1) Find $\displaystyle\int (2-x)^6\,dx$.

☐ We the substitution $u = 2 - x$, we get

$$du = -dx \quad\longrightarrow\quad dx = -du$$

And so,

$$\int (2-x)^6 dx = \int u^6\,(-du) = -\int u^6\,du = -\frac{1}{7}(u)^7 + C$$
$$= -\frac{1}{7}(2-x)^7 + C \quad \blacksquare$$

---

(2) Find $\displaystyle\int \frac{x}{x^2+1}\,dx$.

☐ We use the substitution $u = x^2 + 1$, and so

$$du = 2xdx \quad\longrightarrow\quad x\,dx = \frac{1}{2}du$$

Then,

$$\int \frac{x}{x^2+1}\,dx = \frac{1}{2}\int \frac{du}{u} = \frac{1}{2}\ln u + C = \frac{1}{2}\ln|x^2+1| + C \quad \blacksquare$$

---

(3) Find $\displaystyle\int y^3\sqrt{2y^4-1}\,dy$.

☐ We use the substitution $u = 2y^4 - 1$ so that

$$du = 8y^3\,dy \quad\longrightarrow\quad y^3\,dy = \frac{1}{8}\,du$$

so that

$$\int y^3\sqrt{2y^4-1}\,dy = \int \sqrt{2y^4-1}\cdot y^3\,dy = \int \sqrt{u}\cdot\frac{1}{8}\,du$$
$$= \frac{1}{8}\int \sqrt{u}du = \frac{1}{8}\cdot\frac{2}{3}u^{3/2} + C$$
$$= \frac{1}{12}(2y^4-1)^{3/2} + C \quad \blacksquare$$

(4) Find $\int \sqrt{x}e^{x^{3/2}}\,dx$.

☐ We use the substitution $u = x^{3/2}$ to generate

$$du = \frac{3}{2}\sqrt{x}\,dx \quad \longrightarrow \quad \sqrt{x}dx = \frac{2}{3}\,du$$

so that

$$\int \sqrt{x}e^{x^{3/2}}\,dx = \int e^{x^{3/2}} \cdot \sqrt{x}\,dx = \int e^u \cdot \frac{2}{3}\,du$$
$$= \frac{2}{3}\int e^u\,du = \frac{2}{3}e^u + C = \frac{2}{3}e^{x^{3/2}} + C \quad ■$$

(5) Find $\int (1+\tanh(\theta))^5 \operatorname{sech}^2(\theta)\,d\theta$.

☐ We use the substitution $u = 1 + \tanh(\theta)$, which leads to $du = \operatorname{sech}^2(\theta)\,d\theta$, and that's all we need:

$$\int (1+\tanh(\theta))^5 \operatorname{sech}^2(\theta)\,d\theta = \int u^5\,du$$
$$= \frac{1}{6}u^6 + C = \frac{1}{6}(1+\tanh(\theta))^6 + C$$

See, it looked horrible, but it wasn't so bad! ■

(6) Find $\int \frac{\sin x}{1+\cos x}\,dx$ and $\int \frac{\sinh x}{1+\cosh x}\,dx$.

☐ For the first, we use the substitution $u = 1 + \cos x$ to get

$$du = -\sin x dx \quad \longrightarrow \quad \sin x\,dx = -du$$

and so

$$\int \frac{\sin x}{1+\cos x}dx = -\int \frac{du}{u} = -\ln|u| + C = -\ln|1+\cos x| + C \quad ■$$

The second follows the same way, but with different details. We use $u = 1 + \cosh x$

$$du = \sinh x dx \quad \longrightarrow \quad \sinh x\,dx = du$$

and so

$$\int \frac{\sinh x}{1+\cosh x}dx = \int \frac{du}{u} = \ln|u| + C = \ln|1+\cosh x| + C \quad ■$$

(7) Find $\int \frac{x}{(x^2+1)^2}\,dx$.

☐ We use the substitution $u = x^2 + 1$ to get

$$du = 2x\,dx \quad \longrightarrow \quad x\,dx = \frac{1}{2}\,du$$

and so

$$\int \frac{x}{(x^2+1)^2}\,dx = \frac{1}{2}\int \frac{du}{u^2} = -\frac{1}{2}\cdot\frac{1}{u} + C = -\frac{1}{2(x^2+1)} + C \quad \blacksquare$$

(8) Find $\int \cos x \sin(\sin x)\,dx$.

☐ Well, this one has quite a snarl in it. Let's look at the term $\sin(\sin x)$ and target the inner $\sin x$ with the substitution $u = \sin x$, for which $du = \cos x\,dx$. If we rewrite the integral,

$$\int \cos x \sin(\sin x)\,dx = \int \sin(\sin x)\cdot\cos x\,dx$$

then everything falls into place:

$$\int \sin(\sin x)\cdot\cos x\,dx = \int \sin u du = -\cos u + C = -\cos(\sin x) + C \quad \blacksquare$$

A.5 Chapter 5: Practice Problem Solutions

A.5.1 *Related Rates — Practice — Solved*

---

(1) The volume of a cone increases at $5\,m^3/min$. Assuming the height of the cone stays a constant $10\,m$, how fast is the radius of the cone increasing when the volume of the cone is $500\,m^3$?

□ Let's name the volume, height and radius as the usual suspects V, h and r. We are given that $dV/dt = 5\,m^3/min$, and we want to compute dr/dt at the instant $V = 500\,m^3$.

Our static geometric relationship is the formula for the volume of a cone, $V = \pi r^2 h/3$; because the height is constant at $10\,m$ for the duration of the problem, we can plug that in right away and power up this static relationship.

$$V = \frac{10}{3}\pi r^2$$
$$\frac{d}{dt}(V) = \frac{10}{3}\pi\frac{d}{dt}r^2$$
$$\frac{dV}{dt} = \frac{20}{3}\pi r\,\frac{dr}{dt}$$

The last piece of missing information is r itself. We want r at the moment of our snapshot, which is when $V = 500$:

$$500 = \frac{10}{3}\pi r^2$$
$$r^2 = \frac{10\pi}{3\cdot 500} = \frac{\pi}{150}$$
$$r = \sqrt{\frac{\pi}{150}} = \frac{\sqrt{\pi}}{5\sqrt{6}}$$

And into the related rates equation it goes, with other known information, so we can solve for dr/dt. I'm going to hide the units so the expression is not cluttered up, but make sure you see how they track through the calculations.

$$\begin{aligned}\frac{dV}{dt} &= \frac{20}{3}\pi r \cdot \frac{dr}{dt} \\ 5 &= \frac{20}{3}\pi\left(\frac{\sqrt{\pi}}{5\sqrt{6}}\right) \cdot \frac{dr}{dt} \\ 5 &= \frac{4\pi^{3/2}}{3\sqrt{6}} \cdot \frac{dr}{dt} \\ \frac{dr}{dt} &= \frac{15\sqrt{6}}{4\pi^{3/2}} \approx 1.65\,\frac{m}{min}\end{aligned}$$

The radius is changing at about $1.65\, m/min$. ■

---

(2) A conspiracy theorist trying to get to the bottom of chemtrails is filming a rocket launch with a cell phone. He sits $4000\, ft$ from the launch pad. At launch, the rocket rises vertically at $600\, ft/sec$. How fast is the distance from this spectator to the rocket changing at the instant the rocket is $3000\, ft$ up?

□ The parameters we have are:

- Distance from the cellphone camera to the launch pad, constant 4000 ft.
- Height of the rocket, variable h.
- Distance from the camera to the rocket, variable D.

We are interested in finding dD/dt when $h = 3000$ ft. We need an equation relating D and h. But there is a right triangle between the camera, launch pad, and rocket, so the Pythagorean Theorem will do:

$$D^2 = h^2 + 4000^2$$

We power this up by taking the derivative with respect to t; both h and D vary with time, so we get:

$$\begin{aligned}\frac{d}{dt}(D^2) &= \frac{d}{dt}(h^2 + 4000^2) \\ 2D\frac{dD}{dt} &= 2h\frac{dh}{dt} + 0 \\ \frac{dD}{dt} &= \frac{h}{D}\frac{dh}{dt}\end{aligned}$$

Our snapshot happens when $h = 3000\, ft$, and we also need to know D at that instant; when $h = 3000$, we have $D^2 = (3000)^2 + (4000)^2$, or

$D = 5000$ ft. So,

$$\frac{dD}{dt} = \frac{h}{D}\frac{dh}{dt}$$
$$\frac{dD}{dt} = \frac{3000}{5000}(600) = 360$$

The distance from the camera to the rocket changes at 360 ft/sec. ■

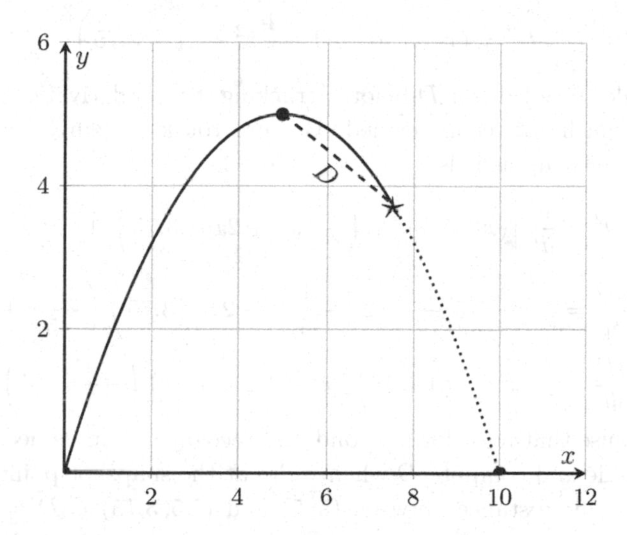

Fig. A.16 A birdie on a birdie.

---

(3) In 2018, golfer Kelly Kraft's tee shot on the 14th hole of a PGA tournament course hit a bird; the deflected shot caused him to miss out on advancing to the next round. Let's say that Kelly's tee is at the origin, and the trajectory of his ball follows the curve $y = -x^2/5 + 2x$. This clearly does not reflect real world distance units. In these new mystery units, the ground speed of the tee shot is $dx/dt = 2$, and the bird was hovering at the point $(7.5, 3.75)$. What was the rate of change of the straight line distance D from the ball to the bird when the ball was at the peak of its trajectory? (See Fig. A.16.)

□ The ball was at its peak at $x = 5$, the half-way point; the corresponding y coordinate comes from the equation of the parabola, and is $y(5) = 5$. So the vertex is $(x, y) = (5, 5)$. Our static equation needs to present the distance from the ball to the bird at $(7.5, 3.75)$, so that we can power it up and then take a snapshot at $x = 5$.

The distance D between any point (x, y) in the universe to $(7.5, 3.75)$ comes from

$$D^2 = (x - 7.5)^2 + (y - 3.75)^2$$

and so the distance D between any point on $\left(x, -\frac{1}{5}x^2 + 2x\right)$ to $(7.5, 3.75)$ is

$$D^2 = (x - 7.5)^2 + \left(-\frac{1}{5}x^2 + 2x - 3.75\right)^2$$

We could solve this for D before attacking it with derivatives, but that is very much not recommended. Be sure you know why! Rather, let's just power it up as it is:

$$\begin{aligned}
\frac{d}{dt} D^2 &= \frac{d}{dt}\left[(x - 7.5)^2 + \left(-\frac{1}{5}x^2 + 2x - 3.75\right)^2\right] \\
2D\frac{dD}{dt} &= 2(x - 7.5)\frac{dx}{dt} + 2\left(-\frac{1}{5}x^2 + 2x - 3.75\right)\left(-\frac{2}{5}x + 2\right)\frac{dx}{dt} \\
2D\frac{dD}{dt} &= \left[2(x - 7.5) + \left(-\frac{1}{5}x^2 + 2x - 3.75\right)\left(-\frac{4}{5}x + 4\right)\right]\frac{dx}{dt}
\end{aligned}$$

Let's pause that mess for a second and recognize that we need to go off to the side and compute D when we're at the snapshot point of $x = 5$. This specific distance between $(5, 5)$ and $(7.5, 3.75)$ is $D = 5\sqrt{5}/4$ (I have complete confidence you can compute this aside). Placing this value, $x = 5$, and $dx/dt = 2$ into the ongoing mess lets us continue with:

$$\begin{aligned}
\frac{10\sqrt{5}}{4}\frac{dD}{dt} &= \left[2(-2.5) + \left(-\frac{1}{5} + 2(5) - 3.75\right)\left(-\frac{4}{5}(5) + 4\right)\right](2) \\
\frac{5\sqrt{5}}{2}\frac{dD}{dt} &= \left[-5 + (-(5)(25) + 10 - 3.75)(-4 + 4)\right](2) \\
\frac{5\sqrt{5}}{2}\frac{dD}{dt} &= (-5)(2) \\
\frac{dD}{dt} &= -10 \cdot \frac{2}{5\sqrt{5}} = -\frac{4}{\sqrt{5}}
\end{aligned}$$

And so, in the strange not-real-world distance units invented for this problem, the rate of change of the straight line distance from the ball's peak to where the bird waited for its doom was $dD/dt = -4/\sqrt{5}$ units per time. ■

A.5.2 *Derivatives and Graphs — Practice — Solved*

(1) Find the critical points and inflection points of $f(x) = x^3 + x^2 + x$.

☐ We find critical points where $f'(x) = 0$, i.e. where $3x^2 + 2x + 1 = 0$. However this equation has no solutions (for the quadratic formula, the discriminant $b^2 - 4ac$ is negative). So, there are no critical points for this function. We find inflection points where $f''(x) = 0$, i.e. where $6x + 2 = 0$, so there is an inflection point at $x = -1/3$. ■

(2) Find the critical points and inflection points of $f(t) = \sqrt{t}(1 - t)$.

☐ We find critical points where $f'(t) = 0$. Rewriting $f(t) = t^{1/2} - t^{3/2}$, we get $f'(t) = 0$ where

$$\frac{1}{2}t^{-1/2} - \frac{3}{2}t^{1/2} = 0$$
$$\frac{1}{t^{1/2}} - 3t^{1/2} = 0$$
$$1 - 3t = 0\ldots$$

so $t = 1/3$ is a critical point. Note that $f'(t)$ is not defined for $t = 0$, so $t = 0$ is also a critical point. We find inflection points where $f''(t) = 0$:

$$-\frac{1}{4}t^{-3/2} - \frac{3}{4}t^{-1/2} = 0$$
$$-\frac{1}{t^{3/2}} - 3\frac{1}{t^{1/2}} = 0$$
$$-\frac{t^{3/2}}{t^{3/2}} - 3\frac{t^{3/2}}{t^{1/2}} = 0$$
$$-1 - 3t = 0\ldots$$

so $t = -1/3$ is an inflection point. And again, $t = 0$ is a point at which the original function is defined, but $f''(x)$ is undefined. So it is also an inflection point. ■

(3) Find the critical points and inflection points of $f(x) = \tanh(x)$ and confirm that these results are consistent with the graph you saw in Practice Problem 1 of Sec. 1.6.

☐ Since $f'(x) = \text{sech}^2(x)$, then there are critical points of $\tanh(x)$ wherever $\text{sech}(x) = 0$. But there are no such points. (Remember,

$\text{sech}(x) = \dfrac{2}{e^x + e^{-x}}$, which is never zero.) Then, we have

$$
\begin{aligned}
f''(x) &= 2\,\text{sech}(x) \cdot \frac{d}{dx}\,\text{sech}(x) \\
&= 2\,\text{sech}(x) \cdot \text{sech}(x)\tanh(x) = 2\frac{\sinh(x)}{\cosh^2(x)}
\end{aligned}
$$

So $f''(x) = 0$ where $\sinh(x) = 0$, and that happens only at $x = 0$. So overall, the graph of $f(x) = tanh(x)$ has no critical points (and in fact will be increasing everywhere since $f'(x) = \text{sech}(x) > 0$ for all x), and has an inflection point at $x = 0$. This matches Fig. A.8. ■

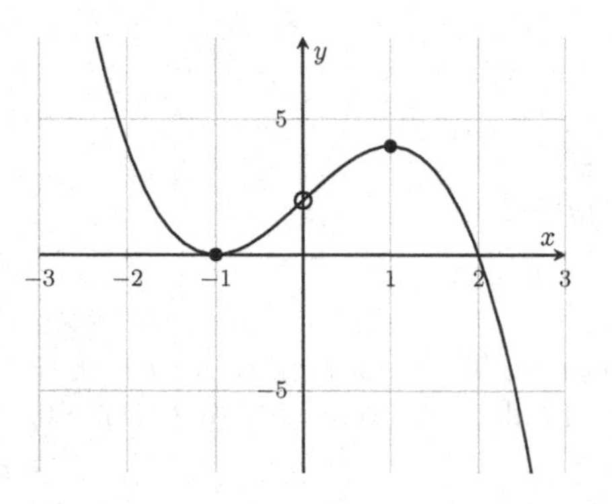

Fig. A.17 Graphical analysis of $f(x) = 2 + 3x - x^3$.

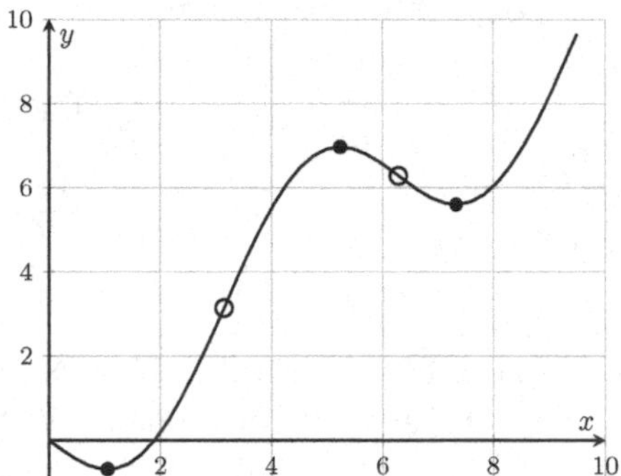

Fig. A.18 Graphical analysis of $f(x) = x - 2\sin(x)$.

(4) Find the critical points of $f(x) = xe^{2x}$.

□ We find critical points where $f'(x) = 0$ or is undefined. Since $f'(x) = (2x + 1)e^{2x}$, it is always defined and the only place it is zero is $x = -1/2$, so this is the only critical point. ■

(5) Provide a complete graphical analysis of $f(x) = 2 + 3x - x^3$ and sketch the graph.

□ There are no asymptotes, and the domain is all real numbers. We have

$$
\begin{aligned}
f'(x) &= 3 - 3x^2 \; ; \; f'(x) = 0 \text{ at } x = -1, 1 \\
f''(x) &= -6x \; ; \; f''(x) = 0 \text{ at } x = 0
\end{aligned}
$$

The function's values at these critical points and inflection points, and resulting categorizations, are:

x	-1	0	1
$f(x)$	0	2	4
type	CP	IP	CP
ext	min		max

Intervals set up by these critical points and inflection points, the signs of the derivatives in each, and the resulting shape are:

int	$(-\infty,-1)$	$(-1,0)$	$(0,1)$	$(1,\infty)$
$f'(x)$	$-$	$+$	$+$	$-$
$f''(x)$	$+$	$+$	$-$	$-$
shape	dec,ccu	inc,ccu	inc,ccd	dec,ccd

Figure A.17 shows a graph of this function with any critical points and inflection points marked. ■

---

(6) Provide a complete graphical analysis of $f(x) = x - 2\sin x$ on $(0, 3\pi)$ and sketch the graph of $f(x)$ on that interval.

□ There are no asymptotes, and the domain is all points within the given interval. We have

$$f'(x) = 1 - 2\cos x \; ; \; f'(x) = 0 \text{ at } x = \pi/3, 5\pi/3, , x = 7\pi/3$$

$$f''(x) = 2\sin x \; ; \; f''(x) = 0 \text{ at } x = \pi, 2\pi$$

The function's values at these critical points and inflection points, and resulting categorizations, are:

x	$\pi/3$	π	$5\pi/3$	2π	$7\pi/3$
$f(x)$	$\pi/3-\sqrt{3}$	π	$5\pi/3+\sqrt{3}$	2π	$7\pi/3-\sqrt{3}$
$\approx$	-0.685	3.14	6.97	6.28	5.60
type	CP	IP	CP	IP	CP
ext	min		max		min

Intervals set up by these critical points and inflection points, the signs of the derivatives in each, and the resulting shape are:

int	$(0,\pi/3)$	$(\pi/3,\pi)$	$(\pi,5\pi/3)$	$(5\pi/3,2\pi)$	$(2\pi,7\pi/3)$	$(7\pi/3,3\pi)$
$f'(x)$	$-$	$+$	$+$	$-$	$-$	$+$
$f''(x)$	$+$	$+$	$-$	$-$	$+$	$+$
shape	dec,ccu	inc,ccu	inc,ccd	dec,ccd	dec,ccu	inc,ccu

Figure A.18 shows a graph of this function with any critical points and inflection points marked. ■

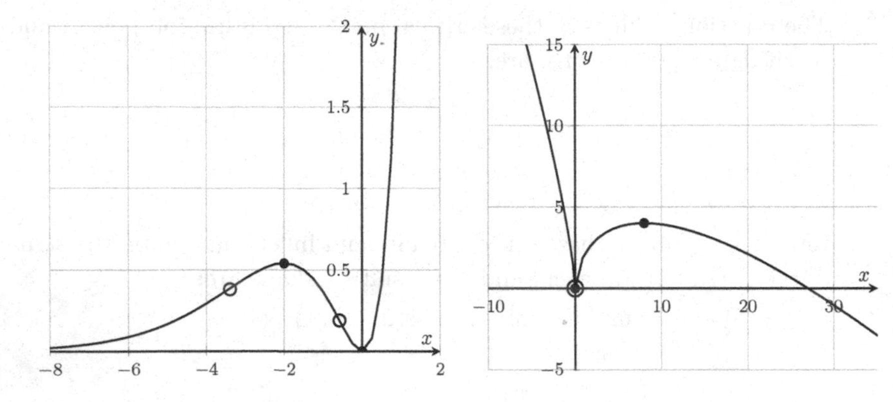

Fig. A.19 Graphical analysis of $f(x) = x^2e^x$.

Fig. A.20 Graphical analysis of $B(x) = 3x^{2/3} - x$.

---

(7) Provide a complete graphical analysis of $f(x) = x^2e^x$ and sketch its graph.

□ There are no asymptotes, and the domain is all real numbers. To find critical and inflection points, we have

$$f'(x) = xe^x(x+2)\ ;\ f'(x) = 0 \text{ at } x = -2, 0$$

$$f''(x) = e^x(x^2 + 4x + 2)\ ;\ f''(x) = 0 \text{ at } x = -2 \pm \sqrt{2}$$

The function's values at these critical points and inflection points, and resulting categorizations, are:

x	$-2-\sqrt{2}$	-2	$-2+\sqrt{2}$	0
$f(x) \approx$	0.3835	0.5413	0.1910	0
type	IP	CP	IP	CP
ext		max	min	

Intervals set up by these critical points and inflection points are, in number-line order,

$$I_1 :(-\infty, -2-\sqrt{2}) \quad I_2 : (-2-\sqrt{2}, -2) \quad I_3 : (-2, -2+\sqrt{2})$$

$$I_4 :(-2+\sqrt{2}, 0) \quad I_5 : (0, \infty)$$

The signs of the derivatives in each, and the resulting shape are:

int	I_1	I_2	I_3	I_4	I_5
$f'(x)$	+	+	−	−	+
$f''(x)$	+	−	−	+	+
shape	inc,ccu	inc,ccd	dec,ccd	dec,ccu	inc,ccu

Figure A.19 shows a graph of this function with any critical points and inflection points marked. ■

(8) Provide a complete graphical analysis of $B(x) = 3x^{2/3} - x$ and sketch its graph. (Hint: if you use tech to view the function, are you sure you're seeing the whole thing? Be sure you've looked at the graph for You Try It 2 in this section.)

☐ There are no asymptotes, and the domain is all real numbers. We have (simplified)

$$B'(x) = \frac{2}{x^{1/3}} - 1\ ;\ f'(x) = 8 \text{ at } x = 0, \text{is undefined at } x = 0$$

$$B''(x) = -\frac{2}{3x^{-4/3}}\ ;\ f''(x) \neq 0 \text{ but is undefined at} x = 0$$

The function's values at these critical points and inflection points, and resulting categorizations, are:

x	0	8
$f(x)$	0	4
type	CP/IP	CP
ext	min	max

Intervals set up by this critical / inflection point, the signs of the derivatives in each, and the resulting shape are:

int	$(-\infty, 0)$	$(0, 8)$	$(8, \infty)$
$f'(x)$	$-$	$+$	$-$
$f''(x)$	$-$	$-$	$-$
shape	dec,ccd	inc,ccd	dec,ccd

Figure A.20 shows a graph of this function with any critical points and inflection points marked. ■

(9) Provide a complete graphical analysis of $f(x) = x/(x^2 - 9)$ and sketch its graph.

☐ There are vertical asymptotes at $x = -3, 3$ and a horizontal asymptote at $y = 0$. The domain is all real numbers. To find critical and inflection points, we have (simplified)

$$f'(x) = -\frac{x^2 + 9}{(x^2 - 9)^2}\ ;\ f'(x) \neq 0 \text{ at } x = 0$$

$$f''(x) = \frac{2x(x^2 + 27)}{(x^2 - 9)^3}\ ;\ f''(x) = 0 \text{ at } x = 0$$

There are no critical points other than the x-coordinates of the asymptotes; there is an inflection point at $(0, 0)$.

Intervals set up by the asymptotes and inflection point, the signs of the derivatives in each, and the resulting shape are:

int	$(-\infty, -3)$	$(-3, 0)$	$(0, 3)$	$(3, \infty)$
$f'(x)$	$-$	$-$	$-$	$-$
$f''(x)$	$-$	$+$	$-$	$+$
shape	dec,ccd	dec,ccu	dec,ccd	dec,ccu

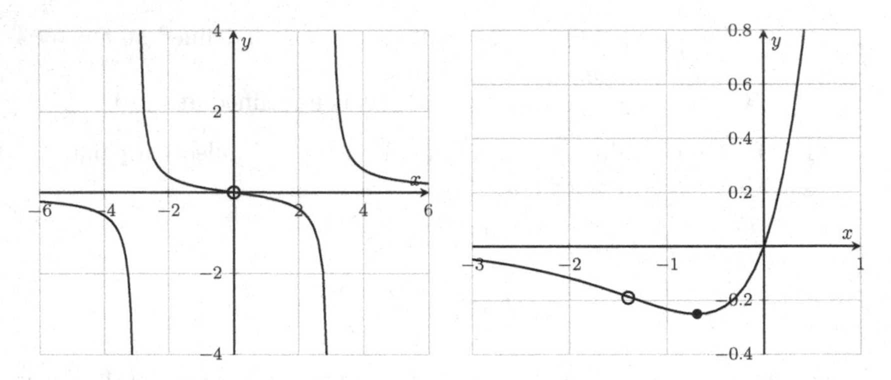

Fig. A.21 Graphical analysis of $f(x) = x/(x^2 - 9)$.

Fig. A.22 Graphical analysis of $f(x) = e^{2x} - e^x$.

Figure A.21 shows a graph of this function with any critical points and inflection points marked. ∎

---

(10) Provide a complete graphical analysis of $f(x) = e^{2x} - e^x$ and sketch its graph.

□ There are no asymptotes and the domain is all real numbers. To find critical and inflection points, we have (simplified)

$$f'(x) = 2e^{2x} - e^x \; ; \; f'(x) = 0 \text{ at } x = -\ln 2$$

$$f''(x) = 4e^{2x} - e^x \text{ at } x = -2\ln 2$$

The function's values at these critical points and inflection points are:

x	$-2\ln 2$	$-\ln 2$
$f(x)$	$-3/16$	$-1/4$
type	*IP*	*CP*
ext		min

Intervals set up by the asymptotes and inflection point, the signs of the derivatives in each, and the resulting shape are:

int	$(-\infty, -2\ln 2)$	$(-2\ln 2, -\ln 2)$	$(-\ln 2, \infty)$
$f'(x)$	$-$	$-$	$+$
$f''(x)$	$-$	$+$	$+$
shape	dec,ccd	dec,ccu	inc,ccu

Figure A.22 shows a graph of this function with any critical points and inflection points marked. ■

A.5.3 *Optimization — Practice — Solved*

(1) Find the global extremes of $f(x) = x^3 - 3x + 1$ on $[0, 3]$.

☐ Since $f'(x) = 3x^2 - 3$, we have $f'(x) = 0$ at $x = \pm 1$. But $x = -1$ is not in the given interval $[0, 3]$ so we toss it out. Candidates for the global extremes thus include the critical point $x = 1$ and the endpoints $x = 0$ and $x = 3$. Now $f(1) = -1$, $f(0) = 1$, and $f(3) = 19$. The points to choose from are thus $(1, -1)$, $(0, 1)$ and $(3, 19)$. On $[0, 3]$, then, the global minimum is at $(1, -1)$ and the global maximum is at $(3, 19)$. ■

(2) Find the absolute extremes of $f(x) = (x^2 - 4)/(x^2 + 4)$ on $[-4, 4]$.

☐ By the quotient rule (details omitted) we have

$$f'(x) = \frac{16x}{(x^2 + 4)^2}$$

So $f'(x) = 0$ at $x = 0$. Candidates for the global extremes thus include the critical point $x = 0$ and the endpoints $x = -4$ and $x = 4$. The full points at these locations are $(0, 0)$, $(-4, 3/5)$, and $(4, 3/5)$. On the interval $[-4, 4]$, then, the global minimum is $(0, -1)$ and the global maximum is at both endpoints, $(-4, 3/5)$ and $(4, 3/5)$. ■

(3) Find the global extremes of $f(x) = (x^2 - 1)^3$ on $[-1, 2]$.

☐ By the chain rule, we have $f'(x) = 6x(x^2 - 1)^2$. So $f'(x) = 0$ at $x = 0, \pm 1$. Note that $x = -1$ is also an endpoint of the given interval $[-1, 2]$. Candidates for the global extremes thus include the points $x = -1$, $x = 0$, $x = 1$, and $x = 2$. The full points at these locations are: $(-1, 0)$, $(0, -1)$, $(1, 0)$, and $(2, 27)$. On the interval $[-1, 2]$, then, the global minimum is $(0, -1)$ and the global maximum is $(2, 27)$. ■

(4) Find a positive number such that the sum of the number and its reciprocal is as small as possible.

☐ We want to minimize $f(x) = x + 1/x$.

$$f(x) = x + \frac{1}{x}$$

$$f'(x) = 1 - \frac{1}{x^2}$$

Then $f'(x) = 1$ when $x = 1$. (FFT: How do we know $x = 1$ does not provide a maximum?) ■

(5) A box with a square base and open top must have a volume of $32000\,cm^3$. Find the dimensions of the box that minimize the amount of material used.

□ Let x be the length of the sides of the square base, and h the height of the box. We want to minimize the amount (area) of material used, $A(x,h) = x^2 + 4xh$. We must reduce this function to one variable x or h. But we know that the total volume must be $32000\,cm^3$, so that $x^2h = 32000$ or $h = 32000/x^2$. Then starting with $A(x,h) = x^2 + 4xh$, we reduce to $A(x)$ by:

$$A(x) = x^2 + 4x\left(\frac{32000}{x^2}\right) = x^2 + \frac{128000}{x}$$

and find / solve $A'(x) = 0$:

$$\begin{aligned} A'(x) &= 2x - \frac{128000}{x^2} \\ 0 &= 2x - \frac{128000}{x^2} \\ x^3 &= 64000 \\ x &= 40 \end{aligned}$$

Now we know $A'(x) = 0$ when $x = 40\,cm$. Then also $h = 128000/(40)^2 = 20\,cm$. ■

(6) Find the point on the line $6x+y=9$ that is closest to the point $(-3,1)$.

□ We want to minimize distance to the point $(-3,1)$. Now for any point (x,y), the distance r to this point is:

$$r^2(x,y) = (x+3)^2 + (y-1)^2$$

Specifically, for points on the line $6x+y=9$, i.e. $y = 9-6x$, this becomes

$$r^2(x) = (x+3)^2 + ((9-6x)-1)^2 = (x+3)^2 + (8-6x)^2$$

To minimize, we find $\dfrac{dr}{dx}$ and then set $\dfrac{dr}{dx} = 0$:

$$\begin{aligned} r^2 &= (x+3)^2 + (8-6x)^2 \\ 2r\frac{dr}{dx} &= 2(x+3) + 2(8-6x)(-6) \\ 2r\frac{dr}{dx} &= (2x+6) + (72x-96) \\ 2r\frac{dr}{dx} &= 74x - 90 \\ 0 &= 74x - 90 \\ x &= \frac{90}{74} \end{aligned}$$

If $x = 90/74$ then $y = 63/37$ (calculations done behind the scenes). This point is the one on the line $6x + y = 9$ that is closest to $(-3, 1)$. ■

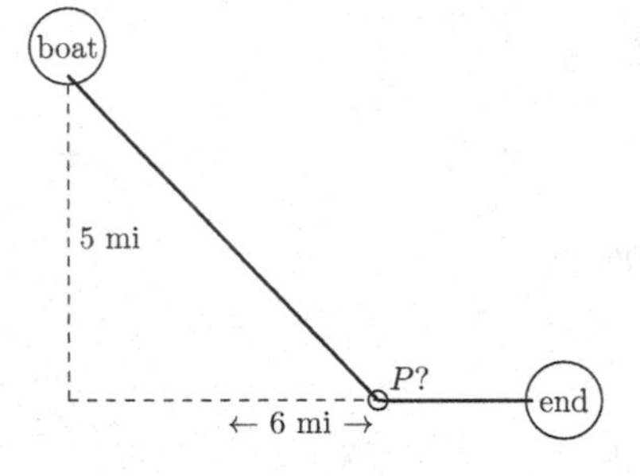

Fig. A.23 Rowing and walking in minimum time.

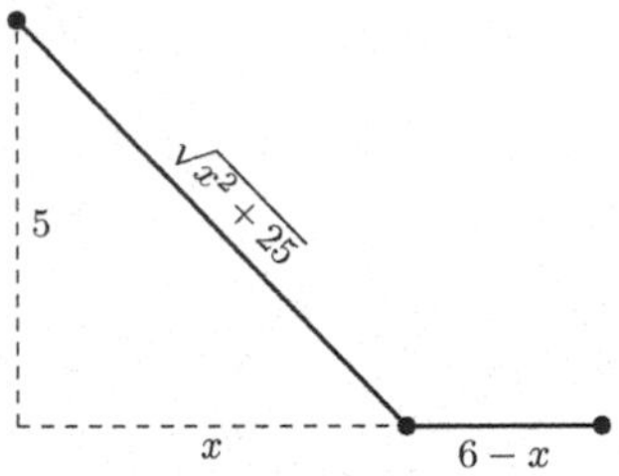

Fig. A.24 Labeling total route with variables.

---

(7) A man in a rowboat is 5 miles from the nearest point on shore. He needs to reach, in the shortest time possible, a second point on shore that is 6 miles east of the current closest landing spot. Where should he land if he can row $2\,mi/hr$ and walk $4\,mi/hr$?

□ Figure A.23 shows the general set-up. We'll name as x the horizontal distance along the shore from the boat's starting point to where it will land; that means the remaining horizontal distance is $6 - x$. The actual distances covered over land and through water are, using the Pythagorean theorem,

$$\sqrt{x^2 + 25} \qquad \text{in water and} \qquad 6 - x \qquad \text{on land}$$

(Figure A.24 shows this geometric labelling.) Now, we need to find an expression for total time taken. We have a distance and rate for each leg of the race, and so we can write a formula for the total time taken using "time is distance over rate":

$$T = T_{water} + T_{land} = \frac{\sqrt{25 + x^2}}{2} + \frac{6 - x}{4}$$

To minimize this time, we'll look for where $\frac{dT}{dx} = 0$ (using tech for the resulting estimate):

$$\begin{aligned} T &= \frac{\sqrt{25 + x^2}}{2} + \frac{6 - x}{4} \\ \frac{dT}{dx} &= \frac{x}{2\sqrt{25 + x^2}} - \frac{1}{4} \\ 0 &= \frac{x}{2\sqrt{25 + x^2}} - \frac{1}{4} \\ t &\approx 2.9 \end{aligned}$$

Therefore, the boater should land about 2.9 mi horizontally along the shore from his current position. ■

A.5.4 *Local Lin Approx & L-Hopital's Rule — Practice — Solved*

(1) Find the local linear approximation of $f(x) = 2/x$ at $x_0 = 1$.

☐ We need these ingredients:

- $f(x_0) = f(1) = 2$
- $f'(x_0) = f'(1) = -2$

so that

$$L(1) = f(1) + f'(1)(x-1) = 2 + (-2)(x-1) = -2x + 4$$

The local linear approximation of $f(x) = 2/x$ at $x = 1$ is $L(1) = -2x + 4$. Figure A.25(A) shows $f(x)$ and $L(1)$. ■

(2) Find the local linear approximation of $f(x) = e^x$ at $x_0 = 0$.

☐ We need these ingredients:

- $f(x_0) = f(0) = e^0 = 1$
- $f'(x_0) = f'(0) = e^0 = 1$

so that

$$L(0) = f(0) + f'(0)(x-0) = 1 + (1)(x-0) = x + 1$$

The local linear approximation of $f(x) = e^x$ at $x = 0$ is $L(0) = x + 1$. Figure A.25(A) shows $f(x)$ and $L(0)$. ■

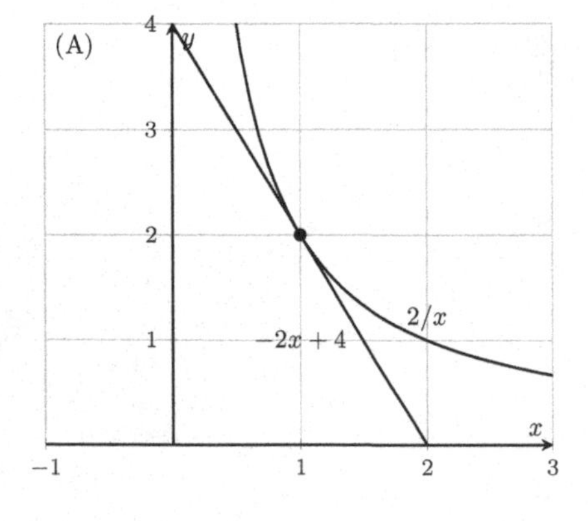

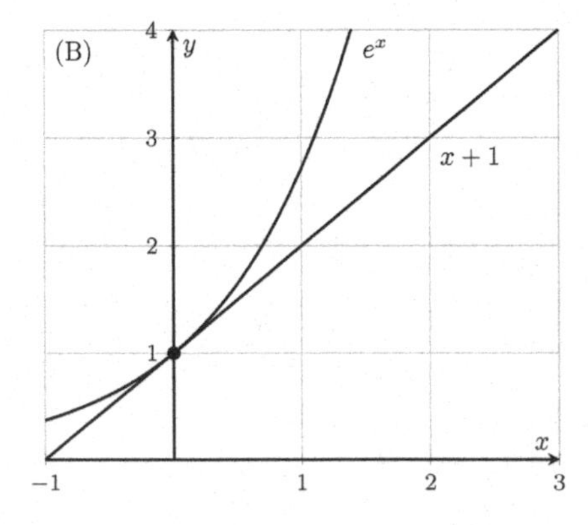

Fig. A.25 Two functions and their local linear approximations; (A) PP1, (B) PP2.

(3) Evaluate $\lim_{x\to 1}\dfrac{x^a-1}{x^b-1}$ $\quad (a, b \neq 0)$.

□ The given limit is of the form 0/0, which is an indeterminate form, so L-Hopital's Rule applies, and

$$\lim_{x\to 1}\frac{x^a-1}{x^b-1}=\lim_{x\to 1}\frac{ax^{a-1}}{bx^{b-1}}=\frac{a}{b}\quad \blacksquare$$

(4) Evaluate $\lim_{t\to 0}\dfrac{e^{3t}-1}{t}$.

□ The given limit is of the form 0/0, which is an indeterminate form, so L-Hopital's Rule applies, and

$$\lim_{t\to 0}\frac{e^{3t}-1}{t}=\lim_{t\to 0}\frac{3e^{3t}}{1}=3\quad \blacksquare$$

(5) Evaluate $\lim_{x\to 0}\dfrac{\sin x - x}{x^3}$.

□ The given limit is of the form 0/0, which is an indeterminate form, so L-Hopital's Rule applies, and

$$\lim_{x\to 0}\frac{\sin x - x}{x^3}=\lim_{x\to 0}\frac{\cos x-1}{3x^2}=\lim_{x\to 0}\frac{-\sin x}{6x}=\lim_{x\to 0}\frac{-\cos x}{6}=-\frac{1}{6}$$

(Note we had to apply L-Hopital's Rule THREE times since the first two application resulted in more 0/0 limits.) ■

(6) Evaluate $\lim_{x\to -\infty} x^2e^x$.

□ The given limit is of the form $\infty\cdot 0$, which is an indeterminate form. It is not yet in a form which L-Hopital's Rule can handle, so we must rearrange things:

$$\lim_{x\to-\infty} x^2e^x=\lim_{x\to-\infty}\frac{x^2}{e^{-x}}$$

which is now an indeterminate form ∞/∞ and we can apply L-Hopital's Rule:

$$\ldots=\lim_{x\to-\infty}\frac{2x}{(-e^{-x})}$$

We need to apply L-Hopital's Rule again,

$$\ldots=\lim_{x\to-\infty}\frac{2}{e^{-x}}=0$$

Together,

$$\lim_{x\to-\infty} x^2e^x=0\quad \blacksquare$$

A.5.5 *Good Value — Practice — Solved*

---

(1) Show that $f(x) = \sin(x)$ meets the conditions of Rolle's Theorem on $[\pi/4, 9\pi/4]$. How many appropriate values $x = c$ are guaranteed by Rolle's Theorem to exist on that interval? How many such values are there, really? Find them all (exactly).

□ With the interval identified as $[a, b] = [\pi/4, 9\pi/4]$, we can check the endpoints:

$$f(a) = \sin\left(\frac{\pi}{4}\right) = \frac{1}{\sqrt{2}} \quad ; \quad f(b) = \sin\left(\frac{9\pi}{4}\right) = \frac{1}{\sqrt{2}}$$

Since $f(x)$ is continuous on $[a, b]$ and differentiable on (a, b) and $f(a) = f(b)$, then the conditions of Rolle's Theorem are met. And so, Rolle's Theorem guarantees one value $x = c$ such that $f'(c) = 0$, although there may be more. In fact, just because we are really smart about trig functions, we know that there are two such places in the given interval: $f'(\pi/2) = 0$ and $f'(3\pi/2) = 0$. ■

---

(2) Does the function $f(x) = x^3 + x^2$ on $[-1, 1]$ satisfy the conditions of the Mean Value Theorem? If so, find a value c which is guaranteed by the MVT.

□ This function is a polynomial, and so is continuous and differentiable everywhere, and so of course is continuous on $[-1, 1]$ and differentiable on $(-1, 1)$. Thus, the MVT guarantees we can find at least one location $x = c$ in this interval where $f'(c) = 0$. Since $f'(x) = 3x^2 + 2x = x(3x + 2)$, then $x = 0$ is such a point. So is $x = -2/3$, and this too is in the interval. The MVT guarantees one such location, we actually found two! ■

---

(3) Does the function $f(x) = |\cos(x)|$ on $[0, 2\pi]$ satisfy the conditions of the Mean Value Theorem? If so, find a value c which is guaranteed by the MVT.

□ Since $f'(x)$ does not exist at $x = \pi/2$ and $x = 3\pi/2$, then $f(x)$ is not differentiable everywhere within $[0, 2\pi]$. So $f(x)$ does not satisfy the conditions of the MVT. Now sure, there are locations on $f(x)$ where $f'(x) = 0$ within this interval, but we can't confirm that in advance using the MVT. The MVT says that if some conditions are met, then a certain thing will happen. We can't flip that around to say that if those conditions are not met, then then the certain thing will *not* happen. ■

(4) Does the function $f(x) = \sec(x)$ on $[0, \pi]$ satisfy the conditions of the Mean Value Theorem? If so, find a value c which is guaranteed by the MVT.

□ The function $f(x) = \sec x$ has a discontinuity at $x = \pi/2$ and so it does not meet hte conditions of the Mean Value Theorem. ■

(5) Does the function $f(x) = \sinh(x)$ on $[0, \ln 2]$ satisfy the conditions of the Mean Value Theorem? If so, find a value c which is guaranteed by the MVT.

Since $f(x)$ is continuous and differentiable everywhere, it certainly satisfies the conditions of the Mean Value Theorem on $[0, \ln 2]$. Since we have $\sinh(0) = 0$ and $\sinh(\ln 2) = 3/4$, then

$$\frac{f(b) - f(a)}{b - a} = \frac{\frac{3}{4} - 0}{\ln 2 - 0} = \frac{3}{4\ln 2}$$

The Mean Value Theorem guarantees there is a point $x = c$ in the interval where $f'(c) = 3/(4\ln 2)$. This value is the solution to

$$\cosh(c) = \frac{3}{4\ln 2}$$

We must get an estimate of this solution value using tech, and that value is $c \approx 0.4$. ■

(6) On January 01, 2020, the price of one bitcoin was $7194 USD. On June 02, 2020, the price was $10,211. What is the average rate of change between those two days? Assuming the change in the conversion rate is a continuous function,[1] did there have to be an instant in between January 01 and June 02 when the price was increasing at a rate exactly equal to the average rate of change?

□ The number of days between Jan 01 and June 02 is 152 days (thanks, Google!). And so the average rate of change in the price of bitcoin was

$$\Delta p = \frac{10211 - 7194}{152} = \frac{3017}{152} \approx 19.849 \text{ USD per day}$$

If we're assuming that the conversion rate is a continuous function of time, then the MVT guarantees there must be at least one instant in between those two days when the instantaneous rate of change was equal to this average rate of change. ■

[1]Narrator: It's not...

(7) Use five iterations of Newton's Method with an initial value of $x_0 = 2$ to estimate a root of $f(x) = -x^4 + x^3 + 2$.

□ We can use a spreadsheet to iterate the Newton's Method formula, $x_{n+1} = x_n - f(x_n)/f'(x_n)$, with $f(x) = -x^4 + x^3 + 2$ and $f'(x) = -4x^3 + 3x^2$. Figure A.26 shows the result. The root is $x \approx 1.54369$. This problem is a good example of why you need to have intuition about the process in order to know when something is going wrong. Spreadsheets can have problems with order of operation, and so if you construct the recipe for $f(x)$ in a cell using, for example, " `= - D2^4 + D2^3 + 2` ", you may very well get the *wrong answer*! Some spreadsheets are very bad with formulas like this, and will construct `-D2` before applying the exponent of 4. To avoid this, you should not be shy about using parentheses to force the proper order of operations into the calculation - such as, `-(D2^4) + D2^3 + 2` for $f(x)$ and `-4*(D2^3) + 3*D2^2` for $f'(x)$. ■

fx =-(D4^4)+D4^3+2

	A	B	C	D
1	Step	f(x)	f'(x)	x_n
2	0			2
3	1	-6	-20	1.7
4	2	-1.4391	-10.982	1.56896
5	3	0.197427893	8.06388966	1.54448
6	4	.0059547943	7.58057875	1.54369
7	5	00005970032	7.56538199	1.54369

Fig. A.26 Spreadsheet cells for Newton's Method, with PP 7.

fx =D3^5-D3+1

	A	B	C	D
1	Step	f(x)	f'(x)	x_n
2	0			-1
3	1	1	4	-1.25
4	2	-0.8017578125	11.20703125	-1.17846
5	3	-0.09440	8.643362543	-1.16754
6	4	-0.00193	8.290801768	-1.16730
7	5	0.00000	8.283377746	-1.16730

Fig. A.27 Spreadsheet cells for Newton's Method, with PP 8.

(8) Use Newton's Method with an initial value of $x_0 = -1$ to estimate a root of $f(x) = x^5 - x + 1$. How many iterations does it take so that your results are no longer changing for the first 5 digits after the decimal?

□ We have $f(x) = x^5 - x + 1$ and $f'(x) = 5x^4 - 1$, with $x_0 = -1$. Figure A.27 shows Newton's Method implemented in a spreadsheet. The root of $x \approx -1.16730$ is identified after the 4th iteration, and confirmed to be fixed to at least five places after the decimal in the 5th iteration. ■

fx | =D5^(3)

	A	B	C	D
1	Step	f(x)	f'(x)	x_n
2	0			-1
3	1	-1	3	-0.6666666667
4	2	-0.296296296	1.333333333	-0.44444
5	3	-0.087791495	0.592592593	-0.29630
6	4	-0.026012295	0.263374486	-0.19753
7	5	-0.007707347	0.117055327	-0.13169
8	6	-0.002283658	0.052024590	-0.08779
9	7	-0.000676639	0.023122040	-0.05853
10	8	-0.000200486	0.010276462	-0.03902
11	9	-0.000059403	0.004567317	-0.02601
12	10	-0.000017601	0.002029918	-0.01734
13	11	-0.000005215	0.000902186	-0.01156
14	12	-0.000001545	0.000400972	-0.00771
15	13	-0.000000458	0.000178210	-0.00514
16	14	-0.000000136	0.000079204	-0.00343
17	15	-0.000000040	0.000035202	-0.00228
18	16	-0.000000012	0.000015645	-0.00152
19	17	-0.000000004	0.000006953	-0.00101
20	18	-0.000000001	0.000003090	-0.00068
21	19	0.000000000	0.000001374	-0.00045
22	20	0.000000000	0.000000610	-0.00030

Fig. A.28 Spreadsheet cells for Newton's Method, with PP 9.

---

(9) Try to use Newton's Method to find the root of $f(x) = x^3$. Obviously the actual root is $x = 0$; pick a number close (but not equal to!) the actual root (like $x_0 = 1$ or $x_0 = -1$, etc.). How many iterations does it take to identify the root correct to three places after the decimal, i.e. $0.000nnn$? Why do you think this problem might be zooming in on the root so much slower than in, say, YTI 6 or PP 7?

□ Let's start with an initial guess of $x_0 = -1$. Figure A.28 shows Newton's Method implemented in a spreadsheet. It takes almost 20 iterations to reach the correct root of $x = 0$ to thousandth-place accuracy $0.000nnn$. This is another example of why knowing the process is crucial. Think about the correction term that adjusts the value of x_n in each step of Newton's Method, $f(x_n)/f'(x_n)$. In most cases, as we approach the root, the values of $f(x_n)$ are getting smaller, whereas the values of $f'(x_n)$ are approaching whatever slope the tangent line at the actual root has — so overall, the correction term $f(x_n)/f'(x_n)$ is shrinking at each step.

For $f(x) = x^3$, though, remember that $x = 0$ is not only a root, it's also a critical point! This means that both $f(x_n)$ *and* $f'(x_n)$ are getting smaller and smaller as x_n gets closer to $x = 0$. As you should remember from considering limits, this means all bets are off in terms of the behavior of the quotient $f(x_n)/f'(x_n)$ as we approach $x = 0$. The adjustment between successive values of x_n is not as effective as it was in YTI 6 or PP 7. If investigating and analyzing this sort of behavior sounds interesting to you, then you are a good candidate for a Numerical Analysis course. ■

A.6 Chapter 6: Practice Problem Solutions

A.6.1 *Sigma Notation — Practice — Solved*

---

(1) Evaluate $\sum_{k=3}^{6}(k^2 - k)$.

□ The value of $a_k = k^2 - k$ for each $k = 3, 4, 5, 6$ is given in this table:

k	3	4	5	6
$\frac{k-1}{k+1}$	6	12	20	30

And so with those values,

$$\sum_{k=3}^{6}(k^2 - k) = 6 + 12 + 20 + 30 = 68 \quad \blacksquare$$

---

(2) Write the following two sum in sigma notation:

$$2 + 4 + 6 + 8 + \ldots + 200 \qquad \text{and} \qquad 1 + 3 + 5 + 7 + \ldots + 201$$

□ The first sum is the sum of all even numbers between 2 and 200; these correspond to $2i$ for $i = 1$ to $i = 100$. There are 100 terms in this sum:

$$2 + 4 + 6 + 8 + \ldots + 200 = \sum_{i=1}^{100}(2i)$$

The second sum is the sum of all odd numbers between 1 and 201; be careful, this has 101 terms in this sum — one more than in the previous one. This sum of odd numbers can be described in two good ways:

- the sum of $2i-1$ for $i = 1$ (which gives the fist term as $2(1)-1 = 1$) to $i = 101$ (which gives the last term as $2(101) - 1 = 201$)
- the sum of $2i+1$ for $i = 0$ (which gives the fist term as $2(0)+1 = 1$) to $i = 100$ (which gives the last term as $2(100) + 1 = 201$)

There is no big reason to pick one over the other of these options right now, so let's just go with the second one because ... well, my dog is sitting next to me and she prefers index values which are even numbers.

$$1 + 3 + 5 + 7 + \ldots + 201 = \sum_{i=0}^{100}(2i + 1) \quad \blacksquare$$

(3) Use trial and error to determine which of the following could be the correct closed form representation of $\sum_{i=1}^{n} i^3$:

$$\left(\frac{n(n+1)}{2}\right)^2 \quad , \quad \left(\frac{(n+1)(n+2)}{2}\right)^2$$

☐ Some sample summations done the long way are:

$$\sum_{i=1}^{5} i^3 = 1+8+27+64+125 = 225$$

$$\sum_{i=1}^{10} i^3 = 1+8+27+\cdots+729+1000 = 3025$$

$$\sum_{i=1}^{20} i^3 = 1+8+27+\cdots+6859+8000 = 44100$$

Let's try the first one with the candidates for the closed form of the sums. For $n = 5$, we get

$$\left(\frac{n(n+1)}{2}\right)^2 = (15)^2 = 225\,, \left(\frac{(n+1)(n+2)}{2}\right)^2 = (21)^2 = 441$$

So we already know the only successful candidate is the first. Let's confirm it with the other two cases. For $n = 10$,

$$\left(\frac{n(n+1)}{2}\right)^2 = \left(\frac{10(11)}{2}\right)^2 = 55^2 = 3025$$

For $n = 20$,

$$\left(\frac{n(n+1)}{2}\right)^2 = \left(\frac{20(21)}{2}\right)^2 = 210^2 = 44100$$

It sure looks like

$$\sum_{i=1}^{n} i^3 = \left(\frac{n(n+1)}{2}\right)^2$$

(I used Wolfram Alpha to do the calculations.) ■

(4) Give equivalent (reindexed) versions of the summation $\sum_{i=1}^{10} \sin\left(\frac{i\pi}{10}\right)^2$ which start at (a) $i = 0$ and (b) $i = -1$.

□ There must be ten terms in each sum, starting with the term $\sin\left(\frac{\pi}{10}\right)^2$ and ending with $\sin\left(\frac{10\pi}{10}\right)^2$.

If we start with $i = 0$, then we have to end with $i = 9$, and the terms are consistent with $\sin\left(\frac{(i+1)\pi}{10}\right)^2$.

If we start with $i = -1$, then we have to end with $i = 8$, and the terms are consistent with $\sin\left(\frac{((i+2)\pi}{10}\right)^2$.

Therefore, the following three sums are equivalent:

$$\sum_{i=1}^{10} \sin\left(\frac{i\pi}{10}\right)^2, \sum_{i=0}^{9} \sin\left(\frac{(i+1)\pi}{10}\right)^2, \sum_{i=-1}^{8} \sin\left(\frac{(i+2)\pi}{10}\right)^2 \quad \blacksquare$$

(5) Do some numerical experiments in a spreadsheet or Wolfram Alpha to determine if the following limit converges or diverges:

$$\lim_{n\to\infty} \sum_{k=1}^{n} \left(\frac{1}{2^k} - \frac{1}{2^{k+1}}\right)$$

□ This table shows several estimated partial sums generated using Wolfram Alpha:

n	5	10	25	50
s_n	0.484	0.4995	0.49999999	0.49999999999999996

I'm going to claim this limit / sum converges, since it looks like there's a cap of 0.5. ■

A.6.2 *Area Under a Curve — Practice — Solved*

(1) Create left-hand and right-hand estimates for the area under the function $f(x)$ tabulated below. Identify which is a lower estimate and which is an upper estimate.

x	0	2	4	6	8	10	12
$f(x)$	9	8.8	8.2	7.25	6	4.1	1

□ The interval we are given is $[a, b] = [0, 12]$. It is natural to make the endpoints of our rectangles at $x = 0, 2, 4, 6, 8, 10, 12$. There are seven endpoints, and therefore six rectangles. The left hand estimate uses left edges as heights:

$$L_6 = 2(9) + 2(8.8) + 2(8.2) + 2(7.25) + 2(6) + 2(4.1) = 2(43.35) = 86.7$$

This will be an overestimate, since the function is decreasing, and therefore the left hand rule will include extra area. The right hand estimate uses right edges as heights:

$$R_6 = 2(8.8) + 2(8.2) + 2(7.25) + 2(6) + 2(4.1) + 2(1) = 2(35.35) = 70.70$$

This will be an underestimate, since the function is decreasing, and therefore the right hand rule will miss some area. ■

(2) Create left-hand and right-hand estimates for the area under $f(x) = \sin(x)$ on the interval $[0, \pi]$ using $n = 4$ rectangles.

□ With 4 rectangles, the endpoints of our rectangles are $x = 0,\pi/4,\pi/2,3\pi/4,\pi$. Then the width of each is $\pi/4$. The function values at each of these endpoints is:

x	0	$\pi/4$	$\pi/2$	$3\pi/4$	π
$f(x)$	0	$1/\sqrt{2}$	1	$1/\sqrt{2}$	0

The left hand estimate uses the left edges of the rectangles as their heights:

$$L_4 = \frac{\pi}{4} \cdot 0 + \frac{\pi}{4} \cdot \frac{1}{\sqrt{2}} + \frac{\pi}{4} \cdot (1) + \frac{\pi}{4}\frac{1}{\sqrt{2}} = \frac{\pi}{4}(1 + \sqrt{2}) \approx 1.896$$

The right hand estimate uses the right edges of the rectangles as their heights:

$$R_4 = \frac{\pi}{4} \cdot \frac{1}{\sqrt{2}} + \frac{\pi}{4} \cdot (1) + \frac{\pi}{4}\frac{1}{\sqrt{2}} + \frac{\pi}{4} \cdot 0 = \frac{\pi}{4}(1+\sqrt{2}) \approx 1.896$$

L_4 and R_4 are the same due to the symmetry of the function on the given interval. ■

(3) Write the left and right hand estimates L_n and R_n in summation form for the area under $f(x) = \sin(x)$ on the interval $[0, \pi]$.

□ The more general expressions for L_n and R_n (for *any* n) are contructed using this information:

- The width of each rectangle will be $\Delta x = \dfrac{b-a}{n} = \dfrac{\pi}{n}$.
- The endpoints are given by $x_i = a + i\Delta x = 0 + i\left(\dfrac{\pi}{n}\right) = \dfrac{i\pi}{n}$
- The heights are given by $f(x_i) = \sin x_i = \sin\left(\dfrac{i\pi}{n}\right)$

Therefore,

$$L_n = \sum_{i=0}^{n-1} f(x_i)\Delta x = \sum_{i=0}^{n-1} \sin\left(\frac{i\pi}{n}\right) \cdot \frac{\pi}{n}$$

and

$$R_n = \sum_{i=1}^{n} f(x_i)\Delta x = \sum_{i}^{n} \sin\left(\frac{i\pi}{n}\right) \cdot \frac{\pi}{n} \quad ■$$

(4) Write the Riemann sum that describes the area under $f(x) = x\cos x$ on $[0, \pi/2]$.

□ We must construct

$$A = \lim_{n\to\infty} \sum_{i=1}^{n} f(x_i)\Delta x$$

For this problem, we have:

- $\Delta x = \dfrac{b-a}{n} = \dfrac{\pi/2 - 0}{n} = \dfrac{\pi}{2n}$
- The individual endpoints are $x_i = a + i \cdot \Delta x = 0 + i\left(\dfrac{\pi}{2n}\right) = \dfrac{i\pi}{2n}$
- The heights of the rectangles are $f(x_i) = \left(\dfrac{i\pi}{2n}\right) \cdot \cos\left(\dfrac{i\pi}{2n}\right)$

So we have

$$A = \lim_{n\to\infty} \sum_{i=1}^{n} f(x_i)\Delta x = \lim_{n\to\infty} \sum_{i=1}^{n} \left[\left(\frac{i\pi}{2n}\right)\cos\left(\frac{i\pi}{2n}\right)\right] \cdot \frac{\pi}{2n}$$

Fortunately, we do not have to evaluate the limit. ■

---

(5) Write the Riemann sum that describes the area under $f(x) = \ln x/x$ on $[3, 10]$.

☐ We must construct

$$A = \lim_{n\to\infty} \sum_{i=1}^{n} f(x_i)\Delta x$$

For this problem, we have:

- $\Delta x = \dfrac{b-a}{n} = \dfrac{10-3}{n} = \dfrac{7}{n}$
- The individual endpoints are $x_i = a + i \cdot \Delta x = 3 + i\left(\dfrac{7}{n}\right)$
- The heights of the rectangles are $f(x_i) = \dfrac{\ln(3 + 7i/n)}{3 + 7i/n}$

So we have

$$A = \lim_{n\to\infty} \sum_{i=1}^{n} f(x_i)\Delta x = \lim_{n\to\infty} \sum_{i=1}^{n} \left(\frac{\ln(3+7i/n)}{3+7i/n}\right) \cdot \frac{7}{n}$$

We are not required to evaluate the limit. Dance for joy! ■

---

(6) The following Riemann sum represents the area under some $f(x)$ on an interval $[a, b]$. What are $f(x)$, a, and b?

$$\lim_{n\to\infty} \sum_{i=1}^{n} \left(5 + \frac{2i}{n}\right)^{10} \cdot \left(\frac{2}{n}\right)$$

☐ By comparing this to

$$\lim_{n\to\infty} \sum_{i=1}^{n} f(x_i)\Delta x$$

we can identify the following:

- Matching the given $\Delta x = \dfrac{2}{n}$ to the generic $\Delta x = \dfrac{b-a}{n}$, we see $b - a = 2$.
- Our function is $f(x_i) = (x_i)^{10}$ where $x_i = 5 + \dfrac{2i}{n}$
- ... matching this to the general $x_i = a + i \cdot \Delta x$, we can spot $a = 5$

Now we know $a = 5$ and $b - a = 2$, so $b = 7$. In all, this Riemann sum computes the area under $f(x) = x^{10}$ on $[5, 7]$. ■

A.6.3 *Definite Integrals — Practice — Solved*

---

(1) Evaluate this definite integral using geometry: $\displaystyle\int_0^{10} |x-5|\,dx$

□ The graph of $|x-5|$ is shaped like an upright V, with the vertex at $(5,0)$. So the area in question contains two equal right triangles to the left and right of $x=5$ on $[0,10]$. The height of each triangle is 5 and the width of each is also 5; the area of each triangle is $A=\dfrac{25}{2}$. But there are two triangles, so the total area is 25:

$$\int_0^{10} |x-5|\,dx = 25 \quad \blacksquare$$

---

(2) Evaluate the following definite integrals using geometry and/or symmetry:

$$\int_0^3 2-x\,dx \quad ; \quad \int_0^{\pi} \cos x\,dx$$

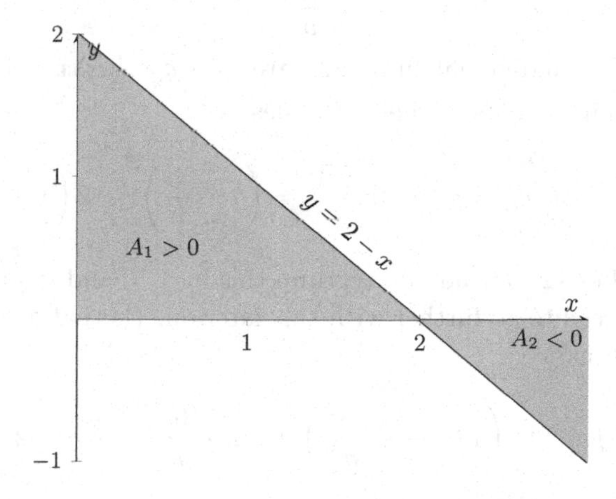

Fig. A.29 The area "under" $y=2-x$ on $[0,3]$.

□ (a) The graph of $2-x$ on the interval $[0,3]$ is shown in Fig. A.29. The total region includes one triangle on $[0,2]$ with a base of 2 and height of 2, and therefore total area $bh/2 = (2)(2)/2 = 2$. The second sub-region, on $[2,3]$ is a triangle of base 1 and height 1, with geometric area $\frac{1}{2}$; however, this triangle is below the axis, and so we assign its contribution to the integral as $-1/2$. Because the whole region is made of two sub-regions, the total area is the sum of the separate contributions,

$$\int_0^3 2 - x\,dx = \int_0^2 2 - x\,dx + \int_2^3 2 - x\,dx = 2 - \frac{1}{2} = \frac{3}{2}$$

(b) A graph of $\cos(x)$ on the interval $[0, \pi]$ shows two equal but opposite areas on $[0, \pi/2]$ and $[\pi/2, \pi]$ and so

$$\int_0^\pi \cos x\,dx = 0 \quad \blacksquare$$

---

(3) Convert this integral into a Riemann sum: $\displaystyle\int_1^4 (x^2 + 2x - 5)\,dx$

□ We must construct $A = \lim_{n\to\infty} \sum_{i=1}^{n} f(x_i)\Delta x$. For this problem, we have:

- $\Delta x = \dfrac{b-a}{n} = \dfrac{4-1}{n} = \dfrac{3}{n}$
- The individual endpoints are $x_i = a + i \cdot \Delta x = 1 + i \cdot \dfrac{3}{n}$
- The heights of the rectangles are

$$f(x_i) = x_i^2 + 2x_i - 5 = \left(1 + \frac{3i}{n}\right)^2 + 2\left(1 + \frac{3i}{n}\right) - 5$$

Let's go ahead and regroup this by 1, i, and i^2 just in case we need to go further with this later on. (Is that a subtle enough hint?)

$$f(x_i) = \left(1 + \frac{6i}{n} + \frac{9i^2}{n^2}\right) + \left(2 + \frac{6i}{n}\right) - 5 = -2 + \frac{12i}{n} + \frac{9i^2}{n^2}$$

Altogether,

$$\begin{aligned}\int_1^4 (x^2 + 2x - 5)\,dx &= \lim_{n\to\infty} \sum_{i=1}^{n} f(x_i)\Delta x \\ &= \lim_{n\to\infty} \sum_{i=1}^{n} \left(-2 + \frac{12i}{n} + \frac{9i^2}{n^2}\right) \cdot \frac{3}{n} \quad \blacksquare\end{aligned}$$

(4) Convert the following Riemann sum into a definite integral.

$$\lim_{n\to\infty}\sum_{i=1}^{n}\frac{e^{1+4i/n}}{1+4i/n}\cdot\frac{4}{n}$$

□ The term at the end of the sum tells us that $\Delta x = 4/n$, so $b-a=4$. The term $1+4i/n$ appears twice, and suggests we have $x_i = 1+4i/n$, which means that $a=1$, and so $b=5$. The fate of the x_i's is that they are plugged into $f(x)=e^x/x$. Therefore,

$$\lim_{n\to\infty}\sum_{i=1}^{n}\frac{e^{1+4i/n}}{1+4i/n}\cdot\frac{4}{n}=\int_1^5\frac{e^x}{x}\,dx$$

Note that we could also assign $x_i=\dfrac{4i}{n}$, so that $a=0, b=4$, and

$$\lim_{n\to\infty}\sum_{i=1}^{n}\frac{e^{1+4i/n}}{1+4i/n}\cdot\frac{4}{n}=\int_0^4\frac{e^{1+x}}{1+x}\,dx$$

Be sure you understand why both are equally valid and would represent the same net area. ■

(5) Compute this definite integral using a Riemann sum: $\displaystyle\int_1^4 (x^2+2x-5)\,dx$

□ The Riemann sum for this integral was created in Practice Problem 3. Picking up there that problem left off, we can expand the terms in the sum and regroup by looking for 1, i and i^2:

$$\begin{aligned}\int_1^4 (x^2+2x-5)dx &= \lim_{n\to\infty}\sum_{i=1}^{n}\left(-2+\frac{12i}{n}+\frac{9i^2}{n^2}\right)\cdot\left(\frac{3}{n}\right)\\ &= \lim_{n\to\infty}\sum_{i=1}^{n}\left(-\frac{6}{n}+\frac{36i}{n^2}+\frac{27i^2}{n^3}\right)\\ &= \lim_{n\to\infty}\left(-\frac{6}{n}\sum_{i=1}^{n}(1)+\frac{36}{n^2}\sum_{i=1}^{n}(i)+\frac{27}{n^3}\sum_{i=1}^{n}(i)^2\right)\ldots\end{aligned}$$

Then we can use the closed form expressions from Eq. (6.3) to give

$$\begin{aligned}\ldots &= \lim_{n\to\infty}\left(-\frac{6}{n}\cdot(n)+\frac{36}{n^2}\frac{n(n+1)}{2}+\frac{27}{n^3}\frac{n(n+1)(2n+1)}{6}\right)\\ &= \lim_{n\to\infty}\left(-6+18\frac{n(n+1)}{n^2}+\frac{27}{6}\frac{n(n+1)(2n+1)}{n^3}\right)\end{aligned}$$

Applying the limit,

$$\ldots = -6 + 18 + \frac{27}{6}(2) = -6 + 18 + 9 = 21$$

Altogether, $\int_1^4 (x^2 + 2x - 5)\,dx = 21$. ■

---

(6) Find the value of $\int_0^1 (1 - 2x^2)\,dx$. (Hint: there's an easy way, and a hard way.)

The area indicated here is exactly half (the right half) of the area found in this section's EX 7. There we found

$$\int_{-1}^{1} (1 - 2x^2)\,dx = \frac{2}{3}$$

so here we have

$$\int_0^1 (1 - 2x^2)\,dx = \frac{1}{3} \quad ■$$

A.6.4 *FTOC — Practice — Solved*

(1) Evaluate $\int_{-2}^{5} 6\,dx$.

$$\square \int_{-2}^{5} 6dx = 6x\Big|_{-2}^{5} = 6(5-(-2)) = 42 \quad \blacksquare$$

(2) Evaluate $\int_{-2}^{0} (u^5 - u^3 + u^2)\,du$.

$$\square \int_{-2}^{0} (u^5 - u^3 + u^2)\,du = (\frac{1}{6}u^6 - \frac{1}{4}u^4 + \frac{1}{3}u^3)\Big|_{-2}^{0}$$

$$= 0 - \left(\frac{1}{6}(64) - \frac{1}{4}(16) - \frac{1}{3}(8)\right) = -\frac{32}{3} + 4 + \frac{8}{3} = -\frac{24}{3} + 4 = -4 \quad \blacksquare$$

(3) Evaluate $\int_{1}^{9} \frac{1}{2x}\,dx$.

$$\square \int_{1}^{9} \frac{1}{2x}dx = \frac{1}{2}\ln x\Big|_{1}^{9} = \frac{1}{2}(\ln 9 - \ln 1) = \frac{1}{2}(2\ln 3 - 0) = \ln 3 \quad \blacksquare$$

(4) Evaluate $\int_{0}^{9} \sqrt{2t}\,dt$.

$$\square \int_{0}^{9} \sqrt{2t}\,dt = \sqrt{2}\int_{0}^{9} t^{1/2}\,dt = \sqrt{2}\left(\frac{2}{3}t^{3/2}\right)\Big|_{0}^{9}$$

$$= \sqrt{2}\left(\frac{2}{3}9^{3/2}\right) = \sqrt{2}\left(\frac{2}{3}(27)\right) = 18\sqrt{2} \quad \blacksquare$$

(5) Evaluate $\int_{0}^{\pi/4} \frac{1+\cos^2\theta}{\cos^2\theta}\,d\theta$.

$$\square \int_{0}^{\pi/4} \frac{1+\cos^2\theta}{\cos^2\theta}\,d\theta = \int_{0}^{\pi/4} (\sec^2\theta + 1)\,d\theta = (\tan\theta + \theta)\Big|_{0}^{\pi/4}$$

$$= \tan\frac{\pi}{4} + \frac{\pi}{4} - (0) = 1 + \frac{\pi}{4} \quad \blacksquare$$

---

(6) Why can't we use the Fundamental Theorem to evaluate $\int_{-2}^{3} x^{-5}\, dx$?

☐ Since x^{-5} is undefined at $x = 0$, which is within the interval of integration $[-2, 3]$, the given integral does not exist. ■

---

(7) Why can't we evaluate $\int_{0}^{\pi/6} \csc\theta \cot\theta \, d\theta$?

☐ Since $\csc\theta \cot\theta$ is undefined at $x = 0$, which is within the interval of integration $[0, \pi/6]$, the given integral does not exist. ■

---

(8) Find $g'(u)$ for $g(u) = \int_{3}^{u} \frac{1}{x + x^2}\, dx$.

☐ Adapting the Fundamental Theorem to these variables, we have

$$\frac{d}{du} \int_{a}^{u} f(x)\, dx = f(u)$$

Setting $f(x) = 1/(x + x^2)$, we get

$$g'(u) = \frac{d}{du} \int_{3}^{u} \frac{1}{x + x^2}\, dx = \frac{1}{u + u^2} \quad ■$$

---

(9) Find dy/dx for $y = \int_{1}^{\cos x} (t + \sin t)\, dt$.

☐ Adapting the Fundamental Theorem (with chain rule) to these variables, we have

$$\frac{d}{dx} \int_{a}^{g(x)} f(t)\, dt = f(g(x))g'(x)$$

Identifying $f(t) = t + \sin(t)$ and $g(x) = \cos(x)$, we have

$$\frac{dy}{dx} = (\cos x + \sin(\cos x))\frac{d}{dx}\cos x = -(\cos x + \sin(\cos x))\sin x \quad ■$$

A.6.5 *Def Ints With Substitution — Practice — Solved*

---

(1) Evaluate $\int_{-3}^{-2} (4-x)^2 \, dx$.

☐ With $u = 4 - x$ we have $-du = dx$ and

$$x = -3 \to u = 7$$
$$x = -2 \to u = 6$$

So that

$$\int_{-3}^{-2} (4-x)^2 \, dx = -\int_{7}^{6} u^2 \, du = -\frac{1}{3}u^3\bigg|_7^6 = -\frac{1}{3}(6^3 - 7^3) = \frac{127}{3} \quad \blacksquare$$

---

(2) Evaluate $\int_{\sqrt{2}}^{\sqrt{3}} \frac{x}{x^2+1} \, dx$.

☐ With $u = x^2 + 1$ we have $du/2 = x\,dx$, and

$$x = \sqrt{2} \to u = 3$$
$$x = \sqrt{3} \to u = 4$$

So,

$$\int_{\sqrt{2}}^{\sqrt{3}} \frac{x}{x^2+1} \, dx = \frac{1}{2}\int_3^4 \frac{du}{u} = \frac{1}{2}\ln|u|\bigg|_3^4 = \frac{1}{2}(\ln 4 - \ln 3) = \frac{1}{2}\ln\frac{4}{3} \quad \blacksquare$$

---

(3) Evaluate $\int_0^1 y^3 e^{y^4} \, dy$.

☐ With $u = y^4$ we have $du = 4y^3\,dy$ or $y^3\,dy = du/4$, and

$$y = 0 \to u = 0$$
$$y = 1 \to u = 1$$

So together,

$$\int_0^1 y^3 e^{y^4} \, dy = \frac{1}{4}\int_0^1 e^u \, du = \frac{1}{4}e^u\bigg|_0^1 = \frac{1}{4}(e^1 - e^0) = \frac{1}{4}(e-1) \quad \blacksquare$$

(4) Evaluate $\int_0^{\pi/4} (1 + \tan\theta)^2 \sec^2\theta \, d\theta$.

☐ With $u = 1 + \tan\theta$ we have $du = \sec^2\theta \, d\theta$, with

$$\theta = 0 \to u = 1$$
$$\theta = \frac{\pi}{4} \to u = 2$$

Altogether,

$$\int_0^{\pi/4} (1 + \tan\theta)^2 \sec^2\theta \, d\theta = \int_1^2 u^2 \, du = \frac{1}{3}u^3 \bigg|_1^2 = \frac{1}{3}(2^3 - 1^3) = \frac{7}{3} \quad \blacksquare$$

(5) Evaluate $\int_0^{\pi/4} \frac{\cos x}{1 + \sin^2 x} \, dx$.

☐ With $u = \sin x$ we get $du = \cos x \, dx$, and

$$x = 0 \to u = \sin 0 = 0$$
$$x = \frac{\pi}{4} \to u = \sin\frac{\pi}{4} = \frac{1}{\sqrt{2}}$$

Then

$$\int_0^{\pi/4} \frac{\cos x}{1 + \sin^2 x} \, dx = \int_0^{1/\sqrt{2}} \frac{du}{1 + u^2} = \tan^{-1}(u) \bigg|_0^{1/\sqrt{2}}$$
$$= \tan^{-1}\left(\frac{1}{\sqrt{2}}\right) - \tan^{-1}(0) = \tan^{-1}\frac{1}{\sqrt{2}} \quad \blacksquare$$

(6) Evaluate $\int_0^{\pi/2} \cos x \sin(\sin x) \, dx$.

☐ With $u = \sin x$ we have $du = \cos x \, dx$, and

$$x = 0 \to u = 0$$
$$x = \pi/2 \to u = 1$$

So,

$$\int_0^{\pi/2} \cos x \sin(\sin x) \, dx = \int_0^1 \sin u \, du = -\cos u \bigg|_0^1$$
$$= -(\cos(1) - \cos 0) = 1 - \cos(1) \quad \blacksquare$$

(7) Which integral has the largest value? Give that value:

$$\int_{-1}^{1} \sinh(x)\,dx \quad , \quad \int_{-1}^{1} \cosh(x)\,dx \ , \quad \text{or} \quad \int_{-1}^{1} \tanh(x)\,dx$$

☐ The sneaky way to answer this is to recall that $\sinh(x)$ and $\tanh(x)$ are odd functions, so that $\displaystyle\int_{-1}^{1} \sinh(x)\,dx$ and $\int_{-1}^{1} \tanh(x)\,dx$ are both zero. Since the whole graph of $\cosh(x)$ is above the x-axis, then $\displaystyle\int_{-1}^{1} \cosh(x)\,dx > 0$ and this will be the largest value:

$$\begin{aligned}\int_{-1}^{1} \cosh(x)\,dx = \sinh(x)\Big|_{-1}^{1} &= \frac{e^1 - e^{-1}}{2} - \frac{e^{-1} - e^{-(1)}}{2} \\ &= 2\left(\frac{e^1 - e^{-1}}{2}\right) = e - \frac{1}{e}\end{aligned}$$

Just to be sure, we can check the others ... it's good practice:

$$\int_{-1}^{1} \sinh(x)\,dx = \cosh(x)\Big|_{-1}^{1} = \frac{e^1 + e^{-1}}{2} - \frac{e^{-1} + e^{-(1)}}{2} = 0$$

$$\int_{-1}^{1} \tanh(x)\,dx = \operatorname{sech}^2(x)\Big|_{-1}^{1} = \frac{2}{e^1 + e^{-1}} - \frac{2}{e^{-1} + e^{-(1)}} = 0 \quad \blacksquare$$

Appendix B

Solutions to All Challenge Problems

B.1 Chapter 1: Challenge Problem Solutions

B.1.1 *Basics of Functions — Challenge — Solved*

(1) Give a function $f(x)$ that does not appear in a prior example or problem for which $f(a+b) = f(a) + f(b)$, and demonstrate that this expression is true for your function. Give another function $f(x)$ that does not appear in a prior example or problem for which $f(a+b) \neq f(a) + f(b)$, and demonstrate that this expression is true for your function.

□ Any linear function of the form $f(x) = mx$ will obey $f(a+b) = f(a) + f(b)$. For example, if $f(x) = 3x$, then

$$f(a+b) = 3(a+b) = 3a + 3b = f(a) + f(b)$$

For all other kinds of functions, we'll have $f(a+b) \neq f(a) + f(b)$. ■

(2) For $g(t) = t^2 + 1$, create and simplify the expressions $g(t+1)$, $g(2t)$, and $\dfrac{g(t) - g(1)}{t-1}$.

□ The first two are direct substitutions for t, the third is a combination of two instances of g:

$$\begin{aligned} g(t+1) &= (t+1)^2 + 1 = t^2 + 2t + 1 + 1 = t^2 + 2t + 2 \\ g(2t) &= (2t)^2 + 1 = 4t^2 + 1 \\ \frac{g(t) - g(1)}{t-1} &= \frac{(t^2+1) - (1^1+1)}{t-1} = \frac{t^2 - 1}{t-1} \\ &= \frac{(t-1)(t+1)}{t-1} = t + 1 \text{ for } t \neq 1 \end{aligned}$$ ■

(3) Find the domain and range of this function, and show whether it is even, odd, or neither: $f(t) = \dfrac{t^4 - 4}{t^2 + 2}$.

□ The domain of this function is all real numbers. Nothing can go wrong no matter what we plug in. We can simplify the function to

$$f(t) = \frac{t^4 - 4}{t^2 + 2} = \frac{(t^2 - 2)(t^2 + 2)}{t^2 + 2} = t^2 - 2$$

and from this easily see that we'll never get a result less than -2, and therefore the range of f is $[-2, \infty)$. Also from the simplified version of the function, we can see that:

$$f(-t) = (-t)^2 + 2 = t^2 + 2 = f(t)$$

and so this is an even function. ■

B.1.2 *Essential Types — Challenge — Solved*

---

(1) Find the equation of the line that connects the cusp of the graph of $f(x) = 4 - |x - 1|$ and the hole in the graph of

$$g(x) = \frac{x^2 + x - 2}{x^2 - x}$$

□ The cusp of $f(x)$ occurs at the point where the two pieces of the graph will join, which is the point at which the absolute value bars turn on/off, which is when $x - 1$ switches from positive to negative, which is at $x = 1$. When $x = 1$, we have $f(1) = 4$, and so the cusp is at $(1, 4)$.

The hole in the graph of $g(x)$ will occur where a term giving a root of the denominator cancels out:

$$g(x) = \frac{x^2 + x - 2}{x^2 - x} = \frac{(x + 2)(x - 1)}{x(x - 1)} = \frac{x + 2}{x}$$

There is a root in the denominator at $x = 1$, and the corresponding term $x - 1$ cancels out — so there is a hole in the graph of the (simplified) function at $x = 1$. The y-value of this point is the value that we would assign to $x = 1$ if we were really allowed to use it; from the simplified form of the function, this is at $(1 + 2)/1 = 3$. So the hole in the graph is at $(1, 3)$.

The line connecting $(1, 4)$ to $(1, 3)$ is the vertical line $x = 1$. ■

---

(2) Find the domain and graphical oddities (asymptotes, holes) of $g(x) = \dfrac{x^2 - 5x + 4}{\sqrt{x} + 1}$.

□ We can immediately see from the denominator that the domain of this function is $x \geq 0$. To find any possible graphical oddities, we can try to rewrite the function. Let's rationalize the denominator (multiply up and down by $\sqrt{x} - 1$) and factor the numerator:

$$\begin{aligned} g(x) &= \frac{x^2 - 5x + 4}{\sqrt{x} + 1} \cdot \frac{\sqrt{x} - 1}{\sqrt{x} - 1} \\ &= \frac{(x - 1)(x - 4)(\sqrt{x} - 1)}{x - 1} = (x - 4)(\sqrt{x} - 1) \end{aligned}$$

This graph has no holes or asymptotes; the denominator has no roots to begin with and there are no other "bad" points in the domain. ■

(3) Put the function $f(x) = |x^3 - 2x^2 - 3x|$ into piecewise form and make a reasonable sketch of its graph.

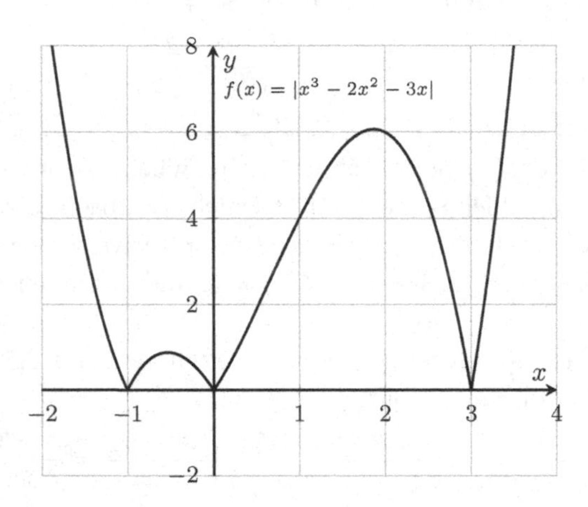

Fig. B.1 The graph of $f(x) = |x^3 - 2x^2 - 3x|$.

□ Let's factor the cubic polynomial inside the absolute value bars:

$$f(x) = |x^3 - 2x^2 - 3x| = |x(x^2 - 2x - 3)| = |x(x+1)(x-3)|$$

The absolute value bars "turn on" when the interior is negative; the polynomial changes from positive to negative at $x = -1, 0, 3$: we can easily check the sign of the polynomial in the resulting intervals:

$$\begin{aligned} x < -1 &\to x(x+1)(x-3) < 0 \\ -1 < x < 0 &\to x(x+1)(x-3) > 0 \\ 0 < x < 3 &\to x(x+1)(x-3) < 0 \\ x > 3 &\to x(x+1)(x-3) > 0 \end{aligned}$$

So there are four pieces in total on this graph; there are two pieces where the absolute value bars do their thing and change the sign of the polynomial. In piecewise form, we can write:

$$f(x) = \begin{cases} -(x^3 - 2x^2 - 3x) & \text{for} \quad x \le -1 \\ x^3 - 2x^2 - 3x & \text{for} \quad -1 < x \le 0 \\ -(x^3 - 2x^2 - 3x) & \text{for} \quad 0 < x \le 3 \\ x^3 - 2x^2 - 3x & \text{for} \quad x > 3 \end{cases}$$

The graph is shown in Fig. B.1. ■

B.1.3 *Exp and Log Functions — Challenge — Solved*

(1) Would you rather have a bank account that starts with one million dollars and receives a deposit of one million dollars per week for a year, or a bank account that starts with once cent and then has its balance doubled every week for that same year? Make sure to give quantitative reasons for your answer.

☐ A bank account that starts with one million dollars and receives a deposit of one million dollars per week will have a balance that goes by $B(t) = 10^6 + 10^6 t$, where t is in weeks. Therefore, it will have 53 million dollars in it at the end of the year.

A bank account that starts with once cent and then has its balance doubled every week will have a balance that goes by $B(t) = 0.01 \cdot 2^t$, where t is in weeks. Therefore, at $t = 52$, we will have $B(52) = 0.01 \cdot 2^{52}$. Now I don't know exactly what this value is, but I know it's a LOT bigger than 52 million. I'll take the second option! ■

(2) Solve the equation $\log_9(1-x) + \log_9(3-x) = 1/2$.

☐ We need to form a single logarithm before proceeding into algebra: Follow the steps in order *(i)* to *(vii)*:

i) $\log_9(1-x) + \log_9(3-x) = \dfrac{1}{2}$

ii) $\log_9((1-x)(3-x)) = \dfrac{1}{2}$

ii) $9^{\log_9((1-x)(3-x))} = 9^{1/2}$

iv) $(1-x)(3-x) = 3$

v) $x^2 - 4x + 3 = 3$

vi) $x^2 - 4x = 0$

vii) $x(x-4) = 0$

and so there are two possible solutions, $x = 0$ and $x = 4$. But since $x = 4$ is not in the domain of the left hand side (do you see why?), it cannot be a solution — so the only good solution is $x = 0$. ■

(3) When first measured, the global population of the purple nosed pointy headed pine warbler (a bird, obviously), was one million. Every year since that time, the population has decreased by a third. Write a function describing the number of birds as a function of time, $B(t)$, since the first measurement, and estimate how many years are needed for the population to drop to 2500.

□ Since there are two-thirds of the previous population remaining after every year, we have

$$B(t) = 10^6 \cdot \left(\frac{2}{3}\right)^t$$

where t is in years. Therefore to drop to 2500, we need to find t such that (in order *(i)* to *(vi)*):

i) $2500 = 10^6 \cdot \left(\frac{2}{3}\right)^t$

ii) $\frac{2500}{1000000} = \left(\frac{2}{3}\right)^t$

iii) $\ln \frac{1}{400} = \ln \left(\frac{2}{3}\right)^t$

iv) $-\ln(400) = t \ln \frac{2}{3}$

v) $t = -\frac{\ln 400}{\ln(2/3)}$

vi) $t = \frac{\ln 400}{\ln(3/2)}$

which is approximately 14.8 years. ■

B.1.4 *Compositions and Inverses — Challenge — Solved*

(1) If $f(x) = \sqrt{2x+3}$ and $g(x) = x^2 + 1$, then create and find the domains of $f(g(x))$, $g(f(x))$, $f(f(x))$, $g(g(x))$. Then, pick the function from $f(x)$ and $g(x)$ which is already one-to-one, and find its inverse. State the domain of that inverse.

□ The domain of $f(x)$ is $x \geq -3/2$; the domain of $g(x)$ is all reals. Then,

- $f(g(x)) = \sqrt{2(x^2+1)+3} = \sqrt{2x^2+5}$; its domain must ensure (1) that we use only members of the domain of g (which is all reals) and that, subsequently, the inside part of the composition stays positive; in each case, we're safe with all reals.
- $g(f(x)) = (\sqrt{2x+3})^2 + 1 = 2x+4$; its domain is $x \geq -3/2$ (points not allowed in f are still not allowed even though the composition appears simplified).
- $f(f(x)) = \sqrt{2\sqrt{2x+3}+3}$; its domain begins with the restrictions already in place on the inside f ($x \geq -3/2$) and we need to ensure that nothing "bad" gets passed along. But note that the output of $\sqrt{2x+3}$ is always positive, so certainly $2\sqrt{2x+3}+3 > 0$ for any value of x allowed. Thus, the domain of $f \circ f$ is also $x \geq -3/2$.
- $g(g(x)) = (x^2+1)^2 + 1$; its domain is all reals.

The function $f(x) = \sqrt{2x+3}$ is one-to-one. It's inverse is found by:

i) $y = \sqrt{2x+3}$

ii) $x = \sqrt{2y+3}$

iii) $x^2 = 2y+3$

iv) $y = \frac{1}{2}(x^2 - 3)$

v) $f^{-1}(x) = \frac{1}{2}(x^2 - 3)$

Since the range of the original $f(x)$ is $y \geq 0$, the domain of the inverse is $x \geq 0$. ■

(2) What three functions $f(x)$, $g(x)$, and $h(x)$ form $H(x) = \sqrt[3]{\sqrt{x}-1}$ when composed as $f(g(h)))$? Also, find $H^{-1}(x)$ and state its domain.

□ $H(x) = \sqrt[3]{\sqrt{x}-1}$ can be decomposed into $f(g(h)))$ with $f(x) = \sqrt[3]{x}$, $g(x) = x - 1$, and $h(x) = \sqrt{x}$. There are other possible answers. The inverse $H^{-1}(x)$ is found by:

i) $y = \sqrt[3]{\sqrt{x} - 1}$

ii) $x = \sqrt[3]{\sqrt{y} - 1}$

iii) $x^3 = \sqrt{y} - 1$

iv) $y = (x^3 + 1)^2$

v) $H^{-1}(x) = (x^3 + 1)^2$

Since the range of the original $H(x)$ is $y \geq -1$, the domain of $H^{-1}(x)$ is $x \geq -1$. ■

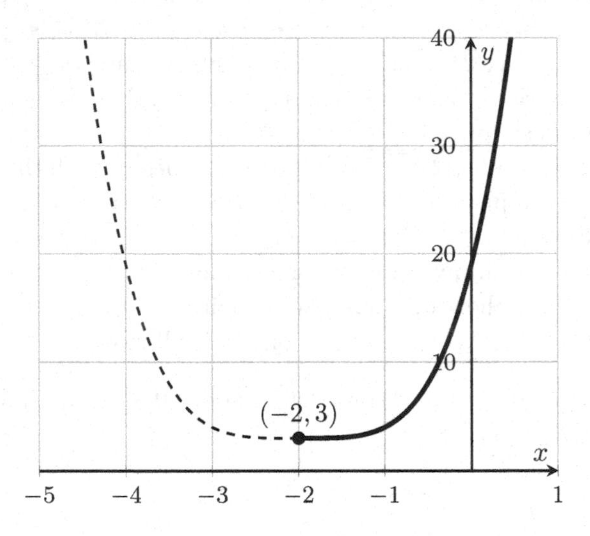

Fig. B.2 Restricting the domain of $y = (x + 2)^4 + 3$ to make it one-to-one.

---

(3) Describe how the graph of $y = x^4$ can be adapted to form the graph of $y = (x + 2)^4 + 3$, and sketch that new graph. Also, state a domain on which $y = (x + 2)^4 + 3$ has an inverse, and find the inverse.

□ The graph of $y = (x + 2)^4 + 3$ can be found by taking the graph of $y = x^4$, shifting to the left by two units and then raising that result by 3 units. In Fig. B.2, this is dashed and solid curves together.

If we restrict the domain of $y = (x + 2)^4 + 3$ to $x \geq -2$ (the solid curve only), then the resulting function is one-to-one, and its inverse is found by:

i) $y = (x+2)^4 + 3$

ii) $x = (y+2)^4 + 3$

iii) $x - 3 = (y+2)^4$

iv) $y + 2 = \sqrt[4]{x-3}$

v) $y = \sqrt[4]{x-3} - 2$ ■

B.1.5 *Trig Functions & Their Inverses — Challenge — Solved*

(1) Sketch the graph of $f(x) = \sec(x)$ on its restricted domain, and $\sec^{-1}(x)$, on the same set of axes.

☐ These curves are shown in Fig. B.3. ■

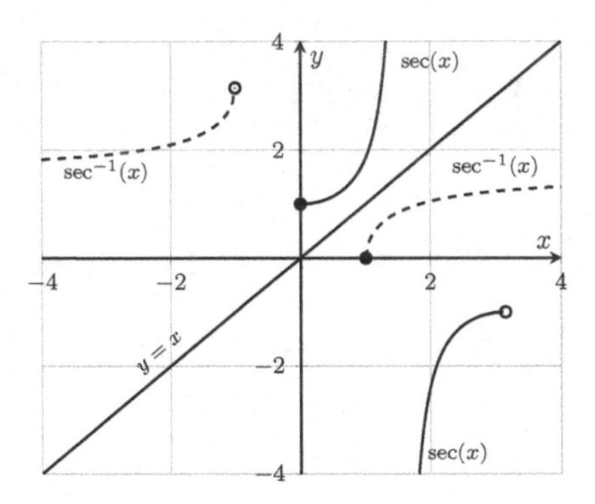

Fig. B.3 *sec*(x) and its inverse.

(2) Evaluate $\sec^{-1}(\sqrt{2})$ and $\arcsin(1)$.

☐ $\sec^{-1}(\sqrt{2})$ represents the angle on $[0, \pi/2) \bigcup (\pi/2, \pi]$ whose secant is $\sqrt{2}$. Note that angle is the same angle as $\cos^{-1}(1/\sqrt{2})$. That angle is $\pi/4$. The expression $\arcsin(1)$ represents the angle between $-\pi/2$ and $\pi/2$ whose sine is 1. That angle is $\pi/2$. Together,

$$\sec^{-1}(\sqrt{2}) = \frac{\pi}{4} \quad \text{and} \quad \arcsin(1) = \frac{\pi}{2} \quad ■$$

(3) Simplify $\cos(\cot^{-1}(1/x))$ into its algebraic form.

☐ The expression given in the problem statement represents the cosine of the angle whose cotangent is $1/x$. This is also the angle whose tangent is x, meaning that we have an opposite side of length x, an adjacent side of length 1, and so a hypotenuse of length $\sqrt{1+x^2}$. All of this information is exactly the same as that laid out in Fig. A.7. So the cosine of this angle is adj / hyp $= 1/\sqrt{1+x^2}$, and

$$\cos\left(\cot^{-1}\left(\frac{1}{x}\right)\right) = \frac{1}{\sqrt{1+x^2}} \quad ■$$

B.1.6 *Hyperbolic Funcs — Challenge — Solved*

---

(1) Categorize each of $\sinh(x), \cosh(x)$, and $\tanh(x)$ as even or odd, and explain how you know.

☐ Remember the definitions: a function $f(x)$ is even if $f(-x) = f(x)$, and is odd if $f(-x) = -f(x)$. Since EX 1 showed that $\sinh(-1) = -\sinh(1)$, and You Try It 3 showed that $\tanh(-1) = -\tanh(1)$, perhaps $\sinh(x)$ and $\tanh(x)$ are odd functions. Since You Try It 1 showed that $\cosh(-1) = \cosh(1)$, perhaps $\cosh(x)$ is even. We have to use the definitions in Def. 1.3 and 1.4 to be sure.

For $\sinh(x)$,

$$\sinh(-x) = \frac{e^{-x} - e^{-(-x)}}{2} = \frac{-(e^x - e^{-x})}{2} = -\frac{e^x - e^{-x}}{2} = -\sinh(x)$$

For $\cosh(x)$,

$$\cosh(-x) = \frac{e^{-x} + e^{-(-x)}}{2} = \frac{e^x - e^{-x}}{2} = \cosh(x)$$

For $\tanh(x)$,

$$\tanh(-x) = \frac{e^{-x} - e^{-(-x)}}{e^{-x} + e^{-(-x)}} = -\frac{e^x - e^{-x}}{e^x - e^{-x}} = -\tanh(x)$$

So, since $\sinh(-x) = -\sinh(x)$ and $\tanh(-x) = -\tanh(x)$ for any x, these are both odd functions. Since $\cosh(-x) = cosh(x)$ for any x, this is an even function. ■

---

(2) From EX 2, we know that there is no direct counterpart to $\sin^2 x + \cos^2 x = 1$, in that $\sinh^2(x) + \cosh^2(x) \neq 1$. But can $\sinh^2(x) + \cosh^2(x)$ be related to *anything* interesting? (Hint: In You Try It 2, we've already seen an identity involving $\sinh(2x)$, so maybe ...)

☐ First, let's see what $\sinh^2(x) + \cosh^2(x)$ looks like:

$$\begin{aligned}
\left(\frac{e^x - e^{-x}}{2}\right)^2 &+ \left(\frac{e^x + e^{-x}}{2}\right)^2 \\
&= \frac{1}{4}(e^{2x} - 2 + e^{-2x}) + \frac{1}{4}(e^{2x} + 2 + e^{-2x}) \\
&= \frac{1}{4}(2e^{2x} + 2e^{-2x}) = \frac{1}{2}(e^{2x} + e^{-2x})
\end{aligned}$$

or,

$$\sinh^2(x) + \cosh^2(x) = \frac{e^{2x} + e^{-2x}}{2}$$

And for the fun of it, let's look at $\cosh(2x)$:

$$\cosh(2x) = \frac{e^{2x} + e^{-2x}}{2}$$

They match! The identity is $\cosh(2x) = \sinh^2(x) + \cosh^2(x)$. ■

---

(3) If there are any solutions to $\sinh^2(x) - 8\cosh(x) + 16 = 0$, find them; write the values in their pure exact form.

□ Let's use the identity from EX 2, $\sinh^2(x) - \cosh^2(x) = -1$ and switch it up a bit to $\sinh^2(x) = \cosh^2(x) - 1$; we can use this to rewrite the equation

$$\begin{aligned}\sinh^2(x) - 8\cosh(x) + 16 &= 0\\(\cosh^2(x) - 1) - 8\cosh(x) + 16 &= 0\\\cosh^2(x) - 8\cosh(x) + 15 &= 0\end{aligned}$$

Now we can factor the equation as $(cosh(x) - 3)(\cosh(x) - 5) = 0$. We have solutions, then, when $\cosh(x) = 3$ or $\cosh(x) = 5$, meaning the solution values are, in pure form, $x = \cosh^{-1}(3)$ and $x = \cosh^{-1}(5)$. ■

B.2 Chapter 2: Challenge Problem Solutions

B.2.1 *Intro to Limits — Challenge — Solved*

(1) Estimate the limit as $x \to 0^+$ for $f(x) = x\ln(x + x^2)$ by examining several values of $f(x)$ in the vicinity of $x = 0$.

☐ Here are values in the vicinity of $x = 0$ (only from the right, since the function is not defined for $x < 0$):

x	1	0.5	0.1	0.05	0.01	0.005	0.001
$f(x)$	0.693	−0.144	−0.221	−0.147	−0.046	−0.026	−0.007

So it looks like $\lim_{x\to 0^+} x\ln(x + x^2) = 0$. ■

(2) Assess the limit $\lim_{x\to -2^+} \frac{x-1}{x^2(x+2)}$.

☐ As x approaches -2 from the right, $x-1$ is negative, $x+2$ is positive, and x^2 is positive. So

$$\lim_{x\to -2^+} \frac{x-1}{x^2(x+2)} = -\infty$$ ■

(3) Investigate the following limit by evaluating the value of the function for a sequence of several x values starting at $x = 1$ and decreasing to a very small positive number: $\lim_{x\to 0^+} \frac{\tan x - x}{x^3}$.

☐ Here is a table of values of $f(x) = (\tan x - x)/x^3$ at some values:

x	1	0.5	0.1	0.05	0.01	0.005
$f(x)$	0.55741	0.37042	0.33467	0.33367	0.33335	0.33334

so it sure looks like

$$\lim_{x\to 0^+} \frac{\tan x - x}{x^3} \approx 0.3333 = \frac{1}{3}$$

Note that in some calculators, if you reduce the test values too much smaller, you can get erroneous results. For example, you might see $f(10^{-8})$ being reported as 0. This is because we can hit the limit of some calculators to properly handle the quotient of such small values. So be careful! ■

B.2.2 *Computation of Limits — Challenge — Solved*

---

(1) If $f(x) = \sqrt{x}$, build and evaluate the limit $\lim_{h\to 0} \dfrac{f(1+h) - f(1)}{h}$.

□ We have $f(1+h) = \sqrt{1+h}$ and $f(1) = \sqrt{1} = 1$, so

$$\begin{aligned}\lim_{h\to 0} \frac{f(1+h) - f(1)}{h} &= \lim_{h\to 0} \frac{\sqrt{1+h} - 1}{h} \\ &= \lim_{h\to 0} \frac{\sqrt{1+h} - 1}{h} \cdot \frac{\sqrt{1+h} + 1}{\sqrt{1+h} + 1} \\ &= \lim_{h\to 0} \frac{(1+h) - 1}{(\sqrt{1+h} + 1)h} = \lim_{h\to 0} \frac{1}{\sqrt{1+h} + 1} = \frac{1}{2} \quad \blacksquare\end{aligned}$$

---

(2) Determine $\lim_{x\to 1.5} \dfrac{2x^2 - 3x}{|2x - 3|}$.

□ Note that with factoring,

$$\lim_{x\to 1.5} \frac{2x^2 - 3x}{|2x - 3|} = \lim_{x\to 1.5} \frac{x(2x - 3)}{|2x - 3|}$$

But the quotient $(2x-3)/|2x-3|$ is equal to 1 for $x > 3/2$ and is equal to -1 for $x < 3/2$, making the function

$$f(x) = \begin{cases} x & (x > 3/2) \\ -x & (x < 3/2) \end{cases} \tag{B.1}$$

So, the left and right hand limits are different,

$$\lim_{x\to 3/2^-} \frac{2x^2 - 3x}{|2x - 3|} = -3/2 \qquad \lim_{x\to 3/2^+} \frac{2x^2 - 3x}{|2x - 3|} = 3/2$$

and the overall limit does not exist. ■

---

(3) Invent an example of two functions f and g such that $\lim_{x\to a} [f(x) + g(x)]$ exists even though neither $\lim_{x\to a} f(x)$ nor $\lim_{x\to a} g(x)$ exist.

□ There are lots of possible examples, how about

$$f(x) = \frac{x}{|x|} \text{ and } g(x) = -\frac{x}{|x|} \text{ at } x = 0$$

Neither limit exists as $x \to 0$ but $f(x) + g(x) = 0$ so the limit of the sum exists and is zero. Other answers to this question don't even need to be represented algebraically, just draw graphs of two functions that behave this way! ■

B.2.3 *Limits Involving Infinity — Challenge — Solved*

(1) Determine all asymptotes of $f(x) = \dfrac{\sqrt{x^2+4}}{(x+2)^2}$.

☐ First, don't make the easy error of writing $\sqrt{x^2+4} = x+2$. That's false. We all wish it was true, but life isn't fair.

There is a vertical asymptote at $x = -2$ due to the denominator.

The x-axis is likely a horizontal asymptote, since the "degree" of the numerator is effectively 1 (because of the square root), while the degree of the denominator is 2. But, let's do the analysis:

$$\frac{\sqrt{x^2+4}}{(x+2)^2} = \frac{\sqrt{x^2(1+4/x^2)}}{x^2+4x+4} = \frac{|x|\sqrt{1+4/x^2}}{x^2(1+4/x+4/x^2)}$$
$$= \frac{1}{|x|}\frac{\sqrt{1+4/x^2}}{(1+4/x+4/x^2)}$$

We see that as $x \to \pm\infty$, the fractional portion collapses to 1 while the leading $1/|x|$ collapses to zero; therefore

$$\lim x \to \pm\infty f(x) = 0$$

and indeed, $y = 0$ is the horizontal asymptote. ■

(2) Determine all asymptotes of $g(x) = \dfrac{x^3-2x+3}{5-2x^2}$.

☐ There are vertical asymptotes at $x = \pm\sqrt{5/2}$ due to the denominator. (Neither of those roots are cancelled by the numerator. If you don't trust me, just use tech to plug those values into $x^3 - 2x + 3$, and you won't get zero.)

Since the degree of the numerator is larger than the degree of the denominator, the numerator "wins" and the size of the function increases as x increases — so there will not be a horizontal asymptote. How about a slant asymptote? Well,

$$\frac{x^3-2x+3}{5-2x^2} = \frac{x^3(1-2/x^2+3/x^3)}{x^2(5/x^2-2)} = x \cdot \frac{(1-2/x^2+3/x^3}{5/x^2-2}$$

So for large values of x, the fractional part collapses to $-1/2$, and so overall $f(x)$ tends towards $-x/2$ as $x \to \pm\infty$. Thus, we have a slant asymptote at $y = -x/2$. This is shown in Fig. B.4. ■

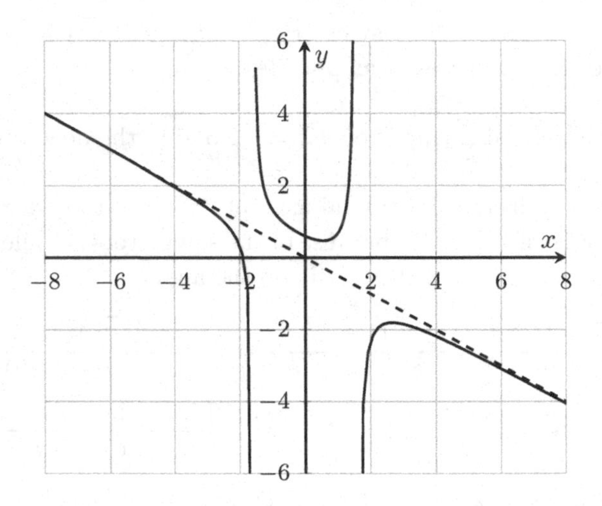

Fig. B.4 A plot of $f(x)$ from CP 2 and the asymptote $y = x/2$.

---

(3) Determine all asymptotes of $y = \dfrac{x^3+1}{x^3+x}$.

□ First, simplify:

$$y = \frac{x^3+1}{x^3+x} = \frac{(x+1)(x^2-x+1)}{x(x^2+1)}$$

So it looked like some cancellation might occur, but no such luck. This function has a vertical asymptote at $x = 0$ due to the denominator. There is a horizontal asymptote at $y = 1$ because

$$\lim_{x\to\infty} \frac{x^3+1}{x^3+x} = \lim_{x\to\infty} \frac{1+1/x^3}{1+1/x^2} = 1 \quad ■$$

B.2.4 *Limits and Continuity — Challenge — Solved*

(1) What value of c makes the following function continuous?

$$f(x) = \begin{cases} x^2 - c^2 \text{ for } x < 4 \\ cx + 20 \text{ for } x \geq 4 \end{cases}$$

□ Since both pieces of this function are continuous by themselves, the only place this function can be discontinuous is at $x = 4$, where the two pieces may or may not connect. To make this function continuous at $x = 4$, the pieces need to join there, i.e. we need:

$$4^2 - c^2 = c(4) + 20$$
$$0 = c^2 + 4c + 4$$
$$0 = (c+2)^2$$

So $c = -2$ will make the function continuous at $x = 4$. ■

(2) Is there a number that is equal to one more than its own cube? (Hint: Use the IVT.)

□ If there is, that number x satisfies the equation $x = x^3 + 1$. This equation has a solution if the function $f(x) = x^3 - x + 1$ has a root (i.e. if $x^3 - x + 1 = 0$). Consider that $f(-2) = -5$ and $f(0) = 1$. Since, then, $f(-2) < 0$ and $f(0) > 0$, and $f(x)$ is continuous everywhere, then by the Intermediate Value Theorem, $f(x)$ has a root somewhere between -2 and 0, and so there is a solution to $x = x^3 + 1$ somewhere between -2 and 0. ■

(3) One plate is in a freezer, the other in a hot oven. The locations of the two plates are then switched. Must there be a moment in time when the plates are at exactly the same temperature at that instant? Why or why not? (Hint: Suppose you have two functions $T_o(t)$ and $T_f(t)$ which describe the temperatures of the individual plates. The new function $T_o - T_f$ would be very interesting, indeed!)

□ Let $T_o(t)$ be the temperature of the plate initially in the oven, and $T_f(t)$ be the temperature of the plate initially in the freezer. Let's say that the oven is at a temperature of 375° and the freezer is at a temperature of 10° (both in Farenheit). These specific values don't really matter that much, any reasonable values (one high, one low) would do.

Since these temperature functions measure physical properties (temperature), they must each be continuous for all times. Temperatures don't just spontaneously leap from one to another. Therefore, if we build a function of their difference, $T_o - T_f$, that function is also continuous at all times.

Let t_1 be the start of the experiment, and t_2 be the end. At the start of the experiment, $T_o(t_1) = 375$ and $T_f(t_1) = 10$. Therefore, $(T_o - T_f)(t_1) = 365$. At the end of the experiment, $T_o(t_2) = 10$ and $T_f(t_2) = 375$. Therefore, $(T_o - T_f)(t_2) = -365$.

The difference function is positive at t_1 and negative at t_2, and it is a continuous function. Therefore, according to tht IVT, there must be some instant when the difference function hits a value of zero.

Since we must find $T_o - T_f = 0$ somewhere during the experiment, we must find an instant when $T_o = T_f$. So, YES, there be a moment in time when the plates are at exactly the same temperature at that instant. ∎

FFT: Note that this is similar of the "mountain climber problem", a brain teaser that poses the following thought experiment: Suppose there are two completely identical mountains, each with an identical trail that goes between the peak and the bottom. One climber starts down from the top of his mountain at 6am and reaches the bottom at 6pm. The other climber starts at the bottom of his mountain at 6am and reaches the top at 6pm. Is there an instant during the day when both climbers must be at exactly the same altitude at exactly the same time? The answer is yes. Why?

B.2.5 *Secant & Tangent Lines — Challenge — Solved*

(1) Use a progression of secant lines and their slopes to estimate the slope of the line tangent to $y = e^x$ at the point $(1, e)$. Then find the equation of that tangent line.

□ We'll form secant lines by connecting $(1, e)$ to several points to its left and right, and see the trend in slope as we get closer to $x = 1$. For any (x, y), the slope of the line connecting (x, y) to $(1, e)$ is

$$m = \frac{y - e}{x - 1}$$

Here is a table of such slope values:

x :	0.5	0.9	0.99	0.999	1	1.001	1.01	1.1	1.5
y :	1.6487	2.4596	2.6912	2.7155	e	2.7210	2.7456	3.0041	4.4816
m :	2.1391	2.5869	2.7047	2.7169	?	2.7196	2.7319	2.8588	3.5268

So it looks like the slope of the line tangent to the curve at $(1, e)$ should be about 2.718 or so. Or, the slope is probably e! The equation of the tangent line is then:

$$y - e = e(x - 1) \quad \blacksquare$$

(2) Use limits to find the slope of the tangent line to $f(x) = 2x/(x + 1)^2$ at (0,0), then find the equation of that line.

□ We're looking for the tangent line at $x = a = 0$, so the slope of the tangent line there is:

$$\begin{aligned} m_{tan} &= \lim_{h \to 0} \frac{f(0 + h) - f(0)}{h} \\ &= \lim_{h \to 0} \frac{[2(0 + h)/(0 + h + 1)^2] - [0/(0 + 1)^2]}{h} \\ &= \lim_{h \to 0} \frac{[2h/(h + 1)^2] - 0}{h} = \lim_{h \to 0} \frac{2}{(h + 1)^2} = 2 \end{aligned}$$

So the equation of the tangent line is $y - 0 = 2(x - 0)$, or $y = 2x$.

■

(3) Use limits to find the form of the slope of the tangent line to $f(x) = 2/(x + 3)$ at an unspecified point $x = a$. Then use that slope form to get the equations of the lines tangent to $f(x)$ at $x = -1$, $x = 0$ and $x = 1$. Plot $f(x)$ and these three tangent lines.

☐ The slope of the tangent line to $f(x) = 2/(x+3)$ at $x = a$ comes from:

$$m_{tan} = \lim_{h\to 0} \frac{f(a+h)-f(a)}{h} = \lim_{h\to 0} \frac{\frac{2}{(a+h)+3} - \frac{2}{a+3}}{h}$$

$$= \lim_{h\to 0} \frac{\frac{2(a+3)}{(a+h+3)(a+3)} - \frac{2(a+h+3)}{(a+h+3)(a+3)}}{h} = \lim_{h\to 0} \frac{\frac{2a+6-2a-2h-6}{(a+h+3)(a+3)}}{h}$$

$$= \lim_{h\to 0} \frac{\frac{-2h}{(a+h+3)(a+3)}}{h} = \lim_{h\to 0} \frac{-2}{(a+h+3)(a+3)} = \frac{-2}{(a+3)^2}$$

The slopes of the tangent lines at $x = -1$, $x = 0$ and $x = 1$ are then, respectively,

$$\frac{-2}{(-1+3)^2} = -\frac{1}{2} \quad ; \quad \frac{-2}{(0+3)^2} = -\frac{2}{9} \quad ; \quad \frac{-2}{(1+3)^2} = -\frac{1}{8}$$

The y-coordinates associated with these x-values are

$$\frac{2}{-1+3} = 1 \quad ; \quad \frac{2}{0+3} = \frac{2}{3} \quad ; \quad \frac{2}{1+3} = \frac{1}{2}$$

So the tangent lines themselves are

$$y - 1 = -\frac{1}{2}(x+1)$$

$$y - \frac{2}{3} = -\frac{2}{9}(x-0)$$

$$y - \frac{1}{2} = -\frac{1}{8}(x-1) \quad ■$$

B.3 Chapter 3: Challenge Problem Solutions

B.3.1 *Rate of Change — Challenge — Solved*

(1) The volume of a spherical balloon is a function of its radius, $V = 4\pi r^3/3$. Use either a succession of slopes of secant lines or a limit to estimate the instantaneous rate of change of the volume with respect to the radius at $r = 1$.

□ It's easier to use a limit:

$$\begin{aligned}\left(\frac{\Delta V}{\Delta r}\right)_{inst} &= \lim_{h\to 0}\frac{V(1+h)-V(1)}{h}\\ &= \lim_{h\to 0}\frac{(4\pi/3)(1+h)^3-(4\pi/3)(1)^3}{h}\\ &= \lim_{h\to 0}\frac{4\pi}{3}\cdot\frac{(1+h)^3-1}{h} = \lim_{h\to 0}\frac{4\pi}{3}\cdot\frac{h^3+3h^2+3h}{h}\\ &= \lim_{h\to 0}\frac{4\pi}{3}\cdot(h^2+3h+3) = \frac{4\pi}{3}\cdot(3) = 4\pi\end{aligned}$$

The instantaneous rate of change of volume with respect to radius at $r = 1$ is 4π. ■

(2) A standard physics formula says that the height in feet of a projectile launched straight upwards with an initial velocity $v_0\, ft/s$ is given by $y = v_0t - 16t^2$. Use a limit to find the instantaneous velocity of the object at $t = 3$ seconds. (Your answer will be in terms of v_0.)

□ Let's find the the pieces we'll need:

$$\begin{aligned}y(3) &= v_0(3) - 16(3)^2 = 3v_0 - 144\\ y(3+h) &= v_0(3+h) - 16(3+h)^2\\ &= 3v_0 + hv_0 - 16(9+6h+h^2)\\ &= 3v_0 + hv_0 - 144 - 96h - 16h^2\end{aligned}$$

Putting these into the definition of the instantaneous rate of change,

$$\begin{aligned}\left(\frac{\Delta y}{\Delta t}\right)_{inst} &= \lim_{h\to 0}\frac{y(3+h)-y(3)}{h}\\ &= \lim_{h\to 0}\frac{(3v_0+hv_0-144-96h-16h^2)-(3v_0-144)}{h}\\ &= \lim_{h\to 0}\frac{hv_0-96h-16h^2}{h} = \lim_{h\to 0}(v_0-96-16h) = v_0-96\end{aligned}$$

The instantaneous rate of change of height with respect to time at $t = 3$ is $v_0 - 96$ ft/s. ■

---

(3) Suppose the cost of making x gizmos is given by $C(x) = x^3 - 200x^2$. The *marginal cost* of making, say, the 100th gizmo is the instantaneous rate of change of the cost function at $x = 100$. Use a limit to find this marginal cost.

□ We're going to find

$$\left(\frac{\Delta C}{\Delta x}\right)_{inst} = \lim_{h\to 0} \frac{C(100+h) - C(100)}{h}$$

Let's find the the pieces we'll need:

$$\begin{aligned} C(100) &= (100)^3 - 200(100)^2 = -10^6 \\ C(100+h) &= (100+h)^3 - 200(100+h)^2 \\ &= (h^3 + 300h^2 + 300h + 10^6) - 200(10000 + 200h + h^2) \\ &= h^3 + 100h^2 - 10000h - 10^6 \end{aligned}$$

Putting these into the definition of the instantaneous rate of change,

$$\begin{aligned} \left(\frac{\Delta C}{\Delta x}\right)_{inst} &= \lim_{h\to 0} \frac{C(100+h) - C(100)}{h} \\ &= \lim_{h\to 0} \frac{(h^3 + 100h^2 - 10000h - 10^6) - (-10^6)}{h} \\ &= \lim_{h\to 0} \frac{h^3 + 100h^2 - 10000h}{h} \\ &= \lim_{h\to 0} (h^2 + 100h - 10000) = -10,000 \end{aligned}$$

The marginal cost of making the 100th gizmo is \$ $-10,000$. ■

B.3.2 *Derivative at a Point — Challenge — Solved*

(1) If $f(x) = \sqrt{3x+1}$, use a limit to determine $f'(a)$, where a is yet to be specified. Is $f'(x)$ undefined for any value of a at which $f(x)$ is not also undefined?

☐ Let's prepare for Def. 3.2 by building,

$$f(a+h) - f(a) = \sqrt{3(a+h)+1} - \sqrt{3a+1}$$

We are now in the zone where we have to apply the old adage (that I've made up), "Sometimes you have to make things look worse before they can look better." You may recall that a strategy for trying to make the best out of something in the form $\sqrt{A} - \sqrt{B}$ is to multiply up and down by $\sqrt{A} + \sqrt{B}$. The starting expression is,

$$f(a+h) - f(a) = \sqrt{3(a+h)+1} - \sqrt{3a+1}$$

then,

$$\begin{aligned}
\left(\sqrt{3(a+h)+1} - \sqrt{3a+1}\right) \cdot \frac{\sqrt{3(a+h)+1} + \sqrt{3a+1}}{\sqrt{3(a+h)+1} + \sqrt{3a+1}} \\
= \frac{3(a+h)+1-(3a+1)}{\sqrt{3(a+h)+1} + \sqrt{3a+1}} \\
\rightarrow f(a+h) - f(a) = \frac{3h}{\sqrt{3(a+h)+1} + \sqrt{3a+1}}
\end{aligned}$$

And we're now set up to divide by h and complete Def. 3.2,

$$\begin{aligned}
f'(a) &= \lim_{h\to 0} \frac{f(a+h) - f(a)}{h} \\
&= \lim_{h\to 0} \frac{3h}{\sqrt{3(a+h)+1} + \sqrt{3a+1}} \cdot \frac{1}{h} \\
&= \lim_{h\to 0} \frac{3}{\sqrt{3(a+h)+1} + \sqrt{3a+1}} \\
&= \frac{3}{\sqrt{3a+1} + \sqrt{3a+1}} = \frac{3}{2\sqrt{3a+1}}
\end{aligned}$$

In general, $f(a)$ is defined for $a \geq -1/3$. However, $f'(a)$ is not defined at $a = -1/3$. ■

(2) The following expression computes $f'(a)$ for some function $f(t)$ at some $t = a$. What are $f(t)$ and a?

$$\lim_{t \to 1} \frac{t^4 + t - 2}{t - 1}$$

☐ This derivative is in the other alternate form. We must match it to the expression

$$\lim_{t \to a} \frac{f(t) - f(a)}{t - a}$$

So evidently, $a = 1$ (from the denominator) and $f(t) = t^4 + t$ (from the first part of the numerator). This means the given expression is the derivative of $f(t) = t^4 + t$ at $a = 1$. (As a check, we do confirm that $f(1) = 1^4 + 1 = 2$.) ■

(3) The quantity (pounds) Q of coffee sold is a function of price p, $Q = f(p)$, where p is in dollars per pound. (a) Interpret the derivative $f'(8)$ in the context of the problem, and state its units. (b) Would you expect $f'(8)$ to be positive or negative? (There's no one right answer to this, but your answer must be consistent with your explanation.)

☐ (a) the derivative $f'(8)$ represents the rate of change of quantity sold Q as the price reaches (or passes) \$8 per pound. The units of this are pounds per dollars-per-pound. And (b), we should probably expect $f'(8)$ to be negative, since the amount sold is likely decreasing as the price is going up past \$8 per pound. ■

B.3.3 *The Derivative Function — Challenge — Solved*

(1) Use a limit to determine $f'(x)$ for $f(x) = x + \sqrt{x}$.

☐ Let's start building $f(x+h) - f(x)$ in preparation for Def. 3.3:

$$f(x+h) - f(x) = [(x+h) + \sqrt{x+h}] - [x + \sqrt{x}] = h + \sqrt{x+h} - \sqrt{x}$$

By now, you might recognize this as another "make it look worse before we make it look better" situation, since we have a difference of square roots. We can set the h aside by itself and write:

$$\begin{aligned} f(x+h) - f(x) &= h + \left(\sqrt{x+h} - \sqrt{x}\right) \cdot \frac{\sqrt{x+h} + \sqrt{x}}{\sqrt{x+h} + \sqrt{x}} \\ &= h + \frac{(x+h) - (x)}{\sqrt{x+h} + \sqrt{x}} \\ &= h + \frac{h}{\sqrt{x+h} + \sqrt{x}} \end{aligned}$$

and so,

$$\begin{aligned} f'(x) &= \lim_{h \to 0} \frac{f(x+h) - f(x)}{h} \\ &= \lim_{h \to 0} \left[h + \frac{h}{\sqrt{x+h} + \sqrt{x}} \right] \cdot \frac{1}{h} \\ &= \lim_{h \to 0} \left[1 + \frac{1}{\sqrt{x+h} + \sqrt{x}} \right] \\ &= 1 + \frac{1}{\sqrt{x} + \sqrt{x}} = 1 + \frac{1}{2\sqrt{x}} \end{aligned}$$

Altogether,

$$f'(x) == 1 + \frac{1}{2\sqrt{x}} \quad \blacksquare$$

(2) Use a limit to determine $f'(x)$ for $f(x) = ax^2 + bx + c$ (in which a, b, and c are unspecified constants).

☐ Here, we have

$$\begin{aligned} f(x+h) - f(x) &= [a(x+h)^2 + b(x+h) + c] - [ax^2 + bx + c] \\ &= [a(x^2 + 2xh + h^2) + bx + bh + c] - [ax^2 + bx + c] \\ &= 2axh + ah^2 + bh = h(2ax + ah + b) \end{aligned}$$

and so,

$$\begin{aligned} f'(x) &= \lim_{h \to 0} \frac{f(x+h) - f(x)}{h} \\ &= \lim_{h \to 0} \frac{h(2ax + ah + b)}{h} \\ &= \lim_{h \to 0} (2ax + ah + b) = 2ax + b \quad \blacksquare \end{aligned}$$

---

(3) Determine and graph $f'(x)$ for $f(x) = |x - 6|$. You don't need to use a limit, but you must describe how you come up with $f'(x)$.

□ The graph of $f(x) = |x - 6|$ is a big "V" with a cusp at $x = 6$; because of the cusp, f' does not exist at $x = 6$. To actually compute f', we can do long calculations, or we can use what we know about derivatives so far. To the right of $x = 6$, the graph of $f(x)$ is simply the line $y = x - 6$. The rate of change of that line is its slope, 1. So this is the value of f' for $x > 6$. Because of the absolute value bars, the graph of $f(x)$ is, to the left of $x = 6$, the line $y = -(x - 6) = -x + 6$. The rate of change of that line is its slope, -1. This is the value of f' for $x < 6$. Thus,

$$f'(x) = \begin{cases} -1 & \text{for } x < 6 \\ 1 & \text{for } x \geq 6 \end{cases}$$

The graph of f' is the pair of horizontal lines $y = 1$ (for $x > 6$) and $y = -1$ (for $x < 6$). ■

B.3.4 *Simple Derivs and Antiderivs — Challenge — Solved*

(1) Find the derivative of $f(x) = (x-1)(x+2)$.

□ Multiply this out to see its quadratic form. Do NOT try to do the derivative of $x-1$ and $x+2$ and then multiply the results, you'll get the wrong answer.

$$\frac{d}{dx}(x-1)(x+2) = \frac{d}{dx}(x^2+x-2) = 2x+1 \quad \blacksquare$$

(2) Use a derivative to help find the vertex of the parabola $y = -x^2+2x-3$.

□ At the vertex, the tangent line will be horizontal. So we need to find where the derivative is zero. Since $f'(x) = -2x+2$, we have $f'(x) = 0$ at $x = 1$. So the vertex is at the point $(1, f(1)) = (1, -2)$. ■

(3) We know that x has an infinite number of antiderivatives. Find the ONE antiderivative of x that goes through the point $(1,2)$. (Hint: Get the general antiderivative of x, then discover which value of the arbitrary constant C will cause the antiderivative to hit this point.)

□ We start with

$$\int x\,dx = \frac{1}{2}x^2 + C$$

We want the antiderivative to use the point $(1,2)$, so we need to see

$$\begin{aligned} f(1) &= 2 \\ \frac{1}{2}(1)^2 + C &= 2 \\ C &= \frac{3}{2} \end{aligned}$$

So the antiderivative of x that goes through $(1,2)$ is $\frac{1}{2}x^2 + \frac{3}{2}$. ■

B.3.5 *Power Rule — Challenge — Solved*

---

(1) (a) Find the locations where the lines tangent to $y = 2(x^2 - 3)^2$ are horizontal. (b) Design a function that has horizontal tangent lines at $x = 1$ and $x = 5$ and provide the formula for this function.

□ (a) The lines tangent to $y = 2(x^2 - 3)^2$ are horizontal wherever $dy/dx = 0$. So,

$$\begin{aligned}\frac{dy}{dx} &= 4(x^2-3)\cdot(2x)\\ 0 &= 8x(x^2-3)\\ x &= 0, \pm\sqrt{3}\end{aligned}$$

The lines tangent to this curve are horizontal at $x = 0$, $x = -\sqrt{3}$ and $x = \sqrt{3}$.

(b) A function with horizontal tangent lines at $x = 1$ and $x = 5$ will have $dy/dx = 0$ at those locations, so we must have something like

$$\frac{dy}{dx} = (x-1)(x-5) = x^2 - 6x + 5$$

Since the derivative of our function is $x^2 - 6x + 5$, then our function can be any antiderivative:

$$f(x) = \int x^2 - 6x + 5\,dx = \frac{1}{3}x^3 - 3x^2 + 5x + C$$

Since we only need one of these, let's pick $C = 0$, so our function with horizontal tangent lines at $x = 1$ and $x = 5$ is $f(x) = x^3/3 - 3x^2 + 5x$. ■

---

(2) Find the derivative and antiverivative of $y = (x^2 + 4x + 3)/\sqrt{x}$.

□ We should simplify the function first,

$$y = \frac{x^2+4x+3}{\sqrt{x}} = x^{3/2} + 4\sqrt{x} + \frac{3}{\sqrt{x}} = x^{3/2} + 4x^{1/2} + 3x^{-1/2}$$

so by the power rule,

$$\frac{dy}{dx} = \frac{3}{2}x^{1/2} + 4\cdot\frac{1}{2}\cdot x^{-1/2} - 3\cdot\frac{1}{2}\cdot x^{-3/2} = \frac{3}{2}\sqrt{x} + \frac{2}{\sqrt{x}} - \frac{3}{2x\sqrt{x}}$$

and also

$$\begin{aligned}\int y\,dx &= \int x^{3/2} + 4x^{1/2} + 3x^{-1/2}\,dx\\ &= \frac{2}{5}x^{5/2} + 4\cdot\frac{2}{3}x^{3/2} + 3\cdot\frac{2}{1}x^{1/2} + C\\ &= \frac{2}{5}x^{5/2} + \frac{8}{3}x^{3/2} + 6\sqrt{x} + C\end{aligned}$$ ■

---

(3) Find the value of A for which the following function has a horizontal tangent line at $x = 4$:

$$f(x) = \frac{3}{4}x^4 - 2Ax - \frac{4}{3}x^{3/2}$$

□ The derivative of this function is

$$f'(x) = \frac{3}{4}(4x^3) - 2A - \frac{4}{3} \cdot \frac{3}{2}x^{1/2} = 3x^3 - 2A - 2x^{1/2}$$

So at $x = 4$, we have

$$f'(4) = 3(4)^3 - 2A - 2(4)^{1/2} = 3(64) - 2A - 4 = 188 - 2A$$

The tangent line at $x = 4$ will be horizontal when $f'(4) = 0$, i.e. when $188 - 2A = 0$; this happens when $A = 94$. ■

B.3.6 *Power Rule Pt 2 — Challenge — Solved*

(1) Analyze the heck out of the graph of $y = 1/((x^2-1)^2+3)$ (see Practice Problem 2 for terminology). Produce a sketch of the curve based on your work — I hereby declare this as the "cowboy hat function".

☐ From the function itself, note that there are no vertical asymptotes since the denominator is never zero. The horizontal asymptote is $y = 0$ because that is the limit of the function as $x \to \pm\infty$. For the derivative, let's rewrite the denominator as

$$(x^2-1)^2+3 = (x^4-2x^2+1)+3 = x^4-2x^2+4$$

so that

$$y = (x^4-2x^2+4)^{-1} \to \frac{dy}{dx} = -1(x^4-2x^2+4)^{-2} \cdot (4x^3-2x^2)$$
$$= \frac{2x^2(2x-1)}{(x^4-2x^2+4)^2}$$

Then we can see that $dy/dx = 0$ at $x - 0$ and $x = \pm 1/2$. So, these are the locations of the horizontal asymptotes. The y-coordinates of these points are (calculations are trivial and omitted)

$$f(0) = \frac{1}{4} \quad ; \quad f\left(-\frac{1}{2}\right) = f\left(\frac{1}{2}\right) = \frac{16}{57}$$

(yuck!). So, there are three horizontal tangent lines, at (in order from left to right):

$$\left(-\frac{1}{2}, \frac{16}{57}\right) \quad , \quad \left(0, \frac{1}{4}\right) \quad , \quad \left(\frac{1}{2}, \frac{16}{57}\right)$$

To form a good graph, select a few "random" x coordinates, get their y coords, and then play connect the dots while placing the tangent lines in the right places. Figure B.5 shows a graph of this function. ■

(2) Find the antiderivative of $f(x) = 2x/(x^2-1)^2$ that goes through the point $(3, 7/8)$.

☐ First rewrite the function as $f(x) = (2x-2)(x^2-1)^{-2}$, so that when matching the integrand to the general form $g'(x)(g(x))^p$ we identify $g(x) = x^2-1$ and hope to find $g'(x) = 2x$. Since that's what is already in the integrand, we have:

$$\int (2x)(x^2-1)^{-2}\, dx = \frac{1}{-1}(x^2-1)^{-1} + C = -\frac{1}{x^2-1} + C$$

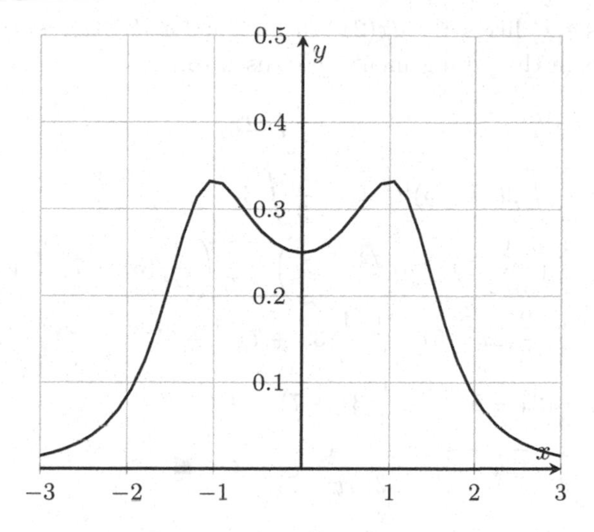

Fig. B.5 A graph of $f(x) = \dfrac{1}{(x^2-1)+3}$.

To find the specific antiderivative that goes through the point $(3, 7/8)$, we find what value of C gives us $y = 7/8$ when $x = 3$:

$$-\frac{1}{3^2-1} + C = \frac{7}{8}$$

$$C = \frac{7}{8} + \frac{1}{8} = 1$$

and so the antiderivative we're after is

$$-\frac{1}{x^2-1} + 1 \quad \text{or} \quad 1 - \frac{1}{x^2-1} \quad \blacksquare$$

---

(3) Find the derivative and antiderivative of $H(x) = \sqrt{2x-5} + \dfrac{1}{(3x+7)^2}$.

□ For the derivative, we can rely on the general rule that $(f+g)' = f' + g'$. With a quick rewriting of the terms to highlight the exponents,

$$H(x) = (2x-5)^{1/2} + (3x+7)^{-2}$$

$$H'(x) = \frac{1}{2}(2x-5)^{-1/2} \cdot (2) + (-2)(3x+7)^{-3} \cdot (3)$$

$$= \frac{1}{\sqrt{2x-5}} - \frac{6}{(3x+7)^3}$$

For the antiderivative, let's start with the rewritten version of the function, $H(x) = (2x-5)^{1/2} + (3x+7)^{-2}$. Then let's pick out the inside functions $g(x)$ like so: $d/dx(2x-5) = 2$ and $d/dx(3x+7) = 3$. Placing the terms in the arrangement we're used to,

$$\begin{aligned}
\int (2x-5)^{1/2} + (3x+7)^{-2}\,dx &= \int (2x-5)^{1/2}\,dx + \int (3x+7)^{-2}\,dx \\
&= \frac{1}{2}\int 2(2x-5)^{1/2}\,dx + \frac{1}{3}\int 3(3x+7)^{-2}\,dx \\
&= \frac{1}{2}\left(\frac{1}{3/2}(2x-5)^{3/2} + C_1\right) + \frac{1}{3}\left(\frac{1}{-1}(3x+7)^{-1} + C_2\right) \\
&= \frac{1}{2}\cdot\frac{2}{3}(2x-5)^{3/2} - \frac{1}{3}(3x+7)^{-1} + C \\
&= \frac{1}{3}(2x-5)^{3/2} - \frac{1}{3}(3x+7)^{-1} + C \\
&= \frac{1}{3}\sqrt{(2x-5)^3} - \frac{1}{3(3x+7)} + C \quad \blacksquare
\end{aligned}$$

B.4 Chapter 4: Challenge Problem Solutions

B.4.1 *Exp & Log Function Derivs — Challenge — Solved*

(1) Find the derivative of $y = \ln(x^2 + e^{x-1}) + \sinh^2(x)$.

□ There is a natural log term and also a term that will need the power rule for $(g(x))^p$:

$$\begin{aligned}\frac{dy}{dx} &= \frac{1}{x^2+e^{x-1}}\frac{d}{dx}(x^2+e^{x-1}) + 2(\sinh(x))^1\frac{d}{dx}(\sinh(x)) \\ &= \frac{1}{x^2+e^{x-1}}\cdot(2x+e^{x-1}) + 2\sinh(x)\cosh(x) \\ &= \frac{2x+e^{x-1}}{x^2+e^{x-1}} + 2\sinh(x)\cosh(x) \quad \blacksquare\end{aligned}$$

(2) Find the antiderivative $\displaystyle\int \frac{1}{x^{2/3}(\sqrt[3]{x}-1)}\,dx$.

□ Let's match the integrand to $g'(x)/g(x)$ in order to use Eq. (4.8). It doesn't *quite* look like it's in that form, but what else do we have to try? Take $g(x) = \sqrt[3]{x} - 1$, i.e. $g(x) = x^{1/3} - 1$, so that

$$g'(x) = \frac{1}{3}x^{-2/3} = \frac{1}{3x^{2/3}}$$

Now we're getting somewhere!

$$\begin{aligned}\int \frac{1}{x^{2/3}(\sqrt[3]{x}-1)}\,dx &= 3\int \frac{1}{3x^{2/3}}\cdot\frac{1}{(\sqrt[3]{x}-1)}\,dx \\ &= 3\ln|\sqrt[3]{x}-1| + C \quad \blacksquare\end{aligned}$$

(3) Find the antiderivative of $f(x)$ that goes through the point $(1, 0)$ for the function

$$f(x) = \frac{1-x}{x^2-2x+2} + x^2e^{-x^3}$$

□ First, note that the general antiderivative is:

$$\int f(x)\,dx = \frac{1}{2}\ln|x^2-2x+2| - \frac{1}{3}e^{-x^3} + C$$

Given the point of interest $(1, 0)$, we know that when we plug in $x = 1$ to the antiderivative, we should get 0 back:

$$\frac{1}{2}\ln|1^2 - 2(1) + 2| - \frac{1}{3}e^{-(1^3)} + C = 0$$

which leads to $C = 1/(3e)$. So the antiderivative of $f(x)$ that goes through $(1, 0)$ is

$$\frac{1}{2}\ln|x^2-2x+2| - \frac{1}{3}e^{-x^3} + \frac{1}{3e} \quad \blacksquare$$

B.4.2 *Trig Function Derivs — Challenge — Solved*

(1) Find dy/dx for $y = \sin(\cos(\tan x))$. (Hint: This requires more than one use of an extended derivative formula.)

☐ $$\begin{aligned}\frac{dy}{dx} &= \cos(\cos(\tan x))\frac{d}{dx}(\cos(\tan x))\\ &= \cos(\cos(\tan x))(-\sin(\tan x))\frac{d}{dx}(\tan x)\\ &= \cos(\cos(\tan x))(-\sin(\tan x))(\sec^2 x)\\ &= -\sec^2 x \sin(\tan x)\cos(\cos(\tan x)) \quad \blacksquare\end{aligned}$$

(2) Follow the procedure in the last part of Example 2 to show that

$$\int \cot x\, dx = \ln|\sin x| + C$$

☐ First, we can use an identity for $\cot x$:

$$\int \cot x\, dx = \int \frac{\cos x}{\sin x}\, dx = \int \cos x \cdot \frac{1}{\sin x}\, dx$$

Next, we can recognize that with $g(x) = \sin x$, we are in the form

$$\int g'(x) \cdot \frac{1}{g(x)}\, dx = \ln|g(x)| + C$$

so that

$$\int \cot x\, dx = \int \cos x \cdot \frac{1}{\sin x}\, dx = \ln|\sin x| + C \quad \blacksquare$$

(3) Follow the procedure shown in this section and the two before it to show why (when $g(x)$ is one-to-one)

$$\frac{d}{dx}\cos(g(x)) = -g'(x)\sin(g(x))$$

☐ The derivative of $\cos(g(x))$ is defined as:

$$\begin{aligned}\frac{d}{dx}\cos(g(x)) &= \lim_{h\to 0}\frac{\cos(g(x+h)) - \cos(g(x))}{h}\\ &= \lim_{h\to 0}\frac{\cos(g(x+h)) - \cos(g(x))}{h}\cdot\frac{g(x+h)-g(x)}{g(x+h)-g(x)}\end{aligned}$$

With regrouping and application of the limit law regarding products, this becomes

$$\frac{d}{dx}\cos(g(x)) = \lim_{h\to 0}\frac{\cos(g(x+h)) - \cos(g(x))}{g(x+h)-g(x)}\cdot\lim_{h\to 0}\frac{g(x+h)-g(x)}{h}$$

The latter limit is just $g'(x)$. The first limit can be redesigned by setting $a = g(x)$, $y = g(x+h)$, and remembering that the limit process $h \to 0$ is now equivalent to $y \to g(x)$, i.e. $y \to a$. Therefore,

$$\lim_{h\to 0} \frac{\cos(g(x+h)) - \cos(g(x))}{g(x+h) - g(x)} = \lim_{y\to a} \frac{\cos(y) - \cos(a)}{y - a}$$

which defines the derivative of $\cos(y)$ at the point $y = a$, that we know to be $-\sin(a)$. But since we have $a = g(x)$, we have, finally, that:

$$\lim_{h\to 0} \frac{\cos(g(x+h)) - \cos(g(x))}{g(x+h) - g(x)} = -\sin(g(x))$$

Together then,

$$\begin{aligned} \frac{d}{dx}\cos(g(x)) &= \lim_{h\to 0} \frac{\cos(g(x+h)) - \cos(g(x))}{g(x+h) - g(x)} \cdot \lim_{h\to 0} \frac{g(x+h) - g(x)}{g(x+h) - g(x)} \\ &= -\sin(g(x)) \cdot g'(x) \end{aligned}$$

Hooray, we did it! ■

B.4.3 *Product & Quotient Rules — Challenge — Solved*

(1) Find and properly name the derivative of $z = w^{3/2}\sin(2w + e^{-w})$.

☐
$$\begin{aligned}\frac{dz}{dw} &= w^{3/2}\frac{d}{dw}\sin(2w+e^{-w}) + \sin(2w+e^{-w})\frac{d}{dw}w^{3/2}\\ &= w^{3/2}\cdot\cos(2w+e^{-w})\frac{d}{dw}(2w+e^{-w})\\ &\quad+\sin(2w+e^{-w})\cdot\frac{3}{2}w^{1/2}\\ &= w^{3/2}\cdot\cos(2w+e^{-w})\cdot(2-e^{-w}) + \frac{3}{2}\sin(2w+e^{-w})\sqrt{w}\\ &= w^{3/2}(2-e^{-w})\cos(2w+e^{-w}) + \frac{3}{2}\sin(2w+e^{-w})\sqrt{w} \quad \blacksquare\end{aligned}$$

(2) Find the equations of all lines tangent to $y = (x-1)/(x+1)$ which are parallel to $x - 2y = 2$.

☐ Using the quotient rule,

$$\begin{aligned}\frac{dy}{dx} &= \frac{(x+1)\frac{d}{dx}(x-1) - (x-1)\frac{d}{dx}(x+1)}{(x+1)^2}\\ &= \frac{(x+1)-(x-1)}{((x+1)^2} = \frac{2}{(x+1)^2}\end{aligned}$$

A line parallel to $x - 2y = 2$ has a slope of $1/2$, so we need

$$\begin{aligned}\frac{2}{(x+1)^2} &= \frac{1}{2}\\ \frac{1}{(x+1)^2} &= \frac{1}{4}\\ (x+1)^2 &= 4\\ x+1 &= \pm 2\\ x = 1 \ &, \ x = -3\end{aligned}$$

These correspond to points $(1,0)$ and $(-3,2)$, so the equations of these tangent lines are:

$$y = \frac{1}{2}(x-1) \quad \text{and} \quad y - 2 = \frac{1}{2}(x+3) \quad \blacksquare$$

(3) At which point will the line tangent to $y = e^{-\sin x}/x - 1/2$ at $x = 1$ cross the x-axis? (If you provide a decimal approximation to the answer, it must have at least 3 decimal place accuracy.)

☐ First, we need the equation of the line tangent to the given function at $x = 1$. The point this tangent line shares with the function is

$$(1, y(1)) = \left(1, e^{-\sin 1} - \frac{1}{2}\right)$$

To find the slope of this tangent line, we need the derivative. Since the derivative of the constant $-1/2$ is zero, the derivative of our function is:

$$\frac{dy}{dx} = \frac{x\frac{d}{dx}(e^{-\sin x}) - e^{-\sin x}\frac{d}{dx}(x)}{(x)^2} = \frac{x(-\cos x e^{-\sin x}) - e^{-\sin x}(1)}{x^2}$$
$$= \frac{e^{-\sin x}(-x\cos x - 1)}{x^2} = -\frac{e^{-\sin x}(1 + x\cos x)}{x^2}$$

and so at $x = 1$, the derivative gives $m_{tan} = -e^{-\sin 1}(1 + \cos(1))$ and the equation of the tangent line is:

$$y - \left(e^{-\sin 1} - \frac{1}{2}\right) = -e^{-\sin 1}(1 + \cos(1)) \cdot (x - 1)$$

Now, this line will cross the x-axis when $y = 0$. To find out the corresponding x coordinate, we solve:

$$0 - e^{-\sin 1} + \frac{1}{2} = -e^{-\sin 1}(1 + \cos(1)) \cdot (x - 1)$$
$$\frac{1}{e^{\sin 1}} - \frac{1}{2} = \frac{1}{e^{\sin 1}}(1 + \cos(1)) \cdot (x - 1)$$
$$1 - \frac{e^{\sin 1}}{2} = (1 + \cos(1)) \cdot (x - 1)$$
$$\frac{2 - e^{\sin 1}}{2} = (1 + \cos(1)) \cdot (x - 1)$$
$$x - 1 = \frac{2 - e^{\sin 1}}{2(1 + \cos(1))}$$
$$x = \frac{2 - e^{\sin 1}}{2(1 + \cos(1))} + 1$$

The decimal approximation of this value is $x \approx 0.896$. ■

B.4.4 *Chain Rule — Challenge — Solved*

(1) Find the derivative of $y = x\sin^{-1}(1/x)$.

☐ We use a combination of the product rule and the chain rule. To prepare for the product rule, we'll need the following derivative; it requires the chain rule, and is messy enough that we should solve it on its own first:

$$\begin{aligned}\frac{d}{dx}\sin^{-1}\left(\frac{1}{x}\right) &= \frac{1}{\sqrt{1-(1/x)^2}}\cdot\frac{d}{dx}\frac{1}{x} = \frac{1}{\sqrt{1-1/x^2}}\cdot\left(-\frac{1}{x^2}\right)\\ &= -\frac{1}{x^2\sqrt{1-1/x^2}} = -\frac{1}{x\cdot\sqrt{x^2}\cdot\sqrt{1-1/x^2}}\\ &= -\frac{1}{x\sqrt{x^2-1}}\end{aligned}$$

and so the full derivative of the given function is

$$\begin{aligned}\frac{dy}{dx} &= x\cdot\frac{d}{dx}\sin^{-1}\left(\frac{1}{x}\right)+\sin^{-1}\left(\frac{1}{x}\right)\cdot\frac{d}{dx}(x)\\ &= x\cdot\left(-\frac{1}{x\sqrt{x^2-1}}\right)+\sin^{-1}\left(\frac{1}{x}\right)\cdot(1)\\ &= -\frac{1}{\sqrt{x^2-1}}+\sin^{-1}\left(\frac{1}{x}\right) \quad \blacksquare\end{aligned}$$

(2) Find the derivative of $y = e^{\sin(x^2)}$ and $z = e^{\sinh(x^2)}$.

☐ This requires the chain rule twice!

$$\begin{aligned}\frac{dy}{dx} = e^{\sin(x^2)}\cdot\frac{d}{dx}\sin(x^2) &= e^{\sin(x^2)}\cdot\cos(x^2)\cdot\frac{d}{dx}(x^2)\\ &= e^{\sin(x^2)}\cdot\cos(x^2)\cdot 2x = 2x\cos(x^2)e^{\sin(x^2)}\end{aligned}$$

The second result is literally the first result with sin replaced by sinh. Be sure you know why!

$$\frac{dz}{dx} = 2x\cosh(x^2)e^{\sinh(x^2)} \blacksquare$$

(3) Find the equation of the line tangent to $y = \cos 2x/(x^2+1)$ at $x = \pi/4$.

☐ The point this tangent line shares with the function is

$$\left(\frac{\pi}{4}, f\left(\frac{\pi}{4}\right)\right) = \left(\frac{\pi}{4}, 0\right)$$

The slope of the tangent line comes from dy/dx at $x = \pi/4$. This derivative starts as a quotient rule.

$$\begin{aligned}\frac{dy}{dx} &= \frac{(x^2+1)\frac{d}{dx}\cos(2x) - \cos(2x)\frac{d}{dx}(x^2+1)}{(x^2+1)^2}\\ &= \frac{(x^2+1)(-2\sin(2x)) - (\cos(2x))(2x)}{(x^2+1)^2}\\ &= \frac{-2(x^2+1)\sin(2x) - 2x\cos(2x)}{(x^2+1)^2}\end{aligned}$$

So at $x = \pi/4$ we have

$$\begin{aligned}\sin(2x) &= \sin\left(2\cdot\frac{\pi}{4}\right) = \sin\left(\frac{\pi}{2}\right) = 1\\ \cos(2x) &= \cos\left(2\cdot\frac{\pi}{4}\right) = \cos\left(\frac{\pi}{2}\right) = 0\end{aligned}$$

Therefore,

$$\begin{aligned}\frac{dy}{dx} &= \frac{-2((\pi/4)^2+1)(1) - 0}{((\pi/4)^2+1)^2} = \frac{-2}{((\pi/4)^2+1)}\\ &= -\frac{2}{\frac{\pi^2}{16}+1} = -\frac{2}{\frac{\pi^2+16}{16}} = -\frac{32}{\pi^2+16}\end{aligned}$$

Altogether, then, the equation of the tangent line is

$$y - 0 = -\frac{32}{\pi^2+16}\left(x - \frac{\pi}{4}\right) \blacksquare$$

B.4.5 *Implicit Differentiation — Challenge — Solved*

(1) Find dy/dx for the inverse secant function $y = \sec^{-1}(x)$ $(x > 0)$.

☐ We can rewrite the function as $\sec(y) = x$, which we can convert to $\cos y = \dfrac{1}{x}$. Then we get started with the derivative operator:

$$\frac{d}{dx}\cos(y) = \frac{d}{dx}\frac{1}{x}$$
$$-\sin(y)\frac{dy}{dx} = -\frac{1}{x^2} \rightarrow \frac{dy}{dx} = \frac{1}{x^2 \sin y}$$

We can encode $\cos y = 1/x$ into a right triangle by putting y in as an acute angle, with 1 and x being the lengths of the adjacent side and hypotenuse, respectively. Then the length of the missing opposite side is known to be $\sqrt{x^2 - 1}$. Now we have the lengths of all three sides, and can immediately find that $\sin y = \sqrt{x^2 - 1}/x$. (This is all shown in Fig. B.6.) So,

$$\frac{dy}{dx} = \frac{1}{x^2 \sin y} = \frac{x}{x^2\sqrt{1 - x^2}} = \frac{1}{x\sqrt{1 - x^2}}$$

(Since we set $x > 0$ we don't need to worry about x/x^2 being anything other than x itself.) ■

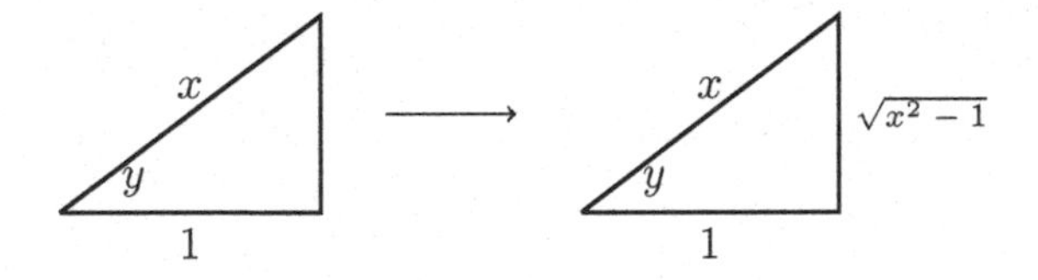

Fig. B.6 Encoding $\cos y = \dfrac{1}{x}$ into a right triangle.

(2) We know that if $y = e^x$ then $dy/dx = e^x$. Also, if $y = \ln x$, then $dy/dx = 1/x$. But when we have a different base b involved, the derivative of $y = b^x$ is not $dy/dx = b^x$ and the derivative of $y = \log_b(x)$ is not $1/x$. Use implicit differentiation to find what the derivatives of b^x and $\log_b(x)$ really are. (Hint: For the first, use implicit differentiation directly. For the second, write $y = \log_b(x)$ in exponential form, *then* use implicit differentiation.)

☐ If $y = b^x$ then $\ln y = x \ln b$. So

$$\frac{d}{dx} \ln y = \frac{d}{dx} x \ln b$$
$$\frac{1}{y} \frac{dy}{dx} = \ln b$$
$$\frac{dy}{dx} = y \ln b = b^x \ln b$$

In the very last step, note the exchange of y back to b^x, which allows us to write the derivative in terms of x only.

If $y = \log_b(x)$ then we can write $b^y = x$ and with the rule just found above,

$$\frac{d}{dx} b^y = \frac{d}{dx} x$$
$$b^y \cdot \ln b \cdot \frac{dy}{dx} = 1$$
$$\frac{dy}{dx} = \frac{1}{b^y \ln b} = \frac{1}{x \ln b} \quad \blacksquare$$

---

(3) Given a cube with side length x and full cross diagonal (from one corner to the opposite corner) D, what is the relationship between the rates of change of x and D (with respect to time)?

☐ If one corner of the cube sits at the origin $(0, 0, 0)$ then the opposite corner is at (x, x, x). So, the length of the diagonal is the distance D between these two points. By the distance formula we have

$$D^2 = (x - 0)^2 + (x - 0)^2 + (x - 0)^2 = 3x^2$$

which then gives $D = \sqrt{3}(x)$. Therefore,

$$\frac{dD}{dt} = \sqrt{3} \frac{dx}{dt} \quad \blacksquare$$

B.4.6 Antiderivs Using Substitution — Challenge — Solved

(1) Find $\int e^{2\cosh(t)} \sinh(t)\, dt$.

□ We use the substitution $u = 2\cosh(t)$ to get

$$du = 2\sinh(t)\, dt \quad \longrightarrow \quad \sinh(t)\, dt = \frac{1}{2}\, du$$

so that

$$\int e^{2\cosh(t)} \sinh(t)\, dt = \frac{1}{2}\int e^u\, du = \frac{1}{2}e^u + C = \frac{1}{2}e^{2\cosh(t)} + C$$

■

(2) Find $\int \frac{4t^3}{t^2-3}\, dt$.

□ It is often better to focus on denominators when picking a substitution, so let's try $u = t^2 - 3$. With this, we have $du = 2t\, dt$. In order for this to work, we have to spot a $t\, dt$ in the numerator. Can we find that? Sure! We have to split off one t from the t^3 in the numerator. Heck, let's split up the 4 into $2 \cdot 2$ as well, to help put a complete $2t\, dt$ into place:

$$\int \frac{4t^3}{t^2-3}\, dt = \int \frac{2t^2}{t^2-3} \cdot 2t\, dt$$

Another consequence of the substitution $u = t^2 - 3$ is that we can write $t^2 = u + 3$, or $2t^2 = 2(u+3)$. In summary, we have three pieces to exchange because of the substitution $u = t^2 - 3$:

(a) $t^2 - 3$ becomes u
(b) $2t^2$ becomes $2(u+3)$
(c) $2t\, dt$ becomes du

Altogether, from beginning to end,

$$\begin{aligned}\int \frac{4t^3}{t^2-3}\, dt = \int \frac{2t^2}{t^2-3} \cdot 2t\, dt &= \int \frac{2(u+3)}{u}\, du \\ &= 2\int 1 + \frac{3}{u}\, du = 2\left(u + 3\ln|u|\right) + C \\ &= 2(t^2-3) + 3\ln|t^2-3| + C\end{aligned}$$

That's good stuff! ■

(3) Find $\int \frac{x^5}{1+x^3}\,dx$.

☐ Let's be clever and split the numerator to write this as:

$$\int \frac{x^5}{1+x^3}\,dx = \int \frac{x^3}{1+x^3} \cdot x^2\,dx$$

Now if we start with $u = 1 + x^3$, we get two exchanges:

(a) $du = 3x^2\,dx$, which can be written in the more helpful form $x^2\,dx = du/3$

(b) The leftover x^3 can be written as $u - 1$

Altogether,

$$\begin{aligned}\int \frac{x^3}{1+x^3} \cdot x^2\,dx &= \int \frac{u-1}{u}\frac{du}{3} = \frac{1}{3}\int 1 - \frac{1}{u}\,du \\ &= \frac{1}{3}\left(u - \ln|u|\right) + C \\ &= \frac{1}{3}\left(1 + x^3 - \ln|1+x^3|\right) + C\end{aligned}$$

We can even do some cleaning up by absorbing the leading constant $\frac{1}{3}$ into C

$$\int \frac{x^3}{1+x^3} \cdot x^2\,dx = \frac{1}{3}\left(x^3 - \ln|1+x^3|\right) + C$$

but this last step is not entirely necessary. ■

B.5 Chapter 5: Challenge Problem Solutions

B.5.1 *Related Rates — Challenge — Solved*

(1) If the length of one of the sides a pentagon is decreasing at $10\,cm/sec$, how fast is the area of the pentagon changing when the five sides of the pentagon are each $50\,cm$ long? (Hint: The area of a pentagon is given by

$$A = \frac{5}{4}\cot\left(\frac{\pi}{5}\right)x^2$$

where x is the length of any one of the five equal sides.)

□ We are given $\frac{dx}{dt} = -10\,\frac{cm}{sec}$ and asked to compute dA/dt. Our static geometry is related to the given formula for the area of a pentagon. Powering it up, we have

$$\frac{dA}{dt} = \frac{5}{4}\cot\left(\frac{\pi}{5}\right)\left(2x \cdot \frac{dx}{dt}\right) = \frac{10}{4}\cot\left(\frac{\pi}{5}\right)x\,\frac{dx}{dt}$$

So when $x = 50\,cm$ we have

$$\frac{dA}{dt} = \frac{10}{4}\cot\left(\frac{\pi}{5}\right)(50\,cm)\left(-10\,\frac{cm}{sec}\right) = -1250\cot\left(\frac{\pi}{5}\right)\frac{cm^2}{sec} \quad ■$$

(2) The surface area of a melting snowball decreases at a rate of $1\,cm^2/min$. Find the rate at which the diameter of the snowball is decreasing at the instant the diameter is $10\,cm$.

□ Assuming the snowball is a sphere, our static geometry is a sphere with diameter D and surface area A. For a sphere, $A = 4\pi r^2$, which can be converted to be in terms of the diameter:

$$A = 4\pi\left(\frac{D}{2}\right)^2 = \pi D^2$$

We are interested in what happens at $D = 10$ cm. The rates of change at that instant are $dA/dt = -1\,cm^2/min$ and dD/dt, which is unknown. Powering up our static equation:

$$\frac{d}{dt}A = \frac{d}{dt}(\pi D^2)$$

$$\frac{dA}{dt} = 2\pi D\frac{dD}{dt}$$

Taking our snapshot when $D = 10$ and $dA/dt = -1$,

$$-1 = 2\pi(10)\frac{dD}{dt}$$

$$\frac{dD}{dt} = -\frac{1}{20\pi}$$

The diameter of the snowball is decreasing at $-1/(20\pi)\, cm/min$. (As usual, we omitted units during the calculation, but we can predict the final units of dA/dt and see that the proper units result from the computation.) ■

---

(3) An object is moving counterclockwise along the unit circle $x^2 + y^2 = 1$. The x coordinate of the object is changing at 1 unit per second. How fast is the object's distance from the point $(1, 0)$ changing when the object is at the point $(0, 1)$?

□ On the upper half of the circle, where this problem takes place, the object will always be at points described by $(x, y) = (x, \sqrt{1 - x^2})$. Thus, the distance D from the object at any point $(x, \sqrt{1 - x^2})$ to the point $(1, 0)$ is given by:

$$D^2 = (x - 1)^2 + (\sqrt{1 - x^2} - 0)^2 = (x^2 - 2x + 1) + (1 - x^2) = 2 - 2x$$

Powering this up,

$$\frac{d}{dt}D^2 = \frac{d}{dt}(2 - 2x)$$
$$2D\frac{dD}{dt} = -2\frac{dx}{dt}$$

At our snapshot, we have $\dfrac{dx}{dt} = -1$ (when moving counterclockwise on the upper half of the circle, the x-coordinate is decreasing). But we still need D at the instant in question. When the object is at $(0, 1)$, the distance from the object to $(1, 0)$ is

$$D = \sqrt{(1 - 0)^2 + (0 - 1)^2} = \sqrt{2}$$

so when the object is at $(0, 1)$,

$$\frac{dD}{dt} = -\frac{1}{\sqrt{2}} \cdot (-1) = \frac{1}{\sqrt{2}}$$

The distance from the object to the point (1,0) at the instant in question is increasing at $1/\sqrt{2}$ units per second. ■

B.5.2 *Derivatives and Graphs — Challenge — Solved*

(1) Provide a complete graphical analysis of $f(x) = \cos^2 x - 2\sin x$ on $(0, 2\pi)$ and sketch the graph of $f(x)$ on that interval.

☐ We have

$$\begin{aligned} f'(x) &= -2\sin x\cos - 2\cos x = -2\cos x(\sin x + 1) \\ &\implies f'(x) = 0 \text{ at } x = \pi/2, 3\pi/2 \\ f''(x) &= 4\sin^2 x + 2\sin x - 2 = 2(\sin x + 1)(2\sin x - 1) \\ &\implies f''(x) = 0 \text{ at } x = \pi/6, 5\pi/6, 3\pi/2 \end{aligned}$$

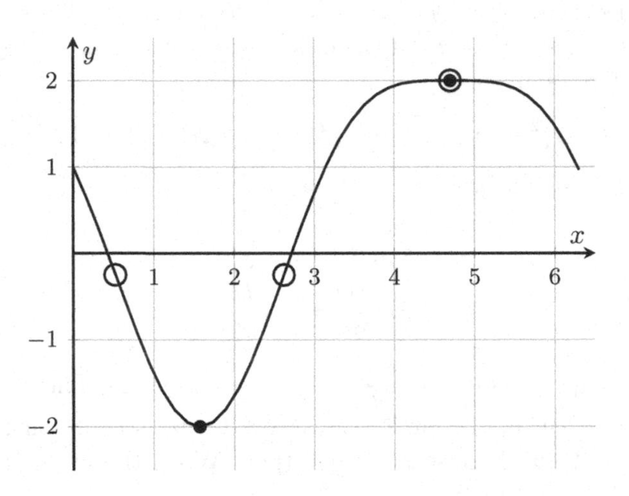

Fig. B.7 Graphical analysis of $f(x) = \cos^2(x) - 2\sin(x)$.

The function's values at these critical points and inflection points are:

x	$\pi/6$	$\pi/2$	$5\pi/6$	$3\pi/2$
$f(x)$	$-1/4$	-2	$-1/4$	2
type	IP	CP	IP	CP/IP
ext		min		max

Intervals set up by these critical points and inflection points, the signs of the derivatives in each, and the resulting shape are:

int	$(0, \pi/6)$	$(\pi/6, \pi/2)$	$(\pi/2, 5\pi/6)$	$(5\pi/6, 3\pi/2)$	$(3\pi/2, 2\pi)$
$f'(x)$	$-$	$-$	$+$	$+$	$-$
$f''(x)$	$-$	$+$	$+$	$-$	$-$
shape	dec,ccd	dec,ccu	inc,ccu	inc,ccd	dec,ccd

Figure B.7 shows a graph of this function with any critical points and inflection points marked. Note that the combined critical point and inflection point at $x = 3\pi/2$ is a rare case of an inflection point where the concavity does not change. ■

---

(2) Provide a complete graphical analysis of $f(x) = (x^2 - 1)^3$ and sketch its graph.

☐ We have (simplified)

$$f'(x) = 6x(x^2 - 1)^2 \; ; \; f'(x) = 0 \text{ at } x = -1, 0, 1$$
$$f''(x) = 6(x^2 - 1)(5x^2 - 1) \; ; \; f''(x) = 0 \text{ at } x = \pm 1, \pm 1/\sqrt{5}$$

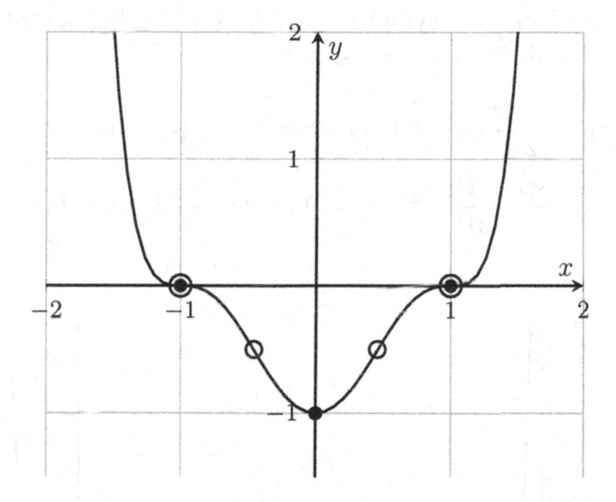

Fig. B.8 Graphical analysis of $f(x) = (x^2 - 1)^3$.

The function's values at these critical points and inflection points are:

x	-1	$-1/\sqrt{5}$	0	$1/\sqrt{5}$	1
$f(x)$	0	$-64/125$	-1	$-64/125$	0
type	CP/IP	IP	CP	IP	CP/IP
ext	neither		min		neither

Intervals set up by these critical points and inflection points, in number-line order, are:

$$I_1 : (-\infty, -1) \quad I_2 : (-1, -1/\sqrt{5}) \quad I3 : (1/\sqrt{5}, 0)$$
$$I_4 : (0, 1/\sqrt{5}) \quad I_5 : (1/\sqrt{5}, 1) \quad I_6(1, \infty)$$

The signs of the derivatives in each, and the resulting shape are:

int	I_1	I_2	I_3	I_4	I_5	I_6
$f'(x)$	$-$	$-$	$-$	$+$	$+$	$+$
$f''(x)$	$+$	$-$	$+$	$+$	$-$	$+$
shape	dec,ccu	dec,ccd	dec,ccu	inc,ccu	inc,ccd	inc,ccu

Figure B.8 shows a graph of this function with any critical points and inflection points marked. ■

(3) Provide a complete graphical analysis of $f(x) = x^2/(x-2)^2$ and sketch its graph.

☐ There is a vertical asymptote at $x = 2$ and a horizontal asymptote at $y = 1$. The derivatives are (simplified):

$$f'(x) = \frac{4x}{(2-x)^3} \; ; \; f'(x) = 0 \text{ at } x = 0, \text{is undefined at } x = 2$$

$$f''(x) = \frac{8(x+1)}{(2-x)^4} \; ; \; f''(x) = 0 \text{ at } x = -1, \text{is undefined at x=2}$$

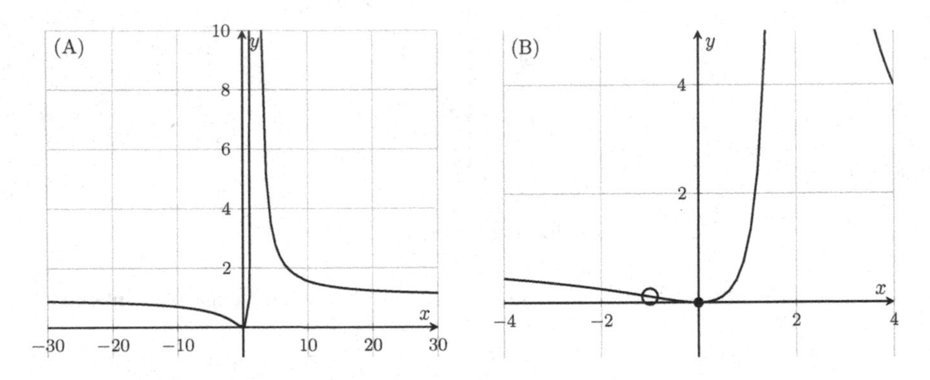

Fig. B.9 The function $f(x) = x^2/(x-2)^2$ on (A) wide, and (B) zoomed scales.

The function's values at the critical points and inflection points are:

x	-1	0
$f(x)$	$1/9$	0
type	IP	CP
ext		min

Intervals set up by the vertical asymptote, critical, and inflection points, the signs of the derivatives in each, and the resulting shape are:

int	$(-\infty,-1)$	$(-1,0)$	$(0,2)$	$(2,\infty)$
$f'(x)$	$-$	$-$	$+$	$-$
$f''(x)$	$-$	$+$	$+$	$+$
shape	dec,ccd	dec,ccu	inc,ccu	dec,ccu

Figure B.9 (A) and (B) show a graph of this function with critical points and inflection points marked. ■

B.5.3 *Optimization — Challenge — Solved*

(1) A box with a rectangular base and open top must have a volume of $10\,m^3$. The length of the base is twice the width. Material for the base costs \$10 / m^2 and the sides cost \$6 / m^2. Find the cost of materials for the cheapest such container.

□ Let the lengths of the sides of the box be l, w (for the base) and h (height). We know $l = 2w$. There are four sides of box, two of area wh and two of area $2wh$. The base has area $2w^2$. We want to minimize the cost of the box, i.e. the cost of the base and the sides:

$$C(w, h) = 10(2w^2) + 6(2(wh) + 2(2wh)) = 20w^2 + 36wh$$

We must reduce this function to one variable. We know the total volume must be 10, so that $10 = 2w^2h$, or $h = 5/w^2$. The total cost function is then,

$$C(w) = 20w^2 + 36w \cdot \frac{5}{w^2} = 20w^2 + \frac{180}{w}$$

So that

$$C'(w) = 40w - \frac{180}{w^2}$$

Finding a critical point,

$$\begin{aligned} 40w - \frac{180}{w^2} &= 0 \\ w^3 &= \frac{9}{2} \\ w &= \sqrt[3]{9/2} \end{aligned}$$

Then $C'(w) = 0$ when $w = \sqrt[3]{9/2}$, and the total cost using this value is

$$C_{min} = 20\left(\sqrt[3]{\frac{9}{2}}\right)^2 + \frac{180}{\sqrt[3]{9/2}} = 90\sqrt[3]{6} \approx 163.54 \quad ■$$

(2) Estimate the point on $y = \tan x$ that is closest to the point $(1, 1)$. (Plan on using tech to estimate the final result.)

□ We want to minimize distance to the point $(1, 1)$. The distance r from any point (x, y) in the universe to $(1, 1)$ is given by:

$$r^2(x, y) = (x - 1)^2 + (y - 1)^2$$

For points chosen from the curve $y = \tan x$ specifically, this becomes

$$r^2(x) = (x-1)^2 + (\tan x - 1)^2$$

To minimize, we search for where $\dfrac{dr}{dx} = 0$:

$$\begin{aligned} r^2 &= (x-1)^2 + (\tan x - 1)^2 \\ 2r\frac{dr}{dx} &= 2(x-1) + 2(\tan x - 1)(\sec^2 x) \\ r\frac{dr}{dx} &= (x-1) + (\tan x - 1)(\sec^2 x) \\ 0 &= (x-1) + (\tan x - 1)(\sec^2 x) \end{aligned}$$

The solution to this equation is (using tech) $x \approx 0.82$. Also, then, $y \approx \tan(0.82) \approx 1.08$. So the point we're looking for is $(0.82, 1.08)$. ■

---

(3) A piece of wire 10 m long is cut into two pieces. One piece is bent into a square and the other is bent into a circle. How should the wire be cut so that the total area enclosed by both shapes is a minimum?

□ Let's take a length x of wire for the circle first. This will be the circumference of the circle, which means $x = 2\pi r$, so the resulting radius of the circle is $r = x/(2\pi)$. The length left for the square is $10 - x$. This must be used for four sides of the square, so each side of the square has length $(10-x)/4$. The total enclosed area and its derivative is thus

$$A(x) = \pi\left(\frac{x}{2\pi}\right)^2 + \left(\frac{10-x}{4}\right)^2 = \frac{x^2}{4\pi} + \left(\frac{10-x}{4}\right)^2$$

Seeking a critical point,

$$\begin{aligned} A'(x) &= \frac{x}{2\pi} - \frac{10-x}{8} = \frac{8x - 2\pi(10-x)}{16\pi} \\ 0 &= \frac{4x - \pi(10-x)}{8\pi} \\ 0 &= 4x - 10\pi + \pi x \\ x &= \frac{10\pi}{4+\pi} \end{aligned}$$

This x value is between 4 and 5. Since $A'(4) < 0$ and $A'(5) > 0$, we know this x value provides a minimum. So to minimize the total enclosed area, take $\dfrac{10\pi}{4+\pi}$ meters of wire for the circle, and the rest for the square. ■

B.5.4 *Local Lin Approx & L-Hopital's Rule — Challenge — Solved*

---

(1) Evaluate $\lim_{x\to 0} \frac{x + \sin x}{x + \cos x}$.

☐ The given limit is not indeterminate, it can be evaluated without L-Hopital's Rule:

$$\lim_{x\to 0} \frac{x + \sin x}{x + \cos x} = \frac{0+0}{0+1} = 0$$

Why was this a Challenge Problem? To make sure you are not making every problem a nail now that you have the hammer of L-Hopital's Rule! ■

---

(2) Evaluate $\lim_{x\to 0} \frac{\tan x - x}{x^3}$.

☐ The given limit presents the indeterimate form $0/0$, so L-Hopital's Rule applies, and

$$\lim_{x\to 0} \frac{\tan x - x}{x^3} = \lim_{x\to 0} \frac{\sec^2 x - 1}{3x^2}$$

This is still $0/0$, so we use L-Hopital's Rule again:

$$\lim_{x\to 0} \frac{\sec^2 x - 1}{3x^2} = \lim_{x\to 0} \frac{2\sec^2 x \tan x}{6x}$$

This is still $0/0$, but we can resolve *part* of it this way:

$$\lim_{x\to 0} \frac{2\sec^2 x \tan x}{6x} = \lim_{x\to 0} 2\sec^2 x \cdot \lim_{x\to 0} \frac{\tan x}{6x} = 2 \lim_{x\to 0} \frac{\tan x}{6x}$$

The remaining piece of the limit is still in form $0/0$ and so back to L-Hopital's Rule:

$$2 \lim_{x\to 0} \frac{\tan x}{6x} = 2 \lim_{x\to 0} \frac{\sec^2 x}{6} = 2 \cdot \frac{1}{6} = \frac{1}{3}$$

and so

$$\lim_{x\to 0} \frac{\tan x - x}{x^3} = \frac{1}{3} \quad ■$$

---

(3) Evaluate $\lim_{x\to 0^+} \sin x \ln x$.

☐ The given limit presents the indeterminate form $0 \cdot -\infty$. It is not yet in a form which L-Hopital's Rule can handle, so we must rearrange things:

$$\lim_{x\to 0^+} \sin x \ln x = \lim_{x\to 0^+} \frac{\ln x}{\csc x}$$

which is now an indeterminate form $-\infty/\infty$ and we can apply L-Hopital's Rule:

$$\ldots = \lim_{x\to 0+} \frac{1/x}{(-\csc x \cot x)} = -\lim_{x\to 0+} \frac{\sin^2 x}{x \cos x}$$

Now we're at the indeterminate form 0/0, and we must apply L-Hopital's Rule again,

$$\ldots = \lim_{x\to 0+} \frac{2\sin x \cos x}{\cos x - x\sin x} = \frac{0}{1-0} = 0$$

Together,

$$\lim_{x\to 0+} \sin x \ln x = 0 \quad \blacksquare$$

B.5.5 *Good Value — Challenge — Solved*

(1) Does the function $f(x) = \ln x$ on $[0, e]$ satisfy the conditions of the Mean Value Theorem? If so, find a value c which is guaranteed by the MVT.

□ The function $\ln x$ is not defined at $x = 0$ and so cannot be continuous on the closed interval $[0, e]$. (It is, however, differentiable on the open interval $(0, e)$, but that's not enough for the Mean Value Theorem to kick in.) ■

(2) Suppose you are driving on a road that has a speed limit of 55mph. At one point, the police clock you driving at 50mph. Five minutes later, you pass another police car five miles away from the first one, and there you are clocked at 55mph. Now, citations are given based on instantaneous speeds — so why should you get a speeding ticket even though you were not speeding either time you passed a police car? (This question is related to the Mean Value Theorem — your answer must state what you know about this situation from the perspective of the MVT and what conclusion you can draw.)

□ This problem might seem to be all about speed, but it's actually all about your position as a function of time, which is naturally a continuous function. Let's call it $x(t)$. Note that the two positions $x = a$ and $x = b$ are five miles apart, and it took you five minutes to cross the distance between them; this means your average rate of change of position with respect to time (i.e. your average speed) was

$$\frac{\Delta x}{\Delta t} = \frac{5 \text{ miles}}{5 \text{ minutes}}$$

which is one mile per minute, or 60 miles per hour. Since the position function is continuous, the Mean Value Theorem guarantees there was at least one point inside the interval where your instantaneous rate of change of position was the same as the overall average rate. In other words, there was a point where your instantaneous speed matched your average speed of 60 mph. Busted!

(This makes you wonder why tickets are not given on toll roads where there is a record of time vs position; if your average speed between toll plazas is calculated to be larger than the speed limit, then at some single location you were actually traveling at that illegal speed.) ■

---

(3) Try to use Newton's Method to find the a root of $f(x) = x^3 - 3x^2 + x - 1$ with an initial guess of $x_0 = 1$. What happens? Can you explain the behavior of the Newton's Method process? Try again with a different (better?) starting value. Are you able to detect a root? Find the value to a precision of *n.nnnn*.

---

fx =D4^3-3*D4^2+D4-1

	A	B	C	D
1	Step	f(x)	f'(x)	x_n
2	0			1
3	1	-2	-2	0
4	2	-1	1	1.00000
5	3	-2.00000	-2	0.00000
6	4	-1.00000	1	1.00000
7	5	-2.00000	-2	0.00000
8	6	-1.00000	1	1.00000

Fig. B.10 Spreadsheet cells for Newton's Method, with CP 3.

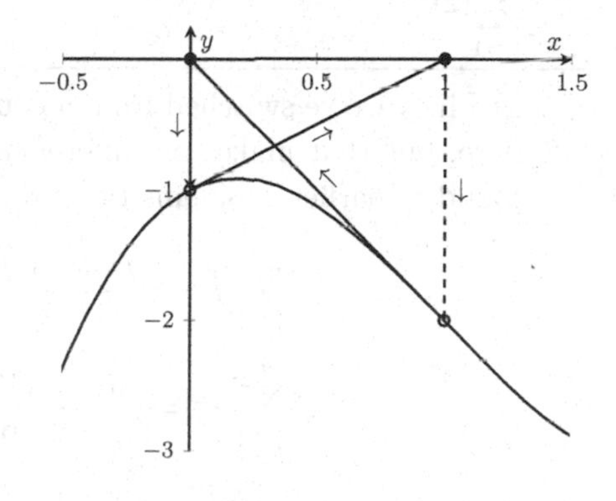

Fig. B.11 Newton's Method caught in a loop, with CP 3.

Since $f(x)$ is continuous and differentiable everywhere, so surely Newton's Method should work! Figure B.10 shows several iterations of Newton's Method starting with $x_0 = 1$. The output is seen to get "caught" oscillating between two values of x with no hope of escape: $x_n = 0$ leads to $x_{n+1} = 1$ and vice versa.

Before explaining this, let's note that a very small shift to $x_0 = 2$ will lead quickly to a root of $f(x)$ at $x \approx 2.76949$.

The first attempt failed because of the shape of $f(x)$. The line tangent to $f(x)$ at $(1, f(1)) = (1, -2)$ is $L_1(x) = -2(x - 1) = 2$, which crosses the x axis at $x = 0$. Then a line tangent to $f(x)$ at $(0, f(0)) = (0, -1)$ is $L_2(x) = -x + 1$, and that crosses the x-axis at $x = 1$. Now we're back where we started. Newton's Method simply repeats this loop over and over. This trajectory is shown in Fig. B.11. ■

B.6 Chapter 6: Challenge Problem Solutions

B.6.1 *Sigma Notation — Challenge — Solved*

(1) Use the closed form values that we're confident with after EX 3, YTI 3, and PP 3 in this section to compute the final value of the sum $\sum_{k=1}^{100}(2k^3 - k^2 + k - 5)$.

□ First, we've switched from a counter of i in the earlier problems to k here, but that makes no difference whatsoever. The closed form sums found in earlier problems (adapted from i to k) are:

$$\sum_{k=1}^{n} k^3 = \left(\frac{n(n+1)}{2}\right)^2 \qquad \text{(PP 3)}$$

$$\sum_{k=1}^{n} k^2 = \frac{n(n+1)(2n+1)}{6} \qquad \text{(YTI 3)}$$

$$\sum_{k=1}^{n} k = \frac{n(n+1)}{2} \qquad \text{(EX 3)}$$

Because the sum posed in this problem has a finite number of terms and is not one of those oddball limit / sum combinations, it's just a really, really long addition / subtraction problem — so we can rearrange terms as needed. Meaning, we can regroup as follows:

$$\begin{aligned}\sum_{k=1}^{100}(2k^3 - k^2 + k - 5) &= \sum_{k=1}^{100}(2k^3) - \sum_{k=1}^{100}(k^2) + \sum_{k=1}^{100}(k) - \sum_{k=1}^{100}(5)\\ &= 2\cdot\sum_{k=1}^{100}(k^3) - \sum_{k=1}^{100}(k^2) + \sum_{k=1}^{100}(k) - \sum_{k=1}^{100}(5)\end{aligned}$$

With $n = 100$ for this sum, the closed form recipes give us

$$\sum_{k=1}^{100} k^3 = \left(\frac{100(101)}{2}\right)^2 = 5050^2 = 25,502,500$$

$$\sum_{k=1}^{100} k^2 = \frac{100(101)(201)}{6} = (50)(101)(67) = 338,350$$

$$\sum_{k=1}^{100} k = \frac{100(101)}{2} = (50)(101) = 5050$$

Also note that $\sum_{k=1}^{100}(5) = 5 + 5 + \ldots 5$ (100 times), and so is equal to 500. Altogether, then,

$$\begin{aligned}\sum_{k=1}^{100}(2k^3 - k^2 + k - 5) &= 2 \cdot (25,502,500) - (338,350) + 5,050 - 500 \\ &= 50,671,200 \quad \blacksquare\end{aligned}$$

---

(2) In PP 5 of this section, we saw that the following limit / sum likely converges to 1/2.

$$\lim_{n\to\infty} \sum_{k=1}^{n} \left(\frac{1}{2^k} - \frac{1}{2^{k+1}} \right)$$

Are we allowed to distribute the limit / sum operation? Is the following statement true?

$$\lim_{n\to\infty} \sum_{k=1}^{n} \left(\frac{1}{2^k} - \frac{1}{2^{k+1}} \right) = \lim_{n\to\infty} \sum_{k=1}^{n} \frac{1}{2^k} - \lim_{n\to\infty} \sum_{k=1}^{n} \frac{1}{2^{k+1}}$$

□ Let's define t_n to be the nth partial sum of $\sum_{k=1}^{n} \frac{1}{2^k}$ and u_n be the nth partial sum of $\sum_{k=1}^{n} \frac{1}{2^{k+1}}$. If we investigate these separately, and then compute their difference, we can build this table and compare the difference $t_n - u_n$ to the original s_n found in PP 5.

n	5	10	25	50
t_n	0.96875	0.999	0.99999997	0.999999999999999
u_n	0.484	0.4995	0.49999999	0.49999999999999996
$t_n - u_n$	0.484	0.4995	0.49999999	0.49999999999999996
s_n	0.484	0.4995	0.49999999	0.49999999999999996

And so in this case, it seems true that

$$\lim_{n\to\infty} \sum_{k=1}^{n} \left(\frac{1}{2^k} - \frac{1}{2^{k+1}} \right) = \lim_{n\to\infty} \sum_{k=1}^{n} \frac{1}{2^k} - \lim_{n\to\infty} \sum_{k=1}^{n} \frac{1}{2^{k+1}} \quad \blacksquare$$

(3) Repeat what you just did in PP5 and CP 2 for the following limit / sum:

$$\lim_{n\to\infty}\sum_{k=1}^{n}\left(\frac{1}{k}-\frac{1}{k+1}\right)$$

That is, investigate whether the following is true:

$$\lim_{n\to\infty}\sum_{k=1}^{n}\left(\frac{1}{k}-\frac{1}{k+1}\right)=\lim_{n\to\infty}\sum_{k=1}^{n}\frac{1}{k}-\lim_{n\to\infty}\sum_{k=1}^{n}\frac{1}{k+1}$$

☐ First, let's just get the former limit / sum, as we did in PP 5. Define s_n to be the nth partial sum of $\sum_{k=1}^{n}\left(\frac{1}{k}-\frac{1}{k+1}\right)$. Using this table generated with tech,

n	10	50	100	1000	10000
s_n	0.909	0.9804	0.99	0.999	0.9999

I put $n = 10,000$ in there just to be safe. It sure looks like the limit / sum may be converging to 1.

However, we can't even start building the right side of our comparison. As we saw in You Try It 5 (see, it really pays off to do all the problems!), the limit / sum

$$\lim_{n\to\infty}\sum_{k=1}^{n}\frac{1}{k}$$

does not converge. So this does not have a value, and it's not possible to build the right side of our mystery statement

$$\lim_{n\to\infty}\sum_{k=1}^{n}\left(\frac{1}{k}-\frac{1}{k+1}\right)=\lim_{n\to\infty}\sum_{k=1}^{n}\frac{1}{k}-\lim_{n\to\infty}\sum_{k=1}^{n}\frac{1}{k+1}$$

Do CP 2 and CP 3 here contradict each other? Is there something else going on? Stay tuned to Chapter 10 (in Volume 2)! ■

B.6.2 *Area Under A Curve — Challenge — Solved*

(1) Write the Riemann sum that describes the area under $f(x) = x/(1-x)$ on $[2, 5]$.

□ We must construct

$$A = \lim_{n\to\infty} \sum_{i=1}^{n} f(x_i)\Delta x$$

For this problem, we have:

- $\Delta x = \dfrac{b-a}{n} = \dfrac{5-2}{n} = \dfrac{3}{n}$
- The individual endpoints are

$$x_i = a + i \cdot \Delta x = 2 + i\left(\frac{3}{n}\right) = 2 + \frac{3i}{n}$$

- The heights of the rectangles are

$$f(x_i) = \frac{2 + 3i/n}{1 - (2 + 3i/n)} = \frac{2 + 3i/n}{-1 - 3i/n}$$

So we have

$$A = \lim_{n\to\infty} \sum_{i=1}^{n} f(x_i)\Delta x = \lim_{n\to\infty} \sum_{i=1}^{n} \frac{2 + 3i/n}{-1 - 3i/n} \cdot \frac{3}{n} \quad ■$$

(2) Write the Riemann sum that describes the area under $f(x) = \cos(x^2)$ on $[0, \pi/2]$.

□ We must construct

$$A = \lim_{n\to\infty} \sum_{i=1}^{n} f(x_i)\Delta x$$

For this problem, we have:

- $\Delta x = \dfrac{\pi/2 - 0}{n} = \dfrac{\pi}{2n}$
- The individual endpoints are $x_i = a + i \cdot \Delta x = 0 + i\left(\dfrac{\pi}{2n}\right) = \dfrac{i\pi}{2n}$
- The heights of the rectangles are

$$f(x_i) = \cos(x_i^2) = \cos\left(\frac{i\pi}{2n}\right)^2$$

So we have

$$A = \lim_{n\to\infty} \sum_{i=1}^{n} f(x_i)\Delta x = \lim_{n\to\infty} \sum_{i=1}^{n} \cos\left(\frac{i\pi}{2n}\right)^2 \cdot \frac{\pi}{2n} \quad ■$$

(3) The following Riemann sum represents the area under some $f(x)$ on an interval $[a, b]$. What are $f(x)$, a, and b?

$$\lim_{n\to\infty} \sum_{i=1}^{n} \ln\left(\left(\frac{i\pi}{n}\right)^2 + 1\right) \cdot \left(\frac{\pi}{n}\right)$$

☐ By comparing this to

$$\lim_{n\to\infty} \sum_{i=1}^{n} f(x_i)\Delta x$$

we can identify the following:

- $\Delta x = \frac{\pi}{n}$; comparing to the general $\Delta x = \frac{b-a}{n}$, we see that $b - a = \pi$.
- $f(x_i) = \ln((x_i)^2 + 1)$ for $x_i = \frac{i\pi}{n}$
- ... which, matching to $x_i = a + i \cdot \Delta x$ shows us $a = 0$

Since $a = 0$ and $b - a = \pi$, then $b = \pi$. This Riemann sum computes the area under $f(x) = \ln(x^2 + 1)$ on $[0, \pi]$. ■

B.6.3 *Definite Integrals — Challenge — Solved*

(1) Convert the following Riemann sum into a definite integral:

$$\lim_{n\to\infty}\sum_{i=1}^{n}\left[4-3\left(\frac{2i}{n}\right)^2+6\left(\frac{2i}{n}\right)^5\right]\cdot\frac{2}{n}$$

□ Breaking this down, we see that $\Delta x = 2/n$. The endpoints in use are $x_i = 2i/n$, and they are being plugged into the function $f(x) = 4 - 3x + 2 + 6x^5$. The endpoint, written as $x_i = 0 + i \cdot (2/n)$, shows us $a = 0$; with $b - a = 2$, then the right endpoint is $b = 2$. Altogether,

$$\lim_{n\to\infty}\sum_{i=1}^{n}\left[4-3\left(\frac{2i}{n}\right)^2+6\left(\frac{2i}{n}\right)^5\right]\cdot\frac{2}{n}=\int_0^2(4-3x^2+6x^5)\,dx \quad ■$$

(2) Evaluate this definite integral using geometry: $\displaystyle\int_{-3}^{0}(1+\sqrt{9-x^2})\,dx$

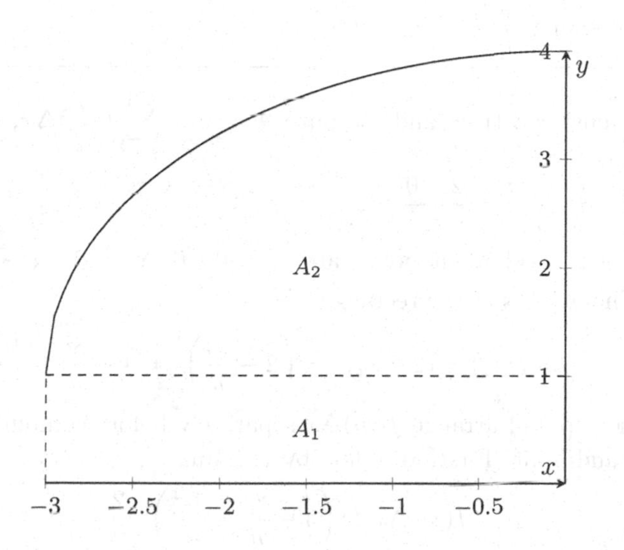

Fig. B.12 The area under $y = 1 + \sqrt{9-x^2}$ on $[-3, 0]$.

□ The graph of $f(x) = 1 + \sqrt{9 - x^2}$ on $[-3, 0]$ is shown in Fig. B.12. This graph suggests we can subdivide the region into two known geometric areas: the rectangular area indicated as A_1 and the semi-circular area indicated as A_2. In other words, we can write

$$\int_{-3}^{0}(1+\sqrt{9-x^2})dx = \int_{-3}^{0}(1)dx + \int_{-3}^{0}\sqrt{9-x^2}\,dx$$

The first of the two new integrals corresponds to the rectangular area under $y = 1$ from $x = -3$ to $x = 0$ — this is just a rectangle of width 3 and height 1, so the area is $A_1 = 3$. The second (new) integral corresponds to the area under the left half of a semicircle of radius 3; a whole circle of radius 3 would have area 9π, so the given fourth of the circle has area $A_2 = 9\pi/4$. When we put these pieces back together,

$$\int_{-3}^{0}(1+\sqrt{9-x^2})dx = \int_{-3}^{0}(1)dx + \int_{-3}^{0}\sqrt{9-x^2}dx = 3 + \frac{9\pi}{4}$$

FFT: Does this result suggest that, in general, that the following relation is true?

$$\int_a^b f(x) + g(x)\,dx = \int_a^b f(x)\,dx + \int_a^b g(x)\,dx$$ ■

---

(3) Convert this definite integral into a Riemann sum, and compute the Riemann sum: $\int_0^2 (2-x)^2\,dx$

☐ We must construct and compute $A = \lim_{n\to\infty}\sum_{i=1}^{n} f(x_i)\Delta x$, where

- $\Delta x = \frac{b-a}{n} = \frac{2-0}{n} = \frac{2}{n}$
- The individual endpoints are $x_i = a + i\cdot\Delta x = 0 + i\cdot\frac{2}{n} = \frac{2i}{n}$
- The heights of the rectangles are

$$f(x_i) = (2-x_i)^2 = \left(2-\frac{2i}{n}\right)^2 = 4 - \frac{8i}{n} + \frac{4i^2}{n^2}$$

Let's set up and arrange $f(x_i)\Delta x$ separately before combining it with a sum and limit. First, grouping by 1, i, and i^2,

$$\begin{aligned} f(x_i)\Delta x &= \left(4 - \frac{8i}{n} + \frac{4i^2}{n^2}\right)\cdot\frac{2}{n} \\ &= \frac{8}{n} - \frac{16i}{n^2} + \frac{8i^2}{n^3} \\ &= \frac{8}{n} - \frac{16}{n^2}\cdot i + \frac{8}{n^3}\cdot i^2 \end{aligned}$$

Now we can put this inside a sum, use the closed forms found in Eq. (6.3), and group the n's together in preparation for the limit:

$$\begin{aligned}\sum_{i=1}^{n} f(x_i)\Delta x &= \left(4 - \frac{4i}{n} + \frac{4i^2}{n^2}\right)\cdot\frac{2}{n}\\ &= \sum_{i=1}^{n}\left(\frac{8}{n} - \frac{16}{n^2}\cdot i + \frac{8}{n^3}\cdot i^2\right)\\ &= \frac{8}{n}\sum_{i=1}^{n}(1) - \frac{16}{n^2}\sum_{i=1}^{n}(i) + \frac{8}{n^3}\sum_{i=1}^{n}(i^2)\\ &= \frac{8}{n}(n) - \frac{16}{n^2}\cdot\frac{n(n+1)}{2} + \frac{8}{n^3}\cdot\frac{n(n+1)(2n+1)}{6}\\ &= 8 - 8\cdot\frac{n(n+1)}{n^2} + \frac{4}{3}\cdot\frac{n(n+1)(2n+1)}{n^3}\end{aligned}$$

And now we're ready for the limit:

$$\begin{aligned}\int_0^2 (2-x)^2 dx &= \lim_{n\to\infty}\sum_{i=1}^{n} f(x_i)\Delta x\\ &= \lim_{n\to\infty}\left[8 - 8\cdot\frac{n(n+1)}{n^2} + \frac{4}{3}\cdot\frac{n(n+1)(2n+1)}{n^3}\right]\\ &= 8 - 8\lim_{n\to\infty}\frac{n(n+1)}{n^2} + \frac{4}{3}\lim_{n\to\infty}\frac{n(n+1)(2n+1)}{n^3}\\ &= 8 - 8(1) + \frac{4}{3}(2) = \frac{8}{3}\end{aligned}$$

Altogether, $\int_0^2 (2-x)^2\,dx = \frac{8}{3}$. ■

B.6.4 *FTOC — Challenge — Solved*

---

(1) Evaluate $\int_1^2 \frac{y+5y^7}{y^3}\,dy$.

$$\square \quad \int_1^2 \frac{y+5y^7}{y^3}\,dy = \int_1^2 (y^{-2}+5y^4)\,dy = (-y^{-1}+y^5)\Big|_1^2$$
$$= \left(-\frac{1}{2}+32\right) - (-1+1) = \frac{63}{2} \quad \blacksquare$$

---

(2) Evaluate $\int_{\pi/4}^{\pi/3} \sec\theta\tan\theta\,d\theta$.

$$\square \quad \int_{\pi/4}^{\pi/3} \sec\theta\tan\theta\,d\theta = \sec\theta\Big|_{\pi/4}^{\pi/3} = \sec\frac{\pi}{3} - \sec\frac{\pi}{4} = 2-\sqrt{2} \quad \blacksquare$$

---

(3) Find dy/dx for $y = \int_1^{x^2} \sqrt{1+r^3}\,dr$.

□ Adapting the Fundamental Theorem (with chain rule) to these variables, we have

$$\frac{d}{dx}\int_a^{g(x)} f(r)\,dr = f(g(x))g'(x)$$

Identifying $f(r) = \sqrt{1+r^3}$ and $g(x) = x^2$, we have

$$\frac{dy}{dx} = \sqrt{1+(x^2)^3}\left(\frac{d}{dx}x^2\right) = 2x\sqrt{1+x^6} \quad \blacksquare$$

B.6.5 *Def Ints With Substitution — Challenge — Solved*

---

(1) Evaluate $\displaystyle\int_4^9 \frac{1}{\sqrt{x}(\sqrt{x}+1)^2}\,dx$.

□ With $u = \sqrt{x}+1$ we have $du = (1/2\sqrt{x})\,dx$ or $2du = (1/\sqrt{x})\,dx$. For endpoints,

$$x = 4 \to u = 3$$
$$x = 9 \to u = 4$$

and so

$$\int_4^9 \frac{1}{\sqrt{x}(\sqrt{x}+1)^2}\,dx = 2\int_3^4 \frac{du}{u^2} = -2\cdot\frac{1}{u}\bigg|_3^4 = -2\left(\frac{1}{4}-\frac{1}{3}\right)$$
$$= -2\left(-\frac{1}{12}\right) = \frac{1}{6} \quad \blacksquare$$

---

(2) Evaluate $\displaystyle\int_0^1 \frac{x}{1+x^4}\,dx$.

□ Note that $u = 1+x^4$ won't work, because we'd need an x^3 to complete the substitution for du. But we can rewrite the integral,

$$\int \frac{x}{1+x^4}\,dx = \int \frac{x}{1+(x^2)^2}\,dx$$

and then with $u = x^2$, we have $\dfrac{1}{2}\,du = x\,dx$ and

$$x = 0 \to u = 0$$
$$x = 1 \to u = 1$$

Altogether,

$$\int_0^1 \frac{x}{1+(x^2)^2}\,dx = \frac{1}{2}\int_0^1 \frac{1}{1+u^2}\,du = \frac{1}{2}\tan^{-1}(u)\bigg|_0^1$$
$$= \frac{1}{2}(\tan^{-1}(1) - \tan^{-1}(0)) = \frac{1}{2}\left(\frac{\pi}{4}-0\right) = \frac{\pi}{8} \quad \blacksquare$$

---

(3) Evaluate $\displaystyle\int_0^1 xe^{-x^2}\,dx$.

□ With $u = -x^2$, we have $du = -2x\,dx$ or $-du/2 = x\,dx$; for the endpoints,

$$x = 0 \to u = 0$$
$$x = 1 \to u = -1$$

and so

$$\int_0^1 xe^{-x^2}\,dx = -\frac{1}{2}\int_0^{-1} e^u\,du = \frac{1}{2}\int_{-1}^0 e^u\,du$$

$$= \frac{1}{2}e^u\Big|_{-1}^0 = \frac{1}{2}(e^0 - e^{-1}) = \frac{1}{2}\left(1 - \frac{1}{e}\right) \quad \blacksquare$$

Index

Printed in the United States
by Baker & Taylor Publisher Services

Printed in the United States
by Baker & Taylor Publisher Services